Real Time Fault Monitoring of Industrial Processes

International Series on
MICROPROCESSOR-BASED AND INTELLIGENT SYSTEMS ENGINEERING

VOLUME 12

Real Time Fault Monitoring of Industrial Processes

by

A. D. POULIEZOS
Technical University of Crete,
Department of Production Engineering and Management,
Chania, Greece

and

G. S. STAVRAKAKIS
Technical University of Crete,
Electronic Engineering and Computer Science Department,
Chania, Greece

KLUWER ACADEMIC PUBLISHERS
DORDRECHT / BOSTON / LONDON

Library of Congress Cataloging-in-Publication Data

Pouliezos, A. D., 1951-
Real time fault monitoring of industrial processes / by A.D. Pouliezos and G.S. Stavrakakis.
p. cm. -- (International series on microprocessor-based and intelligent systems engineering ; v. 12)
Includes bibliographical references and indexes.
ISBN 0-7923-2737-3 (acid-free paper)
1. Fault location (Engineering) 2. Process control. 3. Quality control. I. Stavrakakis, G. S., 1956- . II. Series.
TA169.6.P68 1994
670.42--dc20 94-2137

ISBN 0-7923-2737-3

Published by Kluwer Academic Publishers,
P.O. Box 17, 3300 AA Dordrecht, The Netherlands.

Kluwer Academic Publishers incorporates
the publishing programmes of
D. Reidel, Martinus Nijhoff, Dr W. Junk and MTP Press.

Sold and distributed in the U.S.A. and Canada
by Kluwer Academic Publishers,
101 Philip Drive, Norwell, MA 02061, U.S.A.

In all other countries, sold and distributed
by Kluwer Academic Publishers Group,
P.O. Box 322, 3300 AH Dordrecht, The Netherlands.

Printed on acid-free paper

Printed in the Netherlands

Table of contents

CHAPTER 4

AUTOMATIC EXPERT PROCESS FAULT DIAGNOSIS AND SUPERVISION

CHAPTER 5

FAULT DIAGNOSIS USING ARTIFICIAL NEURAL NETWORKS (ANNs)

CHAPTER 6

IN-TIME FAILURE PROGNOSIS AND FATIGUE LIFE PREDICTION OF STRUCTURES

Preface

This book is basically concerned with approaches for improving safety in man-made systems. We call these approaches, collectively, *fault monitoring*, since they are concerned primarily with detecting faults occurring in the components of such systems, being sensors, actuators, controlled plants or entire strucutures. The common feature of these approaches is the intention to detect an abrupt change in some characteristic property of the considered object, by monitoring the behavior of the system. This change may be a slow-evolving effect or a complete breakdown.

In this sense, fault monitoring touches upon, and occasionally overlaps with, other areas of control engineering such as adaptive control, robust controller design, reliability and safety engineering, ergonomics and man-machine interfacing, etc. In fact, a system safety problem, could be attacked from any of the above angles of view. In this book, we don't touch upon these areas, unless there is a strong relationship between the fault monitoring approaches discussed and the aforementioned fields.

When we set out to write this book, our aim was to include as much material as possible in a most rigorous, unified and concise format. This would include state-of-the-art method as well as more classical techniques, still in use today. As we proceeded in gathering material, however, it soon became apparent that these were contradicting design criteria and a trade-off had to be made. We believe that the completeness vs. compactness compromise that we made, is optimal in the sense that we have covered the majority of available methodologies in such a way as to give to the researching engineer in the academia or the professional engineer in industry, a starting point for the solution to his/her fault detection problem. Specifically, this book may be of value to workers in the following fields:

- Automatic process control and supervision.
- Statistical process control.
- Applied statistics.
- Quality control.
- Computer-assisted predictive maintenance and plant monitoring
- Structural reliability and safety.

The book is structured according to the main categories of fault monitoring methods, as considered by the authors: classical techniques, model-based and parameter estimation methods, knowledge- and rule-based methods, techniques based on artificial neural networks plus a special chapter on safety of structures, as a result of our involvement in this related field. The various methods are complemented with specific applications from industrial fields, thus justifying the title of the book. Wherever appropriate, additional references are summarized, for the sake of completeness. Consequently, it can also be used as a textbook in a postgradute course on industrial process fault diagnosis.

We would like at this point, firstly, to cite our distinguished colleagues, who have before us attempted a similar task, and have in this way guided us in the writing of this book:

Anderson T. and P.A. Lee (1981). Fault tolerance: Principles and practice. Prentice-Hall International.
Basseville M. and A. Benveniste, Eds. (1986). Detection of abrupt changes in signals and dynamical systems, Springer-Verlag.
Basseville M. and I. Nikiforov (1993). Detection of abrupt changes: Theory and application. Prentice Hall, NJ.
Brunet J., Jaume D., Labarrère M., Rault A. and M. Vergé (1990). Détection et diagnostic de pannes: approche par modélisation. Hermes Press.
Himmelblau D.M. (1978). Fault detection and diagnosis in chemical and petrochemical processes. Elsevier Press, Amsterdam.
Patton R.J., Frank P.M. and R.N. Clark, Eds. (1989). Fault diagnosis in dynamic systems: theory and application, Prentice-Hall.
Pau L.F. (1981). Failure diagnosis and performance monitoring. Control and Systems Theory Series of Monographs and Textbooks, Dekker, New York.
Telksnys L., Ed. (1987). Detection of changes in random processes. Optimization Software Inc., Publications Division, New York.
Tzafestas, S. (1989). Knowledge-based system diagnosis, supervision and control. Plenum Press, London.
Viswanadham N., Sarma V.V.S. and M.G. Singh (1987). Reliability of computer and control systems. Systems and Control Series, vol.8, North-Holland, Amsterdam.

Secondly, we would like to cite some very important survey papers, that provided us with useful insights:

Basseville M. (1988). Detecting changes in signals and systems - A survey. *Automatica*, **24**, 309-326.
Frank P.M. (1990). Fault diagnosis in dynamic systems using analytical and knowledge-based redundancy - A survey and some new results. *Automatica*, **26**, 459-474.
Gertler J.J. (1988). Survey of model-based failure detection and isolation in complex plants. *IEEE Control Systems Magazine*, **8**, 3-11.
Iserman R. (1984). Process fault detection based on modeling and estimation methods: A survey. *Automatica*, **20**, 387-404.
Mironovskii L.A. (1980). Functional diagnosis of dynamic systems - A survey. *Automation and remote control*, **41**, 1122-1143.
Willsky A.S. (1976). A survey of design methods for failure detection in dynamic systems. *Automatica*, **12**, 601-611.

Thirdly, we would like to note some important international congresses, devoted to fault monitoring, which show the great importance that this field has recently acquired:
1st European Workshop on Fault Diagnostics, Reliability and related Knowledge-based approaches. Rhodes, Greece, August 31-September 3, 1986. Proceedings appeared in

Tzafestas S., M. Singh and G. Schmidt, Eds. System fault diagnostics and related knowledge-based approaches, D. Reidel, Dordrecht, 1987.
1st IFAC Workshop on fault detection and safety in chemical plants, Kyoto, Japan, September 28th-October 1st, 1986.
2nd European Workshop on Fault Diagnostics, Reliability and related Knowledge-based approaches. UMIST, Manchester, England, April 6-8, 1987. Proceedings appeared in M. Singh, K.S. Hindi, G. Schmidt and S.G. Tzafestas (Eds.). Fault Detection and Reliability: Knowledge-based and other approaches, Pergamon Press, 1987.
IFAC-IMACS Symposium SAFEPROCESS '91, Baden-Baden, Germany, September 10-13, 1991.
International Conference on Fault Diagnosis TOOLDIAG '93, Toulouse, France, April 5-7, 1993.
IFAC Symposium SAFEPROCESS '94, Espoo, Finland, June 13-15, 1994.

Next, we would like to express our sincerest thanks to all those who helped us in this effort: our secretaries Stella Mountogiannaki, Irini Marentaki, Dora Mavrakaki and Vicky Grigoraki, our postgraduate students George Tselentis, Michalis Hadjikiriakos and Eleftheria Sergaki and our wives Olga and Aithra who beared with us through the writing of this book.

Lastly we would like to deeply thank Professor S. Tzafestas, not only because as the Editor of this series, showed trust in us, but also because he has been constantly encouraging and helping us in our career so far.

A.D. Pouliezos

G.S. Stavrakakis

December 1993,
Chania, Greece.

List of figures

List of tables

Introduction

The writing of this book has been motivated from the fact that a very large amount of knowledge, regarding fault monitoring, has been accumulated. This accumulation is the result of two factors: firstly, man has always been interested in preventing catastrophes, being a consequence of his works or of natural causes, and secondly, as technology advances, it seems that unfortunately, the risk of catastrophic events occurring, increases.

The latter fact is the result of bigger and more complicated plants, which makes it impossible for human operators to manage or control them. Thus the need for automatic or operator-aiding fault monitoring systems. Fortunately, results from research into man-made system safety, is equally applicable to protection from undesirable natural phenomena, such as earthquake prediction or meteorological forecasts. Even more, the same results find applications in many diverse scientific disciplines, such as bioengineering (e.g. arrythmia detection), speech processing, traffic control (incident detection) and in any other area where dynamic phenomena with possibly time-varying parameters occur. In this sense, the meaning of the term *failure* can be extended to mean *change*. Thus the following definition can be made:

A *change* is any discrepancy between an assumed value of a monitored parameter of an object and its measured, estimated or predicted value. This change may be the result of a natural operation, assuming many operational modes, or the result of a malfunction.

Since this book is about fault monitoring in industrial, thus man-made, systems, let us concentrate henceforth in this area. Malfunctions can occur in sensors, actuators, controller hardware or software, the process itself and to structures (vessels, pipes, beams etc.). The terms fault and failure are used interchangeably, but a subtle difference does exist: *a fault* occurs when the item in question operates incorrectly in a permanent or intermittent manner; *failure* on the other hand denotes a complete operational breakdown. To avoid confusion these two terms will be used with the same meaning throughout this book. Related to this terminology is the notion of *fault tolerance*, signifying the ability of a system to withstand malfunctions, whilst still maintaning tolerable performance. It is obvious however, that fault tolerance includes fault monitoring and diagnosis and the ability of the system to reorganize or restructure itself, following fault identification.

A fault monitoring system should perform the following tasks:

- Fault detection and isolation (FDI).
- Diagnosis of effect, cause and severity of faults in the components of a system.
- Reconfiguration or restructuring of appropriate control laws, to effect tolerable operation of the system if possible. If not, issue of shutdown or other emergency advice (eg. abandon aircraft).

Additionally, the performance of the above tasks, should meet certain requirements. Stated informally, these are:

- As many as possible true faults should be detected, while as few as possible false alarms should be triggered.
- The delay time between a fault occurrence and a fault declaration should be small.
- The accuracy of the estimated fault parameters (location, size, occurrence time etc.) must be high.
- The employed method must be insensitive (robust) to model inaccuracies (if a mathematical model is used) such as simplification errors resulting from linearization or unmodeled, usually non-linear components, e.g. friction, and external phenomena such as noise, load variation etc.

It is obvious, even to the uninitiated, that simultaneous satisfaction of the above requirements leads to contradiction, which is usually resolved by trade-off methods.

Incorporating a fault monitoring system into an industrial process results in improved reliability, maintanability and survivability. These terms are defined as:

- *Reliability* deals with the ability to complete a task satisfactorily and within the period of time over which that ability is retained.
- *Maintanability* concerns the need for repair and the ease with which repairs can be made, with no premium placed on performance.
- *Survivability* relates to the likelihood of conducting an operation safely (without danger to human operators or the system) whether or not the task is completed.

Furthermore increased system autonomy is achieved.

The main types of failures and errors which endanger the safe operation of a technical process are:

- System component failures caused by physical or chemical faults.
- Energy supply failures, caused for example by power supply faults.
- Environmental disturbances and external interference.
- Human operator errors.
- Maintenance errors and failures caused by wrong repair actions.
- Control system failures.

Since "a chain is not stronger than its weakest link", the task of achieving a high system availability must be performed with a *total system availability* attitude. This concept, however is not easy to be realised in practice. The key point in this practical implementation is the concept of *life-cycle maintenance*. This is defined as those actions that are appropriate for maintaining a facility in a proper condition, so that its required functions are performed throughout its life-cycle.

Two major problems have to be considered in life-cycle maintenance. One is how to cope with unexpected deteriorations and failures. The other is how the maintenance activities are properly adjusted to the various changes inside and outside the facility.

To realize the concept of life-cycle maintenance, in which the optimum maintenance strategies for each component of the facility are selected based on the prediction of deteriorations, enormous amount of information and information processing are required.. Utilization of computers is therefore, the only way to integrate and to process information relevant to the maintenance activities. A general framework for a computer assisted predictive maintenance system (CAPMS) is shown in fig. I.1.

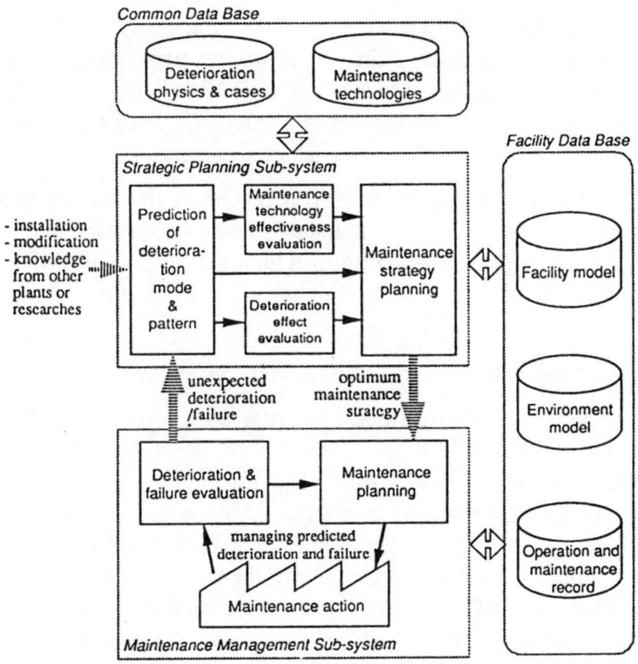

Figure I.1 Architecture of a computer assisted predictive maintenance system (CAPMS) (from Proceedings, IFAC/IMACS Symposium SAFEPROCESS '91 , Baden-Baden, Germany, 10-13 September, 1991)

The system consists of two subsystems and two groups of data bases. The strategy planning subsystem provides the function of selecting the optimum maintenance strategy for each component of the facility. It first predicts deteriorations in each of the components in terms of the mode and the progressive pattern. Deterioration prediction, available maintenance technology and the effect of deterioration are the primary factors for maintenance strategy planning. The effects of deterioration are evaluated in terms of the safety and economic effects of the functional degradations or failures induced by it. The evaluation of maintenace technologies is carried out in terms of the availability of the monitoring, diagnostic and repair techniques. For the prediction of deteriorations, the subsystem needs to refer to specific data about the facility in question. This data is contained in a facility data base which consists of a facility model, an environment model and an operation and maintenance record.

The maintenance management subsystem manages and controls the actual maintenance actions based on the strategy selected in the strategy planning subsystem. From the results of maintenance actions, the deteriorations and failures detected in the facility are analyzed. If the deteriorations and failures correspond to those predicted in the strategy planning subsystem, the maintenance management subsystem keeps the same strategy and makes a plan for the next maintenance cycle. On the other hand, occurrences of unpredicted deteriorations or failures indicate improper predictions in the strategic planning subsystem. In this case, the information is fed back to the strategy planning subsystem, and the prediction of deteriorations is carried out over again for revising the maintenance strategy plan.

To sum up, there are two feed back loops in CAPMS. One is the routine feed back loop to provide the information gathered during maintenance actions to the next maintenance plan. The second is the strategic feed back loop which becomes active when the actual data is reco gnized as inconsistent with the assumed scenario of the maintenance strategy.

To realize a CAPMS the followings major items have to be studied:

- *Prediction of deterioration.* As mentioned already, the prediction of deterioration is an essential function of a CAPMS.
- *Deterioration effect evaluation.* The effects that the deteriorations propagate in the facility cause functional degradations and failures. To evaluate these effects with a computer, one needs functional models of the facility. Significant amount of works on this subject have been done in conjuction with diagnostic expert systems.
- *Monitoring and diagnosis.* Although a number of technologies have been developed for monitoring and diagnosis, one still needs techniques for detecting the progress of various deteriorations at their early stages.
- *Selection of maintenance strategies.* The selection of the optimum maintenance strategy is the key function of a CAPMS.
- *Common data base and facility data base.* CAPMS requires sophisticated data bases. It is necessary to make a study on the structure of the common data base to be effective in deterioration prediction. With regard to the facility data base, the product model should be used as the foundation.

The subject of the present book is the detailed exposition of the first three items.

It is well known that the use of traditional techniques to build a desired control system requires a huge effort. Thus, there is also significant need for better ways to create a system. If the basic tasks of process automation, the feed-forward and feedback control, are dedicated to a first automation level, the various tasks with *supervisory functions* can be considered as forming a second level. These supervisory functions serve to indicate undesired or unpermitted process states and to take appropriate actions in order to avoid a damage of the process or an accident with human beings.

It is assumed that faults affect the technical process and its control. As mentioned earlier, a fault is to be understood as a nonpermitted deviation of a characteristic property of the process itself, the actuators, the sensors and controllers. If these deviations influence the measurable variables of the process, they may be detected by an appropriate signal evaluation. The corresponding ***fault detection and isolation*** (*FDI*) functions, called *monitoring*, consist of checking the measurable variables with regard to a certain tolerance of the normal values (limit or trend checking) and triger alarms if the tolerances are exceeded. Based on these alarms the operator takes appropriate actions. In cases where the limit value violation signifies a dangerous process state, an appropriate action can be initiated automatically. This is called *automatic protection*. Both supervisory functions may be applied directly to the measured signals or to the results of a following signal analysis, as in the case of frequency spectra of vibrations for rotating machines.

These classical ways of limit value checking of some important measurable variables are appropriate for the overall supervision of the processes. However, developing internal process faults are only detected at a rather late stage and the available information does not allow an in-depth fault diagnosis. This is one of the reasons that process operators are still required for the supervision of important processes. These human operators use their own sensors, data records, own reasoning and long term experience to obtain the required information on process changes and its diagnosis.

If the supervision is going to be improved and automated, a natural first step consists of adding more sensors and a second step to transfer the operators' knowledge into computers. Here it is usually desirable to add such sensors which directly indicate faults. Because the number of sensors, transmitters and cables increases, the cost goes up and the overall reliability is not necessarily improved. Furthermore many faults cannot be detected directly by available sensor technology.

In practice, the most frequently used diagnostic approach is the limit checking of individual plant variables. While very simple, this approach has serious drawbacks, namely:

- Since the plant variables may vary widely due to input variations, the test thresholds have to be set quite conservatively;
- Since a single component fault may cause many plant variables to exceed their limits, fault isolation is very difficult (multiple symptoms of a single fault appear as multiple "faults").

Consistency checks for groups of plant variables eliminate the above problems; the price to be paid is the need for an accurate mathematical model. Model-based FDI consists of two stages: residual generation and decision making based on these residuals.

In the first stage, outputs and inputs of the system are processed by an appropriate algorithm (a processor) to generate residual signals which are nominally near zero and which deviate from zero in characteristic ways when particular faults occur. The techniques used to generate residuals differ markedly from method to method.

In the second (decision making) stage, the residuals are examined for the likelihood of faults. Decision functions or statistics are calculated using the residuals, and a decision rule is then applied to determine if any fault has occurred. A decision process may consist of a simple threshold test on the instantaneous values or moving averages of the residuals, or it may be used directly on methods of statistical decision theory, e.g. sequential probability ratio testing.

For a static system, the residual generator is also static; it is simply a rearranged form of an input-output model, e.g. a set of geometric relationships or of material balance equations. For a dynamic system, the residual generator is dynamic as well. It may be constructed by a number of different techniques. These include parity equations or consistency relations, obtained by the direct conversion of the input-output or state-space model of the system, diagnostic observers and Kalman filters.

While a single residual is sufficient to detect a fault, a set of residuals is required for fault isolation. To facilitate isolation, residual sets are usually enhanced, in one of the following ways:

- In response to a single fault, only a fault-specific subset of the residuals becomes nonzero (structured residuals).
- In response to a single fault, the residual vector is confined to a fault specific direction (fixed direction residuals).

Also, to simplify statistical testing in a noisy system, it is useful if the residuals are "white", that is, uncorrelated in time. Residuals need to be insensitive to some disturbance variables. This may be addressed as an explicit disturbance decoupling problem or handled as a special case of structured residuals.

A fundamental issue in the generation of residuals is their robustness (insensitivity) relative to unavoidable modeling errors. Robustness concerns have plagued implementation of detection filters (as well as other failure detection methods) since their introduction. False alarms and incorrect identification of faults due to noise, disturbances, plant parameter uncertainties and unmodelled system dynamics have led to the design of robust detection filters and to determining appropriate thresholds for a given detection filter. Various techniques have been proposed to make the failure detection process more robust. Design methods are proposed with the goal of making the detection filter very much more sensitive to one fault than others. These methods have been shown to be specific cases of the unknown input observer approach. In this approach, noise, disturbances, parameter uncertainties and unmodelled dynamics are modeled as "fault events" of the system, along with the fault events arising from actual system failures. An observer is then designed to be sensitive to a fault event of interest, while insensitive to as many other real and pseudo-fault events as possible.

Increasing automatization of processes means increasing functional dependence on automation systems. The reliability of automation systems is thus becoming more and more important when the overall reliability of the processes is considered. In modern digital automation systems the most unreliable part is the field instrumentation; the reliability of electronics and the man-machine interface devices being very high in normal conditions. The relatively high fault frequency in the field instrumentation is due to the fact that the instruments make contacts with the process often in hard conditions and many of them, especially actuators, are exposed to wearing because of mechanically functioning parts. It is therefore important to look for methods that help detecting faults in this area in as an early phase as possible.

In recent years research efforts have shown that process changes due to faults can be detected in an early stage by using process models and common sensors. Then nonmeasurable quantities like process state variables and parameters may also be used. With this improved knowledge a *process supervision with fault diagnosis* (also called *condition monitoring*) becomes possible. Signal processing now provides *features*, like direct measurable quantities or nonmeasurable quantities in the form of state variables or process parameter estimates. By comparison with normal values, changes are detected resulting in *symptoms*. A knowledge-base fault diagnosis indicates its cause and location. The next step is a *fault evaluation*, that is an assessment is made of how the fault will affect the process. The faults are then divided into different hazard classes according to an incident/sequence or a fault tree analysis. Then the following action can be decided. If the fault is evaluated to be tolerable, the operation may continue and if it is conditionally tolerable, a change of operation, a reconfiguration of process parts or just maintenance has to be performed. However, if the faults are intolerable an immediate stop is required and the fault must be eliminated, e.g. by repair. A looped signal flow exists from the measured signals through the different actions back to the process. It is therefore possible to refer to supervisory loops.

With such advanced supervisory functions it is possible to improve the further automation of technical processes with the following features:

- Early detection and diagnosis of developing faults (also called incipient fault diagnosis).
- Prevention of further fault expansion.
- Preventive maintenance.
- Maintenance on request.
- Telediagnosis by using modern communication nets.

Process control computers can be extended to include automated diagnosis through the integrated use of Artificial Intelligence techniques. Diagnosis, a complex reasoning activity, is first characterized and then decomposed into its constituent information-processing tasks, each being described in terms of input, output, knowledge representation and inferencing strategy. AI-based techniques have been applied throughout the process plant control infrastructure: from the low-end "execution level" to the high-end "supervision and planning level". The execution level includes the use of techniques such as "fuzzy control" or "neural control" for closed loop control. Fuzzy logic is used to express and manipulate ill-defined qualitative terms like "large", "small", "very small", etc. in a well defined mathematical way to mimic the human operator's manual control strategy. Qualitative rules are used to express how the control signal should be chosen in different situations. "Neural control" refers to the use of neural networks in developing process models which are then used to implement robust, model-predictive controllers.

The high-end "supervisory level", on the other hand, seeks to extend the range of conventional control algorithms through the use of knowledge based systems (KBSs) for tuning controllers, performing fault diagnosis and on-line reconfiguration of control systems and process operation.

For diagnosis, these knowledge-driven techniques involve the interpretation of sensor readings and other process observations, detection of abnormal operating conditions, generation and testing of malfunction hypotheses that can explain the observed symptoms and finally resolution of any interactions between hypotheses. Fundamentally, diagnosis is viewed as a decision-making activity that is not numeric in nature. While the governing elements are symbolic, numeric computations still play an important role of providing certain kinds of information for making decisions and drawing diagnostic conclusions.

Neural nets are expected to improve today's automated supervising systems, because complex classifiers can be designed with neural nets. Artificial neural networks, even in their simplest form, are good pattern recognizers. Input vectors are introduced into the network and via supervised or unsupervised learning the weights on the connections of the network are adjusted to achieve certain goals: matching targets for supervised learning or forming clusters with unsupervised learning. Subsequent input vectors of similar types can be classified properly, but, of course, novel input patterns that the network was not trained to recognize, cannot be classified succesfully any more than a clustering code will correctly classify a new cluster on which it was not trained.

As the pattern vectors get to be large and complex, conventional numerical algorithms may not be able to properly handle the task of recognition promptly (with a computer of reasonable cost). For example, in the analysis of faults in rotating equipment, several sensors could be used to collect measurements of the vibrations in the x, y, and z axes. Ultra high-speed data sampling would be applied to get the complex waveforms involved, followed by Fourier analysis to extract the frequency components. Then statistical reduction might be applied to isolate patterns of condition from which a decision could be made as to whether the equipment was operating normally, or not. But all of these calculations take time and computer power. An ANN, once trained, can reach the decision state far more rapidly.

Another desirable feature about ANN is that good models are not required to reach the decision stage. In a typical operation in a chemical plant, the process model may be only approximate and the critical measurements may all be correlated with each other and include non-normally distributed noise. Thus, the assumptions underlying the usual statistical analysis for faults are violated to some unknown extent. An ANN seems to be able to internally map the functional relations that represent the process, filter out the noise, and handle the correlations as well.

Petri nets are suitable for the description of discrete events or processes. An important property of these nets are their capability to model and describe concurrent and asynchronous processes. Petri nets are supported by a rich mathematical theory, enabling one to simulate and analyze them. By this characteristic they can be used for processing in computers and controllers as well as in tool automation.

Since the presentation of the Petri nets in 1962, a range of net classes has been defined. These classes are divided according to the different quantity of inherent information. In Petri nets-based diagnosis, place/transition nets are used on the area level and condition/event nets on the component level. These net classes are, on the one hand, sufficient for modeling events and processes in machines or plants and, on the other hand, they provide high performance during their processing in a computer or a PLC.

Ageing plant (nuclear, conventional power, chemical, offshore, etc.) life management, life-cycle optimization and in-time safety assessment are the integral elements in safe operation and maintenance practices. Life management relies on accurate condition assessment and this can be achieved by integration of on-line monitoring and off-line inspections. The essential role of the maintenance activities is to provide plant operators with all the functions needed for safe and economical plant operation.

A systematic approach to life management has three steps:

- Data management and selection of critical and/or important areas.
- Life assessment in the critical/important areas or condition assessment.
- Control of life and life extension, if needed, including possible refurbishment or upgrading.

During the first step, for example, the economical benefits of the process, the life-limiting factors of components, fault history and safety factors affect the decision of criticality. The second step includes conventional life assessment work. The third step is the actual process of managing plant life by operation, maintenance, refurbishment, training and cost control.

Tools are needed for organizing, analyzing and transferring knowledge between the people involved. Tools must be closely related to the strategies of the company and they must fit the tasks of the users. Therefore, the strategies and the operation models have to be defined before the system definition.

The data to be used in the analyses has to be systematically gathered and saved. Inspection results, maintenance history and fault history are all important when trying to assess the life of a certain component. Data management is a necessity. Before doing any life assessment or action plan, the critical components have to be defined. They are the components that for some reason are suspected as critical and require further analysis.

The criticality of a component can be defined with several methods. Theoretical criticality consists of analysis of economical aspects, like the operational effects of a damage in the component, the delivery time of a component, the statistical fault frequency of the component and safety factors analysis. One way of determining criticality is to find out the components with the lowest remaining life time based on stress calculations and the international standards for life assessment. The effect of creep can be calculated based on estimated (static) temperatures and pressures or actual temperature and pressure history. The calculation gives as a result an estimated remaining life in hours. However, the calculations are very conservative and therefore the results can not be taken as accurate facts. Instead, they offer quite a good picture of the most critical parts of a critical installation, eg. piping. The components with the lowest remaining life time or the components whose usage factor exceeds a certain level are gathered as critical. On the basis of calculations the first plans for more accurate methods like non-destructive tests can be made. Operational history affects the life of components. Also authorities' demands have to be taken into account when planning the components to be checked. NDT test results and fault history give quite an accurate estimate about the condition of a component. There are several kinds of standards, material expert knowledge and company specific directions for determining the criticality.

The planned life time of the utility controls all life management decisions. The decisions made in operational level have to follow the strategy of the company. A schedule of reinspections and other maintenance operations is needed for planning for instance the overhauls and budgeting. It is also needed for the authorities who must be confident on the safe operation of the plant. The schedule gets more tight when the utility gets older. Also, even if a plan for a longer period is needed, it changes over time according to the entering information.

There are several kinds of decisions concerning the life management:

- Reinspection intervals and reinspection methods to be used have to be frequently determined.
- It may be necessary to decide if a component should be repaired or changed (investment costs have to be taken into account).
- If the strategy of the company is to extend the life of the plant, change of the operation parameters of a critical component can be considered.

The following modules are examples of the domain knowledge that is needed for life management:

- Components: the component classification (pipe, bend, weld, etc.) and the relations between the components have to be described.

- Material: material properties, allowable maximum stresses, development speed of a certain kind of fault mechanism (creep, fatigue, etc.) in the material are all important knowledge.
- Material tests: the inspection methods and knowledge of their usage, accuracy, cost, etc. is needed for various reasons (for planning the reinspection methods and intervals and interpreting the accuracy of the test results).

All the modules are represented as object classes and their instances. For example, material standards have been stored in data bases and they can be easily looked at and fetched to the problem solving space when needed.

A lot of work is needed in integrating different systems and in generating a systematic approach for each plant. The life management is a complicated domain. The amount of relations and effects between different input parameters is enormous and getting an accurate, optimized schedule for the operation and maintenance activities is far from simple.

In this book we have tried to analyse the aforementioned aspects by providing theoretical results with appropriate industrial applications. We hope that in this way, the various methods will be clarified and their merits and shortcomings understood.

CHAPTER 1

FAULT DETECTION AND DIAGNOSIS METHODS IN THE ABSENCE OF PROCESS MODEL

1.1 Introduction

Malfunctions of industrial plant equipment and instrumentation increase the operating costs of any plant. Even more serious are the consequences of a gross accident, because of faulty plant operation. A fault may be defined as an abnormal change in the characteristics of a system which gives rise to undesirable performance. The fault may be an actual malfunction or perhaps only a change in an operating parameter related with the structure of the system. Complete malfunction (failure) of equipment is usually relatively easy to detect, but when failure has occurred, considerable damage may have taken place. Therefore, it is desirable for an equipment monitoring system to be able to identify faults of small extent or latent malfunctions, in order to predict a later significant degradation or to locate failures which can very rapidly grow to catastrophic event.

In this chapter fault diagnosis techniques that do not necessitate any process model knowledge are presented. In the first part of the chapter, a hypothesis testing formulation of the fault detection process and some of the more recent statistical tools used in fault decision making are presented. According to this formulation, the time record of one or more process variables or parameters is observed, simple or more complicated statistics of these variables are computed and robust statistical hypothesis tests are carried out to detect faulty operation automatically.

Control charts, as the graphical means of monitoring the state of a dynamical system by observing again the time record of one or more variables, are presented.

Control and monitoring of machine health by means of noise and vibration analysis instrumentation, can also be of considerable help in the early diagnosis of potential failures and are presented in the third part of this chapter. These methods are included in this chapter because no process model is needed for their application and important signal processing is needed for their performance. Illustrative examples from practical fault diagnosis cases for all the above methods are presented.

1.2. Statistical aids for fault occurrence decision making

The diagnosis of the working state of an installation is a treatment lying between the acquisition phase of the information and the control phase of this installation.

The treatment is necessary to have representative information about the process state in order to act on the process operation. This representativeness implies that information must be sufficient (observability concepts), that it must be collected from judiciously positioned sensors and that it has to be free from errors being detrimental to the interpretation of the phenomena represented by these information.

It is also a delicate treatment, because there are various types of information (logical, analogical, deterministic, statistical, fuzzy, ...) and because they are relative to different subsets (the process itself, the actuators, the regulation systems, the chains of measurement acquisition, ...). This treatment comprises of two aspects:

- The first concerns the detection and localization of failures that may affect the process and also the instrumentation set-up, except the sensors. "Failure" may be an alteration in the operation of a proceeding element such as sensor bias, actuator locking, chocking up or significant deviation of the process state variables from the no-fault operation limits. The noises of measurement are not considered as failure.
- The second aspect forms part of the data validation. It consists of detecting failures in sensors and in correcting the doubtful measurement if the case arises. This aspect will be dealt with in Chapter 2.

In this section some of the basic statistical tools used in failure decision making will be described. The strategy proposed is based on exploitation of the information given by sensors and detectors at local level. The results may be synthesized by a "knowledge based system" proposing a diagnosis. Knowledge based diagnosis systems are examined in detail in Chapter 4.

1.2.1 Tests on the statistical properties of process characteristic quantities

A. Univariate statistical aids

The statistical aids presented here can be applied to single output processes or to multioutput processes with outputs mutually statistically independent.

A1. Limit checking

Plant measurements are compared to preset limits; exceeding a limit indicates a failure situation. In many systems, there are two levels of limits: the first level serves for prewarning only, while the second level triggers emergency action. The limits are usually set such that a large enough distance to the appearance of damage is retained on the one hand, while unnecessary fault alarms are avoided on the other.

A2. Constitution of observation windows.

A set of several successive measurements called *window*, is collected for every measured process variable in order to make the following test and treatments possible.

In real time environment the window must be sliding. On this window, statistical properties of the measured signal are studied. During normal operation, the tested parameters develop in an uncertain way around a medium value, according to a partially known law of distribution: the greater the number of samples, the greater the significance of the obtained results.

The adjustment of the window's width is delicate and has to be realized by learning. A narrow window permits the quick detection of failures while a wider window permits the detection of failures of small amplitude, but its utilization is harmful as regards the speed of detection. In order to obtain a satisfactory compromise of detection rapidity-security, several windows of different width can be used together.

Moreover it is judicious, in order to guarantee the significance of the statisical tests, that the minimum number of elements contained in the smallest window be at least equal to 30 (Kendall, (1982) and Basseville, (1980)).

Assume that one observes a piecewise constant signal disturbed by a white noise; in other words, let $y(k)$ be a sequence of independent gaussian variables with variance σ^2, and piecewise constant mean $\mu(k)$, where the jumps in the mean occur at unknown time instants. The problem of the on-line detection of such jumps appears for example in the automatic decoding of striped labels: the detection of the vertical stripes has to be done during the horizontal scanning of the label. In an on-line framework, the basic problem to be solved is the detection of a single jump, as quickly as possible, in order to allow the detection of near successive jumps.

According to the previous paragraph, let $\gamma(k)$ be a white noise sequence with variance σ^2 and let $y(k)$ be the observations sequence such that:

$$y(k) = \mu(k) + \gamma(k) \tag{1.1}$$

where,

$$\mu(k) = \begin{cases} \mu_0 & if \quad k \leq r-1 \ (H_0) \\ \mu_1 & if \quad k \geq r \quad (H_1) \end{cases} \tag{1.2}$$

r being the fault occurrence time instant.

It is first assumed that the means μ_0 and μ_1 before and after the jump are known. Several possible solutions in the (real) situation where μ_0 is known (possibly via a recursive parameter identification) and μ_1 is unknown are indicated in Chapters 2 and 3.

General comments on hypothesis testing. To test any hypothesis on a basis of a random sample of observations, the sample space Ω (i.e. all the possible sets of observations) is divided into two regions. If the observed point, say γ, falls into one of these regions, say ω, the hypothesis is rejected in favour of an alternative hypothesis; if γ falls into the complementary region Ω-ω the hypothesis is accepted. ω is known as the *critical region* of the tests and Ω-ω is called the *acceptance region*.

When making statistical hypothesis tests, the possibility of erroneous inference exists. This falls into two categories for the case where a null hypothesis (H_0) is tested against an alternative hypothesis (H_1):

Type I: (H_0) is rejected when it is true.

Type II: (H_0) is accepted when it is false.

The probability of a type I error is equal to the size of the critical region used, termed the *significance level* of the test and denoted by α. Thus,

$$P[\gamma \in \omega | H_0] = \alpha$$

In the present context α will be defined as the probability, P_f, of a false alarm.

Hence,

$$P[\gamma \in \omega | H_0] = P_f \tag{1.3}$$

The probability of a type II error is a function of the alternative hypothesis (H_1), termed the *operating characteristic* (*OC*) of the test and denoted by β. Hence,

$$P[\gamma \in \Omega\text{-}\omega | H_1] = \beta$$

$$P[\gamma \in \omega | H_1] = 1\text{-}\beta$$

The complementary probability 1-β is called the *power* of the test and in the present context it will be defined as the probability, P_d, of correct fault detection. Thus,

$$\mathrm{P}[\gamma \in \omega | H_1] = P_d$$

For a given P_f solution of (1.3) will generally yield an infinity of sub regions all obeying (1.3). In this case ω is chosen so that P_d is maximum. This is a fundamental principle in statistical decision theory first expressed by J. Neyman and E. S. Pearson (Kendall 1982). A critical region whose power is no smaller than that of any other region of the same size for testing a hypothesis H_0 against an alternative H_1 is called a ***best critical region*** (*BCR*), and a test based on a BCR is called a ***most powerful*** (*MP*) test.

When testing a hypothesis (H_0) against a class of alternatives, i.e. a composite hypothesis (for example, when testing for a zero mean against non zero mean) a MP test could be found for the different members of (H_1) (an infinity for the aforementioned example). If there exists a BCR which is best for every member of (H_1) then this region is called *uniformly most powerful* (UMP) and the test based on it a UMP test.

Fault diagnosis. Any kind of fault occurrence makes the standardized scalar residual stochastic process $\gamma(k) = y(k) - \mu(k)$ depart from its zero mean, σ^2 - variance and/or whiteness properties. Therefore it is useful to perform the following four statistical tests:

a. Sample mean (parametric test).

The test statistic commonly used for testing,

$$(H_0): \overline{\gamma}(k) = 0$$

$$\text{against} \qquad (H_1): \overline{\gamma}(k) = \overline{\gamma}_1(k) \neq 0; k = i, \ldots, j$$

is the sample mean defined by:

$$\hat{\overline{\gamma}} = \frac{1}{n} \sum_{k=i}^{j} \gamma(k) \tag{1.4}$$

Under the null hypothesis, the sample mean is normally distributed with zero mean and variance c/n, where c is the variance of the observations sequence (i.e. σ^2 or the calculated sample variance if σ^2 is unknown) and n is the size of the sample.

The probabilities P_f and P_d are respectively given by:

$$P_f = P\left[|\hat{\overline{\gamma}}| > \frac{\sqrt{c}}{n} z_{P_f/2} \right] \tag{1.5}$$

and

$$P_d = 1 - \left\{ \Phi\left[-\overline{\gamma}_1(k) \frac{\sqrt{n}}{c} + z_{P_f/2} \right] - \Phi\left[-\overline{\gamma}_1(k) \frac{\sqrt{n}}{c} - z_{P_f/2} \right] \right\} \tag{1.6}$$

where z_a is defined by:

$$P[z > z_a] \stackrel{\Delta}{=} \frac{1}{\sqrt{2\pi}} \int_{z_a}^{\infty} e^{-z^2/2} dz = a$$

and

$$\Phi(z) \stackrel{\Delta}{=} \frac{1}{\sqrt{2\pi}} \int_{-\infty}^{z} e^{-y^2/2} dy$$

P_f, P_d and n are functionally related in the two equations defining P_f and P_d. P_d also depends on the unknown $\overline{\gamma}(k)$. Typical values for P_f, are 0.1, 0.05 though this will of course depend on the specific application requirements. Having fixed P_f, then P_d, n and the critical region $|\hat{\overline{\gamma}}|$ can be chosen using equations (1.5), (1.6).

The UCL and LCL values (defined in Section 1.2.2) are given by:

$$\mathrm{UCL} = \frac{\sqrt{c}}{n} z_{P_f/2} ,$$

$$\mathrm{LCL} = -\frac{\sqrt{c}}{n} z_{P_f/2}$$

The graph of 1-P_d, called the operating characteristics (OC) curve is shown in Appendix 1.A (figs.1.A.1 and 1.A.2 of Appendix 1.A) for different values of the sample size n and for P_f = 0.01 and 0.05 respectively. As it can be seen from the graphs, increasing the sample size increases P_d, but at the expense of an increase in the detection delay time, since by averaging a larger number of residuals the effect of a fault is smoothed out.

In the case of a sliding window the sample mean $\hat{\bar{\gamma}}$ given by equation (1.4), can be calculated iteratively, thus reducing the amount of computation in on-line operations. Define the window sample mean using a new notation as follows:

$$\hat{\bar{\gamma}}^{i,j} = \frac{1}{n} \sum_{k=i}^{j} \gamma(k)$$

Then, in the case of a sliding window

$$\hat{\bar{\gamma}}^{i+1,j+1} = \frac{1}{n} \sum_{k=i+1}^{j+1} \gamma(k)$$

$$= \frac{1}{n} \left\{ \sum_{k=i}^{j} \gamma(k) + \gamma(j+1) - \gamma(i) \right\}$$

$$= \hat{\bar{\gamma}}^{i,j} + \frac{1}{n} \{\gamma(j+1) - \gamma(i)\}$$

If the residuals are correlated, the sample mean test may still be used but its control limits have to be modified accordingly. Statistical tests for the mean in the presence of correlated measurements do not appear to exist in the statistical literature. This means that in such cases the robustness of the appropriate tests must be examined when the assumption of independence is violated.

To calculate the effect on the sample mean control limits, consider the variance of the residual sample mean, which is now calculated using the formula (Kendall, 1982):

$$\mathrm{var}\left[\hat{\bar{\gamma}}\right] = \frac{1}{n^2} \{nc + 2(n-1)c\rho_1 + 2(n-2)c\rho_2 + \ldots\}$$

where ρ_i is the ith order correlation between the residuals and c is the variance of the residuals under the null hypothesis. Then,

$$\text{var}\left[\hat{\bar{\gamma}}\right] = \frac{c}{n} + \frac{2c}{n^2}\sum_{k=1}^{n-1}(n-k)\rho_k$$

$$\approx \frac{c}{n}\left\{1 + 2\sum_{k=1}^{\infty}\rho_k\right\}$$

If the process is autoregressive of order 1, then $\rho_k = \rho^k$. Hence, for such processes,

$$\sum_{k=1}^{\infty}\rho_k = \sum_{k=1}^{\infty}\rho^k = \frac{\rho}{1-\rho}$$

Consequently,

$$\text{var}\left[\hat{\bar{\gamma}}\right] \approx \frac{c}{n}\left\{1 + \frac{2\rho}{1-\rho}\right\} = \frac{c}{n}\left\{\frac{1+\rho}{1-\rho}\right\} \tag{1.7}$$

This result implies that in the case of correlated measurements, the limits of the control chart (see section 1.2.2.) for the sample mean have to be modified according to (1.7). If the correlation is negative the limits have to be decreased, whereas if the correlation is positive the limits have to be increased, since,

$$\frac{1+\rho}{1-\rho} \text{ is } \begin{cases} <1 \text{ if } \rho < 0 \\ >1 \text{ if } \rho > 0 \end{cases}$$

In the first stage of the fault monitoring process the correlation is not known, therefore if the occurred fault induces large ρ the mean test will give erroneous results.

b. Sign test (non-parametric test).

This is a non-parametric test used to test hypotheses on the value of the median of a population. Since the residuals are normal under all hypotheses the median is equal to the mean and therefore this test can be applied to test for zero mean.

The sign test procedure is as follows: the number of positive residuals in a batch is calculated and compared to two thresholds which depend on the sample size n and significance level α. Thus if,

$n_1 <$ (number of positive residuals) $< n_2$: accept (H_0)

otherwise :(H_0) is rejected.

Table 1.A.1 of Appendix 1.A is a table of the percentage points of the symmetric binomial distribution for different samplé sizes and significance levels. It is shown in Bennett and Franklin (1954), that it may be used for the sign test as follows:

i. Count the number of values above and below zero, say n^+ and n^-.
ii. Choose the smallest of the two values, say n^+.
iii. Compare n^+ with the table entry for chosen n and α, say n_α.
iv. If $n^+ < n_\alpha$, reject (H_0); otherwise accept it.

The entries in Table 1.A.1 of Appendix 1.A may be modified to indicate percentage points for the number of positive residuals. If for a sample size n the table entry is n_α, it follows that the number of positive residuals can vary from n_α to n-n_α.

The thresholds n_1 and n_2 are then chosen to satisfy:

$$n_1 = n_\alpha$$

$$n_2 = n - n_\alpha$$

Hence n_1 and n_2 represent the UCL and LCL respectively for the sign test.

Dixon gave the Tables 1.A.2 and 1.A.3 of Appendix 1.A, where values of $1\text{-}P_d$ for P_f equal 0.05 and 0.01 respectively are shown.

In the case of a sliding window of observations the number of positive residuals can also be calculated iteratively. Let $n^+_{i,j}$ be the number of positive residuals in the residual vector $\boldsymbol{\gamma}^{i,j} = [\gamma(i) \quad \gamma(i+1) \quad \dots \quad \gamma(j)]^{\mathrm{T}}$ and,

$$n_i = 1 \qquad \text{if} \qquad \gamma(i)>0$$

$$= 0 \qquad \text{if} \qquad \gamma(i)<0$$

The best procedure for residual values that are equal to zero is to disregard them and reduce the sample size by their number. This is also intuitively appealing since a zero value contributes equally to both negative and positive values. Then,

$$n^+_{i+1,j+1} = n^+_{i,j} + n_{j+1} - n_i$$

The robustness of the sign test in the case of correlated residuals, can be investigated similarly to the case of the sample mean test. Let,

$$n_i = 1 \text{ if } \gamma(i) > 0$$

$$= \text{-}1 \text{ if } \gamma(i) < 0$$

Then,

$$\mathrm{E}[n_i] = 0 \text{ , var } [n_i] = 1$$

The random variable n_i, may be associated with the positive and negative residuals. Hence, if the $\gamma(i)$ are correlated,

$$\operatorname{cov}\left[n_i, n_{i+j}\right] = E\left[n_i n_{i+j}\right]$$

$$= \frac{2}{\pi}\sin^{-1}\rho_j$$

where,

$$\rho_j = E[\gamma(i)\gamma(i+j)]$$

If $\rho_j = \rho^j$, then $E\left[n_i n_{i+j}\right] = \frac{2}{\pi}\sin^{-1}\rho^j = \rho_j^{(s)}$

The variance of the sign test statistic will be given by:

$$\operatorname{var}\left[n_{i,j}^{+}\right] = \frac{\left[\sigma^{(s)}\right]^2}{n}\left\{1+\frac{2}{n}\sum_{h=1}^{n-1}(n-h)\rho_h^{(s)}\right\}$$

The sum is equal to,

$$\sum_{h=1}^{n-1}\frac{2}{\pi}\sin^{-1}\rho^h$$

Expanding the inverse sine (i.e. $\sin^{-1}(.)$) in Taylor's series about 0,

$$\sin^{-1}\rho^h = \left(\frac{d}{d\rho}\sin^{-1}\rho^h\right)\Bigg|_{\rho=0}\rho^h + O(\rho^{2h})$$

$$= \rho^h + O(\rho^{2h})$$

Hence if orders >2 are neglected,

$$\operatorname{var}\left[n_{i,j}^{+}\right] = \frac{\left[\sigma^{(s)}\right]^2}{n}\left\{1+\frac{2\rho}{\pi(1-\rho)}\right\} = \frac{\left[\sigma^{(s)}\right]^2}{n}\frac{\pi+\rho(2-\pi)}{\pi(1-\rho)} \tag{1.8}$$

The modifying factor in this case is appreciably less than the factor appearing in (1.7). This suggests that the sign test will be more robust in departures from independence than the corresponding sample mean test.

c. Testing for whiteness

Among the various tests proposed for testing independence, two tests for whiteness are investigated here. These include one parametric and one non- parametric:

i. *First order serial correlation,* r_1
ii. *Rank correlation.*

All of these tests are well documented in the statistical literature and a brief review of each method is given here (Kendall (1982), Anderson 1958, Bennett and Franklin 1954).

i) The *first order serial correlation* of a window of observations is defined by:

$$r_1 = \frac{n}{n-1}\,\frac{\sum_{m=i}^{j-1}\left\{\gamma(m)-\hat{\bar{\gamma}}^{i,j})(\gamma(m+1)-\hat{\bar{\gamma}}^{i,j})\right\}}{\sum_{m=i}^{j}(\gamma(m)-\hat{\bar{\gamma}}^{i,j})^2}$$

where $\hat{\bar{\gamma}}^{i,j}$ is the sample mean. For small sample sizes (<20) more accurate forms may be used (Kendall, 1982).

Under the null hypothesis of whiteness the random variable r_1 is distributed asymptotically normal with mean $E(r_1) = -1/(n-1)$ and variance $var(r_1) = (n-2)^2/(n-1)^3$. Sampling experiments on serial correlation distributions suggest that the null case normal theory remains approximately valid even for n_w = 10 or 20, (Kendall, 1982). Confidence limits for hypothesis testing can be found using normal distribution theory. The probabilities P_f and P_d are respectively given by:

$$P_f = P\left[|\rho_1| > z_{(.5P_f)}\right]$$

$$P_d = 1 - \left\{\Phi\left[-\rho_1 + z_{(.5P_f)}\right] - \Phi\left[-\rho_1 - z_{(.5P_f)}\right]\right\}$$

where z_a is defined by

$$P\left[z > z_\alpha\right] = \frac{1}{\sqrt{2\pi}} \int_{z_\alpha}^{\infty} e^{-.5z^2} dz = \alpha$$

and ρ_1 is the standardized normal variable$(r_1 - E(r_1)) / \mathrm{var}(r_1)^{1/2}$.

ii) The rank correlation coefficient is the non-parametric equivalent of the standard correlation coefficient of two sets of variables. The usual procedure for the calculation of the rank correlation coefficient for a set of values $x_1, y_1; x_2, y_2; \ldots, x_n, y_n$ is to replace each x_j, y_j by their rankings x'_j and y'_j among the x's and y's respectively, and calculate:

$$r' = 1 - \frac{\sum_{j=1}^{n} d_j^2}{n(n^2 - 1)}$$

where $d_j = x'_j - y'_j$.

If serial independence is to be tested, the set $\{ x_j \}$ is replaced by the set $\{ j \}$, while the set $\{y_j\}$ represents the population values (Kendall, 1982). Therefore, to test for whiteness of the residual sequence, calculate,

$$r' = \frac{\sum_{m=i}^{j} \{m - i + 1 - \gamma'(m)\}^2}{n(n^2 - 1)}$$

where $\gamma'(m)$ is the rank of $\gamma(m)$ among the $\gamma(i)$'s. The calculated value of r' is then compared to its LCL and UCL values which are found from Table 1.A.4 of Appendix 1.A, for different P_f.

d. Testing for variance.

The residual population variance is calculated from the sample by the formula,

$$\hat{s}^2 = \frac{1}{n}\sum_{m=i}^{j}(\gamma(m) - \hat{\bar{\gamma}}^{i,j})^2$$

Confidence limits for testing,

$$(H_0): c = \sigma^2$$

$$\text{against} \qquad (H_1): c \neq \sigma^2 \quad ; c = \sigma_1^2$$

are found using the fact that the quantity,

$$\chi^2_{n-1} = \frac{(n-1)\hat{s}^2}{\sigma^2} \tag{1.9}$$

is distributed χ^2 with $(n\text{-}1)$ degrees of freedom. It then follows that the relation,

$$\frac{(n-1)\hat{s}^2}{\chi^2_{n-1,\,\alpha/2}} < c < \frac{(n-1)\hat{s}^2}{\chi^2_{n-1,1-\alpha/2}}$$

will have a probability of $(1\text{-}\alpha)$ of being correct (Benett and Franklin, 1954). Equivalently,

$$\frac{\left(\chi^2_{n-1,1-\alpha/2}\right) c}{n-1} < \hat{s}^2 < \frac{\left(\chi^2_{n-1,\alpha/2}\right) c}{n-1}$$

represent the confidence limits on $\hat{s}^2$ with a probability of error a. Hence, the UCL and LCL are given by,

$$\text{UCL} = \frac{\left(\chi^2_{n-1,\alpha/2}\right) c}{n-1} \tag{1.10}$$

$$\text{LCL} = \frac{\left(\chi^2_{n-1,1-\alpha/2}\right) c}{n-1} \tag{1.11}$$

The power of the test is given by,

$$\pi(\sigma_1^2) = \text{P}\left[\chi^2_{n-1} < \frac{1}{\lambda^2}\chi^2_{n-1,\alpha/2}\right] + \text{P}\left[\chi^2_{n-1} > \frac{1}{\lambda^2}\chi^2_{n-1,1-\alpha/2}\right] \tag{1.12}$$

where,

$$\lambda^2 = \frac{\sigma_1^2}{c}$$

Figure 1.A.3 of Appendix 1.A shows some power curves for P_f=0.05 and for n=3, 10, 30.

For real time applications, the variance and the first order serial correlation, as indeed correlations of higher order can be calculated iteratively. The equations describing the evolution of correlations are developed by Pouliezos, (1980). These are:

$$\hat{c}_0^j = \hat{c}_0^{j-1} + \frac{1}{n}\left\{\gamma^2(j) - \gamma^2(i-1)\right\} - a^j(a^j + 2\hat{\bar{\gamma}}^{i-1,j-1})$$

$$\hat{\bar{\gamma}}^{i,j} = \hat{\bar{\gamma}}^{i-1,j-1} + a^j$$

$$a^j = \frac{1}{n}\{\gamma(j) - \gamma(i-1)\}$$

and c_m^j can be calculated from c_{m-1}^j using:

$$\hat{c}_m^j = \frac{1}{n}\left\{(n-m)\left[\hat{\bar{\gamma}}^{i,j}\right]^2 - \left(q_m^j + p_m^j\right)\hat{\bar{\gamma}}^{i,j} + \sum_{k=k_1}^{j} \gamma(k)\gamma(k-m)\right\}$$

where,

$$k_1 = j - n + m + 2$$

$$q_m^j = q_{m-1}^j - \gamma(j-n-m)$$

$$p_m^j = p_{m-1}^j - \gamma(j-n+1)$$

$$p_0^j = q_0^j = n\,\hat{\bar{\gamma}}^{i,j}$$

and $\hat{c}_m^j$ denotes the sample serial correlation of lag m calculated from the residual sample $\gamma^{i,j}$. Specifically the variance, $\hat{s}^2 = \hat{c}_0^j$, and the first order serial correlation $r_1 = \hat{c}_1^j$ can be calculated iteratively using the above formulae.

Both the tests of mean and variance assume that the residuals sequence is white (Kendall 1982, Mehra 1971). Therefore, it is important to test the residuals sequence for whiteness first, especially using tests which are invariant with respect to the mean and variance of the distribution as those presented in paragraph (c) previously.

A3. Test of steady state

The aim of this test is to determine whether the examined variables are in static or dynamic state, so that appropriate treatments can be realized afterwards. When the distribution law of measurements is postulated to be normal, the test of the mean square of the successive differences is utilized (Commissariat à l' Energie Atomique, 1978).

Suppose the measurements x_i represent a gaussian sequence, then the variable r,

$$r = \frac{1}{2}\frac{\sum_{i=1}^{n-1}(x_{i+1}-x_i)^2}{\sum_{i=1}^{n}(x_i-\overline{x})^2}$$

where $\overline{x}$ is the mean of the sequence and n the window's length, has a probable value equal to 1. If n is bigger than 25, it can be assumed that,

$$u = (1-r)\sqrt{(n-1)\cdot(n+1)/2}$$

follows a zero-mean standardized gaussian law. The decision rule used is:

$u > \varepsilon$: dynamic state case

$u \leq \varepsilon$: static state case

The confidence limits for hypothesis testing can be found in a similar way to that used for the first order serial correlation test r_1 (see paragraph (c) previously).

When the distribution law is anomalous, the test can be substituted by the serial correlation test.

A4. ***Drift test.***

This test is aimed at detecting the slow variations. It is based on the exploitation of the results obtained by the previous test, utilized with the smallest and the largest window. The small window is first considered; if the steady state test, utilizing the small window detects a steady state for a duration equal to the large window and if for the same duration the steady state test utilizing the large window detects a dynamic state, then a drift is present.

A5. ***Robust univariate signal detector***

In the previous paragraphs the case in which the signal was disturbed by a white noise was considered. However, in many situations one may only have inexact knowledge of the underlying univariate noise distribution. In these situations, it is desirable to employ a robust detection scheme which offers some degree of protection against deviations from the "assumed" noise distribution.

The discrete-time detection problem under consideration reduces to a hypothesis test of the following form:

$$H_0 : Y_i = N_i$$

$$H_1 : Y_i = N_i + s$$

for i=1, ..., n, where n denotes the number of samples, the positive signal s is assumed to be known and the N_i are i.i.d. random variables. Based on the realizations $\{y_i\}_{i=1}^{n}$ of the random variables $\{Y_i\}_{i=1}^{n}$ the detector attempts to decide between hypotheses H_0 and H_1.

In practical situations one would not likely know the precise distribution of the noise. Accordingly, one will assume that the noise distribution lies within an appropriately defined neighbourhood of the nominal Laplace distribution and employ a robust test. The nominal noise has a probability density function (pdf) given by,

$$p_0(z) = \frac{\gamma}{2} e^{-\gamma|z|}$$

where γ is a positive parameter. The Laplace noise model exhibits the "heavy tail" behavior which typically characterizes impulsive noise, and it appears in many engineering investigations. Examples where the Laplace pdf is used as a noise model include undersea, atmospheric and speech processing applications.

Let $B(R)$ denote the σ - algebra of Borel sets on the real line R, and let M denote the class of all probability measures on $(R, B(R))$. Let P_0 and P_1 denote the probability measures on $(R, B(R))$. The admissible noise families under H_0 and H_1 are given by P_0 and P_1, respectively.

$$P_0 = \{Q \in M: Q((-\infty, z)) \geq (1-\varepsilon_0)P_0((-\infty, z)) - \delta_0 \quad \text{for all } z \in R\}$$

and,

$$P_1 = \{Q \in M: Q((z, \infty)) \geq (1-\varepsilon_1)P_1((z, \infty)) - \delta_1 \quad \text{for all } z \in R\}$$

where the non negative numbers ε_0, ε_1, δ_0, and δ_1 are sufficiently small to insure the disjointness of P_0 and P_1 . The designer must consider the definitions of the aforementioned admissible noise families when assigning values to these parameters, as this will determine the breadth of P_0 and/or P_1. Smaller choices of the ε and δ parameters lead to a correspondingly smaller class of admissible distributions. It should be noted that several popularly used noise classes such as ε-contamination, total variation, Kolmogorov distance, Levy distance, and Prohorov distance are subsets of P_0 and/or P_1 (see Thompson, 1991).

Loosely speaking, the saddlepoint approach is one which takes the "least favorable" distribution from the class of admissible distributions and then specifies a detector to maximize performance for this "least favorable" distribution. Without specifying all of the details, the robust detector, which is the solution of a specific saddlepoint criterion, has the canonical detector structure of a nonlinearity, followed by an accumulator, followed by a threshold comparator. The difference between the Neyman-Pearson optimal detector for Laplace noise and the robust detector for Laplace noise lies in the nonlinearities. Both the Neyman-Pearson (g_{NP}) and robust (g_k) detector nonlinearities are illustrated in the figure below.

The Neyman-Pearson and robust nonlinearities. The robust nonlinearity is obtained from the Neyman-Pearson nonlinearity by censoring at prescribed vertical heights. The method for determining the censoring height can be found in Thompson, (1991), and because of space considerations cannot be included here. Note that by putting $k=\gamma^S$, the

Neyman-Pearson nonlinearity is obtained from the robust nonlinearity. It follows that the test statistic for the robust detector is given by,

$$T_k = \sum_{i=1}^{n} g_k(Y_i)$$

The distribution function of the test statistic T_k is defined and given in closed form by Thompson, (1991).

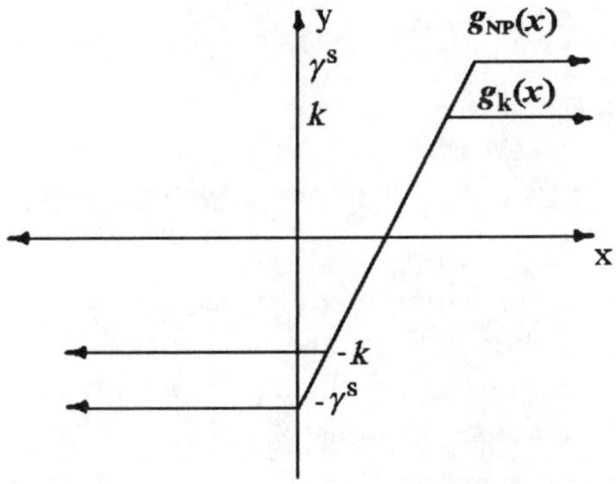

For relatively small sample sizes, the evaluation of this detector revealed that a significant loss in detection probability can occur when an overly cautious approach toward robustness is taken.

For larger sample sizes, the closed form test statistic distribution function proposed by Thompson, (1991), is computationally difficult to use. For this reason, simulation techniques can be used for larger sample sizes. A modified Monte Carlo simulation technique known as *improved importance sampling* has been suggested by Thompson and has been shown to significantly reduce the number of simulation runs required to estimate smaller false alarm probabilities.

B. Multivariate statistical aids

The statistical aids presented in the present section can be applied to multioutput processes with outputs statistically dependent between them. An attempt has been made to give the limiting distribution under the null hypothesis (H_0) in order to be able to calculate P_f for these cases.

B1. Tests of mean

These tests check whether the observed innovation sequence is zero mean or not. Let $\mathbf{x}_j^T = (x_{1j} \; \ldots \; x_{pj}); j = 1, \ldots, n$ denote a sample from a p-dimensional, absolutely continuous residual population with a p-vector mean $\boldsymbol{\theta}$. One wishes to test (H_0): $\boldsymbol{\theta} = \boldsymbol{0}$ versus (H_a): $\boldsymbol{\theta} \neq \boldsymbol{0}$. Here $\boldsymbol{0}$ is used without loss of generality, since (H_0):$\boldsymbol{\theta}=\boldsymbol{\theta}_0$ can be tested by subtracting $\boldsymbol{\theta}_0$ from each observation vector and testing whether these differences are located at $\boldsymbol{0}$.

The classical procedure for this problem is Hotelling's T^2 (Anderson, 1958, Randles, 1989), which assumes that the underlying population is p-variate normal, that is, an $N_p(\boldsymbol{\theta},\boldsymbol{\Sigma})$ distribution with mean vector $\boldsymbol{\theta}$ and variance-covariance matrix $\boldsymbol{\Sigma}_{(pxp)}$. It rejects (H_0) in favour of (H_a) if,

$$T^2 = n\overline{\boldsymbol{X}}^{\mathrm{T}}\boldsymbol{S}^{-1}\overline{\boldsymbol{X}} \geq (n-1)pF_\alpha(p,n-p)/(n-p)$$

where $\overline{\boldsymbol{X}}$ is the sample mean vector,

$$\boldsymbol{S} = \sum_{i=1}^{n}(X_i - \overline{\boldsymbol{X}})(X_i - \overline{\boldsymbol{X}})^{\mathrm{T}} / (n-1)$$

is the unbiased estimator of the variance-covariance matrix and $F_\alpha(n_1, n_2)$ is the upper αth quantile of an F distribution with n_1 and n_2 degrees of freedom. This test is quite effective and has many nice properties, including the intuitive property that it is invariant under all nonsingular linear transformations of the data. That is, if Y_j=$\boldsymbol{D}X_j$; j=1, ..., n, where $\boldsymbol{D}$ is any nonsingular pxp matrix, then $T^2(Y_1 \;\ldots\; Y_n) = T^2(X_1 \;\ldots\; X_n)$

Thus, if the data points are rotated or if they are reflected around a p-1 dimensional hyperplane or if the scales of measurement are altered, the value of T^2 stays the same. This property is intuitively appealing, and it also ensures that the performance of T^2 is the same for any variance-covariance matrix $\boldsymbol{\Sigma}$. Many nonparametric competitors to Hotelling's T^2 have been proposed which concentrate on sign statistics, that is, ones that use the direction of the observations from $\boldsymbol{0}$ rather than the distances from $\boldsymbol{0}$. The most popular such statistic is the component sign test, which uses a sign statistic for each component of the vectors and combines them in a quadratic form. Let $\boldsymbol{S}^{\mathrm{T}} = (S_1 \;\ldots\; S_p)$ where,

$$S_i = \sum_{j=1}^{n} sgn(X_{ij}) \text{ and } \operatorname{sgn}(t) = 1,(0,-1) \text{ as } t > (=,<)0$$

The test rejects (H_0) for large values of,

$$\boldsymbol{S}_n^* = \boldsymbol{S}^{\mathrm{T}}(n\hat{\boldsymbol{W}})^{-1}\boldsymbol{S}, \quad \text{where} \quad \hat{W}_{ii'} = n^{-1}\sum_{j=1}^{n}\operatorname{sgn}(X_{ij})\operatorname{sgn}(X_{i'j})$$

for $1 \leq i$ and $i' \leq p$. The limiting distribution of $\boldsymbol{S}_n^*$ under H_0 is χ_p^2.

The statistic $\boldsymbol{S}_n^*$ is not invariant under nonsingular linear transformations, and it does not have a small-sample distribution-free property unless a null distribution is generated by conditioning on the observed vectors, giving equal probability to each data vector being the observed one or a point on the opposite side of 0. The performance properties of $\boldsymbol{S}_n^{\bullet}$ vary depending on $\boldsymbol{\Sigma}$ and the direction of shift from 0. It has been demonstrated by Randles, (1989), that it may not perform well when there are substantial correlations among variates in the vectors. In an effort to stabilize the performance, this test can be performed on transformed data. This creates invariant (or asymptotically invariant) procedures, that for any $\boldsymbol{\Sigma}$, have significance levels and power comparable with those of $\boldsymbol{S}_n^{\bullet}$ when $\boldsymbol{\Sigma} = \boldsymbol{I}$, (Randles 1989).

Distribution-free tests are also investigated by Randles, (1989), for the one-sample multivariate location problem. Counts, called interdirections, which measure the angular distance between two observation vectors relative to the positions of the other observations, are introduced there. These counts are invariant under nonsingular linear transformations and have a small-sample distribution-free property over a broad class of population models, called distributions with elliptical directions, which include all elliptically symmetric populations and many skewed populations. A sign test based on interdirections is described, including, as special cases, the two-sided univariate sign test and Blumen's bivariate sign test (Blumen, 1958). The statistic is shown to have a limiting χ_p^2 null distribution and, because it is based on interdirections, it is also seen to be invariant and to have a small-sample distribution-free property. Pitman asymptotic relative efficiencies and a Monte Carlo study have shown the test to perform well compared with Hotelling's T^2, particularly when the underlying population is heavy-tailed or skewed. In addition, it consistently outperforms the component sign test, which is often recommended in the nonparametric literature (Anderson (1958), MacNeill (1974), Randles (1989)).

B2. Test of covariance

The unbiased covariance of the p-dimensional residual population sequence is estimated as,

$$\hat{\boldsymbol{S}}_0 = \frac{1}{n-1}\sum_{i=1}^{n}(X_i - \overline{\boldsymbol{X}})(X_i - \overline{\boldsymbol{X}})^{\mathrm{T}}$$

Under the null hypothesis, $\hat{\boldsymbol{S}}_0$ has a Wishart distribution (Kendall 1982, Anderson 1958). The trace of $\hat{\boldsymbol{S}}_0$ has a chi-square distribution with $(n\text{-}1)p$ degrees of freedom. Thus $\hat{\boldsymbol{S}}_0$ can be tested for its null hypothesis covariance, equal to an identity matrix for the case of the standardized residual sequence.

Also here, the parametric tests of mean and covariance assume that the residual sequence is white.

B3. Test for autocorrelation (whiteness)

There exists several tests for autocorrelation and randomness in multiple time series (Chitturi, (1976), MacNeill, (1974), Liggett, (1977), Ali, (1989), Robert, (1985)). The exact distributions of these statistics are generally unknown and the asymptotic distributions that are known, do not provide adequate approximation to the exact ones in small samples. The test statistics were modified by Ali, (1989). Asymptotically, these modified statistics are equivalent to their original counterparts; however, it is found that the asymptotic distributions of these statistics provide adequate approximation to the exact ones in relatively small samples and possibly when the time series are nonnormal. The adequacy of the approximations is examined there by simulation experiments. The original test statistics are based on sample lag cross-covariances, autocovariances, cross-correlations, and autocorrelations standardized by their asymptotic means and covariances. The modified statistics are obtained when the asymptotic means and covariances in the standardization are replaced by the exact means and covariances. The expressions for these moments are derived on the assumption that the time series is Gaussian. These moments (both asymptotic and exact) involve nuisance parameters. In constructing the test statistics, these nuisance parameters are replaced by their sample counterparts, which are consistent estimates of the parameters.

Let $\{\boldsymbol{y}_t\}$ be an m-variate stationary stochastic process with $\mathrm{E}\{\boldsymbol{y}_t\} = \boldsymbol{\mu}$ and,

$$\mathrm{cov}\{\boldsymbol{y}_t, \boldsymbol{y}_{t+k}\} = \mathrm{E}\{(\boldsymbol{y}_t - \boldsymbol{\mu})(\boldsymbol{y}_{t+k} - \boldsymbol{\mu})\}^{\mathrm{T}} = \boldsymbol{\Gamma}_k = (\gamma_k(i,j))$$

$$\boldsymbol{P}_k = (\rho_k(i,j)); \;\; \rho_k(i,j) = \gamma_k(i,j) / \sqrt{\gamma_0(i,i)\gamma_0(j,j)} \; ; \; k = 0,1,2,\ldots$$

Let $\boldsymbol{y}_t = [y_{1t} \;\ldots\; y_{mt}]^{\mathrm{T}}$; $t = 1,2,\ldots,n$, be the sampled time series, $\bar{\boldsymbol{y}}^{\mathrm{T}} = [\bar{y}_1 \;\ldots\; \bar{y}_m)^{\mathrm{T}}$ where $\bar{y}_i = \sum_{t=1}^{n} (y_{it} / n); i = 1,2,\ldots,m$ and $e_{it} = y_{it} - \bar{y}_i$; i=1, 2, ..., m; t=1, 2, ..., n. Then for k=0, 1, 2, ..., n-1,

$$c_k(i,j) = \sum_{t=1}^{n-k} (e_{it} e_{j(t+k)}) / n$$

$$\boldsymbol{C}_k = \frac{1}{n} \sum_{t=1}^{n-k} (\boldsymbol{y}_t - \bar{\boldsymbol{y}})(\boldsymbol{y}_{t+k} - \bar{\boldsymbol{y}})^{\mathrm{T}} = (c_k(i,j))$$

and,

$$r_k(i,j) = c_k(i,j) / \sqrt{c_0(i,i)c_0(j,j)}; \; k = 1,2,\ldots,(n-1); \;\; i,j = 1,2,\ldots,m.$$

A test for lag k; k= 1, 2, . . ., autocorrelation and cross-correlation, $\rho_k(i,j)$, can be based on the statistic $r_k(i,j)$. The null hypothesis of this test is that $\boldsymbol{y}_t$; t=1, 2, ..., n, is a random sample. The alternative is that $\rho_k(i,j) \neq 0$.

It may be desirable to test simultaneously for all autocorrelation and cross-correlations at a specific lag k; k=1, 2, ... A suggested portmanteau statistic to test such a hypothesis is,

$$Q_k = n\sum_{ij} r_k^2(i,j)$$

An analogous test based on the following statistic is suggested by Ali (1989),

$$QS_k = n(\mathrm{vec}\,\boldsymbol{C}_k)^{\mathrm{T}}(\boldsymbol{C}_0^{-1} \otimes \boldsymbol{C}_0^{-1})(\mathrm{vec}\,\boldsymbol{C}_k)$$

where (vec$\boldsymbol{A}$) is the vector obtained by stacking the columns of the matrix $\boldsymbol{A}$ and $\otimes$ is the standard direct product of matrices. The statistic QS_k can be obtained from the statistic proposed by Chitturi, (1976), by defining,

$$\boldsymbol{R}_k = \boldsymbol{C}_k \boldsymbol{C}_0^{-1}$$

Under the null hypothesis of randomness, following Chitturi (1976), $r_k\,(i,j)$ are normal asymptotically and both Q_k and QS_k are distributed as the chi- squared variable with m^2 degrees of freedom. Unfortunately, the distribution of these statistics , $r_k\,(i,j)$, Q_k and QS_k is unknown in small samples.

There is a long history in the investigation of the distribution of the sample lag autocorrelation, $r_k(i,i)$ (Kendall (1982), Anderson, 1971). Except for some partial successes, there has been no practical solution to this problem. Only recently, Ali (1989), has observed that for relatively small sample sizes, the null distribution of $r_k(i,i)$ can be well approximated by a normal distribution matching the first two exact moments. Thus, it is suggested that the null distribution of $r_k\,(i,j)$, $i \neq j$, be approximated by a normal distribution matching the first two moments. Alternatively, the statistic $r_k\,(i,j)$ is modified to,

$$\bar{r}_k(i,j) = (r_k(i,j) - \mathrm{E}\{r_k(i,j)\}) / \sqrt{\mathrm{var}\{r_k(i,j)\}}$$

and the distribution of this modified statistic is approximated by its asymptotic distribution, which is normal. Both $\mathrm{E}\{r_k(i,j)\}$ and $\mathrm{var}\{r_k(i,j)\}$ involve nuisance parameters $\rho_0(i,j)$. These are replaced by their consistent estimates $r_0(i,j)$. Ali, (1989), gave the expressions for the first two exact moments of $r_k\,(i,j)$, $i \neq j$, assuming $\boldsymbol{y}_t$ to be normally distributed.

The statistic QS_k can alternatively be written as,

$$QS_k = (\mathrm{vec}\,\boldsymbol{C}_k - \boldsymbol{E}_k)^{\mathrm{T}} \boldsymbol{A}_k^{-1}(\mathrm{vec}\,\boldsymbol{C}_k - \boldsymbol{E}_k)$$

where $\boldsymbol{E}_k$ is the asymptotic mean vector, which is a vector of zeros under the null hypotnesis of randomness, and $\boldsymbol{A}_k=(\boldsymbol{C}_0 \otimes \boldsymbol{C}_0)\,/\,n$ is the estimated asymptotic covariance matrix for (vec $\boldsymbol{C}_k$). Given experience with approximating the distribution of $r_k(i,j)$ using its exact moments (Ali, 1989) it is suggested that the statistic QS_k be modified to,

$$QA_k = (\mathrm{vec}\,\boldsymbol{C}_k - \hat{\mathrm{E}}\{\mathrm{vec}\,\boldsymbol{C}_k\})^{\mathrm{T}} \hat{\boldsymbol{B}}_k^{-1}(\mathrm{vec}\,\boldsymbol{C}_k - \hat{\mathrm{E}}\{\mathrm{vec}\,\boldsymbol{C}_k\})$$

where $\hat{\mathrm{E}}(\mathrm{vec}\boldsymbol{C}_k)$ is the estimated exact first moment of $(\mathrm{vec}\boldsymbol{C}_k)$, i.e. $\mathrm{E}(\mathrm{vec}\boldsymbol{C}_k)$, and $\hat{\boldsymbol{B}}_k$ is the estimated exact covariance matrix for $(\mathrm{vec}\boldsymbol{C}_k)$, i.e. $\boldsymbol{B}_k$. Both $\mathrm{E}(\mathrm{vec}\boldsymbol{C}_k)$ and $\boldsymbol{B}_k$ involve nuisance parameters $\rho_0(i,j)$, which are replaced by their consistent estimates $r_0(i,j)$ to obtain $\hat{\mathrm{E}}(\mathrm{vec}\boldsymbol{C}_k)$ and $\hat{\boldsymbol{B}}_k$. The exact mean vector and covariance matrix for $(\mathrm{vec}\boldsymbol{C}_k)$ are derived by Ali (1989), assuming $\boldsymbol{y}_t$ to be normally distributed. QA_k can be shown to be asymptotically equivalent to QS_k . The null distribution of QA_k will be approximated by a chi-squared distribution with m^2 degrees of freedom.

The statistic QA_k is a modification of QS_k in that the asymptotic mean and covariance matrix of $(\mathrm{vec}\boldsymbol{C}_k)$ in QS_k are replaced by their exact counterparts. Thus QA_k is the mean- and covariance-corrected QS_k. Similarly, one could modify QS_k by correcting only the mean of $(\mathrm{vec}\boldsymbol{C}_k)$ and the covariance matrix of $(\mathrm{vec}\boldsymbol{C}_k)$. These modified statistics are, respectively,

$$QAM_k = (\mathrm{vec}\boldsymbol{C}_k - \hat{\mathrm{E}}\{\mathrm{vec}\boldsymbol{C}_k\})^{\mathrm{T}} \boldsymbol{A}_k^{-1} (\mathrm{vec}\boldsymbol{C}_k - \hat{\mathrm{E}}\{\mathrm{vec}\boldsymbol{C}_k\})$$

and,

$$QAV_k = (\mathrm{vec}\boldsymbol{C}_k)^{\mathrm{T}} \hat{\boldsymbol{B}}_k^{-1} (\mathrm{vec}\boldsymbol{C}_k)$$

One may also apply a correction to QAM_k to obtain another statistic,

$$QAMH_k = (\, n \,/\, (n - k)\,)\, QAM_k$$

Each of the statistics QAM_k, QAV_k, and $QAMH_k$ is asymptotically distributed as a chi-squared variable with m^2 degrees of freedom. It is expected that this asymptotic distribution will provide adequate approximation to the null distributions of these statistics in small samples.

The statistic QS_k may be modified to $QH_k = (n/(n - k))QS_k$ and to $QM_k = QS_k + m^2/n$. The adequacy of approximating the null distribution of QH_k and QM_k by a chi-squared distribution with m^2 degrees of freedom, were also examined by Ali (1989).

In all, the statistics,

$$\bar{r}_k(i,j),\ Q_k,\ QA_k,\ QAM_k,\ QAV_k,\ QAMH_k,\ QS_k,\ QH_k \text{ and } QM_k$$

may be considered. The null distribution of $\bar{r}_k(i,j)$ is approximated by the standard normal distribution, and the null distribution of each of the rest of the eight statistics is approximated by a chi-squared distribution with m^2 degrees of freedom. These eight statistics are referred to as Q statistics.

1.2.1.1 Limit checking fault monitoring in electrical drives

Several devices have been developed for fault detection in industrial drives, performing limit checking tests.

Limiting states for the healthy or faulty operation of a ***Switched Reluctance Motor*** (*SRM*) drive is given in the table below:

Performance effects of staged faults on a 4-phase switched reluctance motor

Fault Condition	Relative Output	Vibration Increase?	Overcurrent?
Normal operation:	1.0	-	-
Phase disconnected and open circuited:	3/4	No	No
Phase disconnected and short circuited:	3/4	No	No
Phase with shorted pole:	7/8	Yes	No
Phase with midpoint shorted to ground:	>7/8	Yes	Severe
Phase-to-phase short:	1.0	Yes	Yes

One obvious fault detection device is a simple overcurrent detector operating from the current sensor signal, setting a comparator having a threshold above the normal operating range of the phase currents. This detector is easy to implement but is not fast acting, since a fault indication is not set until the current is already very high. Since the detector operates from the current sensor, it will also not detect all kinds of faults. The detectors must be able to operate quickly enough to interrupt a fault in progress before damage to the inverter power switches occurs.

Differential detectors are devices utilizing comparators to provide a logic signal indication of a winding fault and generally they do not exhibit the previously mentioned drawbacks. Current differential, flux differential and rate-of-rise detectors are the most utilized limit checking devices. Stephens, (1991), describes a SRM drive with an implemented limit checking fault detection and management system which should be an ideal selection for reliability-premium drive systems in aerospace, industry and automotive.

Spee and Wallace, (1990), propose failure diagnostics and remedial operating strategies for brushless dc-drives, based on correlations between the predictions of a simulation program and test measurements. Comparison of normal operation with the performance that occurs at the onset of faults has been shown to be capable of predicting post-fault performance to a good level of confidence. The reader is referred to the latter paper for details.

1.2.1.2 Steady-state and drift testing in a grinding-classification circuit

Description of the circuit. Fig. 1.1 gives an outline of a process of grinding-classification currently used in the field of mineral industry for the preparation of the minerals. The role of this circuit is to reduce a mineral to a size small enough to free the useful mineral from its gangue (Cecchin, 1985).

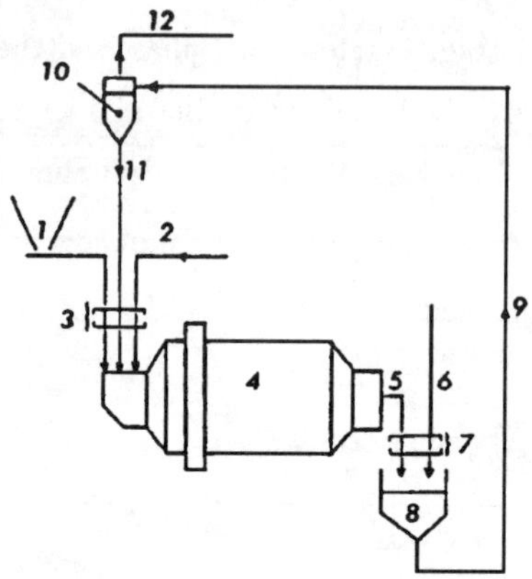

Figure 1.1 Grinding-classification circuit.

The studied circuit is composed of :

- A ball-mill (4), fed with ore (stream 1) and with water (stream 2).
- A recovery sump for the grinded products (8); this sump receives an addition of water (stream 6).
- An hydrocyclone (10) fed with pulp by stream (9). This appliance assures the separation between big particles, which have to be recycled (stream 11) and the fine particles (stream 12).

The following sensors are installed:

- stream 1: sensor of ore flowrate ($W1$).
- stream 2: sensor of water flowrate ($Q2$).
- stream 6: sensor of water flowrate ($Q6$).
- stream 8: sensor of level ($h8$) and density ($d8$).
- stream 9: sensor of pulp flowrate ($Q9$).
- stream 12: sensor of pulp flowrate ($Q12$) and density ($d12$).

The installed actuators are:

- stream 1: conveyer with adjustable speed.
- stream 2 and 6: pneumatic valves.
- stream 9: pump with adjustable speed.

The use of PID analog regulation modules has enabled the setting-up of the following local regulations:

- regulation of ore flowrate ($W1$).

- regulations of water flowrate (*Q*2 and *Q*6).
- regulation of pulp level (*h*8).

Realized decomposition. The circuit is split up into 5 subsystems:

- subsystem 1 includes the whole of mill-sump-hydrocyclone and it uses the totality of the measured variables apart from measurement *h*8.
- subsystems 2, 3 and 4 correspond to the local regulations of *W*1, *Q*2 and *Q*6. These subsystems link the measurement to the set-point.
- subsystem 5 links level *h*8 to the inputs and outputs of the sump.

Implementation of the fault detection strategy.

Test of steady state. Fig. 1.2 shows the evolution of *Q*6, the evolution of the corresponding test variable, *u*, and the result of the test of steady state (see paragraph A3). This test is realized on a window of 50 measurements, that is to say 2.5 min. The discrimination threshold of the test variable, *u*, is taken equal to 6.

Drift test. Fig 1.3 shows the results obtained by the realization of a drift test on the variable 9 (see section A4). On fig. 1.3 it can be seen the evolution of the non filtered variable *Q*9, the mean of this variable realized on 50 measurements, and the test variable *u*, calculated from a window of 50 measurements and from a window of 300 measurements. The threshold of static-dynamic discrimination is taken equal to 3 for the first window and equal to 2 for the second. For the narrow window, it is noticed that the variable *u* is inferior to its discrimination threshold: the variable *Q*9 is therefore considered in static state for that window. For the window of the large width, the variable *u* crosses its threshold very quickly, so a slow drift is present.

Standard deviation test. This test is realized on all variables declared to be in a steady state. The value of the standard deviation of the measurements is compared to thresholds determined by learning. If the standard deviation exceeds the thresholds, there is a modification of the noise due to bad contacts or to a modification of the process noise (see section A2).

Several "visible" failures were created on variable *Q*9 in order to prove the usefulness of a standard deviation test. Fig. 1.4 shows the evolution of variable *Q*9. The functioning of the installation is normal except at the following points:

- From 5.5 to 10.5 mn: changing of the proportional coefficient *P* in the regulation of level.
- From 15.75 to 21.75 mn: suppression of level regulation, the pump works at a constant speed.
- From 26 to 31.5 mn: addition of an integral action I on the level regulation
- from 35 mn to the end: false intermittent contacts on the current output of the sensor Q9.
- The value of the standard deviation calculated on a window of 50 measurements. The arbitrarily negative values of the standard deviation correspond to the points in dynamic state.

- The value of the variable "result of the tests". This variable has 3 states:
  It is zero when the variable Q9 is in static state, the process operation being normal.
  It is negative when the variable Q9 is in dynamic state
  It is positive when the variable Q9 is in static state and when the standard deviation test has detected an anomaly.

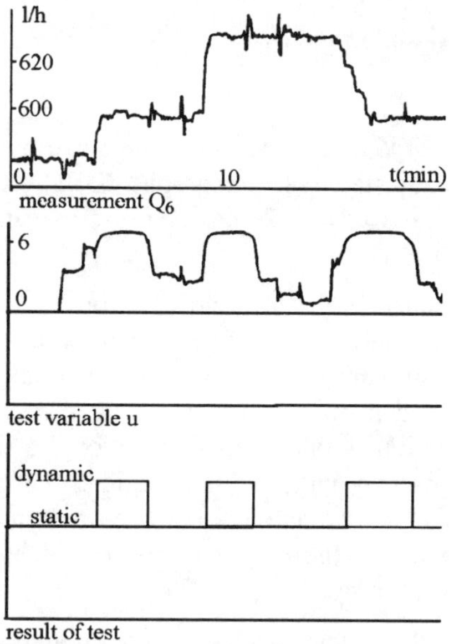

Figure 1.2 Test of steady state applied on *Q6*

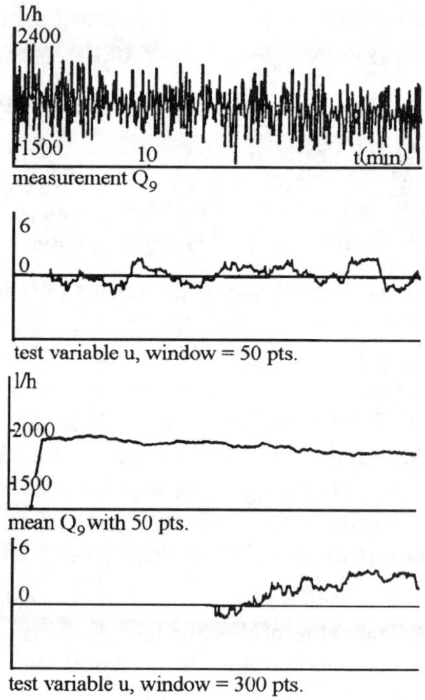

Figure 1.3 Drift test applied on *Q9*

1.2.1.3 Conclusions

The algorithms described in this section may be used to provide all basic functions of the fault monitoring strategy.

The efficiency of the localization and the precision of the diagnosis depend on the structure of the installation and on the knowledge of its operation quality and quantity of information.

To improve a partial diagnosis, a "knowledge based system" realizing this synthesis is needed and may propose some actions to the user: quick verification of a part of the installation, use of extra signals at certain points of the circuit, measurement of variables normally not measured.

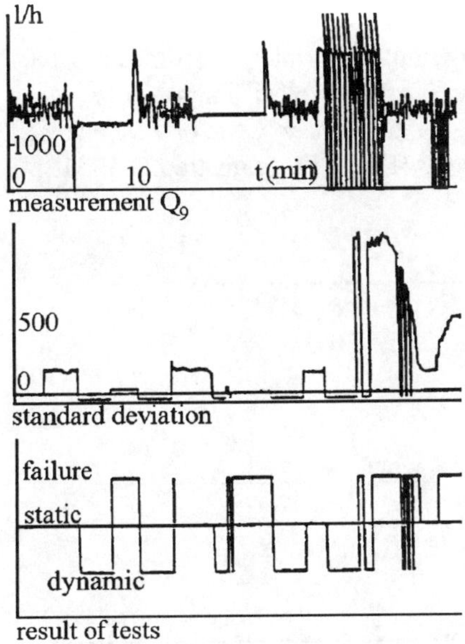

Figure 1.4 Standard deviation test applied on *Q*9

An extension of the diagnosis strategy concerns the study of decision- making, that can be considered from different aspects:

- Proceed to a thorough study of the behaviour of suspect elements.
- Correct defective measurements (in case of a sensor failure).
- Compensate control signals (in case of actuators' failures).
- Reconfigure the control structure (in the case of loss of essential organs) and proceed the substitutions and reparations of organs.
- Put installation in "emergency state" (in case of a non-satisfying reconfiguration).
- Shut down production.

The knowledge- base aspects of the fault diagnosis problem will be investigated in detail in Chapter 4.

1.2.2 Process control charts

Control charts are used in monitoring the statistical state of a process whose measurements are available sequentially in time. Some statistic w (sample mean or sample range etc.) is computed from successive samples of size n and plotted on a graph containing lower and upper limits corresponding to the critical region of the hypothesis on w under test. If the statistic w is distributed normally with mean m_w and variance s_w, where m_w and s_w are calculated a-priori, then typical limits are $m_w + 3s_w$ for the ***Upper Control Limit*** (*UCL*) and $m_w - 3s_w$ for the ***Lower Control Limit*** (*LCL*). Such charts are usually referred to as *Shewhart charts* (Fig. 1.5), originated in 1931.

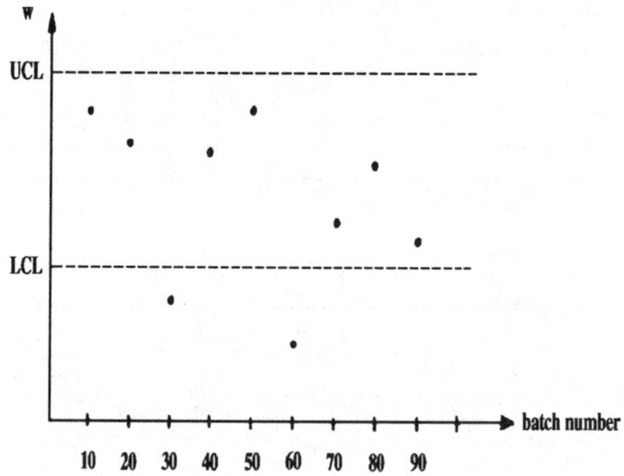

Figure 1.5 Shewhart control chart

Control charts were used by Himmelblau, (1978), to monitor dynamical systems and detect malfunctions in the plant equipment quite recently. An extensive bibliography on control charts is given by Vance, (1983).

Univariate Shewhart control chart techniques.

Univariate control charts can be used for the first stage of fault monitoring (as are the cases of Section 1.2.1, paragraph A), as follows: given successive samples of residuals $\boldsymbol{\gamma}^{i,j}, \boldsymbol{\gamma}^{i+1,j+1}, \dots, \boldsymbol{\gamma}^{i+m,j+m}$, where $\boldsymbol{\gamma}^{j,k} = [\gamma(j) \quad \gamma(j+1) \quad \dots \quad \gamma(k)]^{\mathrm{T}}$, an appropriate statistic is calculated and plotted on a corresponding control chart with the precomputed UCL and LCL. A decision that a fault has occurred is made when the statistic falls outside its normal operation level for a specified subsequent number of times. This procedure will decrease the probability of type I errors (see section 1.2.1., paragraph A). A logical flowchart for a computer-operated control chart is shown in Fig. 1.6. The calculations of the central line, UCL and LCL for the appropriate test statistic of univariate Shewhart charts are summarized in Himmelblau (1978), where the reader is referred for further details. The Shewhart chart essentially treats each sample separately. However, practical experience has shown that by taking into account the information from all collected samples, it is possible to determine in a better way whether the process is in control or out of control.

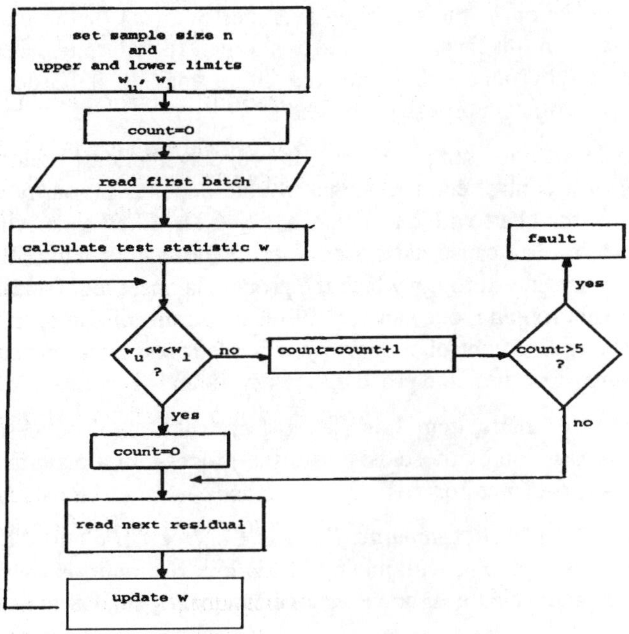

Figure 1.6 Flowchart for computer operated control chart.

Many efforts to use Shewhart charts in industry have been of limited success. This is not because of a technical shortcoming of the method, but typically because of one or more of the following reasons:

1. The formulas used to calculate the limits are incorrect.
2. The sampling and subgrouping plans used to supply and group data for the charts are poorly chosen.
3. The improvement work demanded by the charts is so radical in the context of the organization's culture, that the organization is unable to properly respond.

To one not familiar with industrial environments, the three preceding items may seem uninteresting or even trivial, but they are the critical criteria for successful application.

Item 1 seems on the surface to be beneath one's attention. This problem is so widespread however, that when consulting, one must regularly question clients on how they did their calculations. The root cause in this case is a lack of understanding that: "the essential statistical power of control charts comes from using variation within the subgroups (or in the case of individuals charts, between neighboring measurements) to calculate the width of the control limits". Those who have missed this point have recommended, for example, that limits on a control chart for individuals be based on the sample standard deviation of all of the data thrown together, instead of the average moving range. Similarly, some statistical software allows the user to calculate control limits using the sample standard deviation of the subgroup averages or of all of the individuals. Charts with limits calculated in this fashion are nearly doomed to failure because special causes occurring between subgroups will inflate the limits, making it harder to detect such causes, which is the original purpose of the chart.

Item 2, the problem of poor sampling plans, is probably the least understood of the issues. The ability of a control chart to signal trouble depends primarily on the sampling plan used to supply the chart with data. Indeed, being identified as special or common is not due to a property of a cause itself but rather to the way in which the control chart works as a window or filter through which the process is observed. By carefully selecting the sampling and subgrouping, one can control what sources of variation will show up as special causes. This makes control charts very versatile tools, since one can select several filters with which to view the same process.

Proper response to the charts, item 3, is the most critical issue in successful application. This requires *prompt* action by those closest to the process. In a manufacturing environment, this is typically the operator.

Every one should agree that ***Automatic Process Control*** (*APC*) is the most effective means of maintaining a setpoint with minimal variation. If one deals only with statistical theory and ignores practical issues, however, one can make APC seem like a panacea.

Since APC systems are continually making physical adjustments to a process, there can be increased wear, especially in a typical industrial environment. This, together with the maintenance requirements of the control equipment itself, can substantially increase maintenance costs.

Experts estimate that at any point in time, from 25% to 35% of the world's advanced automatic control systems are on manual. Lack of operator confidence is one critical reason for this. Operators, like most people, have an inherent distrust of a "black box" that makes decisions on a basis that they do not understand. This is not an inherent fault of APC but rather of the way in which it may have been misapplied.

Box and Kramer, (1992), discuss rationales for process monitoring using some of the control chart techniques of ***Statistical Process Control*** (*SPC*) for feedback adjustment. Minimum cost feedback schemes are discussed for some practically interesting models.

The critical question is how to integrate APC and SPC for total system improvement. Hoerl and Palm, (1992), provided a more basic discussion of this topic. APC should be applied to the critical variables that have a direct control knob to minimize variation, manage setpoint changes, and ensure safety. The time and money required should be invested. SPC should be applied to this system to monitor its long-term effectiveness, detect special causes, and monitor the on-line measurement system. Experimental design could be used to tune the control system. SPC or algorithmic statistical process control (Vander Wiel *et al.*, 1992) can be used as a substitute when the cost of APC cannot be justified.

Cumulative sum (*CUSUM*) control charts are very effective in detecting special causes. The CUSUM chart is usually maintained by taking samples at fixed time intervals and plotting a cumulative sum of differences between the sample means and the target value ordered in time, on the chart. The process mean is considered to be on target as long as the CUSUM statistic computed from the samples does not fall into the signal region of the chart. A value of the CUSUM statistic in the signal region should be taken as an indication that the process mean has changed and that the possible causes of the change should be investigated.

The number of observations taken before an out-of-control signal is triggered, is called the ***run length***. The performance of a CUSUM chart is commonly measured by the ***average run length*** (*ARL*) corresponding to various possible choices of the mask parameters.

Cumulative sum (CUSUM) charts are often used instead of standard Shewhart charts when detection of small changes in a process parameter is important. For comparable average run lengths (ARLs) when the process is on-target, CUSUM charts can be designed to give shorter ARLs than Shewhart charts for detecting certain small changes in process parameters. The superiority of the CUSUM chart over the Shewhart chart also holds when the Shewhart chart is augmented with runs rules. Thus, it is only natural to investigate whether the shorter ARLs for the univariate case can be extended to the multivariate case (Hawkins, (1992), Pignatiello, (1990), Blazek, (1987)).

The exponentially weighted moving average control chart. A control chart technique that may be of value to both manufacturing and continuous process quality control engineers is the *exponentially weighted moving average* (*EWMA*) control chart (Hunter, (1986), Lucas and Saccucci, (1990)). The EWMA has its origins in the early work of econometricians, and although its use in quality control has been recognized, it remains a largely neglected tool. The EWMA chart is easy to plot, easy to interpret, and its control limits are easy to obtain. Further, the EWMA leads naturally to an empirical dynamic control equation.

The exponentially weighted moving average (EWMA) is a statistic with the characteristic that it gives less and less weight to data as they get older and older. A plotted point on an EWMA chart can be given a long memory, thus providing a chart similar to the ordinary CUSUM chart, or it can be given a short memory and provide a chart analogous to a Shewhart chart.

The EWMA is very easily plotted and may be graphed simultaneously with the data appearing on a Shewhart chart. The EWMA is best plotted one time position ahead of the most recent observation since it may be viewed as the forecast for the next observation. The immediate purpose is only to plot the statistic. The EWMA is equal to the present predicted value plus λ times the present observed error of prediction. Thus,

$$\begin{aligned} \text{EWMA} &= \hat{y}_{t+1} = \hat{y}_t + \lambda e_t \\ &= \hat{y}_t + \lambda(y_t - \hat{y}_t) \\ &= \lambda y_t + (1-\lambda)\hat{y}_t \end{aligned}$$

where, $\hat{y}_{t+1}$ is the predicted value at time t+1 (the new EWMA), y_t is the observed value at time t, $\hat{y}_t$ is the predicted value at time t (the old EWMA), $e_t = y_t - \hat{y}_t$ is the observed error at time t and and λ is a constant ($0<\lambda<1$) that determines the depth of memory of the EWMA. As shown in Hunter, (1986), the EWMA can be written as,

$$y_{t+1} = \sum_{i=0}^{t} w_i y_i$$

where the w_i are weights defined by,

$$w_i = \lambda(1-\lambda)^{t-i}$$

with sum $\sum_{i=0}^{t} w_i = 1$. The constant λ determines the "memory" of the EWMA statistic. That is, λ determines the rate of decay of the weights and hence the amount of information secured form the historical data. Note, that as $\lambda \to 1$, $w_1 \to 1$ and $\hat{y}_{t+1}$ practically equals the most recent observation y_t.

When the process is under control and λ=1, points plotted on the classical Shewhart chart and those on an EWMA chart are therefore almost equal in their ability to detect signals of departures from assumptions. As $\lambda \to 0$, the most recent observation has small weight, and previous observations near equal (though lower) weights. Thus, as $\lambda \to 0$, the EWMA takes on the appearance of the CUSUM. The EWMA control chart for values of $0<\lambda<1$ stands between the Shewhart and CUSUM control charts in its use of the historical data.

The choice of λ can be left to the judgment of the quality control analyst or can be estimated using an iterative least squares procedure. The analyst, by considering the data as new data arriving sequentially, could for different values of λ, compute the corresponding sequential set of predicted values $\hat{y}$ based on the EWMA. The value of λ which provides the smallest error sum of squares is preferred, based upon this very limited evidence.

As shown in Hunter, (1986), the variance of the EWMA is,

$$\text{Var (EWMA)} = [\lambda / (2 - \lambda)]\ \sigma^2$$

and thus,

$$\sigma_{\text{EWMA}} = [\ \lambda / (2 - \lambda)\ ]^{0.5}\ \sigma$$

An estimate of σ^2 can be obtained from the minimum error sum of squares $\sum e_t^2$ obtained while estimating λ. That is,

$$\hat{\sigma}^2 = \sum_{t=1}^{T} e_t^2 / (T-1)$$

When the Shewhart and EWMA charts are constructed and plotted simultaneously σ_{EWMA} can be estimated from the information used in establishing the 3σ Shewhart control limits. That is,

$$\hat{\sigma}_{\text{EWMA}} = [\lambda / (2-\lambda)]^{0.5}\, \hat{\sigma}_{\text{Shewhart}}$$

The 3σ control limits for the EWMA are,

$$\tau \pm 3\hat{\sigma}_{\text{EWMA}}$$

The Shewhart and the EWMA control limits may be conveniently placed on the same chart.

In practice the classical Shewhart chart (only the last plotted point falling outside the 3σ limits providing a signal) is aided by the use of runs. For example, two out of three successive points falling within the region $\eta \pm 2\sigma$ and $\eta \pm 3\sigma$, or six points in sequence above the chart's center line, are frequently taken to be signals that the process is no longer under control.

Employing runs provides an informal use of the recent history and, in the hands of an experienced analyst, can make the Shewhart chart take on the aspects of the EWMA chart. However, the EWMA chart provides a regular and formal use of the historical data. Runs and all other data configurations are encompassed in the EWMA forecast. Further, once the EWMA has been computed it contains all the information provided by the historical record. There is no need to save or, as in the case of the modified CUSUM, to give occasional zero weights to recorded observations.

The recognition that an EWMA control scheme can be represented as a Markov chain allows its properties to be evaluated more easily and completely than has previously been done. Lucas and Saccucci, (1990), evaluate the properties of an EWMA control scheme used to monitor the mean of a normally distributed process that may experience shifts away from the target value. A design procedure for EWMA control schemes is given. Parameter values not commonly used in the literature are shown to be useful for detecting small shifts in a process. In addition, several enhancements to EWMA control schemes are considered. These include a fast initial response feature that makes the EWMA control scheme more sensitive to start-up problems, a combined Shewhart EWMA that provides protection against both large and small shifts in a process, and a robust EWMA that provides protection against occasional outliers in the data that might otherwise cause an out-of-control signal. An extensive comparison reveals that EWMA control schemes have average run length properties similar to those for cumulative sum control schemes.

The EWMA can be used as a dynamic process control tool. When the process mean η is on the target (i.e., $\eta=\tau$) all three charting procedures, the Shewhart, CUSUM and EWMA, are roughly equivalent in their ability to monitor departures from target. However, the EWMA provides a forecast of where the process will be in the next instance of time. It thus provides a mechanism for dynamic process control.

To control a process it is convenient to forecast where the process will be in the next instance of time. Then, if the forecast shows a future deviation from target that is too large, some electro-mechanical control system or process operator can take remedial action to compel the forecast to equal the target. In modern manufacturing, and particularly where an observation is recorded on every piece manufactured or assembled, a forecast based on the unfolding historical record can be used to initiate a feedback control loop to adjust the process.

Of course, if an operator is part of a feedback control loop he/she must know what corrective action to perform, and care must be taken to avoid inflating the variability of the process by making changes too often. But control engineers long ago learned how to close the feedback loop linking forecast and adjustment to target. The same information feedback loop exists in many situations which only the operator can control. The EWMA chart not only provides the operator with a forecast, but also with control limits to inform when the forecast is statistically significantly distant from the target. Thus, when an EWMA signal is obtained, appropriate corrective action based on the size of the forecast can often be devised.

The EWMA can be modified to enhance its ability to forecast. In situations where the process mean steadily trends away from target the EWMA can be improved by adding a second term to the EWMA prediction equation. That is,

$$\text{modified EWMA: } \hat{y}_{t+1} = \hat{y}_t + \lambda_1 e_t + \lambda_2 \sum e_t$$

where λ_1 and λ_2 are constants that weight the error at time t and the sum of the errors accumulated to time t. The coefficients λ_1 and λ_1 can be estimated from historical data by an iterative least squares procedure similar to that mentioned earlier for the estimation of the constant λ.

A third term can be added to the EWMA prediction equation to give the empirical control equation,

$$\hat{y}_{t+1} = \hat{y}_t + \lambda_1 e_t + \lambda_2 \sum e_t + \lambda_3 \nabla e_t$$

where the symbol ∇ means the first difference of the errors e_t ; that is $\nabla e_t = e_t - e_{t-1}$. Now observe that the forecast $\hat{y}_{t+1}$ equals the present predicted value (zero if the process has been adjusted to the target) plus three quantities: one proportional to e_t, the second a function of the sum of the e_t, and the third a function of the first difference of the

e_t. These terms are sometimes called the "proportional", "integral", and "differential" terms in the process control engineer's basic proportional, integral, differential (PID) control equation. The parameters λ_1, λ_2 and λ_3 weight the historical data to give the best forecast.

The EWMA may thus be viewed as more than an alternative to either the Shewhart or the CUSUM control charts. The EWMA may also be viewed as a dynamic control mechanism to help keep a process mean on target whenever discrete data on manufactured items or on the process operation are sequentially available.

Multivariate control charts.

The quality of the output of a production process is often measured by the joint level of several correlated characteristics. For example, a chemical process may be a function of temperature and pressure, both of which need to be monitored carefully; a particular grade of lumber might depend on correlated characteristics such as stiffness and bending strength. In a geochemical process in coal mining each observation consists of 14 correlated characteristics. In these types of situations, separate univariate control charts for each characteristic are often utilized to detect changes in the inherent variability of the process. When these characteristics are mutually correlated, however, the univariate charts are not as sensitive as multivariate methods that capitalize on the correlation.

One common method of constructing multivariate control charts is based on Hotelling's T^2 statistic. Currently, when a process is in the start-up stage and only individual observations are available, approximate F and chi-square distributions are used to construct the necessary multivariate control limits. These approximations are conservative in this situation. Tracy *et al.* (1992), present an exact method based on the beta distribution, for constructing multivariate control limits at the start-up stage. An example from the chemical industry illustrates that this procedure is an improvement over the approximate techniques.

Note that in the following, monitoring of the mean of a multivariate normal process is required. The term "on-target" is used to indicate that the process is in-control with respect to its mean. Likewise, the term "off-target" is used to indicate that the mean of the multivariate normal process has shifted.

For successive samples, multivariate control chart techniques used for controlling the mean of the multivariate normal process can be interpreted as repeated tests of significance of the form,

$$H_0: \boldsymbol{\mu} = \boldsymbol{\mu}_o$$
$$H_1: \boldsymbol{\mu} \neq \boldsymbol{\mu}_o$$

where $\boldsymbol{\mu}$ represents a multivariate normal process mean, whose true value is unknown and $\boldsymbol{\mu}_0$ is the target value for the parameter vector. For simplicity, it will be assumed that $\boldsymbol{\mu}_0=0$, since the general case can be handled easily by translation.

Let $\boldsymbol{X}_t = [x_{1t} \; \ldots \; x_{pt}]^{\mathrm{T}}$ denote the p-vector of quality characteristic measurements made on a part of a multivariate normal process where $x_{j,t}$ is the observation on variate j at time t. It is assumed that the successive $\boldsymbol{X}_t$ are independent and identically distributed multivariate normal random vectors with known and constant covariance matrix $\boldsymbol{\Sigma}$. That is, the $\boldsymbol{X}_t$ are i.i.d. $N_p(\boldsymbol{\mu}, \boldsymbol{\Sigma})$. Without loss of generality, let $\boldsymbol{X}_t$ denote the (sample) mean vector at time t and let $\boldsymbol{\Sigma}$ denote its covariance matrix.

The Multivariate Shewhart X² Chart.

To test the previous hypothesis, the null hypothesis should be rejected at time t if $\chi^2 > \chi^2_{p;\alpha}$ where,

$$\chi^2 = (\boldsymbol{X}_t - \boldsymbol{\mu}_o)^{\mathrm{T}} \boldsymbol{\Sigma}^{-1} (\boldsymbol{X}_t - \boldsymbol{\mu}_o)$$

and $\chi^2_{p;\alpha}$ is the upper 100α percentage point of the χ^2 distribution with p degrees of freedom. The noncentrality parameter associated with χ^2 is,

$$\lambda^2(\mu) = (\boldsymbol{\mu} - \boldsymbol{\mu}_o)^{\mathrm{T}} \boldsymbol{\Sigma}^{-1} (\boldsymbol{\mu} - \boldsymbol{\mu}_o)$$

Note that $\lambda(\mu)$, the square root of the noncentrality parameter, is often used to represent a measure of the distance of $\boldsymbol{\mu}$ from $\boldsymbol{\mu}_0$. This measure of distance is also called the *Mahalanobis distance* or the *statistical distance*. Note that the *straight line* or *Euclidian* distance assumes an identity covariance matrix instead. Henceforth, the word "distance" will be used to mean the square root of the noncentrality parameter defined above.

A χ^2 control chart operates by plotting χ^2 on a chart with an appropriate UCL. If a point plots above the upper control limit, the process mean is deemed to be out of control and the assignable causes of the variation are sought. The average run length (ARL) of this control scheme can be calculated as $1/P$ where P denotes the probability that χ^2 exceeds the UCL. The on-target value of P is determined from the probability that χ^2 exceeds the UCL under the central χ^2_p distribution while the off-target value of P is the probability that χ^2 exceeds the UCL under the non-central χ^2_p distribution.

The multiple univariate CUSUM scheme.

A p-dimensional multivariate normal process can be monitored by using p two-sided univariate CUSUM charts. The jth two-sided univariate CUSUM is operated by forming the cumulative sums,

$$S_{j,t} = \max(0,\; S_{j,t-1} + x_{jt} - k_j)$$

$$T_{j,t} = \min(0,\; T_{j,t-1} + x_{jt} + k_j)$$

where $S_{j,0} \geq 0,\; T_{j,0} > 0,\; k_j > 0$ and x_{jt} is the sample mean at time t for variate j.

The jth two-sided chart signals that the corresponding process mean has shifted when either $S_{j,t} > k_j$ or $T_{j,t} < -h_j$ for some CUSUM control chart parameters k_j and h_j. The multiple univariate CUSUM scheme signals an off-target condition when any of the p two-sided schemes produces an off-target signal. Therefore, the on-target average run length of the multiple univariate scheme is less than the average run length of any one of the univariate CUSUM charts.

Two new multivariate CUSUM charting procedures are proposed by Pignatiello, (1990). These procedures make direct use of the covariance matrix and are based on quadratic forms of the mean vector. The ARL performance for these charts has shown a superior performance with the ARL of the classical CUSUM charts as compared by Pignatiello, (1990). Overall, these charts appear to be a good control charting procedure for detecting a variety of shifts in the mean of a multivariate normal process.

To introduce the first multivariate CUSUM scheme, the multivariate sum

$$\boldsymbol{C}_t = \sum_{i=t-n_t+1}^{t} (\boldsymbol{X}_i - \boldsymbol{\mu}_o)$$

is considered, where n_t can be interpreted as the number of subgroups since the most recent renewal (i.e. zero value) of the CUSUM, and is formally defined below. Since $(1/n_t)\boldsymbol{C}_t$ may be written as ,

$$\frac{1}{n_t}\boldsymbol{C}_t = \left(\frac{1}{n_t}\sum_{i=t-n_t+1}^{t}\boldsymbol{X}_i\right) - \boldsymbol{\mu}_o,$$

the vector $\boldsymbol{C}_t/n_t$ represents the difference between the accumulated sample average and the target value for the mean. Consequently, at time t, the multivariate process mean can be estimated to be $(\boldsymbol{C}_t/n_t)+\boldsymbol{\mu}_0$. The norm of $\boldsymbol{C}_t$

$$\|\boldsymbol{C}_t\| = \sqrt{\boldsymbol{C}_t^{\mathrm{T}}\boldsymbol{\Sigma}^{-1}\boldsymbol{C}_t}$$

is seen as a measure of the distance of our estimate of the mean of the process from the target mean for the process. A multivariate control chart can then be constructed by defining MC1 as,

$$\mathrm{MC1}_t = \max\{\|\boldsymbol{C}_t\| - kn_t, 0\}$$

and,

$$n_t = \begin{cases} n_{t-1}+1 & \text{if } \mathrm{MC1}_{t-1} > 0 \\ 1 & \text{otherwise} \end{cases}$$

where the choice of the reference value $k>0$ is discussed below. The MC1 chart operates by plotting $\mathrm{MC1}_t$ on a control chart with an upper control limit of UCL_1. If $\mathrm{MC1}_t$ exceeds UCL_1 then the process is deemed to be off-target. Rather than basing a multivariate CUSUM statistic on the square of the distance of the accumulated sample average from $\boldsymbol{\mu}_0$, one could consider the square of the distance of each sample mean from $\boldsymbol{\mu}_0$ and then accumulate those squared distances. Hence, as an alternative to MC1, one could consider the square of the distance D_t^2 of the tth sample mean from the target value of $\boldsymbol{\mu}_0$ where,

$$D_t^2 = (\boldsymbol{X}_t - \boldsymbol{\mu}_0)^{\mathrm{T}} \boldsymbol{\Sigma}^{-1} (\boldsymbol{X}_t - \boldsymbol{\mu}_0)$$

has a central χ^2 distribution with p degrees of freedom when the process is on-target and a non-central χ^2 distribution when the process is off-target. A one-sided univariate CUSUM can now be formed as,

$$\mathrm{MC2}_t = \max\{0,\ \mathrm{MC2}_{t-1} + D_t^2 - k\}$$

with $\mathrm{MC2}_0$=0. The choice of the reference value k is discussed below. To use this multivariate CUSUM, one would declare the process to be off-target if $\mathrm{MC2}_t$ exceeds an upper control limit h_2.

Choice of reference value k. In the (upper) one-sided, univariate CUSUM procedure the reference value k is often taken to be the average of the expected values of the process mean under (H_0): $\boldsymbol{\mu}=\boldsymbol{\mu}_0$ and (H_1): $\boldsymbol{\mu}=\boldsymbol{\mu}_1$ where $\boldsymbol{\mu}_0$ represents the on-target state and $\boldsymbol{\mu}_1$ a specified, unacceptable off-target state. The choice of a value for k follows from the derivation of the CUSUM and from Wald's sequential probability ratio test. Although the (non-) central χ^2-distribution of the observations for the (off-) on-target state of MC2 is not symmetric, this same approach can be used. That is, the k used in MC2 is,

$$p + 0.5\lambda^2(\boldsymbol{\mu}_1)$$

Since the form of MC1 is different from the other CUSUM charts, the choice of k cannot be derived analogously. Instead, one can chose k to be half of the distance of $\boldsymbol{\mu}_1$ from $\boldsymbol{\mu}_0$ (where again, $\boldsymbol{\mu}_1$ is a specified off-target state). That is, one can take k=0.5 $\lambda(\boldsymbol{\mu}_1)$. Note that for both multivariate CUSUM charts, the value for k depends only on the magnitude of the distance of $\boldsymbol{\mu}_1$ from $\boldsymbol{\mu}_0$. It may be possible to improve the performance of the MC1 chart at selected off-target conditions with alternate choices for k.

The procedures presented above for treating quality control data assumes the existence of normality and independence of the data. When either independence and/or normality are not present, as it is usually the case, application of the presented methodology introduces large errors in the analysis of the data and renders conclusions based on them dubious.

Once serial correlation is confirmed, identification techniques can be used to suggest a specific kind of model that might be worth considering. The identification techniques make use of the autocorrelation and partial autocorrelation functions. When the identification is complete, the likelihood function can provide maximum likelihood estimates of the parameters of the identified model (Ljund, 1987).

Suppose that a correlation test revealed the presence of data dependence and the identification techniques suggested as the best model, an autoregressive process AR(p) of order p. Vassilopoulos, (1978), has modified and extended the existing standard control chart methodology by utilizing the time series analysis approach and by introducing dependence via a second order autoregressive process (AR(2) model). Curves of the modified auxiliary quality control factors are presented, showing the substantial effect of dependence on the classical quality control factors.

Yashchin, (1993), discusses a situation in which one is interested in evaluating the run-length characteristics of a cumulative sum control scheme when the underlying data show the presence of serial correlation. In practical applications, situations of this type are common in problems associated with monitoring such characteristics of data as forecasting errors, measures of model adequacy, and variance components. The discussed problem is also relevant in situations in which data transformations are used to reduce the magnitude of serial correlation. The basic idea of analysis involves replacing the sequence of serially correlated observations by a sequence of independent and identically distributed observations for which the run-length characteristics of interest are roughly the same. Applications of this method to several classes of processes arising in the area of statistical process control are discussed in detail, and it is shown that it leads to approximations that can be considered acceptable in many practical situations. The reader is referred to Vasilopoulos, (1978), and Yashchin, (1993), for details.

Statistical process control (SPC) by displaying multivariate data.

Control charts are more valuable in practice especially when used as simple graphical aids to let the process operator, who is untrained in statistical techniques, get a mental picture of the process history and interpret whether or not the quality of the process operation is at a satisfactory level.

Displaying multivariate data is probably the most popular topic of current statistical graphics research (Blazek, 1987). Particularly in the area of process control, where micro- and mini-computers collect, analyze, and store thousands of observations on each phase of a process each day, effective graphical display of these data is a necessity. In one plant the data collected by process control computers can amount to millions of bits of information. Even narrowing down the variables to Juran's "significant few" and using computers to help process the data, graphics is the only means available to present this plethora of data to the operators and supervisors for rapid interpretation of results.

This problem has been most keenly felt with the increasing use of statistical process control (SPC) methods. Although well versed in basic techniques and SPC charting, operators and supervisors feel overwhelmed when they realize that their complex, interdependent processes produce ten to twenty significant variables to be monitored simultaneously instead of just two or three. Although summary measures, like Hotelling's T^2 presented in the previous section, can quantify the overall status of the system, these people also need information on the individual contributors as well. This is especially true in facilities where modernization has produced new measurements for which component interdependencies are unknown.

A graphic which will display the individual and collective relationship sequentially, called *polyplot*, will be presented in the following. The polyplot is a glyph in the shape of a polygon with rays emanating from the vertices. A glyph can represent a single observation, or, as it is often the case in SPC applications, a composite of several observations. Each side is of equal length, and the number of sides of the polygon corresponds to the number of variables being studied. Each vertex and corresponding ray is associated with a specific variable. The rays of the polygon are oriented along invisible line segments passing from the center of the polygon through the appropriate vertex. The length of a ray varies with the value of each observation. The vertex corresponds to the mean of each variable $\overline{\overline{x}}_i$, and the distance from the mean is transformed into standard error units. Thus one can plot,

$$\frac{\overline{x}_i - \overline{\overline{x}}_i}{s / \sqrt{n}}$$

which indicates the number of standard error units the ith data average is from its process mean ($\overline{x}_i$ is the ith average of n data points; s is the estimate of the process standard deviation and n is the sample size). Values less than the mean are mapped by rays starting at the vertex and going toward the center. Rays starting at the vertex and going away from the center correspond to observed values greater than the mean. The distance between the vertex and the center is four standard error units. To ensure that the rays of one glyph do not overwrite a nearby glyph nor extend past the center, rays greater than four standard error units are truncated at that limit.

The use of different line styles or colors for the rays encodes the level of significance of the deviation from the estimated mean. For example, the ray is dotted if the deviation is within two standard error units, dashed between two and three standard error units, and solid beyond three standard error units. If colors can be used, it is suggested to use blue for rays less than two standard error units, green between two and three, and red to signify rays beyond three standard error units.

Figs. 1.7, 1.8 and 1.9 demonstrate the appearance of the glyph for three, five and seventeen variables, respectively. The glyph displays the relative number of variables dramatically as the polygon changes from a triangle to a pentagon to a virtual circle with seventeen variables. Yet the figure still allows the user to identify individual variables easily. The use of different line styles or colors is also effective in this setting. The pattern of blue, green and red rays in addition to the varying ray-length lends itself to effective pattern recognition among related variables. For black and white presentation, it is suggested that the use of varying line styles is a good substitute to color.

Because the initial application of the polyplot was as an SPC tool, the characteristics of the glyph were designed to relate multivariate quality control information. The sequencing of the glyphs displays the production narrative over time, while the number of vertices informs the reader of the number of variables being tracked. The length of each variable's arm away from the polygon is significant information. That is, when a variable is "in control" near its mean value, the arm length is very short.

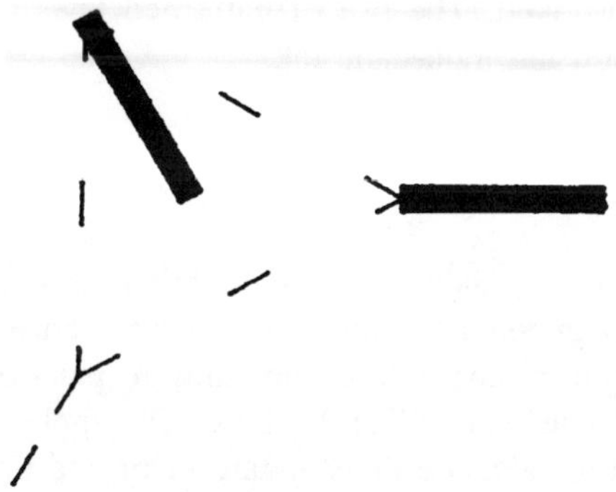

Figure 1.7 Three variable polyplot

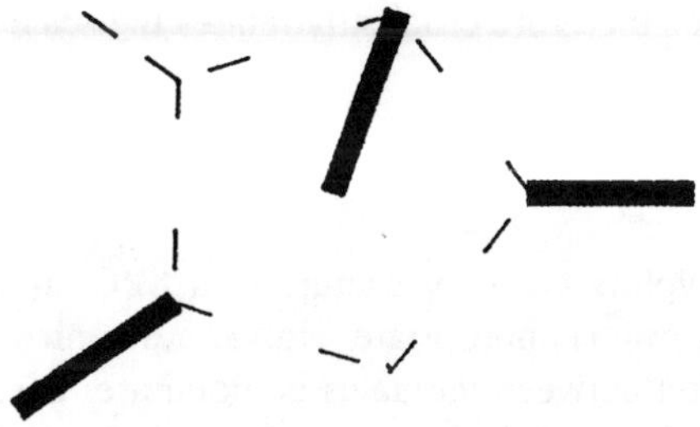

Figure 1.8 Five variable polyplot

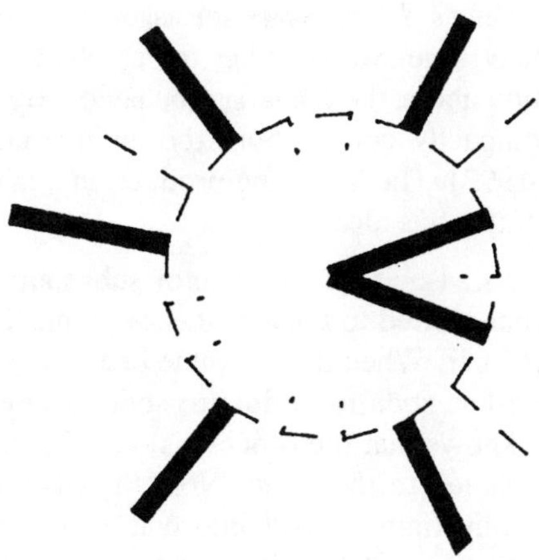

Figure 1.9 Seventeen variable polyplot

The dotted line style of the ray, or alternately the blue color, is unobtrusive. However, when the variable is "out of control", the arm lengthens and the line becomes solid or changes color, thus calling attention to itself visually. The glyphs are arranged in a time sequence in a left to right and top to bottom order, the natural reading sequence in the western world. Also, the use of differing line styles or colors emphasizes the points that SPC personnel are most interested in studying. The shift from dotted to dashed (or blue to green) occurs when the variable is approaching an "out-of-control" condition; the change to a solid (or red) line occurs when that condition is reached. The experience has been that the line pattern or color in combination with the different arm lengths identify many relationships which can be otherwise overlooked. The use of patterned lines or color therefore efficiently relates the scale without using "non-data" ink.

1.2.2.1 An application example for Statistical Process Control (SPC)

Polyplots are very useful in an SPC environment, especially when multivariate control charting is appropriate. Univariate control charts can be misleading even when the correlation between variables is moderate. Bivariate control charts limit the study to pairs of variables and lack time sequence information (Himmelblau, 1978). In these SPC applications, the number of individual observations used to calculate the standard error and average for each glyph are under user control. The rays of each glyph correspond to univariate X-bar chart entries. Each polyplot, then, represents the condition of the process for a given unit of time. By reading consecutive glyphs in normal order, left to right and top to bottom, the sequential history of a process is narrated. Finally, the bar to the left of each glyph represents the relative value of Hotelling's T^2 for the data displayed by each glyph. Charting Hotelling's T^2 provides an assessment of multivariate control in time sequence, but usually one would like to combine the multivariate control information with information about the univariate behavior. By including T^2 information in a polyplot, one can compactly convey both the multivariate and univariate control information (Tracy *et al.*, 1992). The following production quality control example taken from Blazek, (1987), illustrates this idea.

In Fig. 1.10 the values of T^2 associated with vector subgroup averages are represented by the height of a vertical bar plotted to the left of each glyph. The critical value for T^2 is the hash mark above the T^2 bar. When the T^2 value becomes significant and crosses the hash mark, the bar changes line style from dots to solid. Therefore, when T^2 exceeds a control limit, one not only knows that the process is out of control, but also which combination of univariate signals led to the event. Note that this implementation efficiently displays multivariate and univariate control information simultaneously. Figure 1.10 displays six measurements from 50 coils produced at a particular plant. The measurements are identified as variable 1 through variable 6. Each glyph represents the composite statistic on two consecutive coils.

The most interesting variable is Variable 1. After producing four good coils (the first two glyphs), Variable 1 went out of control when the control system lost its calibration. As a result the next 13 coils were out of control in a positive direction. During this time Variables 2 and 4 were also affected and went out of control at several points. The control system was tuned between coils 17 and 18 (and again after coil 20 as it was overcorrected the first time), and good coils were produced until coil 36 (glyph 18). Then, Variable 1 went out of control in a negative direction until the system was repaired again. During this time Variables 2, 4 and 5 also showed out-of-control values. Normal processing then resumed through the end of these records.

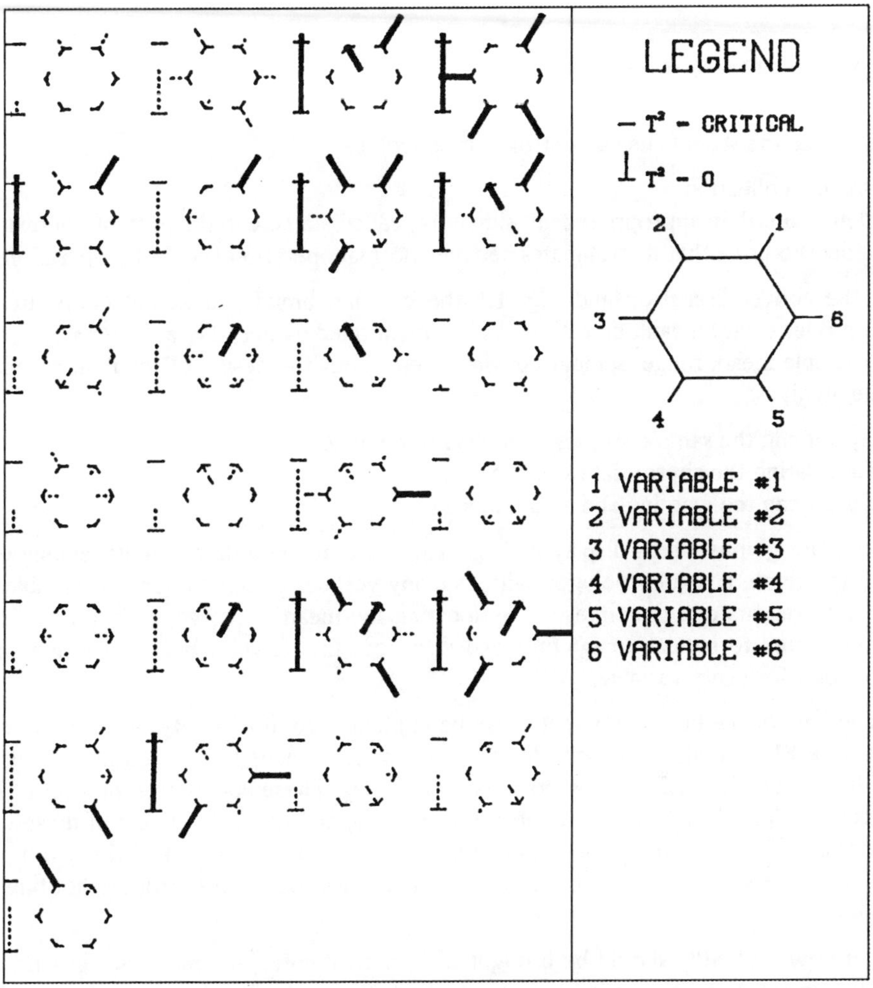

Figure 1.10 Six variable polyplot with Hotelling's T^2 of production data, *2 observations per glyph*

With this technique, both the ray and Hotelling "thermometer" visually indicated when the system went out of control and why. The polyplot also demonstrates the process relationship between Variables 2, 4 and 5 and Variable 1. The nature of the process is such, that only certain types of Variable 1 problems also manifest themselves in the other variables. Therefore, a conventional analysis of the data would probably not correlate the two. With this graphical hint and some process knowledge, conjunctures about the actual relationship can be developed.

1.2.2.2 Conclusions

A fault detection scheme using control charts involves:

(i) sample collection,
(ii) definition of an appropriate fault signature, called statistic in the present context,
(iii) a decision rule that discriminates between normal operation and faulty operation.

From the above discussion and Fig. 1.6 the warning limits and action limits are the decision rules used for fault detection. As it is mentioned earlier, the fault signature could be the sample mean, range, standard deviation etc. Thus the design of the fault detection scheme involves:

a) Specifying the sample size and sampling frequency.
b) Calculating the appropriate statistic.
c) Specifying the warning and action limits.

Polyplots are glyphs which display multivariate and univariate data simultaneously over time. Each glyph, a regular polygon with as many vertices as the number of variables of interest, represents one time interval. By locating the mean at the vertex and using rays from the vertex to indicate magnitude and direction, the polyplot becomes a univariate control chart for every variable.

An advantage of the polyplot is that it can be implemented in a variety of forms. For instance, in a SPC application, consider a quality measurement y modeled as a constant plus a linear combination of k components x_1, x_2,..., x_k. These may represent a set of key process variables which have the greatest impact on y, and the main effects of these variables dominate the variation in y. A polyplot of the x's with the height of the glyph proportional to the value of y could provide valuable real-time process troubleshooting information.

The real power of SPC should be brought to bear to involve all employees in studying their own processes to continuously improve them. This must involve not only manufacturing but marketing, sales, research and development, safety and environment, accounting, procurement, and so forth. As a general principle, one must consider automating routine, highly repetitive tasks, freeing the operators to use their intelligence for higher level problem-solving and prevention. The real future of computer technology in manufacturing and process monitoring is not in automating to replace workers but rather in "informating", that is, in providing workers with unprecedented levels of knowledge about the process, enabling them to make critical, collaborative decisions on the operating floor.

1.3 Fault diagnosis based on signal analysis instrumentation

Machinery condition monitoring is gaining increasing general acceptance throughout industry as a means of achieving reduced maintenance costs, increased machine reliability and increased safety standards. The correct and efficient application of condition monitoring to rotating machinery should enable plant and maintenance engineers to predict the major events affecting machine mechanical condition (i.e. bearing failures) and plan corrective action in advance. This will reduce or avoid the amount of routine maintenance work undertaken at fixed time intervals.

An effective condition monitoring programme will usually offer significant economic advantages over routine or breakdown maintenance by providing advanced warning of maintenance requirements. These can then be incorporated into a scheduled work programme to reduce down-time and lost production.

The major concern of this section is to present methods and instruments to analyse vibration or acoustical signals generated by a machine to reveal its operating conditions. The vibrations or the noise produced by mechanism forces can be used to reveal faults in the mechanism itself or some change in the vibration path. The fault revealed may be an actual malfunction or perhaps only a change in an operating parameter of the machine. In the latter case, the operating condition may be changed by a control system, but if a failure is detected, the machine probably needs to be shut down. Changes in the vibratory path may signal the need for the replacement or repair of structural elements. The use of vibration or noise to detect signal changes in mechanisms or structures is termed by Lyon, (1987) and Baines, (1987), *machinery diagnostics* or *machine health monitoring*.

1.3.1 Machine health monitoring methods

The success of a *condition monitoring* (CM) system is as dependent on its planning and design as on the sensors and signal analysis techniques used. Before the sensor fit and techniques can be practically decided, various features should be considered by the operator to define the requirements:

- Which items of machinery should be monitored ?
- What sort of faults should the system detect ?
- What level of diagnostic/prognostic information is required ?

Generally the answers indicate that the monitoring strategy should aim at reducing the number and severity of failure incidents between overhauls (which have high consequential costs in terms of damage and loss of availability) and increasing the prognostic capability so that maintenance can be planned effectively. The high plant availability and low maintenance cost requirements demanded in the current economic climate necessitate efficient, cost-effective monitoring systems. Their performance can be assessed by the criteria:

- Diagnostic and prognostic capability.
- Maximum information from minimum sensor fit.
- Low false-alarm rate.
- Low missed-alarm rate.

In general the equipment to be monitored must be considered as a number of separate machinery components and it is essential to identify the items which experience most faults and therefore cause most loss of revenue: these often vary between different installations. In many cases it is the auxiliary equipment which cause the largest reduction in machine availability and therefore they need to be incorporated into the condition monitoring system design.

Identification of the items to be included in the CM system and the level of monitoring to be applied can be assessed by careful consideration of the plant operation and its fault history; this process is worthy and will not be dealt with in great detail here.

Many current CM systems are derived from control and alarm instrumentation designed for a supervisory rather than a monitoring capability. This sort of system uses a limited number of sensors in key positions, generally on the main pieces of equipment, to monitor simple parameters for alarm level crossings.

The conventional monitoring techniques which will be presented in the following and in particular traditional vibration analysis methods are essentially energy methods and therefore only detect faults that generate significant amounts of changes in energy. Many faults only generate sufficient energy to trigger the alarm in the later stages of development, so that they are not detected until significant and costly damage has occurred.

Efficient extraction of fault signatures from sensory data is a major concern in fault diagnosis. A general self-tuning method of fault signature extraction that enhances fault detection, minimizes false alarms, improves diagnosability, and reduces fault signature variability will be also briefly presented.

***Vibration analysis*.**

Machine operation involves the generation of forces and motions that produce vibrations. These generating events are called sources. When vibration is measured from a transducer mounted on the casing of a machine, what is actually measured is the original force signal from the source of the signal, modified by the characteristics of the transmission path from the source to the measurement point. Expressed in terms of frequency this modification is a multiplication by the mobility of the transmission path.

A developing fault in a machine will show up as increasing vibration at a frequency associated with the fault. However, the fault might be well developed before it affects the overall vibration level. A frequency analysis of the vibration, on the other hand will give a much earlier warning of the fault, since it is selective, and will allow the increasing vibration at the frequency associated with the fault to be identified (see fig. 1.11).

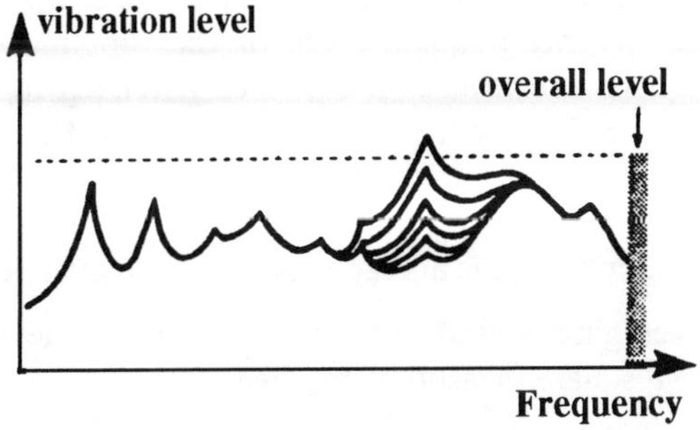

Figure 1.11 Frequency analyzed results give earlier warning.

The vibration from a rotating machine varies as parts wear out and are replaced. However, this variation is over such a long period that the signal can usually be regarded as being stationary. Truly stationary signals do not really exist in nature. Non-stationary signals can exhibit a short-term or a long-term variation. For instance, vibration from a reciprocating machine is stationary when regarded over several cycles, but over a single cycle, which consists of several transients, it is non-stationary. Vibration from a machine which is running up or down in speed, however, is non-stationary on a long term basis.

Stationary deterministic signals show the well-known line spectra. When the spectral lines show a harmonic relationship, the signal is described as being *periodic*. An example of a periodic signal is vibration from a rotating shaft. Where no harmonic relationship exists, the signal is described as being *quasi-periodic*. An example of a quasi-periodic signal is vibration from a turbojet engine, where the vibration signal from the two or more shafts rotating at different frequencies produce different harmonic series bearing no relationship to each other (Lyon 1987, Randall).

The well known Fourier Transform (Lyon 1987) gives the mathematical connection between time and frequency, and vice versa, and given a time signal allows calculation of the spectrum. The Fast Fourier Transform (FFT), see Appendix 1.B, is merely an efficient means of calculating the discrete form of the Fourier Transform (DFT).

A deterministic signal can be analyzed by stepping or sweeping a filter across the frequency span of interest and measuring the power transmitted in each frequency band. A random signal is a continuous signal whose properties can only be described using statistical parameters. Examples of random signals are cavitation and turbulence. Random signals produce continuous spectra. Since random signals have continuous spectra, the amount of power transmitted by the analyzing filter will depend on the filter bandwidth. In all frequency analysis, there is a bandwidth-time limitation. When using a filter, it shows up as the response time of the filter. A filter having a bandwidth of $B(Hz)$ will take approximately 1/B seconds to respond to a signal applied to its input. If the analyzing filter is B (Hz) wide, one has to wait at least 1/B seconds for a measurement. After filtration, the filter output must be detected. One can detect the peak level passed by the filter, the average level, the mean square level, or the root mean square level. Mean square or root mean square detection is used, since it relates to the energy or power content of the signal independent of the phase relationships. Peak detection is relevant when maximum excursions are important. Mean square and root mean square detection require that the output of the analyzing filter be squared and averaged. The period over which the square of the filter output is averaged is called the averaging time, T_A.

With random signals, averaging is used to reduce the standard deviation, σ, of the measured estimate. For a mean square measurement, then:

$$\sigma = \frac{1}{\sqrt{BT_A}}$$

where B is the analyzing filter bandwidth and T_A is the averaging time. For a root mean square measurement:

$$\sigma = \frac{1}{2\sqrt{BT_A}}$$

The above assumes that $BT_A \geq 10$. When FFT analyzers are used, BT will usually be equal to n_d, the number of averages. However, this will depend on the overlap conditions set. Overlap is where overlapping time records are analyzed and averaged. 0% overlap means that only results from statistically independent records are averaged.

As a consequence of the Central Limit Theorem, it can be assumed that any narrow band filtered random signal follows a gaussian distribution. Hence, from the properties of the Gaussian Distribution, there is a 68.3% chance of being within $\pm\sigma$, a 95.5% chance of being within $\pm 2\sigma$, and a 99.7% chance of being within $\pm 3\sigma$ of the true mean value of the signal.

Many FFT analyzers can also average signals in the time domain. Time domain averaging can be used with repetitive signals, for instance, repeated transients or vibration from rotating machines, to suppress extraneous noise. However, there must be a trigger signal synchronous with the signal being averaged. The amount of noise suppression (for random noise) which can be achieved with time domain averaging is equal to $1/\sqrt{n_d}$, where n_d is the number of time domain records averaged. Time domain averaging is also called signal enhancement, or synchronous averaging.

In vibration measurements external measurements of internal effects must be made. However, the transmission path characteristics from the source of vibration to the measuring point will vary from machine to machine, even if the machines are of the same design and construction. This is due to differences in castings, welds, tightness of bolts, etc.. Even in a single machine, the transmission path characteristics will vary with frequency.

Small, insignificant components can be amplified by resonances, and large, significant components damped by anti-resonances. Hence, it is essential to measure the spectra over a large dynamic range, because the largest spectral components are not necessarily the most significant. What can be measured on a machine is the change in vibration, and this will be transferable from machine to machine. If the system is linear, a relative change in the vibration at the source will give the same relative change in the vibration at the measuring point. Hence it is the relative change which is important. When relative changes are important, it is often convenient to express them in dB:

$$\text{Change in dB} = \log\frac{A_1}{A_2}$$

where A_1 is the present level and A_2 the previous level. The same relative change will give the same change in dB, independent of the absolute levels measured. The absolute levels themselves, however, will depend on the transmission path characteristics. This can be extended to making all measurements in dB refer to a common reference. Changes in vibration levels can then be conveniently plotted simply by subtracting the previous vibration level in dB form the present level. The two most commonly used methods of presenting data in the frequency domain are constant bandwidth on a linear frequency scale, and constant relative bandwidth on a logarithmic frequency scale. The two methods have their own different applications. The former gives equal resolution along the frequency axis, making it easier to identify such things as families of harmonics and sidebands. Its limitation is that it can only be used across a frequency range of about 1 and 1/2 decades. The latter can be used across a broader frequency range (3 or 4 decades is typical). Its drawback is that the resolution gets progressively worse at higher frequencies.

In fault detection, it is necessary to use a broad frequency range in order that all machine faults can be detected. Typically this requires a range of under half the slowest shaft speed to more than three times the highest toothmeshing frequency. Also, the possibility of easy speed compensation is desirable, since machine speeds will vary from measurement to measurement. Both these requirements are fulfilled by constant relative bandwidth analysis on a logarithmic frequency axis. Once again, the spectrum should be plotted with logarithmic amplitude.

The most basic level of vibration measurement is to measure the overall vibration level on a broadband basis in a range of, for example, 10-1000 Hz or 10-10000 Hz. Such measurements are also relevant with displacement measurements from proximity probes, where the frequency band of interest is usually from about 30% of the running speed up to about the 4th harmonic. An increasing vibration level is an indicator of deteriorating machine condition. Trend analysis involves plotting the vibration level as a function of time, and using this to predict when the machine must be shut down for repair. Another way of using the measurements is to compare them with published vibration criteria.

One example of a published vibration criterion chart is the *General Machinery Criterion Chart*, which is for displacement or velocity measurements at the bearing cap. In this chart there is always a factor of two involved in movement from one class to the next, that is a constant interval of 6 dB, and logarithmic axes are employed (Mitchell, 1981).

Another example is VDI 2056, from Germany, shown in fig. 1.12. It is for measurements at the bearings of the machine of interest in a frequency range of 10 Hz to 1000 Hz.

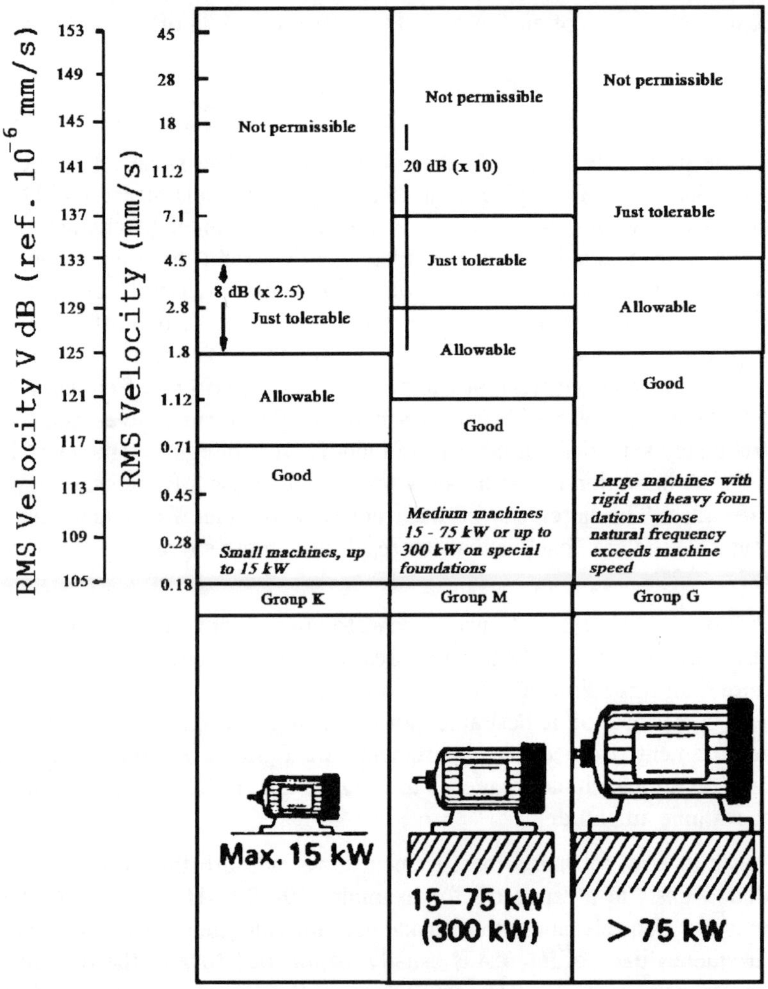

Figure 1.12 Vibration Criterion Chart (from VDI 2056).

This chart differentiates between vibration classes according to machine size. Note again the logarithmic velocity axis and the constant width of the allowable and just tolerable classes, (a change in vibration level by a factor of 2.5 or 8 dB), independent of machine size. This again emphasizes that it is the change in vibration level which is important, and when plotting changes, it is most logical to use a logarithmic axis.

Although overall vibration measurement proves a good starting point for fault detection, far more information can be obtained when a frequency analysis is employed, see fig. 1.13. Firstly, a frequency analysis will usually give far earlier indication of the development of the fault, and secondly, the frequency information can be used to diagnose the fault allowing spare parts to be bought in, etc..

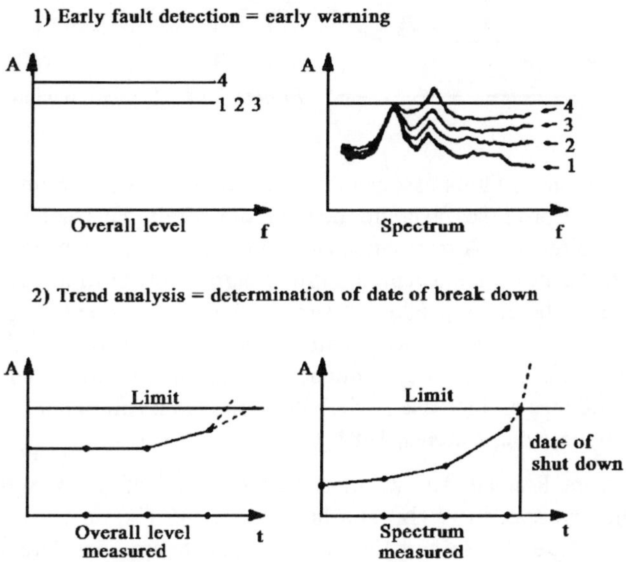

Figure 1.13 Benefits of frequency analysis for fault detection.

The objective of frequency analysis is to break down the vibration signal into its components at various frequencies. It is used in machine health monitoring because a machine running in good condition has a stable vibration spectrum. As parts wear and faults develop, however, the vibration spectrum changes. Since each component in the vibration spectrum can be related to a specific source inside the machine (e.g. unbalanced masses, toothmeshing frequency, blade pass frequency resonances), this then allows diagnosis of the fault.

The basis of fault diagnosis is that different faults in a machine will manifest themselves at different frequencies in the vibration spectrum, as it can be seen in fig. 1.14. The frequency domain information can then be related to periodic events in gears, bearings, etc.. Note that fault diagnosis depends on having a knowledge of the machine in question, that is the shaft frequencies, toothmeshing frequencies, number of teeth on gears, bearing geometries, etc.

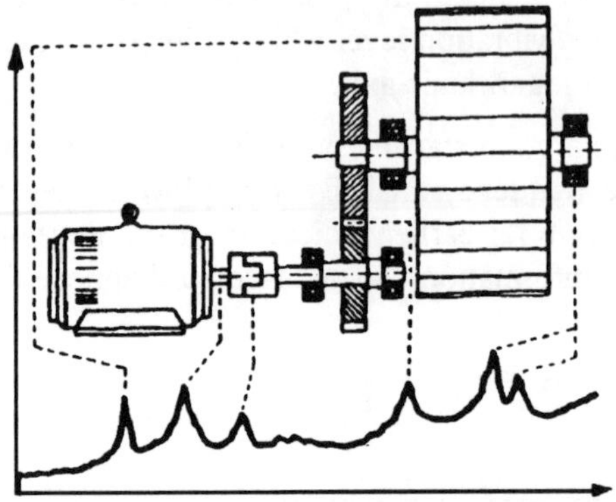

Figure 1.14 Typical machine "signature".

Two of the most common faults associated with rotating shafts are unbalance and misalignment. Unbalance produces a component at the rotational frequency of the shaft, mainly in the radial direction. A misaligned coupling, however, will produce a component at the rotational frequency, plus usually its lower harmonics, both in the axial and radial directions. Misaligned bearings produce a similar symptom, except that the higher harmonics also tend to be excited. A bent shaft is just another form of misalignment, and will produce vibration at the rotation frequency and usually its lower harmonics. Finally. a cracked shaft produces an increase in the vibration at the rotational frequency and the second harmonic (Cue 1990, Mitchell 1981).

Fig. 1.15 taken from Randall, shows an example of the effect of misalignment in a gearbox. Both the low speed (50 Hz) and high speed (85 Hz) shafts are originally misaligned. After repair, the 50 , 85 and 170 Hz components are considerably reduced. The 100 Hz component, however, remains more or less at the same level, which might appear strange until it is realised that it is not only the second harmonic of the shaft speed, but also the second harmonic of the mains frequency, (2-pole synchronous motor). This is a common electromagnetic source of vibration. Note that the higher noise level in the upper spectrum is because it was originally recorded as acceleration and integrated to velocity on playback.

Magnetically induced vibration is an important source of vibration in electrical machines. One source is the rotating magnetic field, which causes alternating forces in the stator. Since there are symmetrical conditions for a north or south pole, this gives rise to vibration at twice the mains frequency, or the "pole passing frequency". Note that in electrical machines, the force is proportional to the current squared, that is the vibration is highly load dependent. In induction motors, the rotational frequency will usually be slightly less than the synchronous frequency. For instance, fig. 1.16 shows the vibration spectra for an induction motor. The lower of the two is a detailed analysis obtained by non-destructive zoom, and shows that the high 100 Hz component is electromagnetic in origin rather than from misalignment.

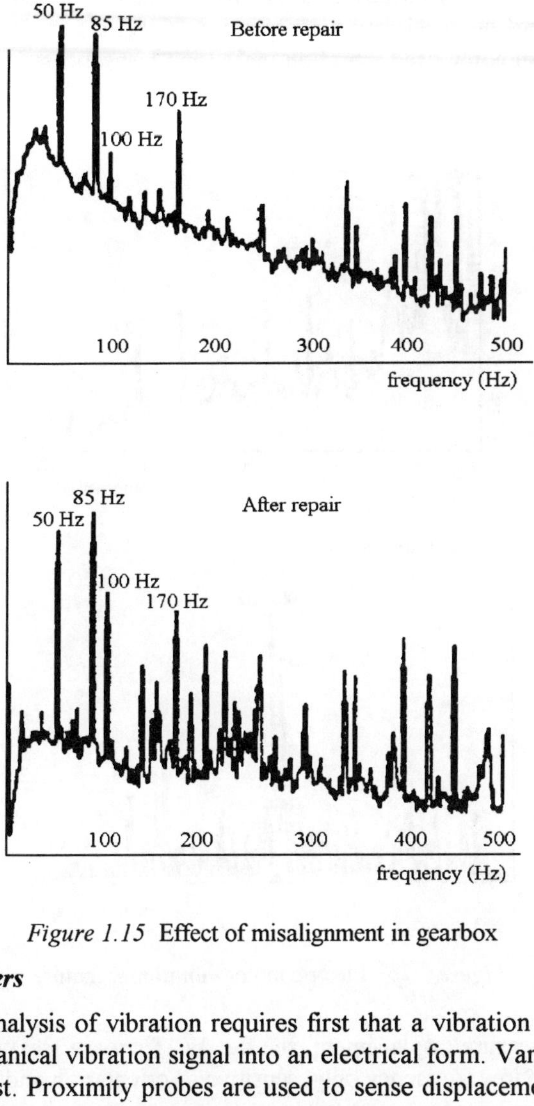

Figure 1.15 Effect of misalignment in gearbox

Vibration transducers

Measurement and analysis of vibration requires first that a vibration transducer be used to convert the mechanical vibration signal into an electrical form. Various types of vibration transducers exist. Proximity probes are used to sense displacement, velocity probes to sense velocity, and accelerometers to sense acceleration.

Displacement, velocity and acceleration are interrelated parameters. Displacement can be differentiated to produce velocity, and velocity can be differentiated to produce acceleration. Likewise, it is possible to integrate from acceleration to velocity and from velocity to displacement. In choosing which parameter to measure, it is usual to choose whichever parameter gives the flattest spectrum, so as to maximize the use of the dynamic range of the measuring instrumentation. As a rule of thumb, this will usually be velocity. Where the velocity spectrum is flat, the displacement spectrum will show a -6 dB/octave slope, and the acceleration spectrum a +6dB/octave slope. A brief presentation of the most commonly used vibration measuring transducers is given in the following.

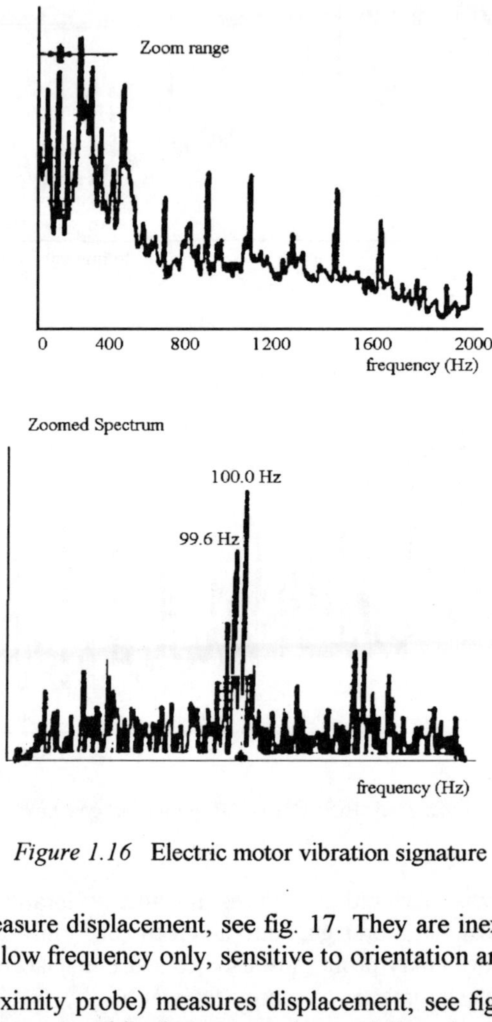

Figure 1.16 Electric motor vibration signature

Mechanical levers measure displacement, see fig. 17. They are inexpensive and self generating but limited to low frequency only, sensitive to orientation and prone to wear.

Eddy current (or proximity probe) measures displacement, see fig. 1.18. There are not moving parts and contacts, resulting in no wear, but variations in the magnetic properties of the shaft give erroneous signal components.

When a force is applied to a piezoelectric material in the direction of its polarisation, an electric charge is developed between its surfaces, giving rise to a potential difference on the output terminals. The charge (and voltage) is proportional to the force applied. The same phenomena will occur if the force is applied to the material in the shear mode. Both modes are used in practical accelerometer design.

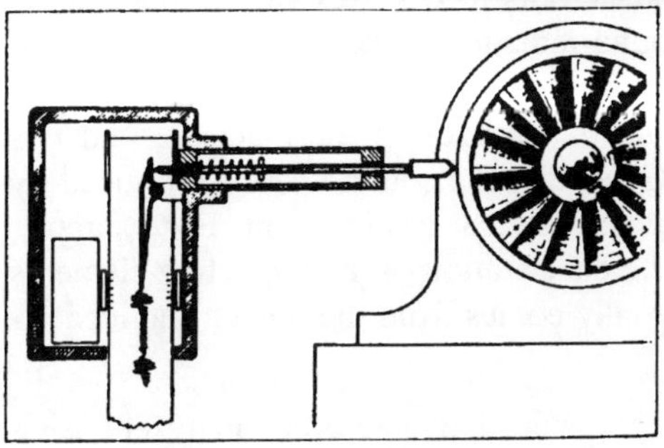

Figure 1.17 Mechanical levers.

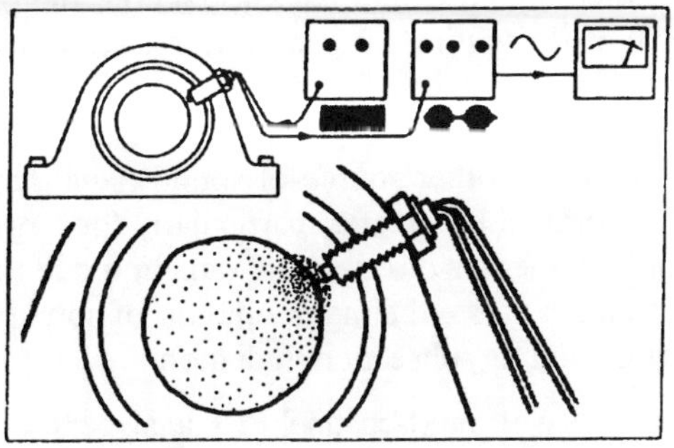

Figure 1.18 Proximity probe.

Accelerometers (compression type or shear type) measure acceleration, see fig. 1.19. Usually they have not moving parts so there is no wear and they have very large dynamic range and wide frequency range making them more suitable for applications.

***Noise analysis*.**

Up to now it has been studied how vibration is transmitted through a machine to its outer surfaces. In the following is considered how that vibration is converted into sound. Sound radiation is inherently a complicated process (Lyon, 1987). It turns out, however, that some fairly simple geometrical and dynamical parameters control sound radiation. These parameters allow to make reasonably good estimates of sound radiation. More specifically, Lyon, (1987), has shown that the sound radiated power of a vibrating machine structure is proportional to the space-time mean square vibration velocity.

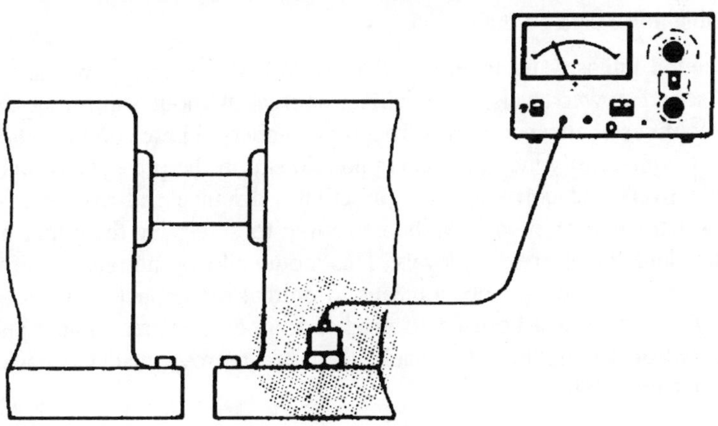

Figure 1.19 Accelerometer.

An example of a sound source is the noise produced by an air jet when it impinges on a rigid obstacle such as a fan blade. When the turbulent flow produces forces on an obstacle, then, by Newton's law of reaction, the obstacle puts forces back on the fluid in the form of fluctuating lift and drag, resulting in sound radiation. Large-scale motions associated with structural vibration are usually much more efficient in radiating sound. Impacting forces also produce a broad spectrum of vibration in the machine, and this represents another source of sound radiation. Generally, the sound energy produced by vibration will be greater, particularly for large machines that are resonant. For example, although there is direct sound radiation due to the deceleration of the impacting elements in a punch press, the major amount of sound usually comes from the impact-induced vibration and its subsequent radiation.

The ability of multi-channel FFT analyzers (see Appendix 1.B) and other analyzers using digital filters to quickly and accurately compute the cross spectrum between microphone signals has been the basis for the very rapid growth in using acoustical intensity measurements to determine the sound power radiated by machines. The usual measurement procedure is to surround the machine with a fixed array of microphone probes or a traverse setup that sweeps over an area surrounding the machine.

Identification and ranking of the noise sources is essential for both new and existing installations. Only the sources which are contributing to the excessive noise levels need to be treated.

Frequently a trial-and-error approach is used. Dominant sources are identified from far-field sound pressure measurements by comparing far-field noise spectra with near-field spectra of probable sources. It is very difficult however to distinguish between spectra of sources when many sources exist in the near-field. It is also difficult to know how much to silence a source and to know whether all the important sources are identified. Often a major source is treated, lowering the near-field noise but reducing the far-field levels only marginally because other sources start to dominate. It is also important that suppliers provide suitable noise data on their products.

The use of sound intensity techniques will provide better sound power information by determining sound power levels of individual sources without subjecting bulky equipment on the confines of anechoic or reverberant chambers. These sound power levels can be calculated from intensity measurements taken in situ in the presence of many sources. Using sound powers and correcting for directivity, distance and excess environmental attenuation, a mathematical model can be generated to determine the effect of the major sources on far-field sound pressure levels. This model allows the ranking of the sources in order of importance and provides a means to predict the impact of a noise abatement programme. The results will be used to predict sources at other similar plants and to provide information to suppliers to enable them to improve machine package design (Laws, (1987), Cue (1990)).

Sound power can be reliably calculated from sound pressure levels in a controlled environment, or in the free-field where sources do not interfere with one another. If ambient noise levels are high and the sound field is reactive, however, only sound intensity measurements will enable calculation of accurate sound power levels.

Sound intensity is the sound energy flux, a vector quantity describing the magnitude and direction of the net flow of acoustic energy. Therefore the dimensions commonly used for sound intensity are W/m^2, By taking ten-times the logarithm of the ratio of sound intensity of a reference value (10^{-12} $Watt/m^2$), the sound intensity level can be expressed in decibels.

The integral of the sound intensity over a surface is the sound power passing through the surface. The sound intensity and sound power levels can be expressed in terms of octave bands, third-octave bands, or overall noise level over any frequency range.

There are different instrumentation packages on the market that can measure sound intensity levels. The instrumentation usually consists of a pair of microphones in conjunction with either a dual channel Fast Fourier Transform (FFT) signal analyser or a real time sound intensity analyser (see Randal). For continuous level noise sources such as gas-turbines, both types of analyser will give similar results. For the study of an unsteady source such as a jack hammer, a real-time instrument should be used to capture the peak levels.

Sound intensity techniques have a number of inherent limitations, such as bias errors resulting from the finite pressure difference approximation for particle velocity, phase mismatch errors due to phase differences in the microphones and analyser channels, and reactivity errors resulting from phase mismatch of both the equipment and the measurements surface. The bias errors limit accuracy in the higher frequencies, while the phase mismatch errors limit the lower frequency capabilities. Reactivity errors could result at any frequency, depending on the location and sound power levels of extraneous sources and the distance between the microphones. The microphone spacing should be selected correctly for the frequency range of interest, in order to minimize the amount of error. These errors are fully discussed in the literature (Randall and Mitchell, 1981).

In addition to sound power levels, an important factor in determining the effect of a source on the far-field is the directivity of the source. There are two components of directivity which can be described as directivity factors; the directivity which a source would exhibit if it was operating in an anechoic chamber or in the air without any reflective surfaces (Q_A), and the directivity effect upon a source due to reflective surfaces, which can be termed spatial directivity (Q_S). Such sources as exhaust ducts, air-intake ducts and vents radiate sound non- uniformly even if there are no reflective surfaces. The spatial directivity factor accounts for reflections from such items as the ground and walls. The spatial directivity factors for spherical, hemispherical and quarter-spherical propagation are one, two and four respectively. The total directivity factor Q_Θ is defined as the product of Q_A and Q_S and this total directivity factor is translated into a directivity index (DI_Θ), expressed in decibels, by the following equation:

$$DI_\Theta = 10 \log Q_\Theta = 10 \log Q_A = 10 \log Q_S$$

Sound intensity measurements for any particular source will use one of three types of control surface; conformal (conforming to the shape of the object), hemisphere (a hemispherical "cover" placed over the source), box (a box-shaped "cover" placed over the source), or a combination of these. The control surfaces are determined using coordinates relative to the object of interest. The physical size of the majority of the sources examined in the case study of fig. 1.20 dictated the use of a box technique. This technique is best explained by describing sound intensity measurements over one of the sources investigated - an inertial air filter extraction fan. A box shape was constructed over the fan as shown in fig. 1.20 and the area of each of the five open sides was determined. The sixth side of the box was covered by the steel plate of the filter house. Since the fan and duct did not radiate sound uniformly through each side of the box, the sound intensity needed to be measured for each of the five sides. Before taking readings of the average sound intensity for the box, the choice was made between the use of a grid or a sweeping technique.

The grid technique involves constructing a real or imaginary grid of equal-area shapes over a surface and taking sound intensity measurements at the centres of each of these shapes. The grid size should be small enough so that the intensity does not vary greatly throughout the shape. An identical grid is set up within the sound intensity computer program (Ikeuchi, 1988) and measurements are then taken systematically.

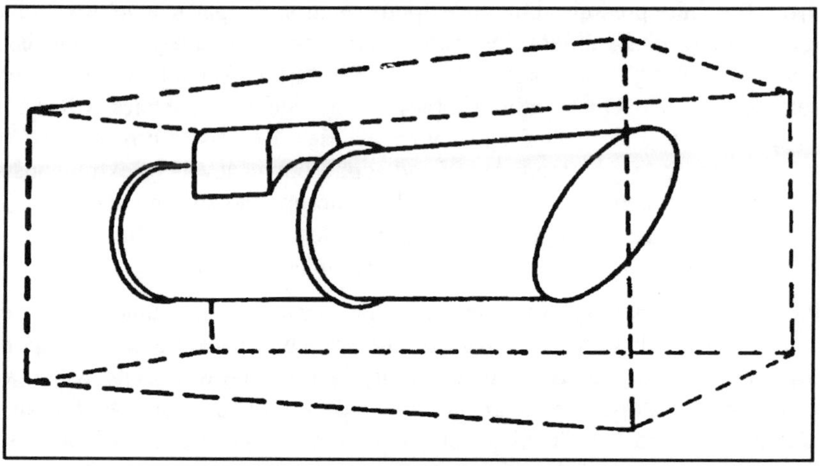

Figure 1.20 Extraction fan control surface.

The operator ensures that the probe is in the correct location in the centre of the shape and perpendicular to the box surface. Once all measurement points are stored, the computer program will display the sound intensity results in tabular or graphical form. The sound power for each of the grid areas, as well as the total sound power of the complete surface, is calculated by the computer.

The sweeping technique is a space-averaging process taken across a complete side. The probe is kept at right angles to the surface and swept uniformly across the surface while the analyser averages the sound intensity.

The sweeping technique was used for each of the five sides of the imaginary box over the extraction fan, using 250 averages over each of the sides or sectors. As described before, the frequency range over which measurements are taken results in various values of bias, phase mismatch and reactivity errors. These errors can be minimized by selection of the proper combination of frequency range and space between the probe microphones (see the case studies below).

Fault signature extraction.

A new general approach to the statistical development of diagnostic models is the use of nonparametric pattern classification techniques, so as not to require knowledge of the probabilistic structure of the system. Recently, Chin and Danai, (1991), introduced a nonparametric pattern classification method with a fast learning algorithm based on diagnostic error feedback that enables it to estimate its diagnosic model based on a small number of measurement - fault data.

This method utilizes a *multi-valued influence matrix* (MVIM) as its diagnostic model and relies on a simple diagnostic strategy ideally suited to on-line diagnosis. The MVIM method can also assess the diagnosability of the system and variability of fault signatures which can be used as the basis for sensor selection and optimization.

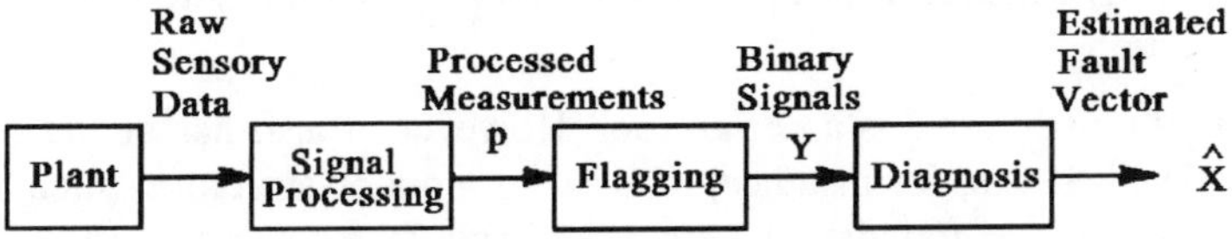

Illustration of the various stages of the fault signature extraction improved diagnosis (see fig. above). In the *flagging unit*, the processed measurements are first flagged by thresholds and then filtered by a single-layer network. A sample batch of measurement-fault vectors is used to tune the flagging unit through iterative learning using a nonparametric pattern classification method. Once all the measurement vectors in the sample batch are flagged, the MVIM is estimated to provide the indices for fault signature variability and system diagnosability. These indices along with the number of false alarms and undetected faults are then fed back to the unit's adaptation algorithm to tune the unit's parameters in its next adaptation iteration. The parameters of the flagging unit are tuned iteratively until its performance indices are extremized.

The effectiveness of this scheme is demonstrated by simulation in Chin and Danai, (1991),where the reader is referred for the detailed mathematical analysis and implementation features of the method. The method is also suitable for automatic tool breakage detection in machining.

System Analysis.

The distinction between signal analysis and system analysis is often made depending on what can be measured.

In practical analysis situations, there is no measurable input but a measurable output or both a measurable input and output. In the first, it is only possible to make a signal analysis, while in the second, because of the presence of information about both the input and the output, it is possible to make an analysis of both the signals and the system.

In signal analysis the input to the system is usually not measured. This can be due to any of three reasons. The first is that the input might be inaccessible. A good example of this is machine health monitoring, where external measurements must be used to monitor internal effects. The second reason is that it might be impossible to define an individual input, as, for instance, in many environment noise measurements. The third reason is that the output might be the only item of interest, as, for instance, in noise dose or whole body vibration measurements.

In system analysis, measurements are made of both the input and the output to the system. It is best to measure the input and output simultaneously (while at the same time taking account of any system delays), so as to maintain the phase relationships, although some limited system analysis (measurement of the magnitude of the frequency response function) is possible using sequential measurements. With system analysis one can obtain the system properties which can then be used to predict how the system will behave under various excitations.

System analysis is mostly used as a design tool. However, it also has applications when systems are installed. It can, for instance, be used to monitor structures such as oil production platforms, machine foundations, etc., for faults. It can also be used for determination of signal sources and signal paths when, for instance, it is necessary to isolate part of a system from vibration. The classical method of system analysis is to use swept sine testing. The system is excited with a sine wave, and feedback from the output is used to hold the input amplitude constant as the sine wave is swept up or down in frequency.

Hence, the amplitude of the sine wave at the output of the system gives the magnitude of the frequency response and the phase difference between the input and output. The advantages of swept sine testing are high signal-to-noise ratio and the possibility of studying non-linearities. The disadvantage is that it is slow. However, the speed limitation has been largely removed by ***Time Delay Spectrometry***, (TDS), where very fast sine sweeps are used to give results almost in real-time.

Dual-channel digital filtering is rarely used for system analysis although it is potentially a very powerful tool, since true real-time measurements can be made. It is a very powerful means of acoustic and vibration intensity measurements.

Dual-channel FFT analysis forms a very powerful and widely used means of system analysis (Randall and Ikeuchi, 1988). Both the input and output of the system are measured simultaneously (taking account of system delays). The basic measured data are the autospectra at the input and output and the cross spectrum between the input and output, from which many other functions can be calculated. The phase information is maintained, and the effects of noise can be reduced.

Some advantages of dual-channel FFT analysis are flexibility and the fact that is easy to use. Also, because the input signal to the system need not be controlled, naturally occurring excitations can be used. Finally, since it is a digital form of analysis, the results can be easily entered into a computer, for example, to carry out a modal analysis.

Dual-channel FFT analyzers are easy to use, but there are several pitfalls that it is necessary to be aware of. Three of them include leakage, the assumption of a linear system, and compensation for system delays.

Leakage is an effect which occurs because FFT analyzers (both single and dual-channel) operate on a time-limited signal. The rectangular weighting introduced produces a (sin$x$)/x filter characteristic, and power "leaks" from the main lobe to the sidelobes, meaning that measured peaks can be too low and measured valleys too high. Leakage can be combated by using higher resolution (zoom), introducing an artificial time window, or where the excitation can be controlled, choosing the right excitation (see Randall).

Linearisation can also be considered an advantage. However, it is important to remember that the dual-channel FFT analyzers impose linearity, even if the system being measured is non-linear.

All physical systems exhibit a propagation delay. When a propagation delay becomes significant, as can frequently happen in mechanical and acoustical (and electrical) systems, it becomes necessary to compensate for it when making a 2-channel FFT analysis, otherwise bias errors will be introduced into the results.

For instance suppose a system has a propagation delay of T seconds, and the analyzer processes data blocks T seconds long. If the analyzer processes simultaneous data blocks at the input and output, the measured frequency response $\hat{H}$ will be lower than the true response H by a factor $(1 - \tau/T)$. Likewise, the measured coherence γ^2 (see below for an exact definition) will be low by a factor $(1 - \tau/T)^2$. System analysis measurements are usually based on the Fourier Transforms, see fig. 1.21, of the input and output time signals $a(t)$ and $b(t)$. The input and output spectra produced (S_A and S_B) are two sided, that is they exist for both positive and negative frequencies. However, since the time functions are real and S_A and S_B will both be conjugate even (that is symmetrical amplitude about f=0, but opposite phase), it is usual to combine the positive and negative frequency halves to form the single sided spectra G_A and G_B, which are zero for negative frequency.

The basic functions usually used for system analysis are the input and output autospectra, (formerly called the input and output power spectra), and the cross (power) spectrum, i.e.

- input autospectrum

$$G_{AA} = \begin{cases} \overline{G_A^* G_A} & (f = 0) \\ \overline{(1/2) G_A^* G_A} & (f > 0) \end{cases}$$

- output autospectrum

$$G_{BB} = \begin{cases} \overline{G_B^* G_B} & (f = 0) \\ \overline{(1/2) G_B^* G_B} & (f > 0) \end{cases}$$

- cross spectrum

$$G_{AB} = \begin{cases} \overline{G_A^* G_B} & (f = 0) \\ \overline{(1/2) G_A^* G_B} & (f > 0) \end{cases}$$

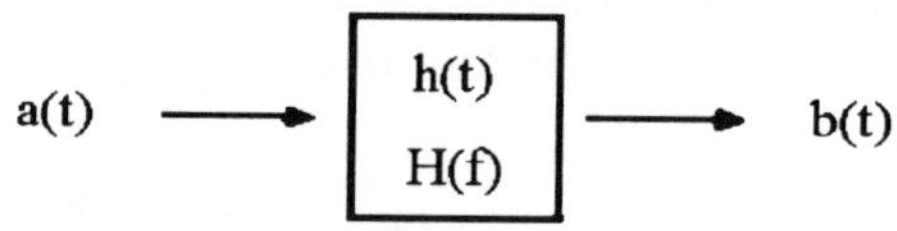

$$S_A = F[a(t)]$$

$$S_B = F[b(t)]$$

$$G_A = \begin{cases} 2S_A & (f > 0) \\ S_A & (f = 0) \end{cases}$$

$$G_B = \begin{cases} 2S_B & (f > 0) \\ S_B & (f = 0) \end{cases}$$

Figure 1.21 System analysis measurements.

The input and output autospectra are the squared and averaged input and output spectra. Note that they contain no phase information. The cross spectrum is the product of the coherent amplitudes at the input and output and the phase difference between the input and output. The cross spectrum is the most important function in system analysis, since it contains the phase information, and since uncorrelated noise at the input and output will be averaged out in the cross spectrum.

Given the three basic functions, many input/output relationships can be calculated by taking various combinations of the three and by using Fourier Transforms. The most important is the system frequency response $H(f)$ and the system impulse response $h(\tau)$. The impulse response is the well known time response of a system to a delta function and it can be calculated by taking the inverse Fourier Transform of the system frequency response (see Appendix 1.B). Cross correlation shows whether the input and output signals are correlated and at what time delays. It can be calculated by taking the inverse Fourier Transform of the cross spectrum.

Three different methods can be used to measure a frequency response function. The first method is based on $|H_a|^2$, the ratio of the output to the input autospectrum as it would be measured using a single-channel analyzer.

Two other methods can be used in a dual-channel analyzer. These are the traditional method, H_1, which is the ratio of the cross spectrum to the input autospectrum, and a newer method, H_2, which is the ratio of the output autospectrum to the inverse cross spectrum, i.e.,

$$H(f) = \frac{B(f)}{A(f)}$$

$$|H_a(f)|^2 = \frac{G_{BB}(f)}{G_{AA}(f)}$$

$$H_1(f) = \frac{G_{AB}(f)}{G_{AA}(f)}$$

$$H_2(f) = \frac{G_{BB}(f)}{G_{BA}(f)}$$

The three methods will behave differently according to whether there is noise at the input, noise at the output, or noise at the input and the output. Noise at the output produces an error in $|H_a|^2$ and H_2. On the other hand, H_1 will be unaffected, since it is a function of the input autospectrum C_{AA}, which is noise free, and the cross spectrum, C_{AB}, where the noise can be averaged out. Hence H_1 will give the correct result. An example where there will be noise at the output is where there are other, unknown inputs to the system. The effects of these other inputs will show up as noise at the output, as shows the figure below:

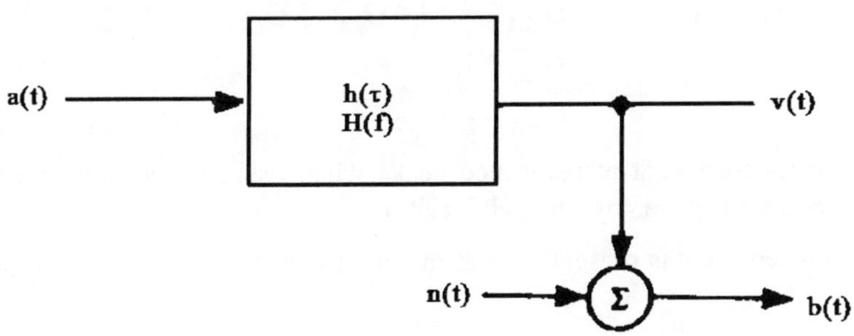

Noise at output: In the case of noise at the output one obtains:

$$|H_a|^2 = \frac{G_{BB}}{G_{AA}} = |H|^2\left[1 + G_{NN} / G_{VV}\right]$$

$$H_1 = \frac{G_{AB}}{G_{AA}} = H$$

$$H_2 = \frac{G_{BB}}{G_{BA}} = H\left[1 + G_{NN} / G_{VV}\right]$$

Noise at the input produces an error in $|H_a|^2$ and H_1. This time, H_2 will give the correct result. An example where there will be noise at the input is where a specimen is being excited with random noise on a shaker. At a resonance of the specimen, the shaker is effectively trying to drive a mechanical short circuit, which drives the input signal down towards the noise floor of the measuring instrumentation. Hence the input signal-to-noise will be low. The output signal-to- noise will be high, however, because of the resonance of the specimen. This situation is shown in the figure below:

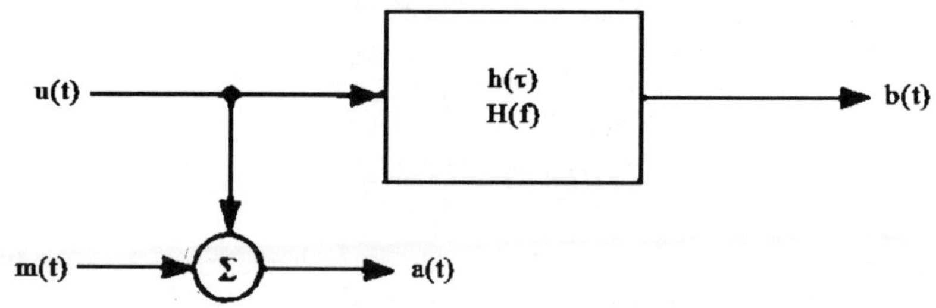

Noise at input: In the case of noise at the input one obtains:

$$|H_a|^2 = \frac{G_{BB}}{G_{AA}} = |H|^2 \frac{1}{1 + G_{MM} / G_{UU}}$$

$$H_1 = \frac{G_{AB}}{G_{AA}} = H \frac{1}{1 + G_{MM} / G_{UU}}$$

$$H_2 = \frac{G_{BB}}{G_{BA}} = H$$

Use of H_2 for measurement at resonance peaks when using broad band random noise excitation was first proposed by Mitchell, (1981).

The situation when noise is present both at the input and the output is shown in the next figure:

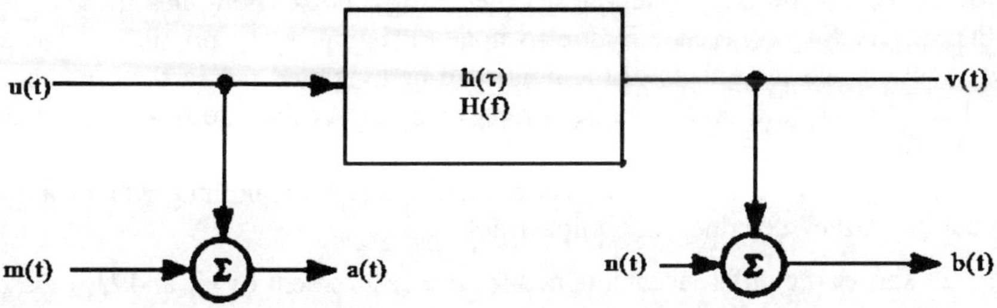

Noise at Input and Output: In the case of noise at the input and output one obtains:

$$\epsilon_i = G_{MM} / G_{UU} \qquad \epsilon_0 = G_{NN} / G_{VV}$$

$$|H_a|^2 = \frac{G_{BB}}{G_{AA}} = |H|^2 \frac{1+\epsilon_0}{1+\epsilon_i}$$

$$H_1 = \frac{G_{AB}}{G_{AA}} = H \frac{1}{1+\epsilon_i}$$

$$H_2 = \frac{G_{BB}}{G_{BA}} = H[1+\epsilon_0]$$

$$\gamma^2 = H_1 / H_2$$

$$|H_a|^2 = |H_1| \cdot |H_2|$$

$$|H_1| \le |H| \le H_2|$$

Note that the true value of the frequency response function will always lie between H_1 and H_2, and that while H_1 tends to give a low estimate, H_2 tends to give a high estimate. The user can choose H_1 or H_2 (after measurement). H_1 is lower bound while H_2 is upper bound. H_2 reduces bias errors for resonance peaks with random excitation. The coherence function relates how much of the measured output signal is linearly related to the measured input signal, i.e.

$$\gamma_{AB}^2 = \frac{|G_{AB}|^2}{G_{AA} \cdot G_{BB}}, \qquad 0 \le \gamma_{AB}^2 \le 1$$

A coherence of 1 indicates a perfect linear relationship, and 0, no relationship. The coherence is always bounded between 0 and 1. It can also be shown that the coherence function is equal to $|H_1|^2$ divided by $|H_a|^2$, indicating that $|H_a|^2$ will always give a high estimate of the frequency response function unless there is a coherence of 1 (Randall and Mitchell, 1981). Likewise, the coherence function is equal to H_1 divided by H_2, again indicating that unless there is a coherence of 1, H_2 will always be greater then H_1.

Low coherence can be due to, amongst other things, noise, non-linearities, or leakage. Note that where low coherence is due to noise, it is still often possible to make good measurements, since the effects of the noise can be averaged out in the cross spectrum. Low coherence due to leakage can be combated by increasing the resolution. Here, it is also important to remember that although the coherence will be the same for H_1 and H_2, H_2 will converge on a resonance peak faster than H_1. There is nothing which can be done to combat low coherence due to non-linearities.

Figure 1.22 shows the differences obtained for a measurement of H_1 and H_2 on a cantilever bar mounted on a shaker and excited with random noise. The resonance peaks in H_1 are about 10dB lower than in H_2.

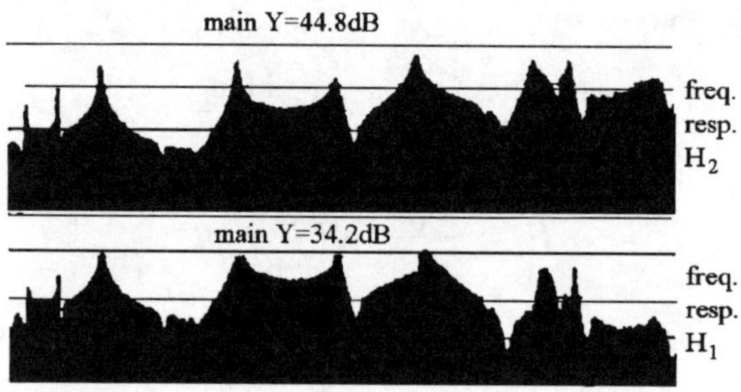

Figure 1.22 Differences between H_1 and H_2 measurements

Another important tool in signal analysis is the *power cepstrum* (see for details Appendix 1.B). The cepstrum is a sort of "spectrum of a spectrum". The distinctive feature of the cepstrum is the logarithmic conversion of the spectrum. The power cepstrum can be applied to the detection of periodic structure in the spectrum (harmonics, side bands, echoes, reflections) and for the separation of source and transmission path effects. The power cepstrum is a sensitive measure of growth of harmonic/sideband family (can be used for separation of different families) and it is insensitive to measurement point, phase combination, amplitude and frequency modulation and loading. An illustration of the use of the cepstrum for both detection and diagnosis of a gear box fault is given in fig. 1.25 of the next section.

1.3.2 Vibration and noise analysis application examples.

A. Gearbox failure diagnosis.

Even gears with perfect involute profiles exhibit vibration due to tooth deflection under load, and in particular the sudden changes in this as the load is shared between different numbers of teeth. The tooth deflection effect is at the toothmeshing frequency (and its harmonics) and is very load dependent, see fig. 1.23.a. (Cue, (1990), Randall).

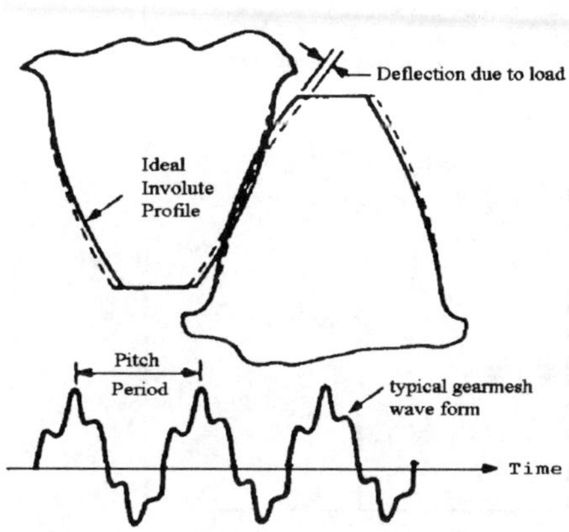

Figure 1.23a Effect of tooth deflection.

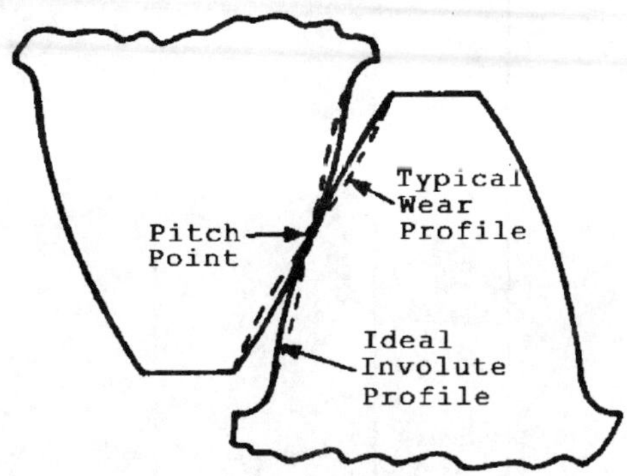

Figure 1.23b Effect of wear

Components also occur at the toothmeshing harmonics due to mean deviation from the ideal profile. These may be a result of initial machining errors, but will eventually be dominated by the effects of uniform wear. Wear tends to be greater on either side of the pitch circle, as illustrated, because of the greater sliding velocity there (with pure rolling at the pitch circle), see fig. 1.23.b. The effects of such geometrical errors are much less load sensitive.

Fig. 1.24 illustrates typical increases in toothmeshing harmonics due to uniform wear. The effect of wear is often first seen in the second harmonic, but usually spreads to the higher harmonics as the profile deteriorates. It is advisable to monitor at least 3 harmonics, as the signal at the first harmonic must first exceed the effects of tooth deflection to be noticeable. Measurements must be made at constant load, for comparisons to be meaningful.

Fig. 1.25 is an illustration of the use of the cepstrum for both detection and diagnosis of a gearbox fault. Sideband family can be clearly seen in spectrum, but in cepstrum it can be detected by monitoring only one component, at 95.9 ms (detection). Measured period (95.9 ms) and corresponding frequency (10.4 Hz) is measured so accurately as to eliminate second harmonic of output shaft speed (5.4 Hz) as a possible source. Source was traced to rotational speed of second gear, even though this was unloaded because first gear was engaged (diagnosis).

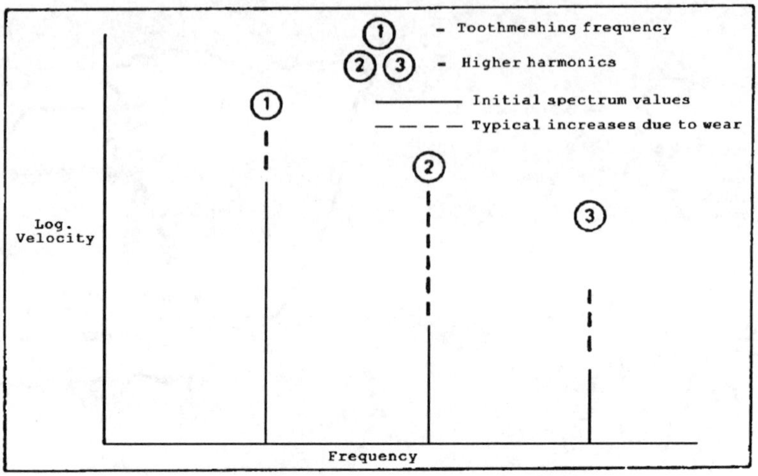

Figure 1.24 Gear toothmeshing harmonics.

B. *Faults in rolling element bearings*

Discrete faults in the elements of a ball or roller bearing give rise to a series of impacts at a frequency determined by the location of the fault, outer race, inner race, etc.. (Randall, Li and Wu (1989), Pengelly and Ast (1988).

The initial impulses are so short, in particular when the faults are still microscopic, that their frequency content extends up to perhaps 300 kHz. The shocks excite structural and other resonances, including the resonance of piezo-electric transducers used to detect them, and produce a series of bursts, as illustrated, with a frequency content dominated by these resonances. This bearing signal is masked by other background vibrations from the machine, and the basic problem is to find a frequency range where the bearing signal is dominant over the background vibration.

Note that the repetition frequency is better indicated by analyzing the envelope of the bursts, rather than the raw signal.

Fig. 1.26 shows how a discrete fault causes a series of bursts with a repetition frequency given by the bearing geometry and rotational speed. The frequency content of the bursts is high (dominated by the resonances excited) and the component at the repetition frequency is small. If the envelope of the bursts is formed, however, its frequency spectrum is dominated by the repetition rate (and its harmonics).

It is possible to calculate the repetition frequency of the bursts using simple classical mechanics, see fig. 1.27. However, note that the relationships assume pure rolling motion, while in reality there is some rolling and some sliding motion. Hence the equations should be regarded as approximate. Also amplitude modulations can produce side-bands.

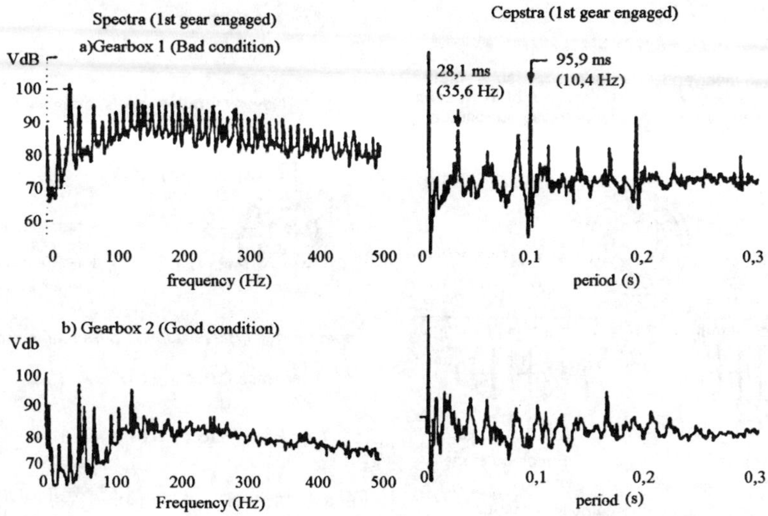

Figure 1.25 The use of the cepstrum for fault detection and diagnosis of a gearbox.

On-line bearing fault monitoring implies automatic data processing without human intervention. Vibrations, picked up by sensors, are transmitted to a monitoring system where they are processed for information extraction. The on-line system comprises a data acquisition stage, where analog bearing signals are converted into digital form and a data processing stage, where modular software algorithms are employed to perform the designed algorithm under the guidance of a supervisor program.

The data acquisition stage comprises (1) an accelerometer, (2) a charge amplifier, (3) a band pass filter, and (4) an analog to digital converter. The data processing stage consists of three functional units: supervisor, defect detection/diagnosis unit, and data base.

The block diagram which illustrates the organization of the complete system is shown in fig. 27.a. The supervisor is responsible for : (1) the proper logic sequence of system operation, (2) the data flow control between the defect detection/diagnosis units and the global data base, and (3) global data base management.

A global data base is constructed to hold all the information to be relayed among the data acquisition stage, the functional units of data processing stage, and the system's human-machine interface. It comprises 3 data files:

(1) general purpose data file, (2) raw bearing signal data file, and (3) pattern vector data file.

One important reason for having a global data base is that the external data files will preserve important data just prior to any unforeseen shutdown of the monitored system or bearing monitoring system itself.

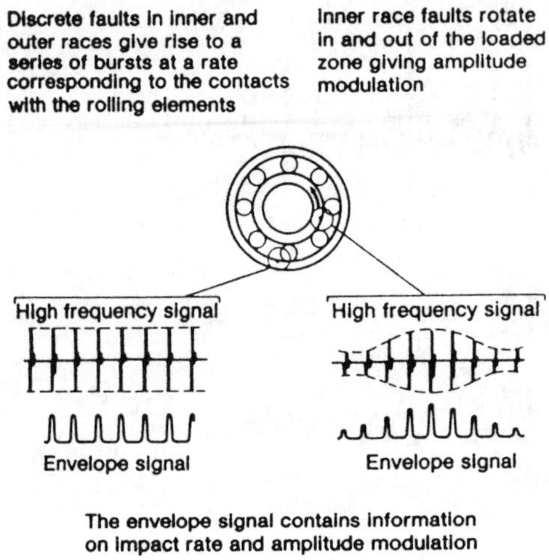

Figure 1.26 Faults in rolling element bearings.

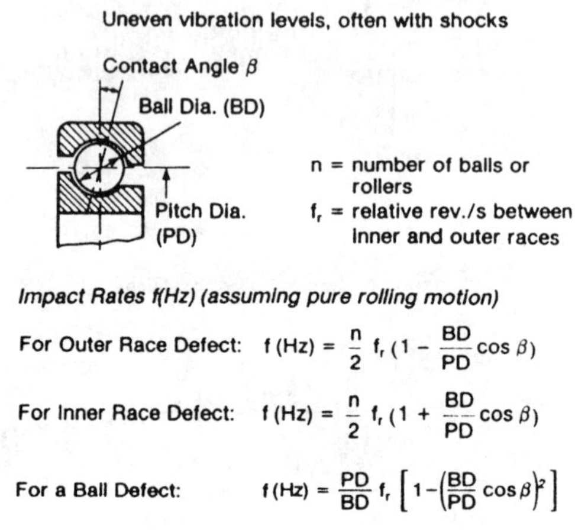

Figure 1.27 Faults in ball and roller bearings

The interface is responsible for the human-monitoring system communication. The necessary input information consists of (a) bearing geometry, (b) bearing rotational speed, and (c) sampling rate. The output quantities through the interface consists of alarm and diagnosis.

Among others, short-time energy function, short-time average zero crossing rate, and median smoothing are employed by the proposed scheme. The definition of the short-time energy function is,

$$E_n = \sum_{m=-\infty}^{\infty} x^2(m)w(n-m)$$

where $x(n)$ is the sampled signal and $w(n) = 1$ if $0 \le n \le N\text{-}1$, and zero otherwise (N is the width of window).

In the context of discrete-time signals, a zero-crossing is said to occur if successive samples have different algebraic signs. The rate at which zero crossings occur is a simple measure of the frequency content of a signal. Its definition is,

$$Z_n = \sum_{m=-\infty}^{\infty} \left| sgn[x(m)] - sgn[x(m-1)] \right| w(n-m)$$

where sgn[$x(n)$] = 1 if $x(n) \geq 0$ and = -1 if $x(n) < 0$, and $w(n) = (1/2)N$ if $0 \leq n \leq N$-1; = 0 otherwise. However, the second equation makes the computation of Z_n appear more complex than it really is. All that is required is to check samples in pairs to determine where the zero-crossings occur with the average being computed over N consecutive samples.

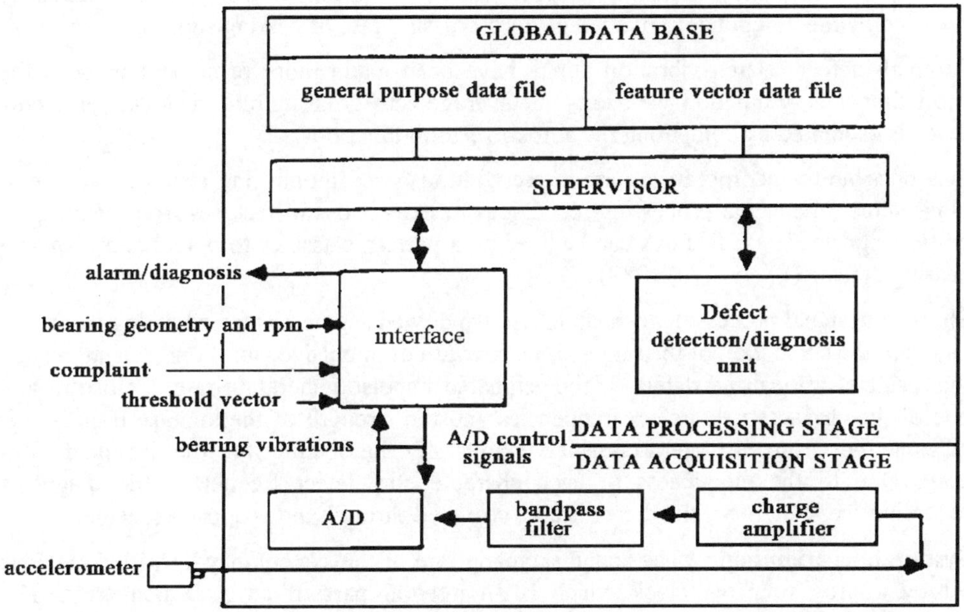

Figure 1.27.a. Block diagram representation of the on-line bearing monitoring system

The definition of median smoothing is

$$M_n = M\left[x(m)w(n-m)\right]\Big|_{m=-\infty}^{\infty}$$

where $w(n) = 1$ if $0 \leq n \leq N$-1, and not valid otherwise. This property makes median smoothing useful for prior processing of the bearing signal because short-time energy function and short-time average zero crossing rate are sensitive to the discontinuities in the signal. The signal discontinuities can be removed by subtracting the output of the median smoother from the signal.

The tactic for feature extraction is based upon the estimation of the rate and the strength of impulse generation from bearing signals. The estimation is performed by highlighting the occurrence of defect related vibration bursts in the signal and computing their rate of occurrence and strength. If this estimated impulse generating rate happens to coincide with any of the characteristic defect frequencies and the strength of vibration bursts is significant, the existence of a localized defect is concluded.

A bearing signal dominated by bearing defect sensitive resonances reveals the occurrence of damaged related impulses in the fluctuations of its amplitude. These amplitude fluctuations are made more prominent by computing their short-time energy functions.

On the other hand, the fluctuations of those defect-excited resonances that have a much higher frequency than the vibration generated by other machine elements will introduce variations in frequency content of bearing signals. Variations of this type may be easily revealed by the computation of average zero-crossing rate of bearing signals.

After all defect related vibration bursts have been made more prominent in both the short-time energy function and short-time average zero-crossing rate, their rate of occurrence is estimated by computing the autocorrelation functions.

It is possible to incorporate a pattern recognition based monitoring scheme which employs short-time signal processing techniques to extract useful features from bearing vibration signals. These features can be used by a pattern classifier to detect and diagnose bearing defects (Li and Wu, 1989).

Short-time signal processing techniques use windowed segments of the bearing signals to facilitate the estimation of the rate and the strength of impulsive vibrations, which may be the result of a localized defect. If the estimated impulse generating rate is close to any one of the characteristic defect frequencies and the strength of the impulse train is significant, the designed pattern classifier will classify the bearing into the damaged category. Due to the uniqueness of each characteristic defect frequency, the diagnosis regarding the location of the defect is also provided through the proposed scheme.

System operation logic. The actual sampling rate of the system may be varied and it is set by a programmable clock which is an integral part of an A/D converter. The supervisor first reads in the necessary inputs from the human operator. Then the data acquisition stage activates the A/D converter according to this information. Once the converter is initiated, it supplies discrete data elements to the system at the selected rate until a specified number of samples has been generated. The data are formulated into records and written onto the peripheral disk memory (data file no. 2).

The data will be processed by the defect detection/diagnosis unit. The loop of data measuring and processing will be continuously carried out until any kind of localized defect is detected and diagnosed. In the case of such an event, the pattern vector, estimated impulse generating rate, and classifier output will be displayed on the CRT to alert the operator.

Details for the on-line implementation of the described technique can be found in Li and Wu (1989).

C. Reciprocating machine and gas turbine fault detection

Vibration and noise signals from reciprocating machinery come, typically, from events which occur at different phases of the machine cycle, see fig. 1.28. This amounts to a signal which, though repeated every cycle, varies during one cycle. Continuous averaging lumps all these signals together and track is lost of this variation which from everyday experience is important in judging the condition of the machine. An FFT Analyzer makes it possible to pick out a short sample length. of the signal which is associated with one particular event and analyze it separately. This is done by triggering the Analyzer from a tacho pulse every cycle and using the variable time delay of the Analyzer to choose that phase of the signal to be analyzed (Randall).

The signal and the tacho pulse are recorded on an instrumentation tape recorder. They are then played back into the FFT Analyzer which is set to trigger on the tacho pulse after a specified delay. A series of spectra for various trigger settings, representative of various phases of the machine cycle, are recorded on the digital cassette recorder. Comparison and data display is handled by a programmable calculator and, optionally, a digital plotter (see Randall).

The basic steps used in the analysis for collecting spectra are presented in fig. 1.29, where:

(a) Represents a typical impulsive signal from a four cylinder diesel engine.
(b) Represents the once per cycle tacho signal.
(c) Represents the positions of the Hanning time window after 2 specified trigger delays, Trig 1 and Trig 2, set up in the FFT Analyzer.
(d) Represents a series of spectra which are used to obtain a representative average spectrum for the phase of the engine cycle corresponding to Trig 1 .
(e) Represents a series of spectra obtained by repeating the process for a new trigger delay setting, Trig 2.

These average spectra can then be stored on the digital cassette recorder. The signals involved vary somewhat from cycle to cycle, and a considerable amount of averaging may be necessary to obtain a reliable result.

A logic diagnosis for the whole engine can be incorporated in the computer memory such as a binary interrogation of the type shown in fig. 1.30, which produces additional prognostic conclusions (see also chapter 4).

By comparing actual values with base values determined for the system theoretically and values measured when the plant was commissioned, the relevant trends can be reported. Additionally, values are theoretically determined for the plant when operating with defects and programmed into the computer; thus when measured conditions fall within those for which the defect applies, the "status" is reported and the defect notified.

Automated fault identification for gas turbines based on spectral features of measurements of various dynamic quantities, such as internal pressure, casing acceleration, acoustic data is presented and applied by Loukis *et al.*, (1992). The examined faults were rotor fouling (fault of all the blades), individual rotor blade fouling (fault of 2 blades of stage 1 rotor), individual rotor blade twisted, stator blade restaggering. The difference pattern (used as fault index) derived from the measured signal of an instrument is defined by the expression :

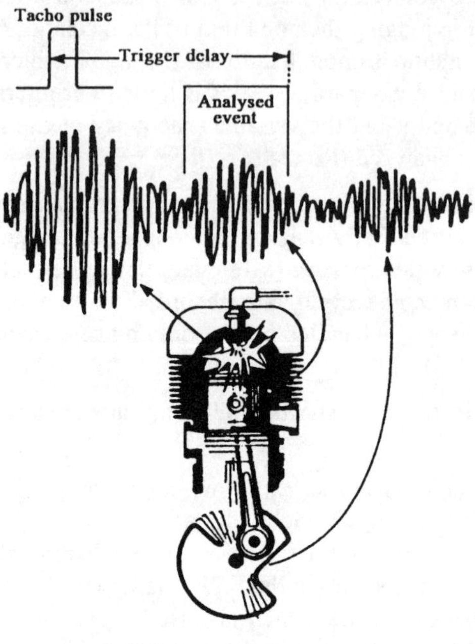

Figure 1.28 Reciprocating machine fault detection.

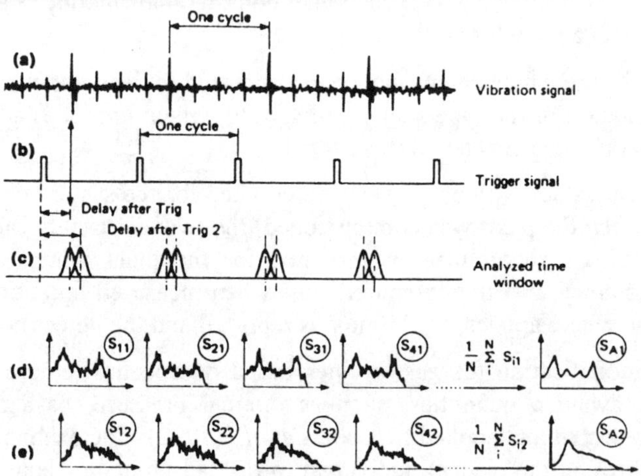

Figure 1.29 Basic steps used in the analysis for collecting spectra.

$$p(f) = 20[\log_{10}(sp(f)) - \log_{10}(sph(f))]$$

where $p(f)$ is the fault index, which is thus a set of values in function of frequency, $sp(f)$ is the power spectrum of the signal of the measuring instrument from the monitored (possibly faulty) engine, $sph(f)$ is the same spectrum from a healthy engine. This index showed high sensitivity to all examined faults and produced distinguishable forms for each fault.

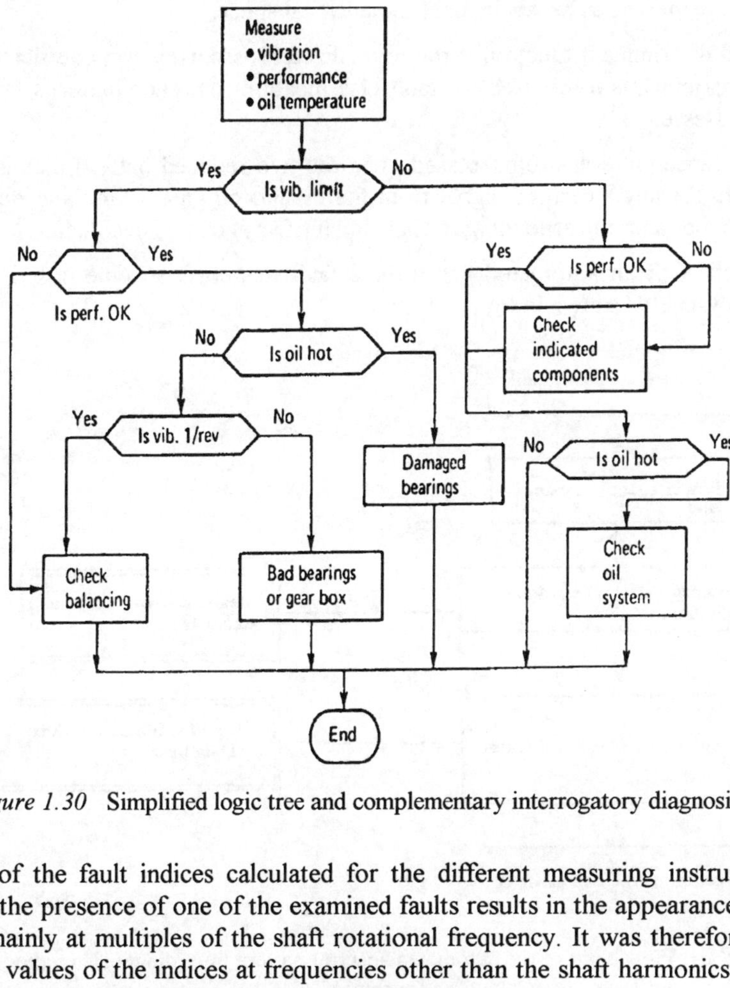

Figure 1.30 Simplified logic tree and complementary interrogatory diagnosis.

Inspection of the fault indices calculated for the different measuring instruments has shown that the presence of one of the examined faults results in the appearance of differentiations mainly at multiples of the shaft rotational frequency. It was therefore decided to filter out values of the indices at frequencies other than the shaft harmonics, since the most useful diagnostic information is contained at these harmonics.

This is done by a filter defined by the following equation:

$$h(f) = \begin{cases} 1 & \text{if } f \text{ is rotational harmonic} \\ 0 & \text{if } f \text{ is not rotational harmonic} \end{cases}$$

The pattern resulting from filtering the difference pattern with the above filter will be refered to as reduced pattern $p_r(f)$.

Two discriminant functions were selected, one of them expressing the quantitative similarity (influenced by both the shape and amplitude of compared patterns) the other expressing the shape similarity (influenced only by the shape of the compared patterns).

The first discriminant function is the usual Euclidean distance between reduced patterns, when they are viewed as points in an N-dimensional space.

The second discriminant function is the normalized crosscorrelation coefficient. In order the two discriminants to be used for fault classification reference patterns for each fault must be possessed.

If such reference patterns are possessed, then the two reduced pattern discriminants can be produced for any measured signal from the monitored gas turbine and depending on their values the fault corresponding to that signal (if any) can be decided.

The flow chart of an automated gas turbine fault diagnosis scheme based on spectral pattern analysis is as shown in fig. 1.31:

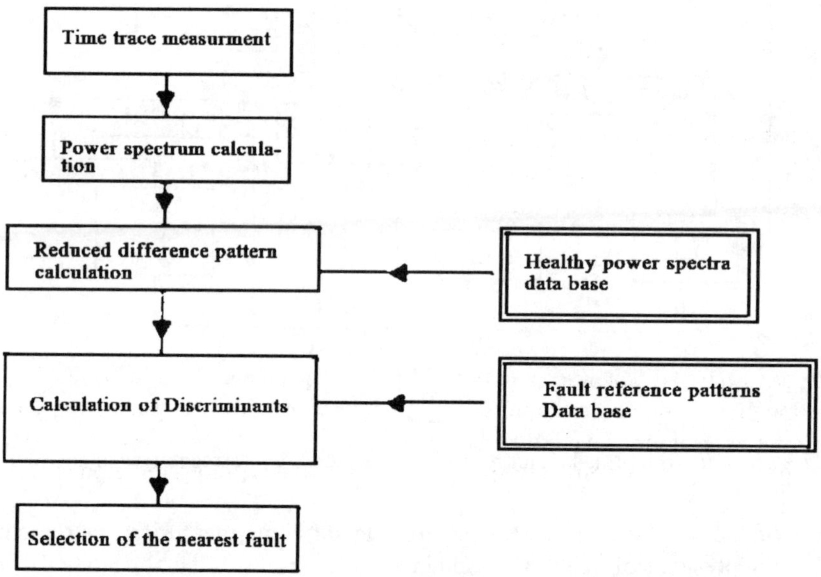

Figure 1.31. Flow chart of the automated spectral pattern fault diagnosis method for gas turbines.

The technique presented briefly previously has been developed on the basis of the data available from experiments in an industrial gas turbine with specific implanted faults

(Loukis *et al.*, 1992). From this point of view part of the findings can be considered of general validity while others will be particular to the specific engine.

D. Induction machine broken bars detection.

The desire to improve the reliability of industrial drive systems has led to concerted research and development activities in several countries to evaluate the causes and consequences of various fault conditions. In particular, ongoing research work is being focused on rotor bar faults and on the development of diagnostic techniques.

Broken bars were shown to produce high localized airgap fields and to degrade mechanical performance. The field perturbation associated with broken bars, which are deliberately disconnected from the endrings by machining, produces, low-frequency components and harmonics in the search coil-induced voltages and gives rise to an oscillatory torque that produces noise and mechanical vibration.

Different techniques for the detection of broken bars were tested and evaluated by Elkasabgy *et al.*, (1992).

A. Search Coil Induced Voltage Detection Techniques. This technique involves an inspection of the time and frequency domain of voltages induced in internal (stator tooth tip and yoke) search coils or an inspection of the time and frequency domain of voltages induced in an external search coil placed against the frame of the motor.

Consider first the voltage induced in an internal stator tooth tip coil for a machine with no rotor faults. The dominant and fundamental frequency will be at the motor excitation frequency of 60 Hz. Higher frequency components will appear due to the periodicity of rotor bars. The nonsinusoidal stator emf distribution, i.e., space harmonics, may also induce time-harmonic voltages.

Consider now the voltage in the same search coil for a machine with one or more adjacent broken bars. An anomalously high airgap local field rotates at rotor speed. This field pulsates at slip frequency and can be considered to be the resultant of two fields, counterrotating at $s\times$(synchronous speed), which are rapidly attenuated away from the fault location. The field associated with the broken bars will, therefore, modulate the coil-induced voltage at a characteristic frequency f_{fault}, given by

$$f_{fault} = \left(\frac{2f}{p}\right)(1-s) \pm sf \text{ Hz}$$

where f is the excitation frequency, s is the slip, and p is the number of poles of the induction motor.

Similar frequency components are anticipated in the yoke and external search coil voltages. These voltages an be captured by a fast data acquisition system and printed out to illustrate their time dependence. Their frequency spectra can also be analyzed. Comparing the tooth-tip search and the yoke-search coils induced voltage frequency spectrum for a fault-free rotor with that for the broken-bar rotor, broken bars can be adequately detected.

Perhaps surprisingly, the external search coil is just as effective as the internal coils in detecting the broken bars. It therefore appears unnecessary to incorporate internal coils to take advantage of this diagnostic technique; an external coil placed against the casing of the machine being entirely adequate.

The use of an external coil placed against the frame of the machine is considered particularly useful in an industrial environment because the motor need not be modified in any way (by the installation of stator search coils) or be taken out of service temporarily. All that is needed is a coil of 10-20 turns with a length equivalent to the active axial length of the machine and of width equal to perhaps half a pole, a power-frequency oscilloscope or low-frequency spectrum analyzer, and the experience of an observant operator.

B. Stator-current detection technique. Each individual rotor bar can be considered to form a short-pitched single-turn single-phase winding. The airgap field produced by a slip-frequency current flowing in a rotor bar will have a fundamental component rotating at slip speed in the forward direction with respect to the rotor, and one of equal amplitude that rotates at the same speed in the backward direction. With a symmetrical rotor, the backward components will sum to zero. For a broken-bar rotor, however, the resultant is nonzero. The field, which rotates at slip frequency back-ward with respect to the rotor, will induce EMFs in the stator side that modulate the mains-frequency component at twice slip frequency.

Under sinusoidal voltage excitation, this effect produces twice slip frequency (2.67 Hz at 1760 r/min) side bands in the spectrum of the phase current, which indicate the existence of the fault.

Thus, the examination of the machine current spectrum provides an important method for detecting rotor-bar faults.

C. Torque-harmonics detection technique. In a balanced three-phase induction machine with no rotor faults, the forward-rotating field interacts with the slip frequency induced rotor currents to produce a steady output torque. For a machine with a rotor fault, a backward rotating field is developed as discussed. This backward rotating field interacts with the rotor currents, induced by the forward rotating field, to produce a torque variation at twice-slip frequency, which is superimposed on the steady output torque.

Rotor faults therefore lead to low-frequency torque harmonics, which result in increased noise and vibration. The torque oscillations can be measured by means of a shaft torque transducer using a data acquisition system while the motor is running on-line at various load conditions. The frequency of the torque oscillation increases as the machine is loaded. The dominant frequency of oscillation corresponds to twice the slip frequency of the operating condition.

1.3.3 Conclusions

Advances in instrumentation and computer information processing technologies offer machinery users an added dimension to their protection systems-diagnostic information for machinery behaviour analysis and predictive maintenance programs. Two decades ago only rudimentary vibration data could be obtained using the electronic technology of the time. Today, computerised vibration instrumentation is a powerful tool for obtaining more specific vibration data, for diagnostic purposes, including steady state data, taken on-line and transient data, taken at start-up and shut-down. The data, properly used, provides the machinery diagnostician with greater insight into the mechanical behaviour of rotating machinery.

As more and more vibration information becomes available, the proper methods for interpreting it are becoming increasingly important for accurately analysing machine behaviour and diagnosing malfunctions. Proper methodology ensures a thorough understanding of the machine's mechanical behaviour and the potential malfunction mechanisms, as well as their symptoms. This knowledge enables one to diagnose and correct machine malfunctions accurately.

Without the ability to reduce and interpret vibration data properly, diagnosing a machine malfunction can be difficult, time consuming and costly. A systematic approach to machine analysis makes it possible to use vibration data for pinpointing the malfunction, despite a myriad of symptoms. Six basic principles create the foundation for accurate machine malfunction diagnosis methodology:

1. Know the basic mechanical characteristics of the machine.
2. Know the types of malfunction mechanism that the machine is likely to incur and their corresponding symptoms.
3. Monitor the key parameters that will indicate a change in the condition of the machine.
4. Reduce diagnostic data to interpretable formats.
5. Understand the historical and related events that may have caused the change in machine condition.
6. Take prompt action, based on the diagnosis.

Any machinery analysis requires that the diagnostician is first familiar with the physical parameters of the machine (size, weight, clearances). However, understanding the basic mechanical factors goes beyond knowing these physical characteristics; it involves understanding the machine's fundamental rotor dynamic behaviour during steady-state and transient conditions. To achieve this, vibration data needs to be acquired during acceptance testing, commissioning, imbalance response testing (synchronous perturbation), and non-synchronous perturbation testing to document the rotor response during steady-state (on-line) and transient (start-up and shut-down) conditions. This process provides baseline data for insight in:

- The machine's natural frequencies.

- The amount of damping in the system.
- The machine's susceptibility to instability.

Analytical models of the machine's dynamic behaviour - particularly models based on a modal approach - are of tremendous value in future diagnostic procedures. It is desirable for critical machines or those that have had a history of vibration problems to incorporate these analytical models into the machine's data base. Critical machines are categorised as those which are indispensable for plant operation; failure of a critical machine results in immediate loss of plant output.

The analytical and testing baseline information is necessary for comparing the machine's previous and present condition when troubleshooting suspects malfunctions. It can also be used after a problem has been corrected to verify the validity of the correction.

References

Anderson T. W. (1958). An Introduction to Multivariable Statistical Analysis. John Willey, New York.

Anderson T. W. (1971). The Statistical analysis of time series. John Willey, New York.

Ali M. M. (1989). Test for autocorrelation and randomness in multiple time series. *Journal of the American Statistical Association*, **84**, 406.

Baines N. (1987). Modern vibration analysis in condition monitoring. *Noise and Vibration control world wide*, p. 148.

Basseville M. and A. Benveniste (1980). Detection of Abrupt Changes in Signals and Dynamical Systems. Springer-Verlag, Berlin.

Bennett C. A. and N.L. Franklin (1954). Statistical analysis in chemistry and the chemical industry. John Wiley.

Blazek L., Novic D. and D. Scott (1987). Displaying Multivariate data using polyplots. *Journal of Quality technology*, **19**, 2, p. 69.

Blumen I. (1958). A new bivariate sign test. *Journal of the American Statistical Association*, **53**, p. 448.

Box G. and T. Kramer (1992). Statistical Process Monitoring and Feedback adjustment - A discussion. *Technometrics*, **34**, 3, p. 251.

Cecchin T. (1985). Conception d' une usine pilote mineralurgique automatisee. Elaboration et application d' une strategie de diagnostic d' etat de fanctionnement de procedes industriels en temps reel. These de doctorat de l' Institut National Polytechnique de Lorraine.

Chin H. and K. Danai (1991). A method of fault signature extraction for improved diagnosis. *ASME Transactions of Dynamic Systems, Measurement and Control*, **113**, p. 635.

Chitturi V. (1976). Distribution of multivariate white noise autocorrelations. *Journal of American Statistical Association*, **71**, 353, p.223.

Commissariat a l' Energie Atomique (1978). Statistique appliquee a l' exploitation des mesures. Editions Masson, Paris.

Cue R.W. and D.E. Muir (1990). Engine performance monitoring and trouble shooting techniques for the CF-18 aircraft. *Proceedings, Gas Turbine and Aeroengine Congress*, Brussels, June 11-14, 1990.

Dixon W. J. Power functions of the sign test and power efficiency against normal alternatives. *Annals of Mathematical Statistics*, **24**, p. 467.

Elkasabgy N.M., Eastham A.R and G.E. Dawson (1992). Detection of broken bars in the cage rotor on an induction machine. *IEEE Transactions on Industry Applications*, **28**, 1, p. 165.

Hawkins D.M. (1992). A fast accurate approximation for average run lengths of CUSUM control charts. *Journal of Quality Technology*, **24**, 1, p. 37.

Himmelblau D. M., (1978). Fault detection and diagnosis in chemical and petrochemical processes. Elsevier, Amsterdam.

Hoerl R.W. and A.C. Palm A.C. (1992). Discussion: Integrating SPC and APC", *Technometrics*, **34**, 3, p. 268.

Hunter J.S. (1986). The Exponentially Weighted Moving Average. *Journal of Quality Technology*, **18**, 4, p. 203.

Ikeuchi T., Shirai M., Nakamachi K., Tanabe S., Ishino K. and T. Fujishima (1988). Computer-assisted noise and vibration analysis system - CANVAS. *Noise and Vibration Control Worldwide*, February, p. 58.

Johns W. D. and R.H. Porter (1988). Ranking of compressor station noise sources using sound intensity techniques. *Noise and Vibration Control Worldwide*, February, p. 70.

Kendall M., Stuart A. and J.K. Ord (1982). The advanced theory of statistics. Vols. 2, 3. Charles Criffin Ltd., London.

Laws W. C. and A. Muszynska (1987)Periodic and continuous vibration monitoring for preventive/predictive maintenance of rotating machinery. *ASME Journal of Engineering for Gas Turbines and Power*, **109**, April, p. 159.

Li C.J. amd S.M. Wu (1989). On line detection of localized detects in bearings by pattern recognition analysis. *ASME Journal of Engineering in Industry*, **111**, November, p. 331.

Liggett W., Jr. (1977). A test for serial correlation in multivariate data. The annals of Statistics, **5**, 2, p. 408.

Loukis E., Mathioudakis K. and K. Papailiou (1992). A precedure for automated gas turbine blade fault identification based on spectral pattern analysis. *ASME Journal of Engineering for Gas Turbines and Power*, **114**, April, p. 201.

Ljund L. and T. Söderstrom (1987). Theory and practice of recursive identification. The MIT Press, London.

Lucas J.M. and M.S. Saccucci (1990). Exponentially Weighted Moving Average Control Schemes: Properties and Enhancements. *Technometrics*, **32**, 1, p.1.

Lyon R. (1987). Machinery noise and diagnostics. Butterworths ed., London.

MacNeill I. B. (1974). Tests for periodic components in multiple time series. Biometrica, **61**, 1, p. 57.

Mehra R.K. and J. Pesčhon (1971). An innovations approach to fault detection and diagnosis in dynamic systems. *Automatica*, 7, p. 637.

Mitchell J. S. (1981). Machinery analysis and monitoring. Penn Well ed., London.

Pnegelly B.W. and G.E. Ast (1988). A computer-based multipoint vibration system for process plant rotating equipment. *IEEE Transactions on Industry Applications*, **24**, 6, p. 1062.

Pignatiello J. and C. Runger (1990). Comparisons of multivariate CUSUM charts. *Journal of Quality Technology*, **22**, 3. p. 173.

Pouliezos A. (1980). An iterative method for calculating sample serial correlation co-efficients. *IEEE Transactions on Automatic Control*, **AC-25**, 4, p. 834.

Proceedings, 15th Symposium "Aircraft Integrated Monitoring Systems", Aachen, Germany, September 12-14, 1989.

Randall R. D. Efficient Machine Monitoring. Bruel and Kjaer publication, code no. 18-212.

Randles R. (1989). A distribution-free multivariate sign test based on interdirections. *Journal of the American Statistical Association*, **84**, 408.

Robert P., Cleroux R. and N. Ranger (1985). Some results on vector correlation. *Computational Statistics and Data Analysis*, **3**, p. 25.

Spee R. and A.K. Wallace (1990). Remedial Strategies for brushless DC drive failures. *IEEE Transactions on Industry Applications*, **26**, 2, p. 259.

Stephens C.M. (1991). Fault detection and management system for fault-tolerant switched raluctance motor drives. *IEEE Transactions on Industry Applications*, **27**, 6, p. 1098.

Thompson R.W. (1991). Importance sampling applied to a robust signal detector. *International Journal of Modeling and Simulation*, **11**, 4, p. 114.

Tracy N.D *et al*. (1992). Multivariate control charts for individual observations. *Journal of Quality Technology*, **24**, 2, p. 88.

Vance L. C. (1983). A Bibliography of Statistical Quality Control Chart Techniques, 1970-1980. *Journal of Quality Technology*, **15**, p. 59.

Vander Wiel S.A *et al*. (1992). Algorithmic Statistical Process Control: Concepts and an application. *Technometrics*, **34**, 3, p. 286.

Vasilopoulos A. and A. Stamboulis A. (1978). Modification of control chart limits in the presence of data correlation. *Journal of Quality Technology*, **10**, 1, p. 20.

Walter T. J., Marchione M. M. and H.C. Shugars (1988). Diagnosing vibration problems in vertically mounted pumps. *ASME Journal of Vibration, Acoustics, Stress and Reliability in Design*, **110**, p. 173.

Yashchin E. (1993). Performance of CUSUM control schemes for serially correlated observations. *Technometrics*, **35**, 1, p. 37.

Appendix 1. A

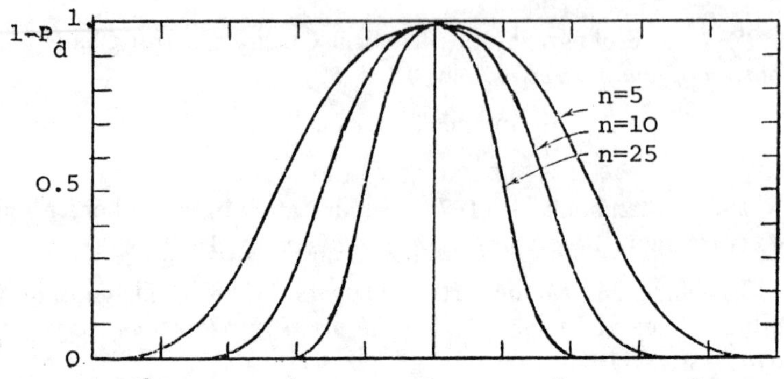

Figure 1.A.1 Operating characteristic curves for the sample mean test, $P_f = 0.01$

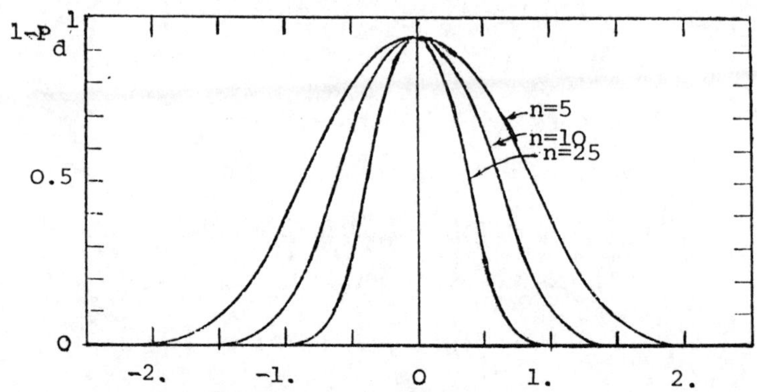

Figure 1.A.2 Operating characteristic curves for the sample mean test, $P_f = 0.05$

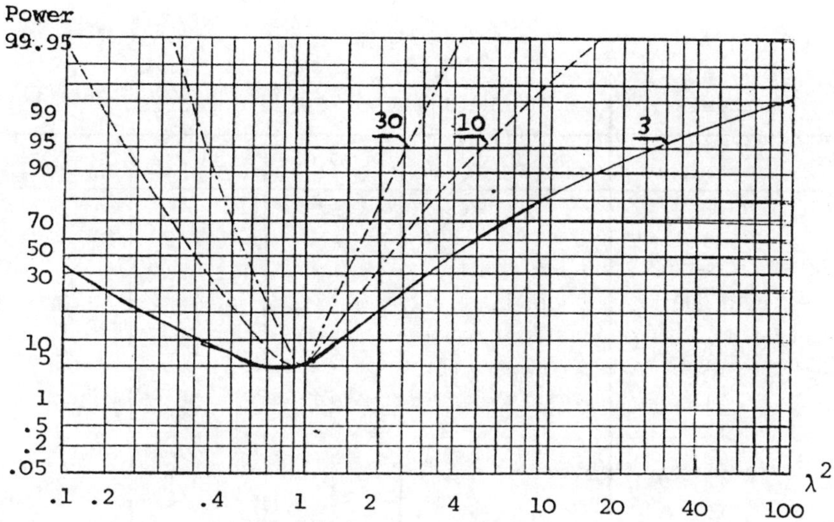

Figure 1.A.3 Power curves for the two-tailed χ^2-test at the 5% level of significance.

Table 1.A.1

Values of k such that $Pr(y \leq k-1) < a/2$ where y has the binomial distribution with p=0.5.

Sample size	Probability of false detection			
n	P_f=0.10	P_f=0.05	P_f=0.02	P_f=0.01
5	1	-	-	-
6	1	1	-	-
7	1	1	1	-
8	1	1	1	1
9	2	2	1	1
10	2	2	1	1
11	3	2	2	1
12	3	3	2	2
13	4	3	2	2
14	4	3	3	2
15	4	4	3	3
16	5	4	4	3
17	5	5	4	3
18	6	5	4	4
19	6	5	5	4
20	6	6	5	4
21	7	6	5	5
22	7	6	6	5
23	8	7	6	5
24	8	7	6	6
25	8	8	7	6
30	11	10	9	8
35	13	12	11	10
40	15	14	13	12
45	17	16	15	14
50	19	18	17	16

Table 1.A.2

Values of $1\text{-}P_d$ for thc sign test

$P_f = 0.05$

n	r	$1-P_d$
8	0	.00781
9	0	.00391
10	0	.00195
11	0	.00098
12	1	.00635
13	1	.00342
14	1	.00183
15	2	.00739
16	2	.00418
17	2	.00235
18	3	.00754
19	3	.00443
20	3	.00258
25	5	.00408
30	7	.00522
35	9	.00599
40	11	.00643
45	13	.00661
50	15	.00660
60	19	.00622
70	23	.00558
80	28	.00968
90	32	.00743
100	36	.00664

Table 1.A.3

Values of $1\text{-}P_d$ for the sign test

$P_f = 0.01$

n	r	$1-P_d$
(5)	(0)	.0625
6	0	.03125
7	0	.01562
8	0	.00781
9	1	.03906
10	1	.02148
(10)	(2)	.10938
11	1	.01172
12	2	.03857
13	2	.02246
14	2	.01294
15	3	.03516
16	3	.02127
17	4	.04904
18	4	.03088
19	4	.01921
(20)	(4)	.01182
20	5	.04139
(20)	(6)	.11532
25	7	.04329
30	9	.04277
35	11	.04096
40	13	.03848
45	15	.03570
50	17	.03284
60	20	.02734
70	26	.04139
80	30	.03299
90	35	.04460
100	39	.0352

Table 1.A.4

Critical values of the rank corelation coefficient

n \ P_f	0.1	0.05	0.02	0.01
5	0.9	–	–	–
6	0.829	0.886	0.943	–
7	0.714	0.786	0.893	–
8	0.643	0.738	0.833	0.881
9	0.6	0.683	0.783	0.833
10	0.564	0.648	0.745	0.794
11	0.523	0.623	0.736	0.818
12	0.497	0.591	0.703	0.78
13	0.475	0.566	0.673	0.745
14	0.457	0.545	0.646	0.716
15	0.441	0.525	0.623	0.689
16	0.425	0.507	0.601	0.666
17	0.412	0.49	0.582	0.645
18	0.399	0.476	0.564	0.625
19	0.388	0.462	0.549	0.608
20	0.377	0.45	0.534	0.591
21	0.368	0.438	0.521	0.576
22	0.359	0.428	0.508	0.562
23	0.351	0.418	0.496	0.549
24	0.343	0.409	0.485	0.537
25	0.336	0.4	0.475	0.526
26	0.329	0.392	0.465	0.515
27	0.323	0.385	0.456	0.505
28	0.317	0.377	0.448	0.496
29	0.311	0.37	0.44	0.487
30	0.305	0.364	0.432	0.478

Appendix 1. B

The Discrete Fourier Transform

The *discrete Fourier transform* (DFT) is a basic operation used in many different signal processing applications. It is used to transform an ordered sequence of data samples, usually from the time domain into the frequency domain, so that spectral information about the sequence can become known explicitly. The DFT is described briefly here.

The DFT is a complex function of frequency, that is, an ordered sequence of complex numbers, each number consisting of a real part and an imaginary part. Usually the data sequence being transformed is real, but it need not be, and in either case the DFT is, in general, complex.

Suppose a real data sequence consisting of N real samples of a signal $x(t)$, given by

$$\text{Samples of } x(t) = [x_k] = [x_0, x_1, x_2, \ldots, x_{N-1}]$$

In this notation, k is usually a time index and ranges from 0 to N-1. When $[x_k]$ is a real data sequence, the computed DFT of $[x_k]$ consists of $(N/2)+1$ complex samples (it is assumed here that N is given), given by,

$$[X_m] = DFT[x_k] = [X_0, X_1, X_2, \ldots, X_{N/2}]$$

The relationship of $[X_m]$ to $[x_k]$ is discussed later; here it is simply noted that $[X_m]$ is the DFT of $[x_k]$ and that the index m designates the frequency of each component X_m. Also, the DFT is complex, so each X_m can be represented in polar form as

$$X_m = |X_m| e^{j\theta_m}$$

In this notation, $|X_m|$ is the amplitude of X_m, and a plot of $|X_m|$ versus the frequency index m is called the *amplitude spectrum* of $[x_k]$. Similarly, a plot of θ_m versus m is called the *phase spectrum* of $[x_k]$.

The relationship implemented by the forward transform between $[x_k]$ and $[X_m]$ can be expressed as,

$$X_m = \sum_{k=0}^{N-1} x_k e^{-j(2\pi mk/N)}; \quad m = 0, 1, \ldots, \frac{N}{2}$$

Again it is assumed here for convenience that N, the number of data samples, is even. In this formula for X_m, the exponential function, $\exp(-j(2\pi mk/N))$,is a *complex sinusoid* and is periodic. If one thinks of $\exp(-j(2\pi mk/N))$ as a function of k, the time index, then its period is seen to be N/m; that is, when k goes through a range of N/m, $\exp(-j(2\pi mk/N))$ goes through one cycle. One can see this even more clearly by separating the real and imaginary parts of the above relation:

$$X_m = \sum_{k=0}^{N-1} x_k \cos\left(\frac{2\pi m}{N}k\right) - j\sum_{k=0}^{N-1} x_k \sin\left(\frac{2\pi m}{N}k\right); \qquad m = 0,1,\ldots,\frac{N}{2}$$

Thus, each part (real or imaginary) of each DFT component X_m is a correlation (summed product) of the data sequence $[x_k]$ with a cosine or sine sequence having a period of N/m data samples.

The periodicity of the DFT just discussed is a very important property. One can see that adding N to the index of X_m does not change the value of X_m, that is,

$$X_{m+N} = \sum_{k=0}^{N-1} x_k e^{-j[2\pi(m+N)k/N]}$$

$$= \sum_{k=0}^{N-1} x_k e^{-j(2\pi mk/N)} e^{-j2\pi k} = X_m$$

Furthermore, this rule holds whether $[x_k]$ is real or complex; X_0 through X_{N-1} is a complete set of DFT components.

Secondly, when $[x_k]$ is real, one can also show that X_m and X_{N-m} are complex conjugates. One can take:

$$X_{N-m} = \sum_{k=0}^{N-1} x_k e^{-j[2\pi(N-m)k/N]}$$

$$= \sum_{k=0}^{N-1} x_k e^{+j(2\pi mk/N)} e^{-j2\pi k} = X_m^*$$

The star (*) denotes the complex conjugate, resulting from the change of sign in the exponential function. Note that if $[x_k]$ is complex, the above relation does not apply.

Thus, when $[x_k]$ is a sequence of real data, all values of X_m outside of the set X_0, X_1, ..., $X_{N/2}$ are redundant. When $[x_k]$ is a sequence of complex data, all values outside of the set X_o, X_1,..., X_{N-1} are redundant.

The inverse transform.

The reverse (or inverse) DFT is used to obtain a data sequence $[x_k]$ from its complex spectrum $[X_m]$. The formula for the inverse DFT, is

$$x_k = \frac{1}{N}\sum_{m=0}^{N-1} X_m e^{j(2\pi mk/N)}; \qquad k = 0,\ 1,\ \ldots,\ N-1$$

Thus the inverse DFT is the same as the forward DFT except for the sign of the exponential and the scaling factor $1/N$.

It can be seen that N values of X_m, not just $(N/2)+1$ values, are required for the inverse transformation. Thus, given X_0 through $X_{N/2}$, one could generate $X_{(N/2)+1}$ through X_{N-1}, using the formula for X_{N-m} before applying the inverse formula. Alternately, the inverse formula can be modified as follows, assuming $[x_k]$ is real and using the above formula for X_{N-m}:

$$x_k = \frac{1}{N}\sum_{m=0}^{N-1} X_m e^{j(2\pi mk/N)}$$

$$= \frac{1}{N}\left[X_0 + \sum_{m=1}^{N/2-1} X_m e^{j(2\pi mk/N)} + X_{N/2}e^{j\pi k} + \sum_{m=N/2+1}^{N-1} X^*_{N-m} e^{j(2\pi mk/N)}\right]$$

$$= \frac{1}{N}\left[X_0 + (-1)^k X_{N/2} + \sum_{m=1}^{N/2-1} X_m e^{j(2\pi mk/N)} + \sum_{n=1}^{N/2-1} X^*_n e^{-j(2\pi nk/N)}\right]$$

$$= \frac{1}{N}\left\{X_0 + (-1)^k X_{N/2} + 2\,\mathrm{Real}\left[\sum_{m=1}^{N/2-1} X_m e^{j(2\pi mk/N}\right]\right\}$$

$$= \frac{2}{N}\sum_{m=0}^{N/2}\left(R'_m \cos\frac{2\pi mk}{N} - I_m \sin\frac{2\pi mk}{N}\right)$$

where $[R_m]$ and $[I_m]$ are, respectively, the real and imaginary parts of $[X_m]$, and $[R'_m]$ is the same as $[R_m]$ but with R_0 and $R_{N/2}$ reduced to half-strength.

The fast Fourier transform.

The *fast Fourier transform* (*FFT*) is not a new kind of transform different from the DFT. Instead, it is simply an algorithm for computing the DFT, and its output is precisely the same set of complex values expressed in X_m. The FFT algorithm eliminates most of the repeated complex products in the DFT, however, so its execution time is much shorter. Specifically, the ratio of computing times is approximately

$$\frac{\text{FFT computing time}}{\text{DFT computing time}} = \frac{1}{2N}\log_2 N$$

Using the FFT, one can also do the computation "in place", so that $[X_m]$ replaces $[x_k]$, with only a limited amount of auxiliary storage needed for work space.

On the other hand, the FFT algorithm is more complicated than the DFT and becomes lengthy when N, the number of data samples, is not a power of two. Thus, in many applications it is simpler and preferable to use a simple DFT algorithm instead of an FFT.

One final but important point about the inverse transform is that most computer routines, in order to preserve symmetry, omit the factor $1/N$. Thus, the sequence $[x_k]$ is scaled to N times its correct amplitude by most inverse DFT and FFT algorithms.

Power spectrum and cepstrum.

The power spectrum of an ergodic signal is derived from the Fourier transform as,

$$Y_T(\omega) = \frac{1}{T}\left|\int_{-T/2}^{T/2} y(t)e^{-j\omega t}dt\right|^2$$

If $y(t)$ is real, then there is also the constraint that $Y(-\omega) = Y^*(\omega)$, which requires that Y_r be an even function of m and Y_i be an odd function of ω:

$$Y(\omega) = Y_r(\omega) \quad [\text{even in } \omega] + jY_i(\omega) \quad [\text{odd in } \omega]$$

Therefore,

$$|Y| = \sqrt{Y_r^2 + Y_i^2}$$

is also an even function of ω, and the phase,

$$\Phi_y = \tan^{-1}\left(\frac{Y_i}{Y_r}\right)$$

is an odd function of ω.

The above estimation of the power spectrum of $y(t)$ is not a satisfactory estimate of $Y(\omega)$, no matter how large T is, because the variance of $Y_T(\omega)$ does not tend to zero. However, it can be used to estimate a smoothed version of $Y(\omega)$ since,

$$\int_{\omega_1}^{\omega_2} Y_T(\omega)d\omega \underset{T\to\infty}{\to} \int_{\omega_1}^{\omega_2} Y(\omega)d\omega$$

With this background, the relationships shown in fig. 1.B.1 are examined. In the upper left corner is shown a linear system with an impulse response $h(t)$ that is driven by a source $x(t)$ and produces a response or output $y(t)$. The output $y(t)$ is computed by convolving the input time record with the impulse response of the system to produce an output waveform. The result of this process is that the input and the system responses are inextricably woven and folded together so that it is not possible to gain direct information regarding the input or the path from an observation of $y(t)$.

In the frequency domain, the transform of the output is the product of the Fourier transform of the input $X(\omega)$ and the system function $H(\omega)$. The magnitude of the output is the product of the magnitudes of the input and system functions, and the phase of the output φ_y is the sum of the phase of the input φ_x and the phase of the system function φ_h.

The magnitude relationship can be used to generate power spectra according to $|Y|^2 = |X|^2 \, |H|^2$. Because of the product relationship of the power spectra, the correlation output is again a convolution of the input correlation function and the convolution of the impulse response. This relationship is also sketched in fig. 1.B.1. Thus, correlation does not solve the problem of mixing between the input and the system response.

This separation problem can sometimes be alleviated, however, by going another step and forming the log of the transform so that, for example,

$$\log \mathrm{Y} = \log|Y| + j\Phi_y$$

Clearly, the logs of the magnitudes of the input and transfer function add to produce the log of the output magnitude. The phase of the output is a linear sum of the phases of the input and of the transfer function. One now has a situation in which the transfer function and the input add their properties to produce an output. This frequency domain process may not result in an effective way to separate the source and transfer function; but if one takes the inverse transform into the time domain, the cepstrum is generated which may provide a way in some cases to separate the individual effects of course and propagation path on the output cepstrum, at least to a degree that is useful for diagnostic purposes.

Since the log magnitude is an even function of frequency and the phase is an odd function of frequency, their inverse transforms are real functions, so the complex cepstrum $C_y(t)$ is a real function of time. The Fourier transform of the log magnitude is called the *power* or ***real cepstrum***, and in situations where the phase is unknown or ignored it may be a useful way to separate source and path effects. The inverse transform of the phase is the phase cepstrum, and the sum of the magnitude and phase cepstra is the complete complex cepstrum of the signal.

It is noted that if one is able to determine the input cepstrum $C_x(t)$, then by a Fourier transformation he could construct the log Fourier transform of x and by exponentiation recreate the Fourier transform itself. Having obtained the Fourier transform of x, both in magnitude and phase, one could use the inverse transform again and get back to $x(t)$. Thus, there is unique and recoverable relationship between the complex cepstrum and the variable from which it is derived. However, it is not possible to recover the initial waveform from the power or real cepstrum because the inverse transform of the real cepstrum only allows to compute the magnitude of the Fourier transform or the power spectrum. Inverse time transformation of the power spectrum reproduces the correlation function, not the initial waveform, because the phase of the signal has been lost.

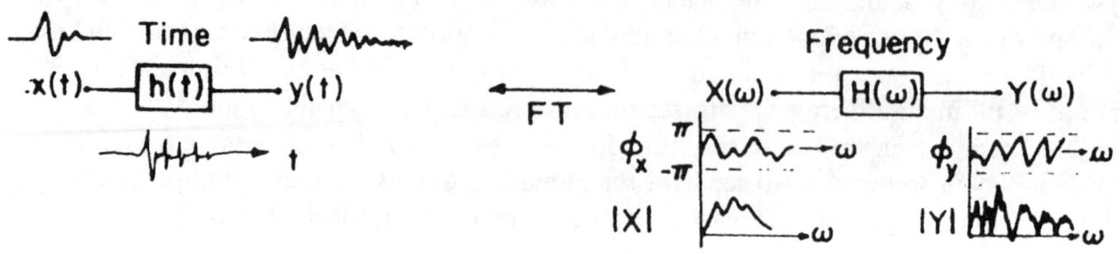

$\leftrightarrow$

Variable	FT	Transform
$y(t) = x(t) * h(t)$		$Y(\omega) = X(\omega)\, H(\omega)$ $\lvert Y\rvert = \lvert X\rvert\ \lvert H\rvert$ $\Phi_y = \Phi_x + \Phi_h$

$\leftrightarrow$

Correlation	FT	Power Spectrum
$R_y(\tau) = R_h(\tau) * R_x(\tau)$		$\lvert Y\rvert^2 = \lvert X\rvert^2 \,.\, \lvert H\rvert^2$

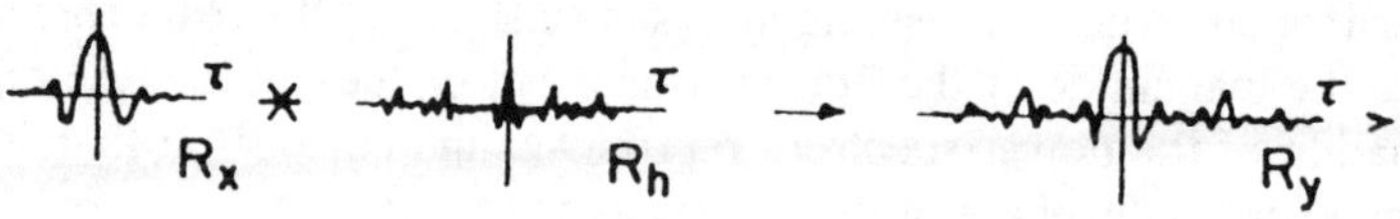

$\leftrightarrow$

Complex Cepstrum	FT	Log Transform
$C_y(\tau) = C_x(\tau) + C_h(\tau)$		$\log Y = \log\lvert Y\rvert + j\Phi_y$ $= \log\lvert X\rvert + \log\lvert H\rvert + j(\Phi_x + \Phi_h)$
Power (Real) Cepstrum		Log Magnitude
$C_{\hat{y}}(\tau) = C_{\hat{x}}(\tau) + C_{\hat{h}}(\tau)$		$\log\lvert Y\rvert = \log\lvert X\rvert + \log\lvert H\rvert$

$\leftrightarrow$

Phase Cepstrum	FT	Phase
$C_{\varphi y}(\tau) = C_{\varphi x}(\tau) + C_{\varphi h}(\tau)$		$\varphi_y = \varphi_x + \varphi_h$

Figure 1. B. 1 Derivation of the power cepstrum.
The cepstrum has an even part (power cepstrum) and an odd part (phase cepstrum), each of which is additive in the time frequency domains for source and path components.

CHAPTER 2

ANALYTICAL REDUNDANCY METHODS

2.1 Introduction

Quantitative model–based failure detection and isolation (FDI) methods rely on the comparison of a system's available measurements, with a–priori information represented by the system's mathematical model. The term *quantitative* is used here as contrary to the term *qualitative,* denoting cruder sustem descriptions. There are two main trends of this approach, namely *analytical redundancy* or *residual–generation* methods and *parameter estimation.* This distinction is not universally adopted and some researchers consider these two approaches as belonging to the same category. However, for reasons of clearer presentation the distinction is adopted here, and parameter estimation methods will be presented in Chapter 3.

The term analytical redundancy arises from the use of analytical relationships describing the dynamical interconnection between various system components. In contrast, *physical* or *hardware redundancy* relies on replication of hardware components (sensors, actuators, computers), thus increasing the reliability of the overall system.

Chow and Willsky, (1984), may be considered the inventors of this terminology, but it is certain that similar ideas were tried before them. Since then, this approach has been extended and evolved into what is currently termed *robust FDI.* Though this term will be elaborated later on, it generally means FDI schemes that are robust with respect to modeling errors and unknown (unmeasured) disturbances. Two main streams of research along this path have been followed in the two sides of the Atlantic. In USA, *parity space* methods have been used in many applications, while in Europe, *observer–based* techniques were developed. However, it has been recently seen, that both approaches are formally equivalent and just use different mathematical tools to achieve the same goal in robustness (Gertler, 1991).

The common denominator of all the approaches that will be presented in this chapter is that decision on whether a specific fault has occurred or not is made according to the values of characteristics quantities called *residuals*. These are generated from the observed input–output history of the system, and the way by which they are generated signifies each different method.

While residuals are zero in ideal situations, in practice, this is seldom the case. Their deviation from zero is the combined result of noise and faults. If the noise is negligible, residuals can be analyzed directly. With any significant noise present, statistical analysis (statistical testing) is necessary. In either case, a logical pattern is generated, showing which residuals can be considered normal and which ones indicate a fault. Such a pattern is called the *signature* of the failure.

As is common in other fields of science, there exist a variety of possible classifications of methods. With respect to the different sectors of a system where faults can occur, one may distinguish between *instrument fault detection* (*IFD*), *actuator fault detection* (*AFD*) and *component fault detection* (*CFD*). While early attempts were not concentrated on any one of these categories separately, increased complexity of robustness requirements has forced researchers to attack each of these in isolation to the others. It is also true to say that most work has been directed towards IFD, since sensor information is very crucial to the safe operation of any system. With respect to noise modeling, analytical redundancy methods may be classified into *stochastic* and *deterministic*. In the former, explicit modeling of the noise is present while in the latter noise is taken into account without any distribution assumptions. Early attempts tend to fall into the first category, while current robust techniques employ the second method.

The main technological areas where the methods of this Chapter have been applied are: aerospace engineering, automotive engineering, machining applications, nuclear engineering, chemical/petrochemical engineering, power plant and power transmission applications.

The structure of this Chapter is as follows: the main concepts of analytical redundancy methods is presented first. This includes modeling and performance criteria. Next, methods of stochastic modeling are briefly considered, mainly for introductory purposes. This is followed by current techniques, aiming at robust FDI, based on deterministic models. Finally, specific examples from industrial applications are presented, which illustrate the various methods which have been considered.

2.2 Plant and failure models

Most model–based failure detection and isolation methods rely on linear discrete–time state–space models. Since most diagnostic computations are performed on sampled data, this represents a reasonable form. This implies that for non–linear plants, any non-linearity is linearized around some operating point. Note, however that some methodologies can be extended to explicitly non–linear models, especially if they can be decomposed into static nonlinearities and linear dynamics (Gertler *et al.*, 1991). Also, continuous–time plants are represented by their discretized model. It must be emphasized however that the type of model used serves the proposed solution method, this being the causc of the many different representations one sees in the fault detection literature. Thus, frequency or z–domain representations have been lately considered, which exploit the additional information carried by the spectrum of the process.

Plant parameters may be varying with time. "Normal" variations are usually small and slow compared to the dynamics of the plant. Such variations will be neglected here for the sake of simplicity. Abrupt and/or significant changes, on the other hand, may and should be considered as *multiplicative* process faults. In addition, *additive* faults, e.g. biases on the different parts of the system are taken into account.

The state–space model relates the state vector $\boldsymbol{x}(k)$ to the input vector $\boldsymbol{u}(k)$ and output vector $\boldsymbol{y}(k)$ using known system matrices $\boldsymbol{A}$, $\boldsymbol{B}$, and $\boldsymbol{C}$. The well–known state equations describing the nominal (fault–free) system are:

$$\boldsymbol{x}(k+1) = \boldsymbol{A}\boldsymbol{x}(k) + \boldsymbol{B}\boldsymbol{u}(k) \tag{2.1}$$

$$\boldsymbol{y}(k) = \boldsymbol{C}\boldsymbol{x}(k) \tag{2.2}$$

The dimensions of the state, input and output vectors are *n, r* and *m* respectively. An equivalent input–output model may be presented in shift–operator form, with matrices $\boldsymbol{G}(z)$ and $\boldsymbol{H}(z)$ consisting of elements that are polynomials in the shift operator z and $\boldsymbol{H}$ being a diagonal matrix:

$$\boldsymbol{H}(z)\boldsymbol{y}(k) = \boldsymbol{G}(z)\boldsymbol{u}(k) \tag{2.3}$$

The matrices of the input–output model are related to those of the state model by,

$$\begin{aligned} \boldsymbol{G}(z) &= \boldsymbol{C}\left[\text{Adjoint}(\boldsymbol{I}z - \boldsymbol{A})\right]\boldsymbol{B} \\ \boldsymbol{H}(z) &= \left[\text{Determinant } (\boldsymbol{I}z - \boldsymbol{A})\right]\boldsymbol{I} \end{aligned} \tag{2.4}$$

Note that (2.3) usually can be simplified by eliminating, row by row, the common factors in the $\boldsymbol{H}(z)$ and $\boldsymbol{G}(z)$ matrices.

Introducing $\boldsymbol{f}_p(k)$ for additive process faults and $\boldsymbol{w}_p(k)$ for (additive) process noise, and $\boldsymbol{P}$ and $\boldsymbol{Q}$ for their coefficient matrices, the state equation (2.1) becomes,

$$\boldsymbol{x}(k+1) = \boldsymbol{A}\boldsymbol{x}(k) + \boldsymbol{B}\boldsymbol{u}(k) + \boldsymbol{P}\boldsymbol{f}_p(k) + \boldsymbol{Q}\boldsymbol{w}_p(k) \tag{2.5}$$

while the input–ouput equation (2.3) becomes,

$$\boldsymbol{H}(z)\boldsymbol{y}(k) = \boldsymbol{G}(z)\boldsymbol{u}(k) + \boldsymbol{L}(z)\boldsymbol{f}_p(k) + \boldsymbol{M}(z)\boldsymbol{w}_p(k) \tag{2.6}$$

Here the matrices $\boldsymbol{L}$ and $\boldsymbol{M}$ are obtained in accordance with (2.4), with $\boldsymbol{B}$ replaced by $\boldsymbol{P}$ and $\boldsymbol{Q}$, respectively. Note that the presence of the new terms in (2.5) may influence $\boldsymbol{H}(z)$ and $\boldsymbol{G}(z)$ since $\boldsymbol{L}(z)$ and $\boldsymbol{M}(z)$ interfere with the simplification of the equations.

Introduce now $\boldsymbol{f}_u(k)$ and $\boldsymbol{f}_y(k)$ for the additive measurement fault (bias) on the input $\boldsymbol{u}(k)$ and output $\boldsymbol{y}(k)$, and $\boldsymbol{w}_u(k)$ and $\boldsymbol{w}_y(k)$ for the respective measurement noise. With these, the *measured* input $\tilde{\boldsymbol{u}}(k)$ and output $\tilde{\boldsymbol{y}}(k)$,

$$\begin{aligned}\tilde{\boldsymbol{u}}(k) &= \boldsymbol{u}(k) + \boldsymbol{f}_u(k) + \boldsymbol{w}_u(k)\\ \tilde{\boldsymbol{y}}(k) &= \boldsymbol{y}(k) + \boldsymbol{f}_y(k) + \boldsymbol{w}_y(k)\end{aligned} \tag{2.7}$$

For controlled inputs, there is no sensory measurement; instead, $\boldsymbol{u}(k)$ is the control signal and $\tilde{\boldsymbol{u}}(k)$ its implementation by the actuators, with $\boldsymbol{f}_u(k)$ representing any actuator malfunction and $\boldsymbol{w}_u(k)$ the actuator noise. Alternatively, inputs may be classified into three groups: measured inputs, $\boldsymbol{u}_m$, controlled inputs $\boldsymbol{u}_c$ and disturbance inputs $\boldsymbol{u}_d$.

Finally, introduce $\Delta\boldsymbol{A}(k)$, $\Delta\boldsymbol{B}(k)$, and $\Delta\boldsymbol{C}(k)$ for the discrepancies between the model matrices $\tilde{\boldsymbol{A}}$, $\tilde{\boldsymbol{B}}$ and $\tilde{\boldsymbol{C}}$ and the true system matrices $\boldsymbol{A}$, $\boldsymbol{B}$, and $\boldsymbol{C}$:

$$\begin{aligned}\tilde{\boldsymbol{A}} &= \boldsymbol{A} + \Delta\boldsymbol{A}(k)\\ \tilde{\boldsymbol{B}} &= \boldsymbol{B} + \Delta\boldsymbol{B}(k)\\ \tilde{\boldsymbol{C}} &= \boldsymbol{C} + \Delta\boldsymbol{C}(k)\end{aligned} \tag{2.8}$$

Such discrepancies may account for multiplicative plant faults. To obtain a complete description of the system with all the possible faults and noises taken into account, the true variables $\boldsymbol{u}(k)$ and $\boldsymbol{y}(k)$ expressed from (2.7) and the true matrices $\boldsymbol{A}$, $\boldsymbol{B}$, $\boldsymbol{C}$ from (2.8) are to be substituted into (2.5) and (2.2). The model becomes,

$$\begin{aligned}\boldsymbol{x}(k+1) &= (\boldsymbol{A} + \Delta\boldsymbol{A}(k))\boldsymbol{x}(k) + (\boldsymbol{B} + \Delta\boldsymbol{B}(k))\left(\boldsymbol{u}(k) + \boldsymbol{f}_u(k) + \boldsymbol{w}_u(k)\right) + \boldsymbol{P}\boldsymbol{f}_p(k) + \boldsymbol{Q}\boldsymbol{w}_p(k)\\ \boldsymbol{y}(k) &= (\boldsymbol{C} + \Delta\boldsymbol{C}(k))\boldsymbol{x}(k) + \boldsymbol{f}_y(k) + \boldsymbol{w}_y(k)\end{aligned} \tag{2.9}$$

or,

$$\begin{aligned}\boldsymbol{x}(k+1) = {} & \boldsymbol{A}\boldsymbol{x}(k) + \boldsymbol{B}\boldsymbol{u}(k) + \Delta\boldsymbol{A}(k)\boldsymbol{x}(k) + \Delta\boldsymbol{B}(k)\boldsymbol{u}(k)\\ & + \boldsymbol{B}\boldsymbol{f}_u(k) + \boldsymbol{B}\boldsymbol{w}_u(k) + \Delta\boldsymbol{B}(k)\boldsymbol{f}_u(k) + \Delta\boldsymbol{B}(k)\boldsymbol{w}_u(k) + \boldsymbol{P}\boldsymbol{f}_p(k) + \boldsymbol{Q}\boldsymbol{w}_p(k)\\ \boldsymbol{y}(k) = {} & \boldsymbol{C}\boldsymbol{x}(k) + \Delta\boldsymbol{C}(k)\boldsymbol{x}(k) + \boldsymbol{f}_y(k) + \boldsymbol{w}_y(k)\end{aligned} \tag{2.10}$$

Now if the various terms are lumped into similar groups one obtains the system,

$$\begin{aligned}\boldsymbol{x}(k+1) &= \boldsymbol{A}\boldsymbol{x}(k) + \boldsymbol{B}\boldsymbol{u}(k) + \boldsymbol{F}_1(k)\boldsymbol{f}_1(k) + \boldsymbol{D}_1\boldsymbol{d}_1(k)\\ \boldsymbol{y}(k) &= \boldsymbol{C}\boldsymbol{x}(k) + \boldsymbol{F}_2(k)\boldsymbol{f}_2(k) + \boldsymbol{D}_2\boldsymbol{d}_2(k)\end{aligned} \tag{2.11}$$

where the meaning of the various terms is clear from comparison of (2.11) and (2.10). The main point to note is that uncertainty (faults and disturbances) is generally split into two categories:

- *structured* uncertainty, acting upon the system as additive faults and disturbances and represented by unknown time–functions multiplied by known distribution matrices, and
- *unstructured* uncertainty, which describes multiplicative (parametric) faults and modeling errors and is represented by unknown matrices multiplying known (observed) variables.

Further specialization of (2.11) is possible in specific situations, where it is clarified what is considered a fault to be detected and what is considerd a disturbance to be ignored.

2.3 Design requirements

Model–based fault diagnosis can be defined as the detection, isolation and characterization of faults in system components from the comparison of available measurements. These three desired functions are stated in order of increased difficulty. Detection is performed by all traditional methods, while isolation and characterization (size and possible time of occurrence of fault) are achieved using more sophisticated algorithms. A fault detection method should usually possess the following characteristics:

- Low *detection delay* time (t_d): this is usually minimised for a fixed false alarm rate.
- High rate of *correct detections* (P_d).
- Low rate of *false alarms* (P_f).
- *Isolability:* is the ability to distinguish (isolate) faults and depends on the statistical tests employed and the structure of the system matrices.
- *Sensitivity*: characterizes the size of faults that can be isolated under certain conditions. It depends on the size of the respective matrices and noise properties and is closely related to the detection delay time. Some researchers use this term in a different context which is related to the robustness requirement but we prefer this definition since it defines a distinct aspect of an FDI algorithm.
- *Robustness:* is the ability to isolate faults in the presence of modeling errors and/or unknown disturbances. This is a most serious requirement, since such errors are practically inevitable and therefore greatly affects the previous properties. Since modeling errors appear as multiplicative faults, false alarms are triggered if the robustness issue is not taken into account in the design process. It has by now become clear that the most essential requirement for a model–based FDI algorithm is robustness to disturbances as well as to model–system mismatches. This is not at all a straightforward problem and as will be seen in the following sections, approximate solutions are found in practice.

Performance measures for the above requirements vary, and will be cited with the specific methods.

2.4 Methods of solution

The procedure of evaluation of the redundancy given by the mathematical model of the system, described by any of the models of Section 2.2 can be roughly divided into the following two steps:

1. Residual generation.
2. Residual analysis: decision and isolation of the faults (time, location, sometimes also type, size, and source).

The analytical redundancy approach requires that the residual generator performs some kind of validation of the nominal relationships of the system, using the actual input $\boldsymbol{u}$, and measured output $\boldsymbol{y}$. The redundancy relations to be evaluated can simply be interpreted as input–output relations of the dynamics. If a fault occurs, the reduncancy relations are no longer satisfied and a residual, $\boldsymbol{r} \neq \mathbf{0}$, occurs. The residual is then used to form appropriate decision functions. They are evaluated in the fault decision logic in order to monitor both the time of occurrence and location of the fault.

A more detailed structural diagram of the overall FDI procedure is depicted in fig. 2.1. Note that for the residual generation three kinds of models are required: nominal, actual (observed) and that of the faulty system. In order to achieve a high performance of fault detection with low false alarm rate, the nominal model should be tracked and updated by the observation model.

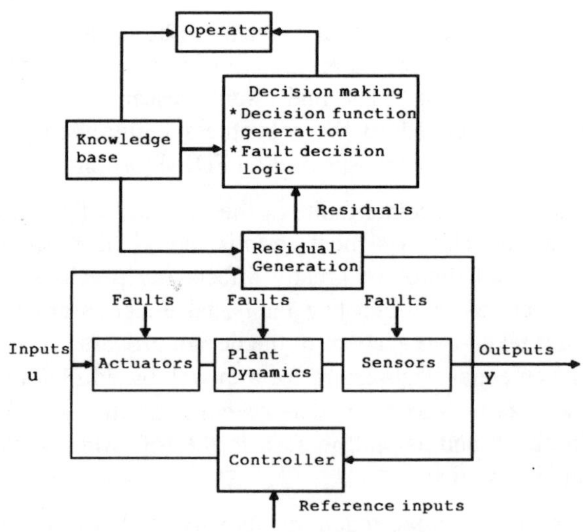

Figure 2.1 General architecture of FDI based on analytical redundancy

Basically, there are three different ways of generating fault–accentuated signals using analytical redundancy: parity checks, observer schemes and detection filters, all of them using state estimation techniques. The resulting signals are used to form decision functions as, for example, norms of likelihood functions. The basis for the decision on the occurrence of a fault is the fault *signature*, i.e. a signal that is obtained from some kind of faulty system model defining the effects associated with a fault.

Deterministic methods for FDI use deterministic state variable methods to generate residual quantities. The detection, isolation and further diagnosis of faults is achieved using these residual quantities. Careful design of the residual can facilitate the use of tighter bounds in the form of threshold levels for detection and isolation.

If the dynamical system with a number of possible faults can be described by the input–output relation,

$$\boldsymbol{y}(s) = \boldsymbol{G_u}(s)\boldsymbol{u}(s) + \boldsymbol{G_f}(s)\boldsymbol{f}(s)$$

where $\boldsymbol{y}(s)$, $\boldsymbol{u}(s)$, $\boldsymbol{f}(s)$ are the Laplace transformed output, input and fault vectors respectively then each component of the residual vector $\boldsymbol{r}(t)$ generated by means of a deterministic model, should satisfy the condition:

$$\boldsymbol{r}(t) = \boldsymbol{0} \text{ if and only if } \boldsymbol{f}(t) = \boldsymbol{0} \tag{2.12}$$

where $\boldsymbol{f}(t)$ is considered to act upon the dynamics of the process in an additive manner. From a practical point of view it is reasonable *not* to make further assumptions about the fault vector $\boldsymbol{f}(t)$ except that it is an unknown time function.

The general structure for all deterministic residual generators, based upon the concept above is shown in fig. 2.2. This structure is expressed mathematically in the frequency domain as:

$$\begin{aligned} \boldsymbol{r}(s) &= \begin{bmatrix} H_u(s) & H_y(s) \end{bmatrix} \begin{bmatrix} \boldsymbol{u}(s) \\ \boldsymbol{y}(s) \end{bmatrix} \\ &= \boldsymbol{H_u}(s)\boldsymbol{u}(s) + \boldsymbol{H_y}(s)\boldsymbol{y}(s) \end{aligned} \tag{2.13}$$

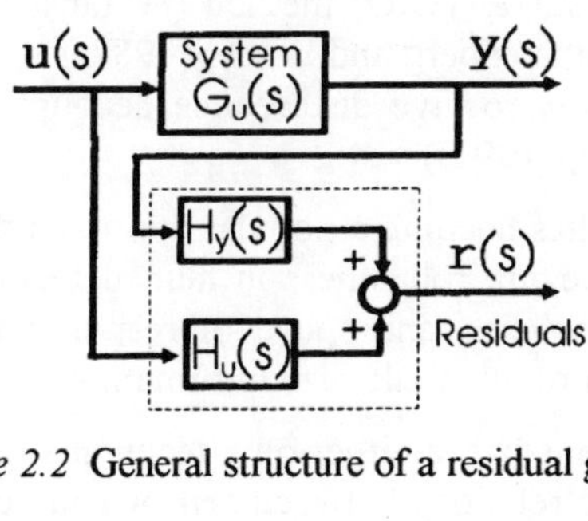

Figure 2.2 General structure of a residual generator

The transfer matrices $\boldsymbol{H}_u(s)$ and $\boldsymbol{H}_y(s)$ are realizable using stable linear systems. In order to make the residual $\boldsymbol{r}(s)$ become zero for the fault–free case (i.e. to achieve requirements in Eq. (2.12)), $\boldsymbol{H}_u(s)$ and $\boldsymbol{H}_y(s)$ must satisfy the null condition:

$$\boldsymbol{H}_u(s) + \boldsymbol{H}_y(s)\boldsymbol{G}_u(s) = \boldsymbol{0}$$

Eq. (2.13) is a generalized representation of all residual generators. The design of the residual generator results simply in the choice of the transfer function matrices $\boldsymbol{H}_u(s)$ and $\boldsymbol{H}_y(s)$ which must satisfy the null condition. The various ways of generating residuals correspond to different parameterizations of $\boldsymbol{H}_u(s)$ and $\boldsymbol{H}_y(s)$. One can obtain different residual generators using different forms for $\boldsymbol{H}_u(s)$ and $\boldsymbol{H}_y(s)$ and using the design freedom, the desired performance of the residual can be achieved.

A fault can be detected by comparing the residual with a *decision* or *threshold function* $D_F(\boldsymbol{r})$ according to the test,

$$\begin{cases} D_F(\boldsymbol{r}) \le T(t) & for \quad \boldsymbol{f}(t) = \boldsymbol{0} \\ D_F(\boldsymbol{r}) > T(t) & for \quad \boldsymbol{f}(t) \ne \boldsymbol{0} \end{cases}$$

If this test is positive (i.e. the decision function is exceeded by the residual), a likely fault is hypothesized. There may also be a likelihood that the decision function will be exceeded even if there is no fault. This would lead to a false alarm in detection as a consequence of modeling errors (in the determination of the residual or in the determination of the decision function) or unknown (i.e. unexpected) disturbances affecting the residual. The simplest method of deciding whether or not there is a fault is to use a fixed threshold applied to the residual signal $\boldsymbol{r}(t)$.

There is a rich variety of methods available for quantitative model–based residual generation, including:

Observer approaches. The underlying idea is the estimation of system outputs from the measurements (or a subset of measurements) by using either full order or reduced order state observers. A suitable weighting of the output estimation error is then defined as a residual, according to the general structure given in Eq. (2.13) and fig. 2.2. Methods for the selection of the observer gains include:

(a) The ***Unknown Input Observer*** (UIO) method (Watanabe and Himmelblau, (1982), Massoumnia, (1986), Wünnenberg and Frank, (1987)).
(b) *Eigenstructure assignment* to give disturbance decoupling (Patton and Kangethe, (1989), Patton and Chen, (1991)).

The state observer approach has become a popular approach due to the flexibility of design, the relative ease in achieving robustness in fault detection and fault isolation, the algorithmic and software simplicity, and speed of response in detecting and isolating faults. However, it does not provide fault size information.

The *parity relations* approach is based either on a technique of *direct redundancy*, making use of the static algebraic relations between sensor and actuator signals, or alternatively upon *temporal redundancy*, when dynamic relations between inputs and outputs are used (differential or difference equations). The term "parity" was first used in computer systems to enable "parity checks" to be performed for error checking. In the FDI field, it has similar meaning in the context of providing an indicator for the presence of a fault (or error) in system components. The key idea is to check the parity (consistency) of the mathematical equations of the system (analytical redundancy relations) by using the actual measurements. A fault is declared to have occurred once preassigned error bounds are surpassed. This method does not provide fault size information either.

In the early developments, parity space methods were applied to *parallel redundancy* schemes (Potter and Suman, (1977), Desai and Ray, (1981), Chow and Willsky (1984)). For such system configurations, the number of measurements is greater than the number of variables to be sensed and the residuals can be obtained directly from the redundant measurements. Inconsistency in the measurement data is then a metric that can be used initially for detecting faults and, subsequently for fault diagnosis. Mironovskii (1980), has independently derived similar relations in the Soviet Union.

Stochastic modeling methods for fault diagnosis are based on statistical testing of the innovations (i.e. the residuals) of Kalman filters or other filters and can be used for both fault detection and isolation, by means of hypothesis testing.

Whilst using a similar structure to the observer, approaches based on the Kalman filter comprise a residual generation mechanism derived by means of a stochastic model of the dynamical system. In normal operation the Kalman filter residual (or innovation) vector (the difference between the measurements and their Kalman filter estimates), is a zero–mean white noise process with known covariance matrix. Mehra and Peschon, (1971) proposed the use of different statistical tests on the innovation to detect a fault of the system. The idea which is common to all these approaches is to test, amongst all possible hypotheses, that the system has a fault or is fault–free. As each fault type has its own *signature*, a set of hypotheses can be used and checked for the likelihood that a particular fault has occurred.

Main methods include:

(i) *Chi–squared testing* (Mehra and Peschon (1971), Willsky *et al.*, (1974a, 1975); Watanabe *et al.*, (1979, 1981)). This is just an alarm procedure, i.e. it does not provide fault location or size information.

(ii) ***Sequential Probability Ratio Testing*** (*SPRT*): The purpose of this test is to check the zero–mean property of the innovations. The decision is based on the value of the likelihood ratio of the p.d.f. of the innovations under the null and altternative hypotheses. Usually the decision space is divided into three regions: fault, no fault and repeat. In this sense the test is sequential. Chien and Adams, (1976), Deckert *et al*, (1977), Yoshimura *et al*, (1979), Bonivento and Tonielli, (1984) and Uosaki, (1985) are the main contributors to this approach. This is also an alarm procedure.

(iii) *Generalized Likelihood Ratio* (*GLR*) testing. The philosophy of this approach is as follows: a Kalman–Bucy filter is implemented on the assumption of no abrupt system changes while a secondary system monitors the measurement residuals of the filter to determine if a change has occurred and adjusts the filter accordingly. Decision of fault occurrence is based on the value of the generalised likelihood ratio of the no–fault and fault hypotheses (Willsky and Jones, (1976), Ono *et al*, (1984), Kumamaru (1984), Pouliezos and Stavrakakis (1987) and Tanaka and Müller (1990)). This method performs all required tasks, i.e. fault detection, isolation and estimation, thus it is possible to perform automatic system reorganization in the case of soft failures. These qualities are however offset by its low robustness.

(iv) *Multiple Model Adaptive Filters* (*MMAF*s). In this early development a bank of linear filters based on different hypotheses concerning the underlying system behaviour is constructed. The innovations of the various filters are monitored and the conditional probability that each system model is correct is computed. The system with the highest probability is declared to be the correct one (Lainiotis (1971), Athans and Willner (1973), Willsky *et al.* (1974b)). The same comments as for the GLR method apply here

2.5 *Stochastic modeling methods*

In this section it is considered the problem of detecting changes in linear, possibly time varying, stochastic dynamical systems described by,

$$\boldsymbol{x}(k+1)=\boldsymbol{A}(k)\boldsymbol{x}(k)+\boldsymbol{B}(k)\boldsymbol{u}(k)+\boldsymbol{w}(k) \tag{2.14}$$

$$\boldsymbol{y}(k)=\boldsymbol{C}(k)\boldsymbol{x}(k)+\boldsymbol{D}(k)\boldsymbol{u}(k)+\boldsymbol{v}(k) \tag{2.15}$$

where $\boldsymbol{w}$ and $\boldsymbol{v}$ are zero-mean, independent, white Gaussian sequences with covariances defined by,

$$\begin{aligned} \mathrm{E}\{\boldsymbol{w}(k)\boldsymbol{w}^{\mathrm{T}}(j)\} &= \boldsymbol{Q}\delta_{kj} \\ \mathrm{E}\{\boldsymbol{v}(k)\boldsymbol{v}^{\mathrm{T}}(j)\} &= \boldsymbol{R}\delta_{kj} \end{aligned} \tag{2.16}$$

where δ_{kj} is the Kronecker delta. Eqs. (2.14)-(2.16) describe the "normal operation" or "no failure" model of the system of interest. If no failures occur, the optimal state estimator is given by the discrete Kalman filter equations,

$$\hat{\boldsymbol{x}}(k+1|k)=\boldsymbol{A}(k)\hat{\boldsymbol{x}}(k|k)+\boldsymbol{B}(k)\boldsymbol{u}(k) \tag{2.17}$$

$$\hat{\boldsymbol{x}}(k|k)=\hat{\boldsymbol{x}}(k|k-1)+\boldsymbol{K}(k)\boldsymbol{\gamma}(k) \tag{2.18}$$

$$\boldsymbol{\gamma}(k)=\boldsymbol{y}(k)-\boldsymbol{C}(k)\hat{\boldsymbol{x}}(k|k-1-\boldsymbol{D}(k)\boldsymbol{u}(k) \tag{2.19}$$

where $\boldsymbol{\gamma}(k)$ is the zero-mean, Gaussian innovation process, and the gain $\boldsymbol{K}(k)$ is calculated from the equations,

$$\boldsymbol{P}(k+1|k)=\boldsymbol{A}(k)\boldsymbol{P}(k|k)\boldsymbol{A}^{\mathrm{T}}(k)+\boldsymbol{Q} \tag{2.20}$$

$$\boldsymbol{V}(k)=\boldsymbol{C}(k)\boldsymbol{P}(k|k-1)\boldsymbol{C}^{\mathrm{T}}(k)+\boldsymbol{R} \tag{2.21}$$

$$\boldsymbol{K}(k)=\boldsymbol{P}(k|k-1)\boldsymbol{C}^{\mathrm{T}}(k)\boldsymbol{V}^{-1}(k) \tag{2.22}$$

$$\boldsymbol{P}(k|k)=\boldsymbol{P}(k|k-1)-\boldsymbol{K}(k)\boldsymbol{C}(k)\boldsymbol{P}(k|k-1) \tag{2.23}$$

Here $\boldsymbol{P}(i|j)$ is the estimation error covariance of the estimate $\hat{\boldsymbol{x}}(i|j)$, and $\boldsymbol{V}(k)$ is the covariance of $\boldsymbol{\gamma}(\mathrm{k})$. Eqs. (2.17)-(2.23) are referrd to as the "normal mode filter" in the sequel.

In addition to the above estimator, one may also have a closed loop control law, such as the linear law

$$\boldsymbol{u}(k)=\boldsymbol{G}(k)\hat{\boldsymbol{x}}(k|k) \tag{2.24}$$

Since the statistics of the innovations sequence is completely known under normal conditions, a number of tests can be devised to check if the observed statistics are the expected. The relevant properties are: zero mean, independence and known covariance. These tests can be used as fault alarms, in situations where this is desirable, or as first level fault detectors in more sophisticated algorithms. Both approaches will be examined in the following.

2.5.1 Simple tests

Mehra and Peschon (1971) were among the first to propose several statistical tests for detection of changes in the system (2.14)-(2.15). Since then various other tests have been proposed and tested in various real applications. It is obvious that the relevant theory is very extended and cannot be described in detail here. The interested reader is referred to the excellent book of Basseville and Nikiforov (1993), in which the statistical change detection theory is presented in detail.

Some of the results presented in this section are specialised versions of general algorithms presented in Chapter 1. However, for reasons of clarity, they will be reintroduced here for the Kalman filter's innovation sequence.

For hypothesis testing purposes, it is more convenient to consider the standardized innovation sequence defined by,

$$\boldsymbol{\eta}(k)=\left(\boldsymbol{C}(k)\boldsymbol{P}(k|k-1)\boldsymbol{C}^{\mathrm{T}}(k)+\boldsymbol{R}(k)\right)^{-\frac{1}{2}}\boldsymbol{\gamma}(k) \tag{2.25}$$

where $(\cdot)^{-\frac{1}{2}}$ denotes the square root of the inverse of a matrix. Then,

$$\mathrm{E}\left\{\boldsymbol{\eta}(i)\boldsymbol{\eta}^{\mathrm{T}}(i)\right\} = \boldsymbol{I}\delta_{ij} \tag{2.26}$$

It is also usual in fault detection applications to use moving windows of data. This results in lower detection delays, since the estimators do not have infinite memory. For this purpose, define,

$$\boldsymbol{\eta}^{\mathrm{T}}(k,n_w) = \left[\boldsymbol{\eta}^{\mathrm{T}}(k-n_w+1) \;\vdots\; \boldsymbol{\eta}^{\mathrm{T}}(k-n_w+2) \;\vdots\; \boldsymbol{\eta}^{\mathrm{T}}(k-n_w+3) \;\vdots\; \ldots \;\vdots\; \boldsymbol{\eta}^{\mathrm{T}}(k)\right] \tag{2.27}$$

to denote a collection of n_w residuals ending at time k.

2.5.1.1 Tests of mean

These tests check whether the observed innovation sequence is zero mean or not. The mean of the innovation sequence is estimated as,

$$\hat{\bar{\boldsymbol{\eta}}}_N = \frac{1}{N}\sum_{i=1}^{N}\boldsymbol{\eta}(i) \tag{2.28}$$

where the subscript is used to signify the dependence on the sample size N, and $\bar{\boldsymbol{\eta}}_N$ denotes the true mean. Under the null (no-failure) hypothesis, $\hat{\bar{\boldsymbol{\eta}}}_N$ has a Gaussian distribution with zero mean and covariance,

$$\mathrm{E}\left\{\hat{\bar{\boldsymbol{\eta}}}_N\hat{\bar{\boldsymbol{\eta}}}_N^{\mathrm{T}}\right\} = \frac{1}{N}\boldsymbol{I} \tag{2.29}$$

Therefore at the 5 per cent significance level, the null hypothesis is rejected whenever,

$$\left|\hat{\bar{\boldsymbol{\eta}}}_N\right| > \frac{1.96}{\sqrt{N}}\boldsymbol{I} \tag{2.30}$$

The above test suffers from the fact that the covariance of $\boldsymbol{\eta}(k)$ is assumed known. A better test is the T^2-test which use the T^2-statistic,

$$T^2 = N\hat{\bar{\boldsymbol{\eta}}}_N\hat{\boldsymbol{C}}_N^{-1}\hat{\bar{\boldsymbol{\eta}}}_N^{\mathrm{T}} \tag{2.31}$$

where $\hat{\boldsymbol{C}}_N$ is the sample covariance calculated by (2.35) (follows). This test is uniformly most powerful among all the tests for zero mean which are invariant with respect to scaling (or covariance), see Anderson (1958).

If a sliding window is used, recursive calculation of the window mean is also possible, using,

$$\boldsymbol{\zeta}(k) = \boldsymbol{\eta}(k-n_w) - \boldsymbol{\eta}(k) \tag{2.32}$$

$$\hat{\bar{\boldsymbol{\eta}}}_{n_w}(k) = \hat{\bar{\boldsymbol{\eta}}}_{n_w}(k-1) - \frac{1}{n_w}\zeta(k) \tag{2.33}$$

where k denotes the current time and n_w the window length..

The above tests assume residual independence, a condition that may be violated if certain faults (non-additive) have occurred. In this situation, non-parametric tests offer greater robustness. Such a test is the multivariable *component sign test* (Bickel, 1965). This test uses a sign statistic for each component of the vectors and combines them in a quadratic form. Define,

$$\boldsymbol{S}_{n_w}(k) = \left[S_{1,n_w}(k) \quad \dots \quad S_{p,n_w}(k)\right]$$

where,

$$S_{i,n_w}(k) = \sum_{t=k-n_w+1}^{k} \operatorname{sgn}\, \eta_i(t) \tag{2.34}$$

and the sgn function is defined as,

$$\operatorname{sgn}(z) = \begin{cases} 1 & z > 0 \\ 0 & z = 0 \\ -1 & z < 0 \end{cases}$$

Now form,

$$\boldsymbol{S}^*_{n_w}(k) = \boldsymbol{S}^{\mathrm{T}}_{n_w}(k)\hat{\boldsymbol{W}}^{-1}_{n_w}(k)\boldsymbol{S}_{n_w}(k) \text{ where}$$

$$\hat{w}^{i,\ell}_{n_w}(k) = \sum_{t=k-n_w+1}^{k} \operatorname{sgn}\eta_i(t)\operatorname{sgn}\eta_\ell(t) \text{ for } 1 \le i \le m \text{ and } 1 \le \ell \le m$$

The test rejects H_0 (detects change) for values of $S^*_{n_w}(k)$ greater than χ^2_m.

2.5.1.2 *Tests of covariance*

The covariance of the innovation sequence is estimated by,

$$\hat{\boldsymbol{C}}_N = \frac{1}{N-1}\sum_{i=1}^{N}(\boldsymbol{\eta}(i) - \hat{\bar{\boldsymbol{\eta}}}_N)(\boldsymbol{\eta}(i) - \hat{\bar{\boldsymbol{\eta}}}_N)^{\mathrm{T}} \tag{2.35}$$

Division by (N-1) instead of N produces unbiased estimates for smaal samples. Under the null hypothesis of no change, $\hat{\boldsymbol{C}}_N$ has a Wishart distribution (Anderson, 1958). The trace of $\hat{\boldsymbol{C}}_N$ has a χ^2 distribution with $(N-1)m$ degrees of freedom. Thus $\hat{\boldsymbol{C}}_N$ can be tested for its null hypothesis covariance equal to an identity matrix.

If moving windows are used, relevant recursive expressions for scalar signals are (Pouliezos, 1980a),

$$\hat{c}_{n_w}(k) = \hat{c}_{n_w}(k-1) + \frac{1}{n_{w-1}}\left(2\zeta(k)\hat{\bar{\eta}}_{n_w}(k-1) - \frac{1}{n_w}\zeta^2(k) - \eta^2(k-n_w) + \eta^2(k)\right)$$

where $\zeta(k)$ and $\hat{\bar{\eta}}_{n_w}(k)$ are defined by (2.32), (2.33).

- ***The sequential probability ratio test***

A more sophisticated approach for detecting changes in the covariance of the innovation is to use Wald's ***Sequential Probability Ratio Test***, *SPRT* (Wald, 1947).

In this context the two hypotheses to be decided upon are (for scalar outputs):

Normal mode (H_0): $E\{\eta_n\} = 0,\ \mathrm{Var}\{\eta_n\} = 1$

Failure mode (H_1): $E\{\eta_n\} = 0,\ \mathrm{Var}\{\eta_n\} = \Delta^2 > 1.$

In order to carry out the test for these two hypotheses by conventional SPRT formulation, the logarithm of the joint likelihood ratio function (LLR) IS computed,

$$\lambda_n = \log\frac{p(\eta_1,\ \eta_2,\ \dots,\ \eta_n|H_1)}{p(\eta_1,\ \eta_2,\ \dots,\ \eta_n|H_0)} \tag{2.36}$$

and compare with two threshold values A^* and B^* ($A^*>0>B^*$) derived from the specified error probabilities for false alarm (α) and miss alarm (β), i.e.,

$$A^* = \log\frac{1-\beta}{\alpha}$$

$$B^* = \log\frac{\beta}{1-\alpha} \tag{2.37}$$

If the LLR exceeds the boundary A^* or falls below the boundary B^*, the observation is terminated with acceptance of the hypothesis H_1 (failure mode) or the hypothesis H_0 (normal mode), respectively. Otherwise, the observation is continued and decision is deferred. It is noted that LLR can be computed recursively by,

$$\lambda_n = \lambda_{n-1} + \log\frac{p(\eta_n|H_1)}{p(\eta_n|H_0)} = \lambda_{n-1} + \frac{\Delta^2-1}{2\Delta^2}\eta_n^2 - \log\Delta \tag{2.38}$$

and the test can be performed recursively as new observations come in.

A failure detection system based on the conventional Wald's SPRT formulation given above, on the average, minimizes the time to reach a decision for specified error probabilities if the system is either in the failure mode or in the normal mode from beginning of the test. However, the failure process considered here is characterized by the system being initially operated in the normal mode and then transition occurs to the failure mode at a random instant during observations. The LLR (2.36) will show, on the average, a negative drift when the system is in the normal mode and a positive drift when it is in the failure mode, and thus the detection system suffers an extra time delay in compensating for a negative quantity accumulated under the normal mode before transition to the failure mode (fig. 2.3). To improve on the performance of the conventional SPRT, Uosaki and Kawagoe, (1988) proposed a modified version called backward SPRT.

Taking into account the change time θ, the two hypotheses can be restated as:

H_0 (normal mode): $E\{\eta_{n-i+1}\} = 0$, $\mathrm{Var}\{\eta_{n-i+1}\} = 1$

H_1 (normal mode): $E\{\eta_{n-i+1}\} = 0$, $\mathrm{Var}\{\eta_{n-i+1}\} = \Delta^2 > 1$, $i=1, 2, \ldots, n-\theta+1$ (2.39)

Define a backward LLR (logarithm of likelihood ratio function computed in reverse from the current observation to the past ones) by,

$$\lambda_{n,k}^{B} = \log \frac{p(\eta_n, \eta_{n-1}, \ldots, \eta_{n-k} | H_1)}{p(\eta_n, \eta_{n-1}, \ldots, \eta_{n-k} | H_0)}$$

$$= \frac{\Delta^2 - 1}{2\Delta^2} \sum_{i=n-k+1}^{N} \left(n_i^2 - \frac{2\Delta^2}{\Delta^2 - 1} \log \Delta \right); k = 1, 2, \ldots, n$$

When the backward LLR is applied to test the hypotheses (2.39) as in SPRT (called *backward SPRT*), no extra time delay will be introduced, since in the backward SPRT, the variance of the normalized innovation is not unity from the beginning for hypothesis H_1 corresponding to the failure mode. Furthermore a negative quantity accumulated under the initial normal mode in the conventional SPRT scarcely appears (fig. 2.3).

Backward LLR can be expressed with the conventional LLR as,

$$\lambda_{n,k}^{B} = \frac{\Delta^2 - 1}{2\Delta^2} \sum_{i=n-k+1}^{n} \left(\eta_i^2 - \frac{2\Delta^2}{\Delta^2 - 1} \log \Delta \right)$$

$$= \frac{\Delta^2 - 1}{2\Delta^2} \sum_{i=1}^{n} \left(\eta_i^2 - \frac{2\Delta^2}{\Delta^2 - 1} \log \Delta \right) - \frac{\Delta^2 - 1}{2\Delta^2} \sum_{i=1}^{n-k} \left(\eta_i^2 - \frac{2\Delta^2}{\Delta^2 - 1} \log \Delta \right)$$

$$= \lambda_n - \lambda_{n-k}; \; k = 1, 2, \ldots, n \qquad (2.40)$$

Since, we have interest only in detecting the failure mode, the decision rule is defined as follows:

> If the backward LLR statistic $\lambda_{n,k}^{B} > K$ for some $k=1, 2, \ldots, n$, where K is a suitable constant, observation is terminated with acceptance of the hypothesis that system is in the failure mode. Otherwise, observation is continued as system is not likely in the failure mode.

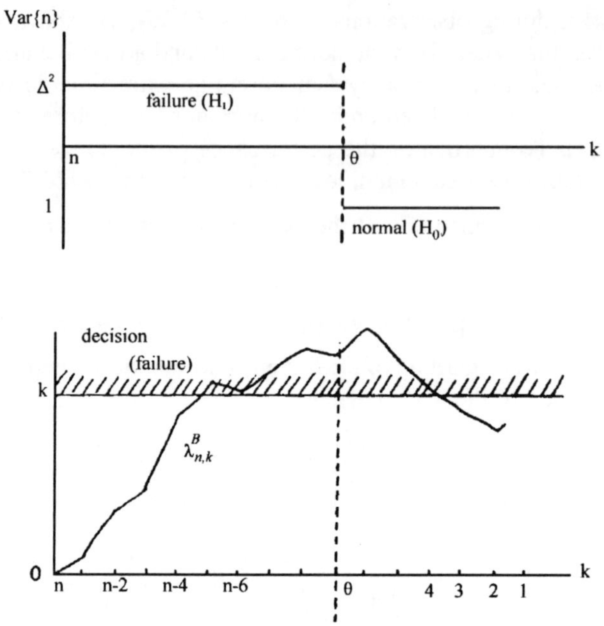

Figure 2.3 Backward SPRT failure detection system and trajectory of backward LLR

The alternative expression of the backward LLR given by (2.40), leads to the decision rule for acceptance of the hypothesis that system is in the failure mode as,

$$\lambda_n - \lambda_{n-k} = \lambda^B_{n,k} > K \text{ for some } k = 1,\ 2,\ \ldots,\ n$$

or,

$$\lambda_n - \min_{1<k<n} \lambda_k > K \tag{2.41}$$

It is easy to show that this decision rule is equivalent to the following decision rule:

If,

$$S_n = \max\left[0,\ \frac{\Lambda^2 - 1}{2\Lambda^2}\eta_n^2 - \log\Lambda + S_{n-1}\right] > K \tag{2.42}$$

with S_0=0, then decide that system is in the failure mode.

Mean detection time of backward SPRT. To determine suitable reference values Λ and decision boundary K, the *mean detection time* (*MDT*) of the backward SPRT should be considered. Here MDT is defined as the mean number of observations until S_n exceeds the decision boundary K when the innovations variance changes to Λ_η^2. To obtain the relation of MDT to Λ and K, first divide $(-\infty, \infty)$ into N subintervals, $(-\infty, 0]$, $(0, W]$, ..., $((N-3)W, K]$, and (K, ∞), with W=$K/(N-2)$, and define states of a stationary finite

Markov chain as $E_1=(-\infty, 0]$, $E_i=((i-2)W, (i-1)W]$; $i=2, \ldots, N-1$, and $E_N=(K, \infty)$. It should be noted that the states E_1 and E_N are reflecting and absorbing, respectively. Let P be the transition probability matrix of this Markov chain with components $p_{ij}=Pr(S_{n-1} \in E_i$ and $S_n \in E_j)$; $i,j=1, \ldots, N)$ which are given by,

$$p_{11} = \Pr\{z-h \le 0\} = F(Dh)$$

$$p_{1i} = \Pr\{(i-2)W \le z-h < (i-1)W\}$$

$$= F\big(D((i-1)W-h) - F((i-2)W+h)\big);\ i=2,\ldots,N-1$$

$$p_{iN} = \Pr\{z-h > K\} = 1-\big(D(k+h)\big)$$

$$p_{i1} = \Pr\left\{z-h \le \left(-i+\frac{3}{2}\right)W\right\} = F\left(D\left(\left(-i+\frac{3}{2}\right)W+h\right)\right);\ i=2,\ldots,N-1)$$

$$p_{ij} = \Pr\left\{\left(j-i-\frac{1}{2}\right)W \le z-h < \left(j-i+\frac{1}{2}\right)W\right\}$$

$$= F\left(D\left(\left(j-i+\frac{1}{2}\right)W+h\right)\right) - F\left(D\left(\left(j-i-\frac{1}{2}\right)W+h\right)\right);\ i,j=2,\ldots,N-1$$

$$p_{iN} = \Pr\left\{z-h > \left(N-i-\frac{1}{2}\right)W\right\} = 1-F\left(D\left(\left(N-i-\frac{1}{2}\right)W+h\right)\right);\ i=2,\ldots,N-1$$

$$p_{Ni} = 0;\ i=1,2,\ldots,N-1$$

$$p_{NN} = 1 \tag{2.43}$$

where,

$$z = \frac{\Lambda^2-1}{2\Lambda^2},\quad h = \log\Lambda,\quad D = \frac{2\Lambda^2}{\left(\Lambda^2-1\right)\Lambda_\eta^2} \tag{2.44}$$

and $F(x)$ is the cumulative distribution function of the χ^2 distribution with one degree of freedom, i.e.,

$$F(x) = \frac{2}{\sqrt{2\pi}}\int_{-\infty}^{x}\frac{1}{\sqrt{v}}\exp\left(-\frac{v}{2}\right)dv = 2\Phi\left(\sqrt{x}\right)-1$$

($\Phi(x)$ is the c.d.f. of the standard normal distribution). Then, the transition probability matrix $\boldsymbol{P}$ can be expressed by,

$$P = \begin{bmatrix} p_{11} & p_{12} & \cdots & p_{1,N-1} & p_{1,N} \\ p_{21} & p_{22} & & p_{2,N-1} & p_{2,N} \\ \cdot & \cdot & \cdot & \cdot & \cdot \\ p_{N-1,1} & p_{N-1,2} & \cdot & p_{N-1,N-1} & p_{N-1,N} \\ 0 & 0 & \cdot & 0 & 1 \end{bmatrix}$$

$$= \begin{bmatrix} \boldsymbol{R} & \boldsymbol{p} \\ \boldsymbol{0} & 1 \end{bmatrix} \tag{2.45}$$

and the mean absorption time vector starting from state E_i is given by,

$$\boldsymbol{\mu} = (\boldsymbol{I} - \boldsymbol{R})^{-1}\boldsymbol{\xi}$$

where $\boldsymbol{\xi} = \begin{bmatrix} 1 & 1 & \dots & 1 \end{bmatrix}^{\mathrm{T}}$.

Since the system starts from the normal mode, the first component of $\boldsymbol{\mu}$ gives the MDT by an N–discrete state Markov chain approximation. The MDT can be obtained by simple extrapolation, as the number of states of discrete state approximation N, goes to infinity. It is assumed that the MDT for a large number of states N, is expressed by,

$$\mu(N) = \mu(\infty) + \frac{A}{N}$$

and $\mu(\infty)$ is determined by a least squares method from $\mu(N)$ for several values of N. To determine the reference value Δ and decision boundary K, it is possible for example, to determine the reference value as the greatest tolerable innovation variance change, and then find the decision boundary K referring the value of the MDT of normal mode ($\Delta_\eta^2 = 1$), which corresponds to the inverse of the probability of false alarm.

2.5.1.3 Tests of whiteness

The most important property of the innovation sequence is whiteness or independence at different time instants. Usual tests of mean and covariance assume that the innovation sequence is white. Therefore, it is important to test the innovation sequence for whiteness first, using tests which are invariant with respect to the mean and covariance of the distribution.

Most of the tests of independence are based on the autocorrelation matrix $\boldsymbol{C}(\tau)$ of a stationary process for lag $\tau = 1, 2, \dots$ defined by,

$$\boldsymbol{C}(\tau) = \mathrm{E}\left\{(\boldsymbol{\eta}(i) - \bar{\boldsymbol{\eta}})(\boldsymbol{\eta}(i-\tau) - \bar{\boldsymbol{\eta}})^{\mathrm{T}}\right\} \tag{2.46}$$

$\boldsymbol{C}(\tau)$ is usually estimated as,

$$\hat{C}_N(\tau) = \frac{1}{N}\sum_{i=\tau}^{N}(\eta(i) - \hat{\bar{\eta}}_N)(\eta(i-\tau) - \hat{\bar{\eta}}_N)^{\mathrm{T}} \tag{2.47}$$

Now, $\hat{C}_N(\tau)$ is an asymptotically unbiased and consistent estimate of $C_N(\tau)$ (Jenkin and Watts, 1968). Under the null hypothesis $\hat{C}_N(\tau)$, τ=1, 2, ... are asymptotically independent and normal with zero mean and covariance of I/N. Thus they can be regarded as samples from the same normal distribution and must lie in the band $\pm 1.96/\sqrt{N}$ more than 95 per cent of the times for the null hypothesis.

Another statistic that can be used for testing independence between the components of the innovation vector is the sample correlation coefficient defined as,

$$\rho_N^{\alpha\beta} = \frac{\sum_{i=1}^{N}(\eta^\alpha(i) - \bar{\eta}^\alpha)(\eta^\beta(i) - \bar{\eta}^\beta)}{\sqrt{\sum_{i=1}^{N}(\eta^\alpha(i) - \bar{\eta}^\alpha)^2}\sqrt{\sum_{i=1}^{N}(\eta^\beta(i) - \bar{\eta}^\beta)^2}} \tag{2.48}$$

where the superscripts α and β indicate the components of the vector η. Anderson, (1958) shows that under the null hypothesis the distribution of $\rho_N^{\alpha\beta}$ is,

$$\frac{\Gamma\left[\frac{1}{2}(N-1)\right]}{\Gamma\left[\frac{1}{2}(N-2)\right]\sqrt{\pi}}\left(1-(\rho_N^{\alpha\beta})^2\right)^{\frac{1}{2}(N-4)}$$

where Γ $[\cdot]$ denotes the gamma function. This statistic is invariant with respect to the mean and the covariance of $\eta(k)$. It is particularly useful in the present case since the true mean and covariance of $\eta(k)$ are unknown. This statistic can also be used for testing whiteness by defining $\rho_N^{\alpha\beta}$ for different lags.

2.5.1.4 Two-stage methods

The previously presented simple tests provide fault alarms and are useful when more sophisticated algorithms cannot be used, because of speed or other limitations. An alternative procedure in such situations would be to implement a two-stage method (Pouliezos and Stavrakakis, 1991). In this approach partial fault isolation is achieved by performing a combination of simple tests, and then use a fault table to limit the possible faults. Such a fault table is the following:

	Zero mean	Whiteness	Stationarity
Plant bias	No	Yes	Yes
Change in A	No	No	
Change in Q	Yes	No	No (Yes*)
Measurement bias	No	Yes	Yes
Change in C	No	No	
Change in R	Yes	No	No(Yes*)

* After fault effects have settled.

If for example, correlated residuals, but of zero mean, are detected, then an increase in the covariance matrix of the noise sequences has taken place. Following this, estimation of the additional noise is possible by appropriate algorithms. In the case of additional sensor noise, Pouliezos *et al.* (1993), estimate the additional noise covariance matrix $\boldsymbol{S}$ for the system described by the time-invariant equivalent of (2.14)-2.23), using the following expressions:

$$\boldsymbol{S} = \hat{\boldsymbol{C}}_{\infty} - \boldsymbol{V} - \boldsymbol{CA\Sigma}(\boldsymbol{CA})^{\mathrm{T}}$$

$$\boldsymbol{\Sigma} = (\boldsymbol{I} - \boldsymbol{KC})\boldsymbol{A\Sigma}[(\boldsymbol{I} - \boldsymbol{KC})\boldsymbol{A}]^{\mathrm{T}} + \boldsymbol{K}(\hat{\boldsymbol{C}}_{\infty} - \boldsymbol{V})\boldsymbol{K}^{\mathrm{T}} - \boldsymbol{KCA\Sigma}(\boldsymbol{KCA})^{\mathrm{T}}$$

where $\boldsymbol{V}$ and $\boldsymbol{K}$ denote the state steady values of the residual covariance and filter gain respectively, as given by (2.21), (2.22). $\hat{\boldsymbol{C}}_{\infty}$ denotes an estimate of the residual covariance obtained from a large number of samples.

This is an algebraic equation in the elements of $\boldsymbol{\Sigma}$, which can be solved directly for its distinct $n(n+1)/2$ elements. Let,

$$\boldsymbol{\sigma} = [\ \sigma_{11}\ \sigma_{12} \ldots \sigma_{1n}\ \sigma_{21} \ldots \sigma_{2n} \ldots \sigma_{nn}\ ]$$

be the vector of the unknown elements of $\boldsymbol{\Sigma}$. Then using brute force algebra,

$$\boldsymbol{\sigma} = (\boldsymbol{I} - \boldsymbol{T})^{-1}\ell$$

$$\boldsymbol{T} = \begin{bmatrix} t_{11}^{11} & t_{11}^{12} & \cdots & t_{11}^{nn} \\ t_{12}^{11} & t_{12}^{12} & \cdots & t_{12}^{nn} \\ \cdots & \cdots & \cdots & \cdots \\ t_{nn}^{11} & t_{nn}^{12} & \cdots & t_{nn}^{nn} \end{bmatrix}$$

$$\ell = \begin{bmatrix} \ell_{11} & \ell_{12} & \cdots & \ell_{1n} & \ell_{21} & \cdots & \ell_{2n} & \cdots & \ell_{nn} \end{bmatrix}$$

$$t_{ij}^{xy} = A_{j,y}A_{i,x} - A_{i,x}\sum_{k=1}^{n} M_{j,k}A_{k,y} - A_{j,y}\sum_{k=1}^{n} M_{i,k}A_{k,x}$$

$$\boldsymbol{L} = \boldsymbol{K}(\hat{\boldsymbol{C}}_{\infty} - \boldsymbol{V})\boldsymbol{K}^T$$

$$\boldsymbol{M} = \boldsymbol{K}\boldsymbol{C}$$

(suffices denote respective matrix elements). In particular, if the system is scalar, its solution is,

$$s = (\hat{c}_{\infty} - v)\left(1 - \frac{(cak)^2}{[(1-kc)a]^2 - 1}\right)^{-1}$$

2.5.2 *The Multiple Model (MM) method*

The MM method was originally developed for problems of system identification and adaptive control (Athans and Willner (1973), Lainiotis (1971)).

The basic MM method deals with the following problem: the inputs $\boldsymbol{u}(k)$; k=0, 1, 2, ... and outputs $\boldsymbol{y}(k)$; k=1, 2, ... of a system which is assumed to obey one of a given finite set of linear, possibly time-varying, stochastic models, indexed by i=1, ..., N, are observed:

$$\boldsymbol{x}_i(k) = \boldsymbol{A}_i(k)\boldsymbol{x}_i(k) + \boldsymbol{B}_i(k)\boldsymbol{u}(k) + \boldsymbol{w}_i(k) + \boldsymbol{g}_i(k)$$

$$\boldsymbol{y}(k) = \boldsymbol{C}_i(k)\boldsymbol{x}_i(k) + \boldsymbol{v}_i(k) + \boldsymbol{b}_i(k)$$

where $\boldsymbol{w}_i(k)$ and $\boldsymbol{v}_i(k)$ are independent, zero–mean Gaussian white noise processes, with,

$$\mathrm{E}\{\boldsymbol{w}_i(k)\boldsymbol{w}_i^{\mathrm{T}}(j)\} = \boldsymbol{Q}_i(k)\delta_{jk}$$

$$\mathrm{E}\{\boldsymbol{v}_i(k)\boldsymbol{v}_i^{\mathrm{T}}(j)\} = \boldsymbol{R}_i(k)\delta_{jk}$$

As usual, the initial state $\boldsymbol{x}_i(0)$ is assumed to be Gaussian, independent of $\boldsymbol{w}_i$ and $\boldsymbol{v}_i$, with mean $\hat{\boldsymbol{x}}_i(0|0)$ and covariance $\boldsymbol{P}_i(0|0)$. The matrices $\boldsymbol{A}_i(k)$, $\boldsymbol{B}_i(k)$, $\boldsymbol{C}_i(k)$, $\boldsymbol{Q}_i(k)$, and $\boldsymbol{R}_i(k)$ are assumed to be known. Also, $\boldsymbol{b}_i(k)$ and $\boldsymbol{g}_i(k)$ are given deterministc functions of time (corresponding to biases, linearizations about different operating points, etc.). In addition, the state vectors $\boldsymbol{x}_i(k)$ may be of different dimensions for different values of i (corresponding to assuming that the different hypothesized models represent different orders for the dynamics of the real system). Note that this is a discrete–time formulation of the MM method. Continuous–time versions can be found in the literature (Greene, 1978), and they differ from their discrete–time counterparts only in a technical and not in a conceptual or structural manner.

Assuming that one of these N models is correct, we now have a standard multiple hypothesis testing problem. That is, let H_i denote the hypothesis that the real system corresponds to the ith model, and let $p_i(0)$ denote the a-priori probability that H_i is true. Similarly, let $p_i(k)$ denote the probability that H_i is true based on measurements through the kth measurement, i.e. given $\boldsymbol{I}_k = \{u(0), ..., u(k-1), y(1), ..., y(k)\}$. Then Bayes' rule yields the following recursive formula for the $p_i(k)$:

$$p_i(k+1) = \frac{p\big(\boldsymbol{y}(k+1)\big|H_i, \boldsymbol{I}_k, \boldsymbol{u}(k)\big)\, p_i(k)}{\sum_{j=1}^{N} p\big(\boldsymbol{y}(k+1)\big|H_j, \boldsymbol{I}_k, \boldsymbol{u}(k)\big)\, p_j(k)} \tag{2.49}$$

Thus, the quantities that must be produced at each time are the conditional probability densities $p\big(\boldsymbol{y}(k+1)\big|H_i, \boldsymbol{I}_k, \boldsymbol{u}(k)\big)$; $i = 1, ..., N$. However, conditioned on H_i, this probability density is precisely the one step prediction density produced by a Kalman filter based on the ith model.

That is, let $\hat{\boldsymbol{x}}_i(k+1|k)$ be the one–step predicted estimate of $\boldsymbol{x}_i(k+1)$ based on $\boldsymbol{I}_k$ and $\boldsymbol{u}(k)$, assuming that H_i is true. Also let $\hat{\boldsymbol{x}}_i(k+1|k+1)$ denote the filtered estimate of $\boldsymbol{x}_i(k+1)$ based on $\boldsymbol{I}_{k+1}=\{\boldsymbol{I}_k, \boldsymbol{u}(k), \boldsymbol{y}(k+1)\}$ and the ith model. Then these quantities are computed sequentially from the following equations:

$$\hat{\boldsymbol{x}}_i(k+1|k) = \boldsymbol{A}_i(k)\hat{\boldsymbol{x}}_i(k|k) + \boldsymbol{B}_i(k)\boldsymbol{u}(k) + \boldsymbol{g}_i(k) \tag{2.50}$$

$$\hat{\boldsymbol{x}}_i(k+1)\big|k+1) = \hat{\boldsymbol{x}}_i(k+1|k) + \boldsymbol{K}_i(k+1)\boldsymbol{\gamma}_i(k+1) \tag{2.51}$$

where $\boldsymbol{\gamma}_i(k+1)$ is the usual innovations process,

$$\boldsymbol{\gamma}_i(k+1) = \boldsymbol{y}(k+1) - \boldsymbol{C}_i(k)\hat{\boldsymbol{x}}_i(k+1)\big|k)$$

and $\boldsymbol{K}_i(k+1)$ is calculated off–line from the following set of equations:

$$\boldsymbol{P}_i(k+1|k) = \boldsymbol{A}_i(k)\boldsymbol{P}_i(k|k)\boldsymbol{A}_i^{\mathrm{T}}(k) + \boldsymbol{Q}_i(k) \tag{2.52}$$

$$\boldsymbol{V}_i(k+1) = \boldsymbol{C}_i(k)\boldsymbol{P}_i(k+1|k)\boldsymbol{C}_i^{\mathrm{T}}(k) + \boldsymbol{R}_i(k) \tag{2.53}$$

$$\boldsymbol{K}_i(k+1) = \boldsymbol{P}_i(k+1|k)\boldsymbol{C}_i^{\mathrm{T}}(k)\boldsymbol{V}_i^{-1}(k+1) \tag{2.54}$$

$$\boldsymbol{P}_i(k+1|k+1) = \boldsymbol{P}_i(k+1|k) - \boldsymbol{K}_i(k+1)\boldsymbol{C}_i(k)\boldsymbol{P}_i(k+1|k) \tag{2.55}$$

Here $\boldsymbol{P}_i(k+1|k)$ denotes the estimation error covariance in the estimate $\hat{\boldsymbol{x}}_i(k+1|k)$ (assuming H_i to be true), and $\boldsymbol{P}_i(k+1|k+1)$ is the covariance of the error $\boldsymbol{x}_i(k+1) - \hat{\boldsymbol{x}}_i(k+1|k+1)$, again based on H_i. Also under hypothesis H_i, $\boldsymbol{\gamma}_i(k+1)$ is zero mean with covariance $\boldsymbol{V}_i(k+1)$, and it is normally distributed (since we have assumed that all noises are Gaussian). Furthermore conditioned on H_i, $\boldsymbol{I}_k$, and $\boldsymbol{u}(k)$, $\boldsymbol{y}(k+1)$

is Gaussian, has mean $C_i(k)\hat{x}_i(k+1|k)$ and covariance $V_i(k+1)$. Thus, it is deduced that,

$$p\big(y(k+1)\big|H_i,I_k,u(k)\big)=\frac{1}{(2\pi)^{m/2}\left[\det V_i(k+1)\right]^{1/2}}\exp\left\{-\frac{1}{2}\gamma_i^{\mathrm{T}}(k+1)V_i^{-1}(k+1)\gamma_i(k+1)\right\} \tag{2.56}$$

(recall m is the dimension of y).

Equations (2.49)–(2.51) and (2.56) define the MM algorithm. The input to the procedure are the $y(k)$ and $u(k)$, and the outputs are the $p_i(k)$. The implementation of the algorithm can be viewed as consisting of a bank of N Kalman filters, one based on each of the N possible models. The outputs of these Kalman filters are the innovations sequences $\gamma_i(k+1)$, which effectively measure how well each of the filters can track and predict the behavior of the observed data. Specifically, if the ith model is correct, then the one–step prediction error $\gamma_i(k)$ should be a white sequence, resulting only from the intrinsic uncertainty in the ith model. However if the ith model is not correct, then $\gamma_i(k)$ will not be white and will include errors due to the fact that the prediction is based on an erroneous model. Thus the probability calculations (2.49), (2.56) basically provide a quantitative way in which to assess which model is most likely to be correct by comparing the performance of predictors based on these models.

Several questions arise in understanding how the MM algorithm should be used. Clearly a very important question concerns the use of MM in problems in which the real system is nonlinear and/or the noises are non–Gaussian. The answer to this problem is application–dependent. The Gaussian assumption is basically used in one place, i.e. in the evaluation of $p\big(y(k+1)\big|H_i,I_k,u(k)\big)$ in (2.56). It may turn out that using this formula, even when $\gamma_i(k+1)$ is non–Gaussian, causes essentially no performance degradation.

As far as the nonlinearity of the real system is concernced, an obvious approach is to linearize the system about a number of operating points for each possible model and use these linearized models to design extended Kalman filters which would be used in place of Kalman filters in the MM algorithm. Again the utility of this approach depends very much on the particular application. Essentially the issue is whether the tracking error from the extended Kalman filter corresponding to the linearized model closest to the true, nonlinear system, is markedly smaller than the errors from filters based on less correct models. This is basically a signal–to–noise ratio problem, similar to that seen in the idealized MM algorithm in which everything is linear. In that case the noise is measured by the $V_i(k+1)$. The larger these are, the harder it will be to distinguish the models. In the nonlinear case, the inaccuracies of the extended Kalman filters effectively increase the $V_i(k+1)$ thus reducing their tracking capabilities and making it more difficult to distinguish among them. Therefore, the performance of MM in this case will depend upon how far apart the different models are, as compared to how well each of the trackers tracks.

The further apart the models are, the more signal we have; the poorer the tracking performance is, the more difficult it is to distinguish among the hypotheses.

Even if the true system is linear, there is clearly the question of the utility of MM given the inevitability of discrepancies between the actual system and any of the N hypothesized models. Again this is a question of signal–to–noise ratio, but in the linear case a number of results and approaches have been developed for dealing with this problem. For example, Baram (1976), has developed a precise mathematical procedure for calculating the distance between different linear modes, and he has shown that the MM procedure will converge to the model closest to the real model (i.e. $p_i(k)\to 1$ for the model nearest the true system). This can be viewed as a technique for testing the robustness of MM or as a tool that enables us to decide what models to choose. That is, if the real system is in some set of models that may be infinite or may in fact represent a continuum of models (corresponding to the precise values of certain parameters), then Baram's results can be used to decide upon a finite set of these models that span the original set and that are far enough apart so that MM can distinguish among them. Willsky (1986) further elaborates on implementation issues of the MM algorithm.

2.5.3 The Generalized Likelihood Ratio (GLR) method

The starting point for the GLR method is a model describing normal operation of the observed signals or of the system which generated them. Abrupt changes are then modeled as *additive* or *multiplicative* disturbances to this model that begin at unknown times. Additive disturbances typify biases, while multiplicative disturbances model parameter changes. As just discussed for MM, the case of a single such change will be considered, the assumption being that abrupt changes are sufficiently separated to allow for individual detection and compensation. The solution to the problem just described and applications of the method can be found amongst others in Willsky and Jones (1976), Pouliezos and Stavrakakis (1987), Tanaka and Müller (1990).

2.5.3.1 Additive changes

The system under consideration is modeled as,

$$\boldsymbol{x}(k+1)=\boldsymbol{A}(k)\boldsymbol{x}(k)+\boldsymbol{B}(k)\boldsymbol{u}(k)+\boldsymbol{w}(k)+\boldsymbol{f}_i(k,\theta)\nu \qquad (2.57a)$$

$$\boldsymbol{y}(k)=\boldsymbol{C}(k)\boldsymbol{x}(k)+\boldsymbol{v}(k)+\boldsymbol{g}_i(k,\theta)\nu \qquad (2.57b)$$

where the normal model consists of these equations without the $\boldsymbol{f}_i$ and $\boldsymbol{g}_i$ terms. These terms represent the presence of the ith type of abrupt change, $i = 1, \ldots, N$. Here θ is the unknown time at which the failure occurs (so $\boldsymbol{f}_i(k,\theta) = \boldsymbol{g}_i(k,\theta) = \boldsymbol{0}$, for $k<\theta$), and $\boldsymbol{f}_i$ and $\boldsymbol{g}_i$ are the specified *dynamic profiles* of the ith change type. For example, if $\boldsymbol{f}_i = \boldsymbol{0}$ and $\boldsymbol{g}_i =$ a

= a vector whose components are all zero except for the jth one which equals 1 for $k \geq \theta$, then this corresponds to the onset of a bias in the jth component of $\boldsymbol{y}$. Finally, the scalar v denotes the magnitude of the failure (e.g. the size of a bias) which we can model as known (as in MM and as in what is called *simplified GLR* (*SGLR*)) or unknown.

Assume that a Kalman filter based on normal operation is designed, i.e. by neglecting $\boldsymbol{f}_i$ and $\boldsymbol{g}_i$. This filter is given by,

$$\hat{\boldsymbol{x}}(k+1|k) = \boldsymbol{A}(k)\hat{\boldsymbol{x}}(k|k) + \boldsymbol{B}(k)\boldsymbol{u}(k) \tag{2.58}$$

$$\hat{\boldsymbol{x}}(k+1|k+1) = \hat{\boldsymbol{x}}(k+1|k) + \boldsymbol{K}(k+1)\boldsymbol{\gamma}(k+1) \tag{2.59}$$

$$\boldsymbol{\gamma}(k+1) = \boldsymbol{y}(k+1) - \boldsymbol{C}(k)\hat{\boldsymbol{x}}(k+1|k) \tag{2.60}$$

where $\boldsymbol{K}$, $\boldsymbol{P}$, and $\boldsymbol{V}$ are calculated as in (2.52)–(2.55). Suppose now that a type i change of size v occurs at time θ. Then, because of the linearity of (2.57)–(2.60),

$$\boldsymbol{x}(k) = \boldsymbol{x}_N(k) + \boldsymbol{\alpha}_i(k,\theta)v \tag{2.61}$$

$$\hat{\boldsymbol{x}}(k|k) = \hat{\boldsymbol{x}}_N(k|k) + \boldsymbol{\beta}_i(k,\theta)v \tag{2.62}$$

$$\hat{\boldsymbol{x}}(k+1|k) = \hat{\boldsymbol{x}}_N(k+1|k) + \boldsymbol{\mu}_i(k+1,\theta)v \tag{2.63}$$

$$\boldsymbol{\gamma}(k) = \boldsymbol{\gamma}_N(k) + \boldsymbol{\rho}_i(k,\theta)v \tag{2.64}$$

where $\boldsymbol{x}_N$, $\hat{\boldsymbol{x}}_N$ and $\boldsymbol{\gamma}_N$ are the responses if no abrupt change occurs, and the other terms are the response due solely to the abrupt change. Straightforward calculations yield recursive equations for these quantities:

$$\begin{aligned} &\boldsymbol{\alpha}_i(k+1,\theta) = \boldsymbol{A}(k)\boldsymbol{\alpha}_i(k,\theta) + \boldsymbol{f}_i(k,\theta) \\ &\boldsymbol{\alpha}_i(\theta,\theta) = \boldsymbol{0} \end{aligned} \tag{2.65}$$

$$\begin{aligned} &\boldsymbol{\beta}_i(k+1,\theta) = [\boldsymbol{I} - \boldsymbol{K}(k+1)\boldsymbol{C}(k+1)]\boldsymbol{\mu}_i(k+1,\theta) + \boldsymbol{K}(k+1)[\boldsymbol{C}(k+1)\boldsymbol{\alpha}_i(k+1,\theta) + \boldsymbol{g}_i(k+1,\theta)] \\ &\boldsymbol{\beta}_i(\theta-1,\theta) = \boldsymbol{0} \end{aligned} \tag{2.66}$$

appropriate to determine its identity i and estimate its time of occurrence θ and size v, if the latter is modeled as being unknown. The solution to this problem involves matched filtering operations. First, define the precomputable quantities,

$$a(k,\theta,i) = \sum_{j=\theta}^{k} \boldsymbol{\rho}_i^{\mathrm{T}}(j,\theta)\boldsymbol{V}^{-1}(j)\boldsymbol{\rho}_i(j,\theta) \tag{2.67}$$

This has the interpretation as the amount of information present in $y(\theta)$, ..., $y(k)$ about a type i change occurring at time θ.

The on–line GLR calculations consist of the calculation of,

$$d(k,\theta,i)=\sum_{j=\theta}^{k}\rho_i^{\mathrm{T}}(j,\theta)V^{-1}(j)\gamma(j) \tag{2.68}$$

which are essentially correlations of the observed residuals with the abrupt change signatures $\rho_i(j,\theta)$ for different hypothesized types i, and times θ. If v is known (the SGLR case), then the likelihood of a type i change having occurred at time θ given data $y(1)$, ..., $y(k)$ is,

$$\ell_s(k,\theta,i)=2vd(k,\theta,i)-v^2a(k,\theta,i) \tag{2.69}$$

If v is unknown, the GLR for this change is,

$$\ell(k,\theta,i)=\frac{d^2(k,\theta,i)}{a(k,\theta,i)} \tag{2.70}$$

and the MLE of v assuming a change of type i at time θ is,

$$\hat{v}(k,\theta,i)=\frac{d(k,\theta,i)}{a(k,\theta,i)} \tag{2.71}$$

Thus the GLR algorithm consists of the single Kalman filter (2.58)-(2.60), the matched filter operations of (2.68), and the likelihood calculation of (2.69). The outputs of the method are these likelihoods and the estimates of eq. (2.71) if v is modeled as unknown. The basic idea behind GLR is that different types of abrupt changes produce different kinds of effects on the filter innovations - i.e. different signatures - and GLR calculates the likelihood of each possible event by correlating the innovations with the corresponding signature.

As with the MM method, a number of issues can be raised about GLR. Some of these, such as the effect of nonlinearities and robustness to model errors, are very similar to the MM case. Essentially it still can be viewed as a signal-to-noise ratio problem: in the nonlinear case the additive decomposition of (2.64) is not precisely valid, but it may be approximately correct. Also, different failure modes can be distinguished even in the presence of modeling errors if their signatures are different enough. Again these issues depend very much on the particular application. The reader is referred to Basseville and Benveniste (1986), for discussions of several applications of GLR to which these issues had to be addressed.

In the case of the GLR algorithm it is simpler to discuss performance measures, as one can use standard detection-theoretic ideas. Specifically, a direct measure of the detectability of a particular type of change is the information $a(k,\theta,i)$ defined in (2.67). This quantity can be viewed as the correlation of $\rho_i(j,\theta)$ with itself at zero lag. Similarly, the relative distinguishability of a type i change at two times θ_1 and θ_2 can be determined as the correlation of the corresponding signatures,

$$a(k,\theta_1,\theta_2,i) = \sum_{j=\max(\theta_1,\theta_2)}^{k} \boldsymbol{\rho}_i'^{\mathrm{T}}(j,\theta_1)\boldsymbol{V}^{-1}(j)\boldsymbol{\rho}_i(j,\theta_2) \qquad (2.72)$$

and the relative distinguishability of type i and m changes at times θ_1 and θ_2 similarly:

$$a(k,\theta_1,\theta_2,i,m) = \sum_{j=\max(\theta_1,\theta_2)}^{k} \boldsymbol{\rho}_i'^{\mathrm{T}}(j,\theta_1)\boldsymbol{V}^{-1}(j)\boldsymbol{\rho}_m(j,\theta_2)$$

These quantities provide extremely useful information. For example, in some applications, the estimation of the time θ at which the change occurs is critical, and the above equations provide information about how well one can resolve the onset time. In failure detection applications these quantities directly provide information about how system redundancy is used to detect and distinguish failures and can be used in deciding whether additional redundancy (e.g. more sensors) are needed. Also, they directly give the statistics of the likelihood measure (2.69). For the SGLR case of (2.69), ℓ_s is Gaussian, and its mean under no failure is $-v^2 a(k, \theta, i)$, while if a type m failure occurs at time φ, its mean is,

$$\mathrm{E}\{\ell_2(k,\theta,i)|(m,\varphi)\} = v^2[2a(k,\theta,\varphi,i,m) - a(k,\theta,i)]$$

In the case of full GLR, under no failure $\ell(k,\theta,i)$ is a χ^2 random variable with 1 degree of freedom, while if a failure (m,φ) of size v occurs, $\ell(k,\theta,i)$ is a non-central χ^2 with mean,

$$\mathrm{E}\{\ell(k,\theta,i)|(m,\varphi)\} = 1 + \frac{v^2 a^2(k,\theta,\varphi,i,m)}{a(k,\theta,i)}$$

These quantities can be very useful in evaluating the performance of GLR detection algorithms and for determining decision rules based on the GLR outputs. If one were to follow the precise GLR philosophy (Van Trees, 1968), the decision rule one would use is to choose at each time k the largest of the $\ell(k,\theta,i)$ over all possible change types i and onset times θ. This largest value would then be compared to a threshold for change detection, and if the threshold is exceeded the corresponding maximizing values of θ and i are taken as the estimates of change type and time. For greater confidence, persistance tests (i.e. ℓ must exceed the threshold over some time period) are often used to cut down on false alarms due to spurious and unmodeled events. Basseville (1981), further discusses the threshold selection procedure.

A final issue to be mentioned is the pruning of the tree of possibilities. As in the MM case, in principle a growing number of calculations have to be performed, as $d(k,\theta,i)$ must be calculated for every possible fault case and all possible change times up to the present, i.e. $\theta = 1, \ldots, k$. As discussed previously, an appropriate procedure is look only over a sliding window of possible times:

$$k - M_1 \le \theta \le k - M_2$$

where M_1 and M_2 are chosen based on the a's (eq. 2.69), i.e. on detectability and distinguishability considerations. Basically after M_2 times steps from the onset of change, enough information is collected, so that a detection may be made with a reasonable amount of accuracy. Further, after M_1 time steps, a sufficient amount of information will be collected so that detection performance is as good as it can be (i.e. there is no point in waiting any longer). Clearly, M_1, M_2 must be large to allow for maximum information collection, but also small enoough for fast response and for computational simplicity. This is a typical tradeoff that arises in all change detection problems.

2.5.3.2 Non-additive changes

GLR has been successfully applied to a wide variety of applications, such as geophysical signal analysis (Basseville and Benveniste, 1983), detecting arrhythmias in electrocardiograms (Gustaffson *et al.*, 1978a and 1978b), freeway incident detection (Willsky *et al.*, 1980), and manoeuver detection (Dowdle *et al.*, 1983). Recall that the model used in (2.57) for such changes is an additive model. Thus it appears on the surface that the types of abrupt changes that can be detected by GLR are a special subset of those that can be detected by MM, since in this method parametric changes are allowed (in $\boldsymbol{A}$, $\boldsymbol{B}$, $\boldsymbol{C}$, $\boldsymbol{Q}$, $\boldsymbol{R}$) as well as additive ones. However, a GLR system based on the detection of additive effects can often also detect parameter failures. For example, a gain change in a sensor does look like a sensor bias, albeit one that is modulated by the value of the variable being sensed. That is, any detectable change will exhibit a systematic deviation between what is observed and what is predicted to be observed. Obviously, the ability of GLR to detect a parametric change when it is looking for additive ones is again a question of robustness. If the effect of the parametric change is "close enough" to that of the additive one, the system will work. For example, Deckert *et al.* (1977), describe an additive-failure-based design that does extremely well in detecting gain changes in sensors. Note of course that in this mode GLR, is essentially only indicating an alarm, but in detection problems where the primary interest is in simply identifying which of several types of changes has occurred, this is acceptable.

Direct application of the GLR methodology to parametric changes has also been reported by some researchers. In this context, the model is described in any of the following modes:

$$(a)\begin{cases}\boldsymbol{x}(k+1)=[\boldsymbol{A}(k)+\Delta\boldsymbol{A}\sigma_{k+1,\theta}]\boldsymbol{x}(k)+\boldsymbol{B}(k)\boldsymbol{u}(k)+\boldsymbol{w}(k)\\ \quad\boldsymbol{y}(k)=\boldsymbol{C}(k)\boldsymbol{x}(k)+\boldsymbol{v}(k)\end{cases}$$

$$(b)\begin{cases}\boldsymbol{x}(k+1)=\boldsymbol{A}(k)\boldsymbol{x}(k)+\boldsymbol{B}(k)\boldsymbol{u}(k)+\boldsymbol{w}(k)\\ \quad\boldsymbol{y}(k)=[\boldsymbol{C}(k)+\Delta\boldsymbol{C}\sigma_{k+1,\theta}]\boldsymbol{x}(k)+\boldsymbol{v}(k)\end{cases}$$

$$(c)\begin{cases} \boldsymbol{x}(k+1)=\boldsymbol{A}(k)\boldsymbol{x}(k)+\boldsymbol{B}(k)\boldsymbol{u}(k)+\boldsymbol{w}(k)+\boldsymbol{\zeta}_x(k)\sigma_{k+1,\theta} \\ \boldsymbol{y}(k)=\boldsymbol{C}(k)\boldsymbol{x}(k)+\boldsymbol{v}(k) \end{cases}$$

$$(d)\begin{cases} \boldsymbol{x}(k+1)=\boldsymbol{A}(k)\boldsymbol{x}(k)+\boldsymbol{B}(k)\boldsymbol{u}(k)+\boldsymbol{w}(k) \\ \boldsymbol{y}(k)=\boldsymbol{C}(k)\boldsymbol{x}(k)+\boldsymbol{v}(k)+\boldsymbol{\zeta}_y(k)\sigma_{k+1,\theta} \end{cases}$$

where $\boldsymbol{\zeta}_x(k)$, $\boldsymbol{\zeta}_y(k)$ are additional noises of covariances $\boldsymbol{S}_x$, $\boldsymbol{S}_y$, independent of the plant and sensor noise sequences.

Model (a) is used in cases where a transition matrix change is to be detected, model (b) is used when a measurement matrix change is to be detected, while models (c) and (d) are used whenever additional noise in the state or measurements must be monitored. For these models, appropriate equations for GLR-based fault detection are (Pouliezos and Stavrakakis, 1987):

For model (a):

$$\boldsymbol{\gamma}(k)=\boldsymbol{\gamma}_N(k)+\sum_{i=\theta}^{k}\boldsymbol{G}_a(k,i,\Delta\boldsymbol{A})\Delta\boldsymbol{A}\boldsymbol{x}_N(i-1)$$

where the matrices $\boldsymbol{G}_a(\,)$ are calculated from,

$$\boldsymbol{G}_a(k,\theta,\Delta\boldsymbol{A})=\boldsymbol{C}(k)\left(\sum_{q=0}^{\eta(=k-\theta)} C_q^{\eta}\boldsymbol{\Phi}(\theta+\eta,\theta+q)(\Delta\boldsymbol{A})^q-\boldsymbol{A}(k-1)\boldsymbol{F}_a(k-1,\theta,\Delta\boldsymbol{A})\right)$$

$$\boldsymbol{F}_a(k,\theta,\Delta\boldsymbol{A})=\boldsymbol{K}(k)\boldsymbol{G}_a(k,\theta,\Delta\boldsymbol{A})+\boldsymbol{A}(k-1)\boldsymbol{F}_a(k-1,\theta,\Delta\boldsymbol{A});\quad k\ge\theta$$

$$\boldsymbol{G}_a(k,\theta,\Delta\boldsymbol{A})=\boldsymbol{F}_a(k,\theta,\Delta\boldsymbol{A})=\boldsymbol{0};\quad k<\theta;$$

C_q^{η} is the binomial coeeficient, $\boldsymbol{\Phi}(i,j)$ is the state transition matrix and the subscript N denotes normal operation quantities defined as in (2.61)-(2.64). As seen these equations have a similar form to Eqs. (2.64), (2.65)-(2.68) but the nice linear structure is lost, since there is modulation of the residuals by the unknown ΔA.

Similarly for model (b):

$$\boldsymbol{\gamma}(k)=\boldsymbol{\gamma}_N(k)+\sum_{i=\theta}^{k}\boldsymbol{G}_b(k,i)\Delta\boldsymbol{H}\boldsymbol{x}_N(i)$$

where the matrices $\boldsymbol{G}_b(\,)$ are calculated from,

$$\boldsymbol{G}_b(k,\theta)=-\boldsymbol{C}(k)\boldsymbol{A}(k-1)\boldsymbol{F}_b(k-1,\theta);\quad k>\theta$$

$$\boldsymbol{F}_b(k,\theta)=\boldsymbol{K}(k)\boldsymbol{G}_b(k,\theta)+\boldsymbol{A}(k-1)\boldsymbol{F}_b(k-1,\theta);\quad k\ge\theta$$

$$\boldsymbol{G}_b(\theta,\theta)=\boldsymbol{I};$$

$$\boldsymbol{G}_b(k,\theta)=\boldsymbol{F}_b(k,\theta)=\boldsymbol{0};\quad k<\theta$$

while for model (c),

$$\boldsymbol{\gamma}(k)=\boldsymbol{\gamma}_N(k)+\sum_{i=\theta}^{k}\boldsymbol{G}_c(k,i)\boldsymbol{\zeta}_x(i)$$

where,

$$\begin{aligned}
&\boldsymbol{G}_c(k,\theta)=-\boldsymbol{C}(k)[\boldsymbol{\Phi}(k,\theta)-\boldsymbol{A}(k-1)\boldsymbol{F}_c(k-1,\theta)] \\
&\boldsymbol{F}_c(k,\theta)=\boldsymbol{K}(k)\boldsymbol{G}_c(k,\theta)+\boldsymbol{A}(k-1)\boldsymbol{F}_c(k-1,\theta);\quad k\geq\theta \\
&\boldsymbol{G}_c(k,\theta)=\boldsymbol{F}_c(k,\theta)=\boldsymbol{0};\qquad\qquad\qquad\qquad k<\theta
\end{aligned}$$

Finally for model (d),

$$\boldsymbol{\gamma}(k)=\boldsymbol{\gamma}_N(k)+\sum_{i=\theta}^{k}\boldsymbol{G}_d(k,i)\boldsymbol{\zeta}_y(i)$$

where,

$$\begin{aligned}
&\boldsymbol{G}_d(k,\theta)=-\boldsymbol{C}(k)\boldsymbol{A}(k-1)\boldsymbol{F}_d(k-1,\theta);\qquad\qquad k>\theta \\
&\boldsymbol{F}_d(k,\theta)=\boldsymbol{K}(k)\boldsymbol{G}_d(k,\theta)+\boldsymbol{A}(k-1)\boldsymbol{F}_d(k-1,\theta);\quad k\geq\theta \\
&\boldsymbol{G}_d(k,\theta)=\boldsymbol{I}; \\
&\boldsymbol{G}_d(k,\theta)=\boldsymbol{F}_d(k,\theta)=\boldsymbol{0};\qquad\qquad\qquad\qquad k<\theta
\end{aligned}$$

These equations can be used to derive GLR-based detection algorithms. However, their implementation is not at all straightforward, because of their complexity. Liu (1977), considered the case of additional plant noise and proposed some approximate solutions to this problem. Pouliezos *et al.* (1993), considered the problem of additional sensor noise and solved it for the time-invariant case. Their approach utilised the steady-state effect additional sensor noise has on the filter innovations. Tanaka and Müller (1990), have used a pattern recognition approach to detect parametric changes in the state and measurements matrices, using a GLR statistic. Their method is robust and relies on the recognition of the pattern of the curve of the maximum GLR calculated by the conventional step-hypothesised GLR.

2.6 Deterministic methods

2.6.1 Observer–based approaches

The basic idea in the observer based approach is to reconstruct the states (and the outputs) of the system with the aid of observers. Observers are dynamic systems that are aimed at reconstructing the state $\boldsymbol{x}$ of a state-space model on the basis of the measured inputs $\boldsymbol{u}$ and outputs $\boldsymbol{y}$. The state estimation error is then used as the residual for the detection of the faults.

The particular type of observer used depends on the particular application's needs, i.e. what kind of failures need to be detected (component, sensor, actuators), what criteria must be set (robustness, isolability etc.) and on the system structure (observability, controllability). To introduce the concept, consider the linear, time–invariant, controllable and observable system described by equations (2.1), (2.2) in its continuous time version:

$$\begin{aligned}\dot{\boldsymbol{x}}(t) &= \boldsymbol{A}\boldsymbol{x}(t) + \boldsymbol{B}\boldsymbol{u}(t) \\ \boldsymbol{y}(t) &= \boldsymbol{C}\boldsymbol{x}(t)\end{aligned} \tag{2.73}$$

It can easily be seen that this mathematical model with the assumptions made, contains all the existing relations among the measured variables $\boldsymbol{y}_i$ (i.e., the "redundancy relations") in the form of differential equations. To reconstruct the states or measured variables from other measured variables and inputs one can use the *linear full–order estimator*,

$$\dot{\hat{\boldsymbol{x}}}(t) = \boldsymbol{A}\hat{\boldsymbol{x}}(t) + \boldsymbol{B}\boldsymbol{u}(t) + \boldsymbol{L}(\boldsymbol{y}(t) - \boldsymbol{C}\hat{\boldsymbol{x}}(t)) \tag{2.74a}$$

$$\hat{\boldsymbol{y}}(t) = \boldsymbol{C}\hat{\boldsymbol{x}}(t) \tag{2.74b}$$

where $\hat{\boldsymbol{x}}(t)$ and $\hat{\boldsymbol{y}}(t)$ denote the estimated state and output vector and $\boldsymbol{L}$ the gain matrix, by the choice of which desired dynamics of the estimator can be achieved.

Let's consider a simple, continuous–time version of equations (2.11):

$$\dot{\boldsymbol{x}}(t) = \boldsymbol{A}\boldsymbol{x}(t) + \boldsymbol{B}\boldsymbol{u}(t) + \boldsymbol{f}(t) + \boldsymbol{w}(t) \tag{2.75}$$

$$\boldsymbol{y}(t) = \boldsymbol{C}\boldsymbol{x}(t) + \Delta C(t) + \boldsymbol{v}(t) \tag{2.76}$$

where $\boldsymbol{f}(t)$ denotes faults in the process, $\Delta C(t)$ sensor faults, and $\boldsymbol{w}(t)$, $\boldsymbol{v}(t)$ process and sensor noise respectively.

Substracting (2.74a) from (2.75) and defining the state estimation error,

$$\boldsymbol{\varepsilon}(t) = \boldsymbol{x}(t) - \hat{\boldsymbol{x}}(t)$$

the equations for the ouput estimation error,

$$\boldsymbol{e}(t) = \boldsymbol{y}(t) - \hat{\boldsymbol{y}}(t)$$

become,

$$\dot{\boldsymbol{\varepsilon}}(t) = (\boldsymbol{A} - \boldsymbol{L}\boldsymbol{C})\boldsymbol{\varepsilon}(t) + \boldsymbol{w}(t) + \boldsymbol{f}(t) \tag{2.77}$$

$$\boldsymbol{e}(t) = \boldsymbol{C}\boldsymbol{\varepsilon}(t) + \Delta C + \boldsymbol{v}(t) \tag{2.78}$$

It can be seen that if all the disturbances $\boldsymbol{f}(t)$, ΔC, $\boldsymbol{w}(t)$, $\boldsymbol{v}(t)$ are zero and the matrix $\boldsymbol{A}$–$\boldsymbol{LC}$ has leftside eigenvalues, the estimation error $\boldsymbol{e}(t)$ goes to zero after any initial condition tracking error has died out. If, however, at least one of these disturbances occurs,

$e(t)$ is driven by it. Thus its effect on $e(t)$ can be evaluated to discover the disturbance. Therefore $e(t)$ can be used as a residual for *FDI*.

Since specific designs depend on design requirements let us now consider three dinstinct cases:

- instrument fault detection (IFD)
- component fault detection (CFD)
- actuator fault detection (AFD)

A. Instrument fault detection

The goal of IFD is to detect and locate faults of the *sensors* of the process with sufficient robustness to system parameter changes and noise. There are two extremes of application, depending on the *purpose* of failure detection and the *type* of failures. One extreme is the detecting of *hard–failures* as, for example, in safety relevant systems. Typically, the failures are then large and the admissible detection time is small (e.g. seconds or fractions of seconds). The other extreme is the detecting of *soft failures* as in the case of *signal validation*, such as repeated (periodic) checking of instrumentation in a nuclear reactor. In this case the failures to be discovered are typically small (a few percent) but a large detection time is available (days or weeks). Hence, in the first case, one can get along with the deterministic approach using observers since noise (and parameter uncertainties) do not play a large role. In the second case noise cannot be neglected and hence Kalman filters and stochastic decision procedures including correlation techniques are indicated. In a concrete situation one has to decide for either (or, in poorly conditioned situations, for non) of the two approaches.

Over the years a number of different estimator schemes for deterministic IFD have been proposed in the literature. They differ in the number of estimators and the number of measured variables.

In the *Dedicated Observer Scheme (DOS)*, (Frank, 1987), it is assumed that the full state or a subset of it can be observed from each measured variable of the process (fig. 2.4). Each sensor (instrument) output, y_i, is used to drive a dedicated observer of full or reduced order to observe as many of the remaining measured variables as possible. If the process is completely observable, an m–fold redundancy of the measurement vectors is achieved. In the non–fault case,

$$\hat{y}_{ik} = \hat{y}_{ij} = y_i; \; i = 1,2,\ldots,n; \; k,j = 1,2,\ldots,m \tag{2.79}$$

If a fault occurs in the q–th instrument then,

$$\hat{y}_{iq} \neq \hat{y}_{ik} = y_i \tag{2.80}$$

holds for i=1, 2, ..., n and k=1, 2, ..., m, and $k \neq q$, i.e., the observer that is driven by the faulty instrument provides a wrong estimate of all measured variables whereas the estimates of all the other observers match the corresponding y_i except that of the erroneous instrument. The output estimation error,

$$e_k = y_k - \hat{y}_k \tag{2.81}$$

represents now the residual vector which, in principle, allows a unique FDI even in the case of several faulty instruments at the same time.

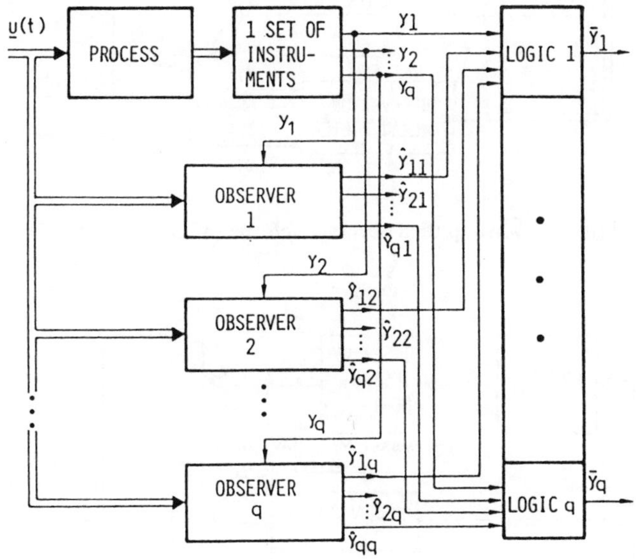

Figure 2.4 Dedicated observer scheme (after Frank, 1987)

The simplest version of an observer scheme for IFD is the so–called ***Simplified Observer Scheme (SOS)*** which includes only a single (full or reduced order) observer driven by only one of the measured variables, see fig. 2.5 (Clark, 1978). As the driving variable one selects the most appropriate sensor output that provides full state (and output) estimation. Though only single redundancy is used this scheme allows a *unique isolation of a single faulty sensor*. If one of the sensors, q, that does not drive the observer fails, the reconstructed output vector is correct and $y_i - \hat{y}_i = 0$ holds for all i except i=q.

If the sensor s that drives the observer fails, all estimates are erroneous so that $y_i - \hat{y}_i \neq 0$ for all i except i=s. An advantage of this scheme is that it applies to processes with very limited observability and that it is easy to implement. However, the price to pay is reduced redundancy and a loss of detection reliability.

At the other end, if DOS is generalized in that each observer is driven with more than one measured variable, say with overlapping subsets of y, the ***Generalized Observer***

Scheme (GOS) is arrived at, see fig. 2.6 (Frank, 1987). It is evident that the decision logic is of the same type as in the case of DOS or SOS. However, this scheme provides more degrees of freedom for the observer design and can be used for the increase of robustness to parameter variations.

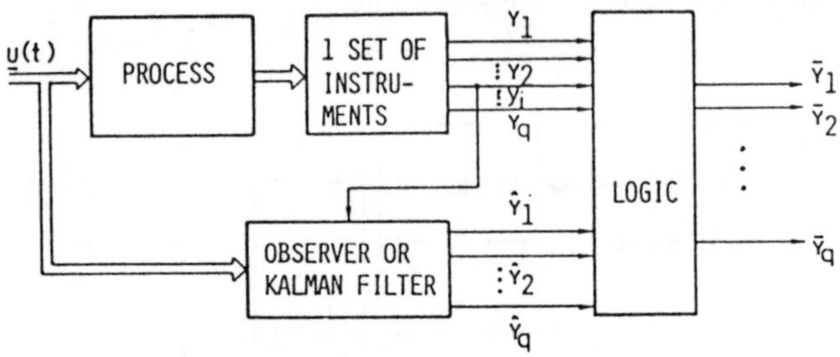

Figure 2.5 Simplified observer scheme (after Frank, 1987)

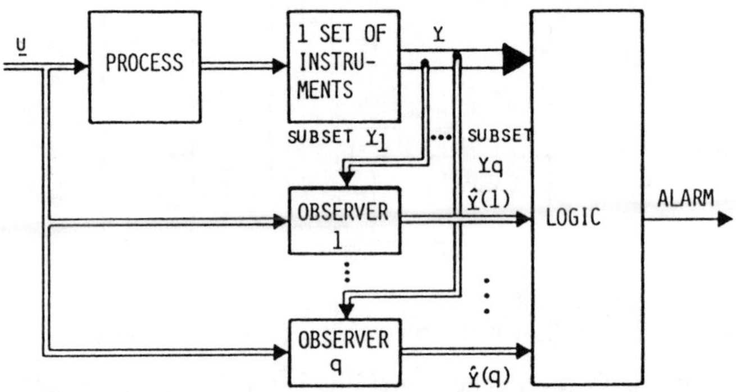

Figure 2.6 Generalized observer scheme (after Frank, 1987)

B. Component fault detection (CFD)

The goal of CFD is to detect and locate faults of components in dynamic processes with sufficient robustness to sensor faults, system parameter changes and noise. While the IFD schemes described in the previous section are also suitable for the alarm task of CFD they are generally not able to solve the problem of location of faulty components. In this Section, observer schemes that allow to detect and locate component faults are discussed.

The most natural approach to the component fault detection and isolation problem is to decompose the system and apply a hierarchical scheme of local observers. One of the advantages of a local observer scheme is that even if the order of the overall system is rather high, the order of each subsystem and consequently the order of the corresponding observers may be rather low. Furthermore, only local observability is required. The problem with this approach lies, however, in the interaction between the subsystems. If they are low, a malfunction in any of the components affects only the estimate of the corresponding local observer. It is thus possible to identify the faulty component uniquely. However, if the interactions are large and not measurable, a fault in one component propagates to observers of other components. The observer scheme would then fail to identify the faulty component. What it could achieve is to identify a fault in *the complex of components* that are largely interconnected. This is the basis for the ***Hierarchical Observer Scheme*** (*HOS*) introduced by Janssen and Frank (1984).

In the HOS, shown in fig. 2.7, the overall system is divided into two levels of components: an upper level including all components with measurable couplings and a lower level of components with unknown couplings. For each of the resulting configurations in both levels a scheme of local observers is designed, which includes the reconstruction of the measured outputs of the components. The observer scheme used for the upper level is called *ASCOS* (***A****vailable*¬***S****tate* ***C****oupled* ***O****bserver* ***S****cheme*) and the one for the lower level *ESCOS* (***E****stimated*¬***S****tate* ***C****oupled* ***O****bserver* ***S****cheme*). The difference between ASCOS and ESCOS is the way of performing the couplings in the observer part of fig. 2.7.

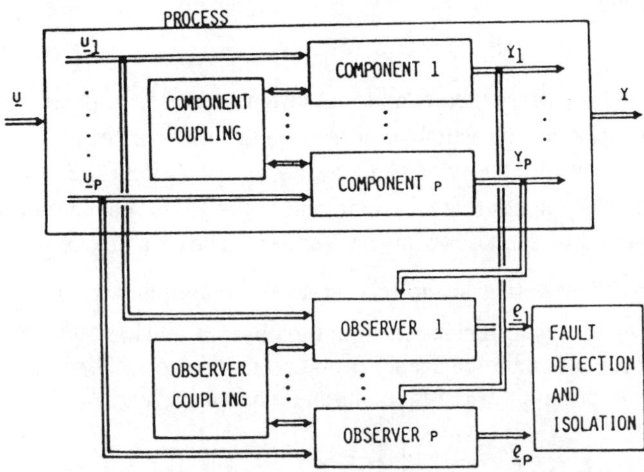

Figure 2.7 Local observer scheme of an HOS (after Frank, 1987)

For a brief development of the main ideas of ASCOS and ESCOS, consider a system represented by equations (2.1), (2.2) in component form:

$$\dot{\boldsymbol{x}}_i(t) = \boldsymbol{A}_{ii}\boldsymbol{x}_i(t) + \sum_{\substack{j=1 \\ j\neq i}}^{N} \boldsymbol{A}_{ij}\boldsymbol{x}_j + \boldsymbol{B}_i\boldsymbol{u}_i \qquad (2.82)$$

$$\boldsymbol{y}_i(t) = \boldsymbol{C}_i\boldsymbol{x}_i(t) \tag{2.83}$$

where $\boldsymbol{x}_i$, $\boldsymbol{y}_i$, $\boldsymbol{u}_i$ are the states, outputs and inputs, respectively, of the *i*th component and $\boldsymbol{A}_{ij}$, $\boldsymbol{B}_i$, $\boldsymbol{C}_i$ are matrices of appropriate dimensions. Note that $\boldsymbol{A}_{ij}$ characterizes the coupling gains with other components.

Assuming that the coupling terms $\sum \boldsymbol{A}_{ij}\boldsymbol{x}_j$ in (2.82) are measurable, the state equations of the corresponding *i*th observer are given by,

$$\dot{\hat{\boldsymbol{x}}}(t) = (\boldsymbol{A}_{ii} - \boldsymbol{L}_i\boldsymbol{C}_i)\hat{\boldsymbol{x}}_i(t) + \sum_{\substack{j=1 \\ j\neq i}}^{N}\left(\boldsymbol{A}_{ij}\boldsymbol{x}_j(t)\right) + \boldsymbol{B}_i\boldsymbol{u}_i(t) + \boldsymbol{L}_i\boldsymbol{y}_i(t) \tag{2.84}$$

$$\hat{\boldsymbol{y}}_i(t) = \boldsymbol{C}_i\hat{\boldsymbol{x}}_i(t) \tag{2.85}$$

where $\boldsymbol{x}_i$ and $\boldsymbol{u}_i$ are measurable and N is the total of components of the upper level. In the nominal case the state estimation error $\boldsymbol{\varepsilon}_i(t) = \boldsymbol{x}_i(t) - \hat{\boldsymbol{x}}_i(t)$ obeys,

$$\dot{\boldsymbol{\varepsilon}}_i(t) = \left(\boldsymbol{A}_{ii} - \boldsymbol{L}_i\boldsymbol{C}_i\right)\boldsymbol{\varepsilon}_i(t) + \Delta\boldsymbol{f}_i + \boldsymbol{v}_i \tag{2.86}$$

where $\Delta\boldsymbol{f}_i$ and $\boldsymbol{v}_i$ denote parameter variations and component faults of the *i*th component. The corresponding output estimation error $\boldsymbol{e}_i(t) = \boldsymbol{y}_i(t) - \hat{\boldsymbol{y}}_i(t)$ of the *i*th component then becomes,

$$\boldsymbol{e}_i(t) = \boldsymbol{C}_i\boldsymbol{\varepsilon}_i(t) + \boldsymbol{N}_i\boldsymbol{n}_i + \boldsymbol{\mu}_i \tag{2.87}$$

where $\boldsymbol{n}_i$ and $\boldsymbol{\mu}_i$ denote sensor faults and measurement noise, respectively. In the ideal case, all coupling terms are measurable, all state and output estimation errors are perfectly decoupled from each other, and $\Delta\boldsymbol{f}_i$, $\boldsymbol{v}_i$, $\boldsymbol{n}_i$, $\boldsymbol{\mu}_i$ only affect the error equations of the *i*th component. This implies that a system fault in the *k*th component only affects $\boldsymbol{e}_k(t)$ leaving the remaining errors $\boldsymbol{e}_i(t)$; $i\neq k$, unchanged. This allows a unique failure isolation.

To derive ESCOS, observe that in the second level the coupling terms $\sum \boldsymbol{A}_{ij}\boldsymbol{x}_j(t)$ are not measurable. The idea is to replace them in the observe scheme by their estimates and estimation errors which both are available from other observers. Then the state equation of the corresponding observer of the *i*th component becomes,

$$\dot{\hat{\boldsymbol{x}}}_i(t) = \left(\boldsymbol{A}_{ii} - \boldsymbol{L}_i\boldsymbol{C}_i\right)\hat{\boldsymbol{x}}_i(t) + \sum_{\substack{j=1 \\ j\neq i}}^{N}\left[\boldsymbol{A}_{ij}\hat{\boldsymbol{x}}_j(t) + \boldsymbol{L}_j\boldsymbol{e}_j(t)\right] + \boldsymbol{B}_i\boldsymbol{u}_i(t) + \boldsymbol{L}_i\boldsymbol{y}_i(t)$$

This leads to the following equation for the state estimation error:

$$\dot{\boldsymbol{\varepsilon}}_i(t) = \left(\boldsymbol{A}_{ii} - \boldsymbol{L}_i\boldsymbol{C}_i\right)\boldsymbol{\varepsilon}_i(t) + \sum_{\substack{j=1 \\ j\neq i}}^{N}\left(\boldsymbol{A}_{ij} - \boldsymbol{L}_j\boldsymbol{C}_j\right)\boldsymbol{\varepsilon}_j(t) \tag{2.88}$$

From (2.88), the output estimation error $e_i(t)$ can again be determined in the same way as above.

2.6.2 Parity space approach

The parity space approach originated by Deckert *et al.* (1977), yields a systematic exploitation of the analytical redundancy provided by the mathematical model of the system.

The basic idea used in this work is to identify the analytical redundancy relations of the system that were known well and those that contained substantial uncertainties. An FDI system (i.e., its residual generation process) is then designed based primarily on the well–known relationships (and only secondarily on the less well–known relations) of the system behavior. Chow, (1980), and Chow and Willsky, (1984), extracted and extended the practical idea underlying this application and developed a general approach to the design of FDI algoritthms. Lou *et al.* (1986), and Massoumnia *et al.* (1988), further developed these ideas.

In addition to its use in specifying residual generation procedure, this approach is also useful as it can provide a quantitative measure of the attainable level of robustness in the early states of a design. This allows the designer to assess overall performance.

The basis for residual generation in analytical redundancy essentially takes two forms:

1. Direct redundancy: the relationship amongst instantaneous outputs of sensors; and,
2. Temporal redundancy: the relationship amongst the histories of sensor outputs and actuator inputs. Based on these relationships, outputs of (dissimilar) sensors (at different times) can be compared. The residuals resulting from these comparisons are then measures of the discrepancy between the behaviour of observed sensor outputs and the behaviour that should result under normal conditions.

In order to develop a clear picture of redundancy, consider the system described by equations (2.1), (2.2) in the following form:

$$\boldsymbol{x}(k+1) = \boldsymbol{A}\boldsymbol{x}(k) + \sum_{j=1}^{r} \boldsymbol{b}_j u_j(k) \tag{2.89}$$

$$y_j(k) = \boldsymbol{c}_j \boldsymbol{x}(k); \quad j = 1, \ \ldots, \ m \tag{2.90}$$

where c_j is an n-row vector. Direct redundancy exists among sensors whose outputs are algebraically related, i.e., the sensor outputs are related in such a way that the variable one sensor measures can be determined by the instantaneous outputs of the other sensors. For the system (2.90), this corresponds to the situation where a number of the c_j's are linearly dependent. In this case, the value of one of the observation scan be written as a linear combination of the other outputs. For example, one might have,

$$y_1(k) = \sum_{i=2}^{m} a_i y_i(K) \tag{2.91}$$

where the a_i's are constants. This indicates that under normal conditions the ideal output of sensor 1 can be calculated from those of the remaining sensors. In the absence of a failure in the sensors, the residual $y_1(k) - \sum_{i=2}^{m} a_i y_i(k)$ should be zero. A deviation from this behavior provides the indication that one of the sensors has failed. Define,

$$C_j(k) = \begin{bmatrix} c_j \\ c_j A \\ \cdot \\ \cdot \\ \cdot \\ c_j A^k \end{bmatrix}; \; k = 0,\ 1,\ \ldots,\ ; \; j = 1,\ \ldots,\ m$$

The well–known Cayley–Hamilton theorem implies that there is an n_j, $1 \le n_j \le n$, such that,

$$\text{rank } C_j(k) = \begin{cases} k+1 & k < n_j \\ n_j & k \ge n_j \end{cases} \tag{2.92}$$

The null space of the matrix $C_j(n_j-1)$ is known as the *unobservable subspace* of the jth sensor. The rows of $C_j(n_j-1)$ span a subspace of $\boldsymbol{R}^n$ that is the orthogonal complement of the unobservable subspace. Such a subspace will be referred to as the *observable subspace* of the jth sensor, and it has dimension n_j.

Let $\boldsymbol{\omega}$ be a row vector of dimension $N = \sum_{i=2}^{m} (n_i + 1)$ such that $\boldsymbol{\omega} = [\boldsymbol{\omega}^1 \ \ldots \ \boldsymbol{\omega}^m]$, where $\boldsymbol{\omega}^j$; j=1, ..., m is a (n_j+1)–dimensional row vector. Consider a nonzero $\boldsymbol{\omega}$ satisfying,

$$\begin{bmatrix} \boldsymbol{\omega}^1 & \ldots & \boldsymbol{\omega}^m \end{bmatrix} \begin{bmatrix} C_1(n_1) \\ \cdot \\ \cdot \\ \cdot \\ C_m(n_m) \end{bmatrix} \boldsymbol{x}(k) = \boldsymbol{0} \tag{2.93}$$

Assuming that the system (2.89)-(2.90) is observable, there are only $N-n$ linearly independent $\boldsymbol{\omega}$'s satisfying (2.93). Let $\boldsymbol{\Omega}$ be an $(N-n) \times n$ matrix with a set of such independent $\boldsymbol{\omega}$'s as its rows. ($\boldsymbol{\Omega}$ is not unique). Assuming all the inputs are zero for the moment, yields,

$$p(k) = \boldsymbol{\Omega} \begin{bmatrix} y_1(k, n_1) \\ \cdot \\ \cdot \\ \cdot \\ y_m(k, n_m) \end{bmatrix} \tag{2.94}$$

where,

$$y_j(k, n_j) = \begin{bmatrix} y_j(k) \\ \cdot \\ \cdot \\ \cdot \\ y_j(k + n_j) \end{bmatrix}; \; j = 1, \; ..., \; m$$

The $(N\text{–}n)$–vector $\boldsymbol{p}(k)$ is called the *parity vector*. In the absence of noise and failures, $\boldsymbol{p}(k)=\boldsymbol{0}$. In the noisy, no–fail case, $\boldsymbol{p}(k)$ is a zero–mean random vector. Under noise and failures, $\boldsymbol{p}(k)$ will become biased. Moreover, different failures will produce different (biases in the) $\boldsymbol{p}(k)$'s. Thus, the partity vector may be used as the signature–carrying residual for FDI.

The matrix $\boldsymbol{\Omega}$ may be generated by making direct use of (2.93). Let,

$$\boldsymbol{T} = \begin{bmatrix} C_1(n_1) \\ \cdot \\ \cdot \\ \cdot \\ C_m(n_m) \end{bmatrix}$$

From (2.93), it is seen that the rows of $\boldsymbol{\Omega}$ span the orthogonal complement of the range space of $\boldsymbol{T}$. This suggests that $\boldsymbol{\Omega}$ can be generated by subtracting the orthogonal projection onto $\boldsymbol{T}$ from the identity operator. That is, $\boldsymbol{\Omega}$ can be chosen to consist of the $(N\text{–}n)$ independent rows of $\boldsymbol{I}-\boldsymbol{T}(\boldsymbol{T}^T\boldsymbol{T})^{-1}\boldsymbol{T}^T$.

When the actuator inputs are not zero, (2.94) must be modified to take into account this effect. In this case,

$$p(k) = \Omega\left\{\begin{bmatrix} y_1(k,n_1) \\ . \\ . \\ . \\ y_m(k,n_m) \end{bmatrix} - \begin{bmatrix} B_1(n_1) \\ . \\ . \\ . \\ B_m(n_m) \end{bmatrix} U(k,n_0)\right\} \tag{2.95}$$

where,

$$B_j(n_j) = \begin{bmatrix} 0 & 0 & . & . & . & . & . & . & 0 \\ c_jB & 0 & . & . & . & . & . & . & . \\ . & c_jB & . & . & . & . & . & . & . \\ . & . & . & . & . & . & . & . & . \\ . & . & . & . & . & . & . & . & . \\ c_jA^{n_j-1}B & c_jA^{n_j-2}B & . & . & c_jB & 0 & . & . & 0 \end{bmatrix}$$

$$B = [b_1 \ \ldots \ b_r]$$

$$u(k) = [u_1(k) \ \ldots \ u_r(k)]^{\mathrm{T}}$$

$$n_0 = \max(n_1, \ \ldots, \ n_m)$$

$$U(k,n_0) = \left[u^{\mathrm{T}}(k) \ \ldots \ u^{\mathrm{T}}(k+n_0)\right]^{\mathrm{T}}$$

and $B_j(n_j)$ is an $(n_j+1)\times n_0 r$ matrix.

The quantity $p(k)$ is the *generalized parity vector*, which is nonzero (or of nonzero mean if noise is present) only if a failure is present. The $(N-n)$ dimensional space of all such vectors is called the *generalized parity space*. Under the no–fail situation ($p(k)=0$), (2.95) characterizes all the analytical redundancies for the system (2.89)-(2.90) because it specifies all the possible relationships amongst the actuator inputs and sensor outputs. Any linear combination of the rows of (2.95),

$$\sum_{j=1}^{m} \omega^j \left[y_j(k,n_j) - B_j(n_j)U(k,n_0)\right] = 0 \tag{2.96}$$

is called a *parity equation* or a *parity relation*; any linear combination of the RHS of (2.95) is called a *parity function*.

To provide insight into the nature of parity relations consider two simple examples.

Example 2.1: A Single sensor. It is always possible to find a nonzero ω^j such that

$$\boldsymbol{\omega}^j\left[\boldsymbol{y}_j(k,n_j)-\boldsymbol{B}_j(n_j)\boldsymbol{U}(k,n_0)\right]=0 \tag{2.97}$$

or,

$$y_j(k)=-\left(\omega_{n_j}^j\right)^{-1}\left[\sum_{t=1}^{n_j}\omega_{n_j-t}^j y_j(k-t)-\sum_{t=1}^{n_j}\boldsymbol{\sigma}_{n_j-t}^j\boldsymbol{u}(k-t)\right] \tag{2.98}$$

where,

$$\left[\boldsymbol{\sigma}_0^j \quad \ldots \quad \boldsymbol{\sigma}_{n_j}^j \quad \boldsymbol{0} \quad \ldots \quad \boldsymbol{0}\right]=\boldsymbol{\omega}^j\boldsymbol{B}_j(n_j)$$

$\boldsymbol{\sigma}_t^j$; $t=0,\ \ldots,\ n_j-1$ is a r–dimensional row vector and ω_t^j; $t=0,\ \ldots,\ n_j-1$ is the (t+1)st component of $\boldsymbol{\omega}^j$. Equation (2.98) represents a reduced–order ARMA model for the jth sensor alone. That is to say, the output of sensor j can be predicted from its past outputs and past actuator inputs according to (2.98). Based on the ARMA model, several methods of residual generation are possible.

Example 2.2: Temporal redundancy between two sensors. A temporal redundancy exists between sensor i and sensor j if there are,

$$\boldsymbol{\omega}^i=\left[\omega_0^i \quad \ldots \quad \omega_{n_i-1}^i \quad 0\right]$$

$$\boldsymbol{\omega}^j=\left[\omega_0^j \quad \ldots \quad \omega_{n_j-1}^j \quad 0\right]$$

satisfying the redundancy relation,

$$[\boldsymbol{\omega}^i \quad \boldsymbol{\omega}^j]\left\{\begin{bmatrix}\boldsymbol{y}_i(k_1n_1)\\ \boldsymbol{y}_j(k,n_j)\end{bmatrix}-\begin{bmatrix}\boldsymbol{B}_i(k,n_j)\\ \boldsymbol{B}_j(k,n_j)\end{bmatrix}\boldsymbol{U}(k,n_0)\right\}=0 \tag{2.99}$$

Clearly, (2.99) holds if and only if,

$$\left[\omega_0^i \quad \ldots \quad \omega_{n_i-1}^i\right]\boldsymbol{C}_i(n_i-1)=\left[\omega_0^j \quad \ldots \quad \omega_{n_j-1}^j\right]\boldsymbol{C}_j(n_j-1)$$

and, this implies that a redundancy relation exists between two sensors if their observable subspaces overlap. Furthermore, when the overlap subspace is of dimension $\tilde{n}$, there are $\tilde{n}$ linearly independent vectors of the form $\left[\boldsymbol{\omega}^i \quad \boldsymbol{\omega}^j\right]$ that will satisfy (2.99).

Assuming $\omega_\rho^j \neq 0$, (2.99) can be rewritten in an ARMA representation for sensor j as in (2.98),

$$y_j(k)=-\left(\omega_\rho^j\right)^{-1}\left[\sum_{t=1}^{\rho}\omega_{\rho-t}^j y_j(k-t)+\sum_{t=0}^{\rho}\boldsymbol{\omega}_{\rho-t}^i y_i(k-t)-\sum_{t=0}^{\rho}(\boldsymbol{\sigma}_{\rho-t}^i+\boldsymbol{\sigma}_{\rho-t}^j)\boldsymbol{u}(k-t)\right]$$

Example 2.3. To illustrate the residual generation procedure consider a simple second–order system with the following parameters:

$$\boldsymbol{A} = \begin{bmatrix} \alpha_{11} & \alpha_{12} \\ 0 & \alpha_{22} \end{bmatrix}, \quad \boldsymbol{b} = \begin{bmatrix} 0 \\ 1 \end{bmatrix}$$

$$\begin{aligned} \boldsymbol{c}_1 &= [1 \quad 0] \\ \boldsymbol{c}_2 &= [0 \quad 1] \end{aligned} \qquad (2.100)$$

In this case n_1=2, n_2=1, and N–n=3. Therefore, there are only three linearly independent parity equations which may be written as,

$$\begin{aligned} y_1(k) - (\alpha_{11} + \alpha_{22})y_1(k-1) + \alpha_{11}\alpha_{22}y_1(k-2) - \alpha_{12}u(k-2) &= 0 \\ y_1(k) - \alpha_{11}y_1(k-1) - \alpha_{12}y_2(k-1) &= 0 \\ y_2(k) - \alpha_{22}y_2(k-1) - u(k-1) &= 0 \end{aligned} \qquad (2.101)$$

Note that these represent temporal redundancies. Just as with direct redundancy relations, the parity function itself can be taken as a residual. For our specific example, this would be,

$$r_1(k) = y_1(k) - \alpha_{11}y_1(k-1) - \alpha_{12}y_2(k-1) \qquad (2.102)$$

Such a residual is a moving average process, i.e., it is a function of a sliding window of the most recent sensor output and (possibly) actuator input values. It is useful to note the effect of noise and failures on the residual. Specifically, if the sensor outputs are corrupted by white noise, the parity function values will be correlated over the length of the window. In this example, $r_1(k)$ is correlated with $r_1(k-1)$ and $r_1(k+1)$ but not with any of its values removed by more than one time step.

The effect of a failure on a parity function depends, of course, on the nature of the failure. To illustrate what typically occurs, consider the case in which one sensor develops a bias. Since the parity function is a moving average process it also develops a bias, taking at most ρ steps to reach the steady–state value. In this example, if $y_2(k)$ develops a bias of size β at time θ, $r_1(k)$ will have a bias of size $-\alpha_{12}\beta$ from time θ+1 onwards.

How can the residuals generated using parity functions can be used for failure detection? Firstly, sensor FDI using direct redundancy will be examined.

Consider a set of m sensors with output vector $\boldsymbol{y}(k)=[y_1(k) \; \ldots \; y_m(k)]^{\mathrm{T}}$ and a parity vector,

$$\boldsymbol{p}(k)=\boldsymbol{\Omega}\boldsymbol{y}(k) \qquad (2.103)$$

where $\boldsymbol{\Omega}$ is a matrix with m columns and a number of rows (the specification of which will be discussed later). $\boldsymbol{\Omega}$ is not unique, and for any choice of $\boldsymbol{\Omega}$ such that (2.103) is a parity vector, $\boldsymbol{p}(k)$ will be zero in the absence of a failure (and no noise). However, the nature of failure signatures contained in the parity vector depends heavily on the choice of $\boldsymbol{\Omega}$. Clearly $\boldsymbol{\Omega}$ should be chosen so that failure signatures are easily recognizable.

A possible approach which uses information about how failures affect the residuals has been examined by Potter and Suman (1977), and Daley *et al.* (1979). This method exploits the following phenomenon. A faulty sensor, say the jth one, contains an error signal $\nu(k)$ in its output

$$y_j(k) = \boldsymbol{c}_j \boldsymbol{x}(k) + \nu(k) \tag{2.104}$$

The effect of this failure on the parity vector defined by (2.102) is,

$$\boldsymbol{p}(k) = \boldsymbol{\Omega}_j \, \nu(k)$$

where $\boldsymbol{\Omega}_j$ is the jth column of $\boldsymbol{\Omega}$. That, is, no matter what $\nu(k)$ is, the effect of a sensor j failure on the residual always lies in the direction $\boldsymbol{\Omega}_j$. Thus, a sensor j failure can be identified by recognizing a residual bias in the $\boldsymbol{\Omega}_j$ direction. $\boldsymbol{\Omega}_j$ is referred to as the *failure direction in parity space* (FDPS) corresponding to sensor j.

It is now clear that $\boldsymbol{\Omega}$ should be chosen to have distinct columns, so that a sensor failure can be inferred from the presence of a residual bias in its corresponding FDPS. In principle, an $\boldsymbol{\Omega}$ with as few as two rows but m distinct columns is sufficient for detecting and identifying a failure among the m sensors. In practice, however, increasing the row dimension of $\boldsymbol{\Omega}$ can help to separate the various FDPS's and increase the distinguishability of the different failures under noise conditions.

Now, consider the extention of this detection method to temporal redundancy relations. In this case, it is generally not possible to find an $\boldsymbol{\Omega}$ to confine the effect of each component failure to a fixed direction in parity space. To see this, consider the parity relations (2.101). The parity vector can be written as,

$$\boldsymbol{p}(k) = \begin{bmatrix} 1 & -(\alpha_{11}+\alpha_{22}) & \alpha_{11}\alpha_{22} & 0 & 0 \\ 1 & -\alpha_{12} & 0 & 0 & -\alpha_{12} \\ 0 & 0 & 0 & 1 & -\alpha_{22} \end{bmatrix} \begin{bmatrix} y_1(k) \\ y_1(k-1) \\ y_1(k-2) \\ y_2(k) \\ y_2(k-2) \end{bmatrix} + \begin{bmatrix} 0 & -\alpha_{12} \\ 0 & 0 \\ 1 & 0 \end{bmatrix} \begin{bmatrix} u(k-1) \\ u(k-2) \end{bmatrix}$$

When sensor 2 fails [with output model (2.104)], the residual vector develops a bias of the form,

$$\boldsymbol{p}(k) = \begin{bmatrix} 0 \\ 0 \\ 1 \end{bmatrix} \nu(k) + \begin{bmatrix} 0 \\ -\alpha_{12} \\ -\alpha_{22} \end{bmatrix} \nu(k-1) \tag{2.105}$$

Unless $v(k)$ is a constant, the effect (signature) of a sensor 2 failure is only confined to a two–dimensional subspace of the parity space. In fact, generally when temporal redundancy is used in the parity function method for residual generation, failure signatures are generally constrained to multidimensional subspace in the parity space. These subspaces may in general overlap with one another, or some may be contained in others. If no such subspace is contained in another, identification of the failure is still possible by determining which subspace the residual bias lies in.

2.7 Robust detection methods

The theory developed in the preceding sections works well as long as the adopted system models represent adequately the monitored physical system and no noise or unexpected disturbances are present. These requirements are rather stringent however, and are met in few real systems. As a consequence, following initial efforts in devising FDI algorithms for idealised situations, current research focuses on *robust* algorithms. There are a number of different, but somewhat overlapping approaches to this problem, and in the following sections they will be briefly described.

2.7.1 Robust observer-based methods

Eigenstructure assignment methods.

Eigenstructure assignment aims at decoupling the observer estimation from the structured type of uncertainty in the following way: assume that all uncertainties of a system can be summarised as *unknown inputs* (disturbances) with known distribution matrix acting on the system model on which the observer is designed. Patton and Willcox (1986) and Patton *et al.* (1987) first demonstrated the eigenstructure assignment approach to robust observer-based failure detection. They have shown in continuing work an approach to solving this problem using the assignment of suitable eigenvectors and eigenvalues (*eigenstructure assignment*) as a way of providing robustness through disturbance decoupling. Furthermore, Patton *et al.* (1992) have studied ways by which the disturbance distribution matrix is obtained in an optimal way. The main results of their work follows.

To approach the problem from the most general and practical point of view, consider the discrete–time representation of Equation (2.11) with structured uncertainty only:

$$\boldsymbol{x}(k+1) = \boldsymbol{A}\boldsymbol{x}(k) + \boldsymbol{B}\boldsymbol{u}(k) + \boldsymbol{D}_1\boldsymbol{d}_1(k)$$
$$\boldsymbol{y}(k) = \boldsymbol{C}\boldsymbol{x}(k) + \boldsymbol{D}\boldsymbol{u}(k) + \boldsymbol{D}_2\boldsymbol{d}_2(k)$$

Now if the state structured uncertainty is decomposed into the unknown input disturbance and state faults, the system becomes,

$$\boldsymbol{x}(k+1)=\boldsymbol{A}\boldsymbol{x}(k)+\boldsymbol{B}\boldsymbol{u}(k)+\boldsymbol{E}\boldsymbol{d}(k)+\boldsymbol{Q}\boldsymbol{f}_a(k) \tag{2.106a}$$

$$\boldsymbol{y}(k)=\boldsymbol{C}\boldsymbol{x}(k)+\boldsymbol{D}\boldsymbol{u}(k)+\boldsymbol{f}_s(k) \tag{2.106b}$$

Recall that $\boldsymbol{x}(k)$ is the $n\times 1$ state vector, $\boldsymbol{A}$ the open–loop system dynamics matrix, $\boldsymbol{u}(k)$ the $r\times 1$ known input vector with the corresponding input distribution matrix $\boldsymbol{B}$. The term $\boldsymbol{E}\boldsymbol{d}(k)$ characterizes a $q\times 1$ *unknown input (disturbance)* vector $\boldsymbol{d}(k)$ with known distribution matrix $\boldsymbol{E}$ acting directly onto the system dynamics, used to represent the *structured uncertainties*, i.e. although the values of the uncertainty are unknown, its distribution matrix (structure) is known. $\boldsymbol{C}$ is the measurement matrix of the system and $\boldsymbol{y}(k)$ the $m\times 1$ output vector that is assumed to be available for further treatment. The sensor faults that corrupt the measurements, are described by the vector $\boldsymbol{f}_s(k)$. The term $\boldsymbol{Q}\boldsymbol{f}_a(k)$ represents the faults acting on the system dynamics, such as actuator or component faults.

The purpose of the robust residual generator is to use an observer which generates estimates of the system states and measurements, and provides residual signals which are independent of uncertainties (fig. 2.8).

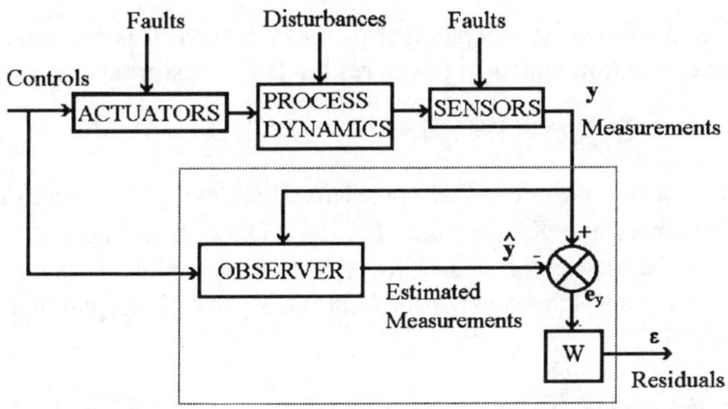

Figure 2.8 General structure of an observer–based residual generation approach

The observer dynamics are described by the following:

$$\hat{\boldsymbol{x}}(k+1)=(\boldsymbol{A}-\boldsymbol{L}\boldsymbol{C})\hat{\boldsymbol{x}}(k)+(\boldsymbol{B}-\boldsymbol{L}\boldsymbol{D})\boldsymbol{u}(k)+\boldsymbol{L}\boldsymbol{y}(k) \tag{2.107}$$

$$\hat{\boldsymbol{y}}(k)=\boldsymbol{C}\hat{\boldsymbol{x}}(k)+\boldsymbol{D}\boldsymbol{u}(k) \tag{2.108}$$

(compare with (2.74)). The state estimation error dynamics are then given by:

$$\boldsymbol{e}(k+1)=\boldsymbol{A}_c\boldsymbol{e}(k)+\boldsymbol{E}\boldsymbol{d}(k)+\boldsymbol{Q}\boldsymbol{f}_a(k)-\boldsymbol{L}\boldsymbol{f}_s(k) \tag{2.109}$$

where $\boldsymbol{A}_c$=$\boldsymbol{A}$–$\boldsymbol{L}\boldsymbol{C}$. A p–dimensional residual vector is generated from the difference between the actual and estimated measurements having the form:

$$\boldsymbol{\varepsilon}(k)=\boldsymbol{W}\big(\boldsymbol{y}(k)-\hat{\boldsymbol{y}}(k)\big)=\boldsymbol{W}\boldsymbol{e}_y(k) \tag{2.110}$$

where W is a $p\times m$ weighting matrix. Substituting (2.107), (2.108) and (2.109) into (2.110) yields,

$$\varepsilon(k) = WCe(k) + Wf_s(k) = He(k) + Wf_s(k) \tag{2.111}$$

where,

$$H = WC \tag{2.112}$$

From equations (2.109) and (2.111), the complete response of the residual vector in shift operator form is,

$$\varepsilon(z) = \left[W - WC(zI - A_c)^{-1}L\right]f_s(k) + WC(zI - A_c)^{-1}Qf_a(z) + WC(zI - A_c)^{-1}Ed(z) \tag{2.113}$$

One can see that the residual is not zero, even if no faults occur in the system. Indeed, it can be difficult to distinguish the effects of faults from the effects of disturbances acting on the system. The effects of disturbances obscure the performance of fault detection and act as a source of false alarms. Therefore, in order to minimize the false alarm rate, one should design the residual generator such that the residual itself becomes decoupled with respect to disturbances. This is essentially the principle of a robust residual generator.

In order that the residual $\varepsilon(k)$ be independent of uncertainties, it is necessary to null the entries in the transfer function matrix between residuals and disturbances, i.e.

$$G_{rd}(z) = WC\left[zI - (A - LC)\right]^{-1}E = 0 \tag{2.114}$$

This is a special case of the *output–zeroing* problem which is well known in multivariable control theory (Karkanias and Kouvaritakis (1979)). Once E is known, the remaining problem is to choose the matrices L and W to satisfy this equation. This can be achieved with the aid of the invariant subspace theory (Antsaklis, 1980). Expanding the inverted matrix one obtains:

$$H\left[zI - A_c\right]^{-1}E = H\left\{a_1(z)I_n + a_2(z)A_c + \ldots + a_n(z)A_c^{n-1}\right\}E \tag{2.115}$$

$$= \left[a_1(z)I_p \quad a_2(z)I_p \quad \ldots \quad a_n(z)I_p\right]\begin{bmatrix} H \\ HA_c \\ \cdot \\ \cdot \\ HA_c^{n-1} \end{bmatrix}E$$

$$= H\begin{bmatrix} E & A_c E & \dots & A_c^{n-1}E \end{bmatrix}\begin{bmatrix} a_1(z)I_q \\ a_2(z)I_q \\ \cdot \\ \cdot \\ a_n(z)I_q \end{bmatrix} \quad (2.116)$$

Hence, (2.115) can be satisfied in the following two cases:

1. If the (H, F_c)–invariant subspace (observable subspace) lies in the left zero space of E, or,
2. If the (F_c, E)–invariant subspace (controllable subspace) lies in the right zero space of H.

These two goals can be achieved by the assignment of either left or right eigenvectors of the observer as proposed by Patton and Chen (1991). To briefly expose these ideas, express $G_{rd}(z)$ in dyadic form as:

$$G_{rd}(z) = \frac{R_1}{z-\beta_1} + \dots + \frac{R_n}{z-\beta_n} \quad (2.117)$$

where $R_i = Hv_i l_i^{\mathrm{T}} E$ and v_i and l_i^T are, respectively, the right and left eignevectors associated with an eigenvalue β_i of A_c. It is well known that, a given left eigenvector l_i^T (corresponding to eigenvalue β_i) of A_c is always orthogonal to the right eigenvectors v_i corresponding to the remaining $(n-1)$ eigenvalues β_j of A_c where $\beta_i \neq \beta_j$. Now, for (2.117) to be satisfied, all eigenvectors must be appropriately scaled so that $VL^{\mathrm{T}} = L^{\mathrm{T}}V = I_n$, where:

$$V = \begin{bmatrix} v_1 & v_2 & \dots & v_n \end{bmatrix}$$
$$L = \begin{bmatrix} l_1 & l_2 & \dots & l_n \end{bmatrix}$$

Thus, disturbance decoupling is possible if and only if $R_i = Hv_i l_i^T E = 0$ for all i=1 to n. This implies that,

$$R_1 + \dots + R_n = Hv_1 l_1^{\mathrm{T}} E + \dots + Hv_i l_i^{\mathrm{T}} E + \dots + Hv_n l_n^{\mathrm{T}} E$$
$$= HVL^{\mathrm{T}} E = HE = WCE = 0 \quad (2.118)$$

Hence, $WCE=HE=0$ is a necessary condition for achieving disturbance decoupling design. Furthermore, it may be proved (Patton and Chen, 1991) that,

Theorem 2.1 If $WCE=0$, and all rows of $H=WCE$ are left eigenvectors of A_c corresponding to any eigenvalues, equation (2.118) is satisfied.

The algorithm then becomes:

1. Compute the weighting matrix $\boldsymbol{W}$ to satisfy equation (2.118). The necessary and sufficient condition for this is rank($\boldsymbol{CE}$)<m.
2. Assign the left eigenvectors of the observer as the rows of $\boldsymbol{H}$ (corresponding to suitable eigenvalues).

Step (2) of the algorithm can be done by a transformation of the dual control problem. Obtaining the right eigenvectors of the dual control problem is equivalent to computing the left eigenvectors of the observer. The assignment of right eigenvectors for a controller is well developed (Moore, 1976). The assignability condition is that for each β_i, the corresponding left eigenvector $\boldsymbol{l}_i^{\mathrm{T}}$ of $\boldsymbol{A}_c$ must belong to the row subspace spanned by the rows of $\boldsymbol{C}(\beta_i\boldsymbol{I}-\boldsymbol{A})^{-1}$.

If the left eigenvector assignability condition is not satisfied, a similar approach can be followed, that is to assign the right eigenvectors of the observer as columns of matrix $\boldsymbol{E}$. In this case the corresponding conditions are:

Theorem 2.2 If $\boldsymbol{WCE}=\boldsymbol{0}$ and all columns of $\boldsymbol{E}$ are right eigenvectors of $\boldsymbol{A}_c$ corresponding to any eigenvalues, equation (2.118) is satisfied.

The assignment of right eigenvectors of the observer (left eigenvector of dual controller) is a relatively new problem. Patton and Chen (1991) derived the following conditions:

Theorem 2.3 For a vector $\boldsymbol{r}_i$ to ba a right eigenvector of $\boldsymbol{A}-\boldsymbol{KC}$ corresponding to the eigenvalue β_i either:

- $\boldsymbol{r}_i$ is a right eigenvector of $\boldsymbol{A}$ corresponding to β_i and $\boldsymbol{Cr}_i=\boldsymbol{0}$ or
- $\boldsymbol{r}_i$ is not a right eigenvector of $\boldsymbol{A}$ corresponding to β_i and $\boldsymbol{Cr}_i \neq \boldsymbol{0}$.

If a number of right eigenvectors must be assigned, the gain matrix $\boldsymbol{L}$ must satisfy a set of equations like,

$$\boldsymbol{L}\,\boldsymbol{C}\,\boldsymbol{r}_i = (\boldsymbol{A} - \boldsymbol{\beta}_i\boldsymbol{I})\,\boldsymbol{r}_i$$

If all columns $\boldsymbol{e}_i$ of $\boldsymbol{E}$ must be assigned as the right eigenvectors of $\boldsymbol{A}_c=\boldsymbol{A}-\boldsymbol{LC}$ corresponding to eigenvalues β_i the following equation must be satisfied:

$$\boldsymbol{LCe}_i = (\boldsymbol{A} - \beta_i\boldsymbol{I})\boldsymbol{e}_i;\ i=1,\ \ldots,\ q \tag{2.119}$$

i.e.

$$\boldsymbol{LCE} = \boldsymbol{A}_\beta \tag{2.120}$$

where,

$$\boldsymbol{A}_\beta = \left[(\boldsymbol{A}-\beta_1\boldsymbol{I})\boldsymbol{e}_1 \quad \ldots \quad (\boldsymbol{A}-\beta_q\boldsymbol{I})\boldsymbol{e}_q\right]$$

Therefore the right eigenvector assignment problem is to solve (2.120) and at the same time ensure that the observer is stable.

Theorem 2.4 The necessary and sufficient condition for solution of equation (2.120) to exist is:

$$\text{rank}\,(A_\beta) = \text{rank}\,(CE)$$

Subject to this conditon, the general form of the solution to (2.120) is:

$$L = A_\beta (CE)^* + K_1[I_m - CE(CE)^*] \tag{2.121}$$

where K_1 is a $n \times m$ arbitrary design matrix and $(CE)^*$ is the pseudo–inverse of CE. When rank$(CE)=q$, $(CE)^*$ is given by:

$$(CE)^* = [(CE)^T CE]^{-1}(CE)^T$$

The dynamic matrix of the observer is thus:

$$A - KC = A - A_\beta (CE)^* C - K_1[I_m - CE(CE)^*]C = A_1 - K_1 C_1$$

where $$\mathbf{A}_1 = \mathbf{A} - \mathbf{A}_\beta(\mathbf{CE})^*\mathbf{C};\;\; C_1 = [I_m - CE(CE)^*]C$$

The necessary and sufficient condition for the observer dynamic matrix $A-LC$ to be stable is that $\{C_1, A_1\}$ is the *dual of a stabilizable pair*. When this condition is satisfied, the assignment problem of the right eigenvectors is to choose the matrix K_1 such that the observer is stable. This problem can be handled by using the traditional pole assignment methods. As β_1, ..., β_q have been assigned as the eigenvalues of $A-LC = A_1-K_1C_1$, only the maximum $(n-q)$ eigenvalues of A_1-LC_1 can be moved by changing the design matrix K_1.

Unknown Input Observer (UIO) method

An alternative procedure for desigining robust fault detection procedures is the *Unknown Input Observer* (*UIO*). In the sequel, following Wünnenberg and Frank (1987), a brief description of how to obtain a robust observer is given.

Assume a simple, discrete-time, system model,

$$\mathbf{x}(k+1) = \mathbf{A}_0\mathbf{x}(k) + \mathbf{B}_0\mathbf{u}(k) + \mathbf{E}\mathbf{d}(k) \tag{2.122}$$

$$\mathbf{y}(k) = \mathbf{C}\mathbf{x}(k) \tag{2.123}$$

where the unknown input distribution matrix $\mathbf{E}$ is of the form

$$\mathbf{E} = \left[\Delta \mathbf{A} \;\vdots\; \Delta \mathbf{B} \;\vdots\; \mathbf{G}\right] \tag{2.124}$$

and, as previously, ΔA represents the difference between the nominal system matrix A_0 and the actual system matrix A and similarly for ΔB; G represents a matrix that distributes external disturbances.

The problem is how to design an observer such that the estimation error is decoupled from any unknown input signal $d(k)$. Starting from the system description (2.122) and (2.123), a regular transformation of the state vector is performed:

$$x(k) = \left[M \ \vdots \ C^R \right] \begin{bmatrix} x^*(k) \\ y(k) \end{bmatrix} \tag{2.125}$$

where,

$$C\left[M \ \vdots \ C^R \right] = \left[0 \ \vdots \ I_m \right] \tag{2.126}$$

This means that the state vector $x(k)$ is separated into the measurable part $y(k)$ and the unmeasurable $x^*(k)$ which has to be estimated by the observer. From Eqs. (2.122), (2.123) and (2.125) it is obtained,

$$Mx^*(k+1) - A_0Mx^*(k) = B_0u(k) + Ed(k) - C^Ry(k+1) + A_0C^Ry(k) \tag{2.127}$$

Multiplying (2.127) from the left with the regular matrix,

$$\begin{bmatrix} N \\ \cdots \\ E^L \end{bmatrix}$$

where,

$$\begin{bmatrix} N \\ \cdots \\ E^L \end{bmatrix} E = \begin{bmatrix} 0 \\ \cdots \\ I_s \end{bmatrix} \tag{2.128}$$

yields,

$$NMx^*(k+1) - NA_0Mx^*(k) = NB_0u(k) - NC^Ry(k+1) + NA_0C^Ry(k) \tag{2.129}$$

On the left hand side of (2.129) there is an expression with the unknown $x^*(k)$. All elements on the right hand side are known or measurable. By substituting,

$$u^*(k) = NB_0u(k) - NC^Ry(k+1) + NA_0C^Ry(k) \tag{2.130}$$

(2.129) reduces to,

$$NMx^*(k+1) - NA_0Mx^*(k) = u^*(k) \tag{2.131}$$

which is a system of difference equations that has to be solved. Using the shift operator z, (2.131) can be rewritten as,

$$(zNM - NA_0M)x^*(k) = u^*(z) \tag{2.132}$$

$$H(z)x^*(k) = u^*(z) \tag{2.133}$$

Now by proper choice of N and M, the so called matrix pencil $H(z)$ can be transformed into the block diagonal Kronecker canonical form (Gantmacher, 1974):

$$H(z) = \mathrm{diag}\left\{ 0_{\mu_0,\varepsilon_0}; L_{\varepsilon_1}(z), \ldots, L_{\varepsilon_s}(z); zI_{\beta_I} - J_e^+; zI_{\beta_s} - J_e^-; zJ_0 - I_\zeta; L_{\mu_1}^{\mathrm{T}}(z), \ldots, L_{\mu_p}^{\mathrm{T}}(z) \right\} \tag{2.134}$$

The ε_i; i=0, ..., s, are the column indices and the μ_i; i=0, ..., p, are the row indices. The expression $\boldsymbol{0}_{\mu_0,\varepsilon_0}$ corresponds to zero rows or columns. The matrix $\boldsymbol{L}_{\varepsilon_i}$ is of dimension $\varepsilon_i \times (\varepsilon_i + 1)$ and has the form,

$$\boldsymbol{L}_{\varepsilon_i} = \begin{bmatrix} z & -1 & 0 & \dots & 0 \\ 0 & z & -1 & \dots & 0 \\ \dots & \dots & \dots & \dots & \dots \\ 0 & \dots & 0 & z & -1 \end{bmatrix} \tag{2.135}$$

and the corresponding matrix for the row indices is,

$$\boldsymbol{L}_{\mu_i}^{\mathrm{T}} = \begin{bmatrix} z & 0 & \dots & 0 \\ -1 & z & \dots & 0 \\ 0 & \dots & \dots & 0 \\ \dots & \dots & \dots & \dots \\ 0 & 0 & \dots & -1 \end{bmatrix} \tag{2.136}$$

The matrix $\boldsymbol{J}_e^+$ is a β_I–dimensional matrix that has only unstable eigenvalues. $\boldsymbol{J}_e^-$ represents a β_s–dimensional Jordan matrix with stable eigenvalues only. $\boldsymbol{J}_0$ is a Jordan matrix with all eigenvalues identical to zero. Consider now that part $\overset{*}{\boldsymbol{x}}_0(k)$ of the state vector that corresponds to the $\boldsymbol{J}_0$ block; this is determined by the difference equations.

$$\begin{aligned} -\overset{*}{x}_{02}(k+1) + \overset{*}{x}_{01}(k) &= \overset{*}{u}_{01}(k) \\ -\overset{*}{x}_{03}(k+1) + \overset{*}{x}_{02}(k) &= \overset{*}{u}_{02}(k) \\ &\dots\dots\dots\dots \\ -\overset{*}{x}_{0\zeta}(k+1) + \overset{*}{x}_{0\zeta-1}(k) &= \overset{*}{u}_{0\zeta-1}(k) \\ \overset{*}{x}_{0\zeta}(k) &= \overset{*}{u}_{0\zeta}(k) \end{aligned} \tag{2.137}$$

which are directly derived with the aid of Eqs. (2.133) and (2.134). It is easily seen, that all components $\overset{*}{x}_{0i}$ are completely determined by the known signals $\overset{*}{u}_{0i}$; $\overset{*}{x}_{0i}$ is then calculated with a maximum delay of ζ time shifts where ζ is the dimension of the $\boldsymbol{J}_0$–matrix. Next, define,

$$\varepsilon = \varepsilon_o + \sum_{i=1}^{s} \varepsilon_i + s \tag{2.138}$$

$$\mu = \sum_{i=1}^{p} \mu_i \tag{2.139}$$

and partition the appropriately chosen matrix $\boldsymbol{M}$ from Eq. (2.125) into the matrices

$$M = \left[M_{\varepsilon} \;\vdots\; M_{\varepsilon}^{+} \;\vdots\; M_{\varepsilon}^{-} \;\vdots\; M_{\infty} \;\vdots\; M_{\mu} \right] \tag{2.140}$$

where M_{ε} containts the first ε columns, M_{e}^{+} the following β_1 columns, M_{e}^{-} the next β_s, M_{∞} the next ζ and M_{μ} the last μ columns of $\dot{M}$.

Therefore, a linear combination of the state variables,

$$z(k) = Tx(k) \tag{2.141}$$

can be reconstructed,

i) Without delay but with free choice in the eigenvalues of the estimations error dynamic matrix if,

$$T\left[M_{\varepsilon} \;\vdots\; M_{\varepsilon}^{+} \;\vdots\; M_{\varepsilon}^{-} \;\vdots\; M_{\infty} \right] = 0 \tag{2.142}$$

ii) Without delay and without free choice of all eigenvalues of the estimation error dynamic matrix if,

$$T\left[M_{\varepsilon} \;\vdots\; M_{\varepsilon}^{+} \;\vdots\; M_{\infty} \right] = 0 \tag{2.143}$$

iii) With delay of a finite number of samples and with free choise of the dynamics of the estimation error dynamics matrix if,

$$T\left[M_{\varepsilon} \;\vdots\; M_{\varepsilon}^{+} \;\vdots\; M_{e}^{-} \right] = 0 \tag{2.144}$$

The above constitute conditions for the existence, structure and eigenvalues of the resulting observers as well as the basis for all possible matrices R. Numerically stable algorithms for the computation of an upper triangular form that contains all information of the Kronecker canonical form are available in Konik and Engell, (1986).

Next, assume that the observer used is expressed by,

$$z(k+1) = Rz(k) + Sy(k) + Ju(k)$$

with the residual,

$$r(k) = L_1 z(k) + L_2 y(k)$$

This observer must fulfill the following robustness requirements:

i) $\lim_{k\to\infty} r = 0$ for all u and d and for all initial conditions x_0 and z_0.

ii) A matrix T must exist, such that $Tx_0 = z_0$ implies $Tx_k = z_k$, for all k.

These conditions lead to the well-known observer equations:

$$
\begin{aligned}
TA_0 - RT &= SC \\
TE &= 0 \\
J &= TB \\
\left[L_1 \;\vdots\; L_2\right]\begin{bmatrix} T \\ \cdots \\ C \end{bmatrix} &= 0
\end{aligned}
\tag{2.145}
$$

Now, if (2.122), (2.123) are enriched to include failure modes, the system is given by,

$$
\begin{aligned}
x(k+1) &= A_0 x(k) + B_0 u(k) + Ed(k) + Kf(k) \\
y(k) &= Cx(k) + Fd(k) + Gf(k)
\end{aligned}
\tag{2.146}
$$

For these model, the set of equations that must be fulfilled in order to achieve disturbance <u>and</u> fault decoupling are Eqs. (2.145) and,

$$
\begin{aligned}
SF &= 0 \\
SG &\neq 0 \\
TK &\neq 0 \\
L_2 F &= 0 \\
L_2 G &= 0
\end{aligned}
\tag{2.147}
$$

These equations can be solved with the Kronecker canonical form, outlined previously. Conditions for existence of solutions can be found in Patton et al. (1989b).

Modeling of uncertainty and approximate decoupling.

A prerequisite for solving Eqs. (2.114), (2.145) and (2.147) and is the a-priori knowledge of the distribution matrix E. Furthermore, certain rank conditions, for example the condition rank(E)<n, must hold for exact decoupling. In the sequel, a procedure for modeling this distribution matrix will be presented, which is due to Patton *et al.* (1992). Approximate decoupling methods, in the cases where the physical conditions for the existence of a robust observer are not met, will be also discussed.

The disturbance term $Ed(k)$ in Eq. (2.106a) or (2.146) can be used to describe: the interconnection term in large scale systems, linearized errors, extra non–linear or time–varying terms and model reduction errors. For example, consider the (continuous time) dynamic system:

$$
\dot{x}(t) = (A + \Delta A)x(t) + (B + \Delta B)u(t) + A_c x_h(t) + G\mu(t) + Qf(x,u,t) \tag{2.148}
$$

where $\mu \in R^q$ is a noise or external disturbance vector, ΔA, ΔB represent the parameter variations and plant uncertainty, whilst A_c, x_h represent the dynamic errors when using a reduced order to approximate full–order systems or represent the interaction term in a large scale system. In this case, the disturbance distribution matrix $\boldsymbol{E}$ can be directly computed as:

$$\boldsymbol{E} = \left[\Delta\boldsymbol{A} \;\vdots\; \Delta\boldsymbol{B} \;\vdots\; \boldsymbol{A}_{c0} \;\vdots\; \boldsymbol{G}_0 \;\vdots\; \boldsymbol{Q}\right] \in^{n\times q} \tag{2.149}$$

Now, consider the situation where the system matrices are functions of a parameter vector $\boldsymbol{\alpha}\in\boldsymbol{R}^g$:

$$\dot{\boldsymbol{x}}(t) = \boldsymbol{A}(\boldsymbol{\alpha})\boldsymbol{x}(t) + \boldsymbol{B}(\boldsymbol{\alpha})\boldsymbol{u}(t) \tag{2.150}$$

If the parameter can be perturbed around a nominal condition $\boldsymbol{\alpha}=\boldsymbol{\alpha}_0$, (2.150) can be expanded as:

$$\dot{\boldsymbol{x}}(t) = \boldsymbol{A}(\boldsymbol{\alpha}_0)\boldsymbol{x}(t) + \boldsymbol{B}(\boldsymbol{\alpha}_0)\boldsymbol{u}(t) + \sum_{i=1}^{g}\left\{\frac{\partial \boldsymbol{A}}{\partial \alpha_i}\delta\alpha_i\boldsymbol{x} + \frac{\partial \boldsymbol{B}}{\partial \alpha_i}\delta\alpha_i\boldsymbol{u}\right\} \tag{2.151}$$

In this case, the distribution matrix and unknown input vector is:

$$\boldsymbol{E} = \left[\frac{\partial \boldsymbol{A}}{\partial \alpha_1} \;\vdots\; \frac{\partial \boldsymbol{B}}{\partial \alpha_1} \;\vdots\; \cdots \;\vdots\; \frac{\partial \boldsymbol{A}}{\partial \alpha_g} \;\vdots\; \frac{\partial \boldsymbol{B}}{\partial \alpha_g}\right] \tag{2.152}$$

$$\boldsymbol{d}(t) = \left[\delta\alpha_1\boldsymbol{x}^{\mathrm{T}} \;\vdots\; \delta\alpha_1\boldsymbol{u}^{\mathrm{T}} \;\vdots\; \ldots \;\vdots\; \delta\alpha_g\boldsymbol{x}^{\mathrm{T}} \;\vdots\; \delta\alpha_g\boldsymbol{u}^{\mathrm{T}}\right]^{\mathrm{T}} \tag{2.153}$$

Now, if rank($\boldsymbol{E}$)=n, (2.118) has no solutions and exact decoupling is impossible. Therefore, some kind of approximate decoupling must be applied. The procedure will be to compute a matrix $\boldsymbol{E}^*$ that is as close as possible to $\boldsymbol{E}$ but has rank($\boldsymbol{E}^*$)$\leq n-1$, i.e. to find the solution to the following optimization problem:

$$\min\left\|\boldsymbol{E} - \boldsymbol{E}^*\right\|_F^2 \quad \text{subject to rank}(\boldsymbol{E}^*)\leq n-1 \tag{2.154}$$

Here $\|\cdot\|_F^2$ denotes the Frobenius norm, defined as the root of the sum of squares of the entries of the associated matrix. This optimization problem can be solved by the *Singular Value Decomposition* (*SVD*) of $\boldsymbol{E}$:

$$\boldsymbol{E} = \boldsymbol{S}\left[\mathrm{diag}\{\sigma_1,\ldots,\sigma_n\},\;\; 0\right]\boldsymbol{T} \tag{2.155}$$

where $\boldsymbol{S}$ and $\boldsymbol{T}$ are othogonal matrices, $\sigma_1\leq\sigma_2\leq \ldots \leq\sigma_n$ are the singular values of $\boldsymbol{E}$. As shown in Lou *et al.* (1986), the matrix $\boldsymbol{E}^*$ that minimizes (2.154) is given by:

$$\boldsymbol{E}^* = \boldsymbol{S}\left[\mathrm{diag}\left\{0,\ldots,0,\sigma_{n_1},\ldots,\sigma_n\right\},\;\; 0\right]\boldsymbol{T} \tag{2.156}$$

where n_1 is the rank of $\boldsymbol{E}^*$ which is less than n.

The operating point of the system varies according to different plant conditons, and different operating points correspond to different unknown input matrices, $\boldsymbol{E}_i$ (i=1, 2, ..., M). It is attractive to be able to design a single FDI scheme for a whole range (or a set) of operating points. The success of the single FDI design depends on its robustness properties. In order to make the disturbance decoupling hold for all operating points, one must make:

$$HE_i = 0, \quad \text{for} \quad i = 1,2,\ldots,M \tag{2.157}$$

or,

$$H\begin{bmatrix} E_1 & E_2 & \ldots & E_M \end{bmatrix} = HP = 0 \tag{2.158}$$

If rank(P)≤n–1, Eq. (2.158) has solutions and exact decoupling at all operating points is achievable. Otherwise, approximate decoupling must be used. This is equivalent to the solution of Eq. (2.144) and can be solved by defining an optimization problem:

$$\min\|P - P^*\|_F^2 \quad \text{subject to: } \operatorname{rank}(P^*) \le n-1 \tag{2.159}$$

Now, consider the case when the full–order system model is <u>not</u> available. A possible approach would be to obtain the nominal model $\{A_0, B_0, C_0, D_0\}$ via identification, with the estimation error $\{\Delta A, \Delta B, \Delta C, \Delta D\}$. Normally, ΔA and ΔB are unknown but bounded:

$$A_1 \le \Delta A \le A_2 \tag{2.160}$$

$$B_1 \le \Delta B \le B_2 \tag{2.161}$$

where A_1, A_2, B_1 and B_2 are known and $\Delta A \le A_2$ denotes that each element of ΔA is not larger than the corresponding element of A_2. This typifies an unstructured but bounded uncertainty. Consider ΔA and ΔB in a finite set of possibilities, say $\{\Delta A_i, \Delta B_i\}$; i = 1, 2 , ..., M within the interval $A_1 \le \Delta A \le A_2$ and $B_1 \le \Delta B \le B_2$. This might involve choosing representative points, reflecting desired weighting on the likelihood or importance of particular sets of parameters. In this situation, a set of unknown input distribution matrices is obtained:

$$E_i = \begin{bmatrix} \Delta A_i & \Delta B_i \end{bmatrix}; \; i = 1,\; 2,\; \ldots,\; M \tag{2.162}$$

In order to make the disturbance decoupling valid for a wide range of model parameter variations, an optimal matrix E^* should be made to be near all E_i; i = 1, 2 , ..., M as closely as possible. The optimization problem is thus defined as:

$$\min_{\{s.t.\, rank(E^*) \le n-1\}} \left\| E^* - \begin{bmatrix} E_1 & \ldots & E_M \end{bmatrix} \right\|_F^2 \tag{2.163}$$

In most cases, not enough knowledge about the state space model of the system is available. What is usually at hand, is the linearized low order model matrices (A, B, C, D). In order to account for modeling errors, the system is assumed to be in the form:

$$\dot{x}(t) = Ax(t) + Bu(t) + d_1(t) \tag{2.164}$$

$$y(t) = Cx(t) + Du(t) \tag{2.165}$$

where $d_1(t)$ represents modeling errors. If $d_1(t)$ can be obtained, it may be decomposed into $Ed(t)$ with E a structured matrix so as to apply the disturbance decoupling concept. Firstly, assume that $d_1(t)$ is slowly time–varying, so that the system model can be re–written in augmented form as:

$$\begin{bmatrix} \dot{x}(t) \\ \dot{d}_1(t) \end{bmatrix} = \begin{bmatrix} F & I \\ 0 & 0 \end{bmatrix} \begin{bmatrix} x(t) \\ d_1(t) \end{bmatrix} + \begin{bmatrix} G \\ 0 \end{bmatrix} u(t) \tag{2.166}$$

$$y(t) = \begin{bmatrix} C & 0 \end{bmatrix} \begin{bmatrix} x(t) \\ d_1(t) \end{bmatrix} + Du(t) \tag{2.167}$$

Using the true system input and output data, an observer based on eqs. (2.166) and (2.167) can be used to estimate $d_1(t)$. Then it is possible to obtain some information about the distribution matrix E. Further details of this method can be found in Patton and Chen (1992). Patton *et al.* (1992) further explored this idea and proposed the *deconvolution* method to estimate the vector $d_1(t)$.

As far as the UIO method is concerned, appropriate optimal approximate solutions have been proposed by Wünnenberg and Frank (1987). In this approach, the "best" residual is found by solving the following minimisation problem:

$$P = \min_{w} \frac{\left\| w^{\mathrm{T}} V_0 H_2 \right\|}{\left\| w^{\mathrm{T}} V_0 H_3 \right\|} \tag{2.168}$$

where,

$$H_2 = \begin{bmatrix} F & 0 & \cdots & \cdots & 0 \\ CE & F & 0 & \cdots & 0 \\ CAE & CE & F & \cdots & 0 \\ \cdots & \cdots & \cdots & \cdots & \cdots \\ CA^{s-1}E & CA^{s-2}E & CA^{s-3}E & \cdots & F \end{bmatrix}$$

$$H_3 = \begin{bmatrix} G & 0 & \cdots & \cdots & 0 \\ CK & G & 0 & \cdots & 0 \\ CAK & CK & G & \cdots & 0 \\ \cdots & \cdots & \cdots & \cdots & \cdots \\ CA^{s-1}K & CA^{s-2}K & CA^{s-3}K & \cdots & G \end{bmatrix}$$

and V_0 is used to ensure that,

$$v^{\mathrm{T}} = \begin{bmatrix} C \\ CA \\ \dots \\ CA^s \end{bmatrix}$$

is satisfied. Differentiating (2.168) leads to the relation,

$$w^{\mathrm{T}}\left(V_0 H_2 H_2^{\mathrm{T}} V_0^{\mathrm{T}} - P V_0 H_3 H_3^{\mathrm{T}} V_0^{\mathrm{T}}\right) = 0$$

This is a generalised eigenvalue-eigenvector problem. The minimal eigenvalue is the optimal performance index, while the corresponding eigenvector is the optimal residual generator v. With this the optimal residual sequence is calculated by,

$$r(k) = v^{\mathrm{T}}\left(\begin{bmatrix} y(k-s) \\ y(k-s+1) \\ \dots \\ y(k) \end{bmatrix} - H_1 \begin{bmatrix} u(k-s) \\ u(k-s+1) \\ \dots \\ u(k) \end{bmatrix}\right)$$

where,

$$H_1 = \begin{bmatrix} 0 & 0 & \dots & \dots & 0 \\ CB & 0 & 0 & \dots & 0 \\ CAB & CB & 0 & \dots & 0 \\ \dots & \dots & \dots & \dots & \dots \\ CA^{s-1}B & CA^{s-2}B & CA^{s-3}B & \dots & 0 \end{bmatrix}$$

These conditions have been obtained in the time domain. Ding and Frank (1989), have produced corresponding results in the frequency domain.

2.7.2 Parity relations for robust residual generation

In this section an approach for obtaining robust parity relations in the face of noise and parameter uncertainty will be given. The exposition follows Chow and Willsky (1984).

The starting point is a model that has the same form as (2.89) but includes noise disturbance and parameter uncertainty:

$$x(k+1) = A(\gamma)x(k) + \sum_{j=1}^{q} b_j(\gamma)u_j(k) + w(k) \qquad (2.169)$$

$$y_j(k) = c_j x(k) + v_j(k) \qquad (2.170)$$

where γ is the vector of uncertain parameters taking values in a specified subset Γ of R^M. This form allows the modeling of elements in the system matrices as uncertain quantities that may be functions of a common quantity. The vectors w and v are independent, zero–mean, white Gaussian noise vectors with constant covariance matrices Q(≥ 0) and R(>0), respectively.

A parity function is essentially a weighted combination of a (time) window of sensor outputs and actuator inputs. The *structure* of a parity function defines which input and output elements are included in this window, and the *coefficients* are the (nonzero) weights corresponding to these elements. A scalar parity function $p(k)$ can be written as,

$$p(k) = \alpha Y(k) + \beta U(k) \tag{2.171}$$

where $Y(k)$ and $U(k)$ denote the vectors containing the output and input elements in the parity function, respectively. Together, $Y(k)$ and $U(k)$ specify the parity structure, and the row vectors α and β contain the parity coefficients. Consider, for example, the first parity function of (2.101). Its corresponding $Y(k)$, $U(k)$, α and β are:

$$Y(k) = \left[y_1(k-2),\ y_1(k-1),\ y_1(k)\right]^{\mathrm{T}}$$

$$U(k) = u(k-2)$$

$$\alpha = [\alpha_{11}\alpha_{22},\ -(\alpha_{11}+\alpha_{22}),\ 1]$$

$$\beta = -\alpha_{12}$$

Under model (2.170), $Y(k)$ has the form,

$$Y(k) = C(\gamma)x(k-\rho) + \Phi(\gamma)\tilde{w}(k) + B(\gamma)U(k) + \tilde{v}(k) \tag{2.172}$$

where ρ is the order of the parity function, and

$$\tilde{w}(k) = \left[w^{\mathrm{T}}(k-\rho)\ \ \dots\ \ w^{\mathrm{T}}(k-1)\right]^{\mathrm{T}}$$

The component of $\tilde{v}(k)$ and $U(k)$, and the rows of $C(\gamma)$, $\Phi(\gamma)$, the B are determined from (2.170) and the structure of $Y(k)$. If, specifically, the ith component of $Y(k)$ is $y_s(k-\sigma)$, then the ith component of $\tilde{v}(k)$ is,

$$\tilde{v}_i(k) = v_s(k-\sigma).$$

The vectors $\tilde{w}$ and $\tilde{v}$ are independent zero–mean Gaussian random sequenses with constant covariances $\tilde{Q}$ and $\tilde{R}$ respectively. The matrix $\tilde{Q}$ is block diagonal with Q on the diagonal; $\tilde{R}_{i,j} = R_{s,t}\delta_{\sigma,t}$, and the ith element of $Y(k)$ is $y_s(k-\sigma)$, while the jth element is $y_t(k-\tau)$. The ith row of $C(\gamma)$, i.e., $C(i, \gamma)$ is,

$$C(i,\gamma) = c_s A^{\rho-\sigma}$$

The ith row, $\Phi(i, \gamma)$ of $\Phi(\gamma)$ (which has ρN columns) is,

$$\boldsymbol{\Phi}(i,\boldsymbol{\gamma}) = \left[\boldsymbol{c}_s\boldsymbol{A}^{\rho-\sigma-1},\ \boldsymbol{c}_s\boldsymbol{A}^{\rho-\sigma-2},\ \ldots,\ \boldsymbol{c}_s,\ \boldsymbol{0},\ \ldots,\ \boldsymbol{0}\right]$$

Note that $\boldsymbol{x}(k-\rho)$ is a random vector that is uncorrelated with $\tilde{\boldsymbol{w}}$ and $\tilde{\boldsymbol{v}}$, and

$$\mathrm{E}\{(\boldsymbol{x}(k-\rho)\} = \boldsymbol{x}_0(k-\rho)$$
$$\mathrm{cov}\{\boldsymbol{x}(k-\rho)\} = \boldsymbol{\Sigma}(\boldsymbol{\gamma})$$

where $\boldsymbol{\Sigma}(\boldsymbol{\gamma})$ is the (steady–state) covariance of $\boldsymbol{x}(k-\rho)$ and it is dependent on $\boldsymbol{\gamma}$ through $\boldsymbol{A}(\boldsymbol{\gamma})$ and $\boldsymbol{B}(\boldsymbol{\gamma})$.

The matrix $\boldsymbol{B}$ and the vector $\boldsymbol{U}(k)$ are determined as follows: first, collect into a matrix $\tilde{\boldsymbol{B}}$ all the rows in $\boldsymbol{B}_j(k,\ \boldsymbol{\gamma})$ corresponding to $\boldsymbol{C}(i,\ \boldsymbol{\gamma})$ (Wald, 1947). Then, collect all the nonzero columns of $\tilde{\boldsymbol{B}}$ into $\boldsymbol{B}$ and the corresponding components of $\boldsymbol{u}$ in the window into $\boldsymbol{U}(k)$.

It is clear from this exposition that when parameter uncertainties are included, it is not possible in general to find any parity functions in this narrow sense. In particular, with reference to the function $p(k)$ defined by (2.171) and (2.172) this condition would require that $\boldsymbol{\alpha}\boldsymbol{C}(\boldsymbol{\gamma})=0$ for all $\boldsymbol{\gamma}\in\boldsymbol{\Gamma}$. Consequently, the notion of a useful parity relation must be modified. Intuitively, any given parity structure will be useful for failure detection if a set of parity coefficients can be found that will make the resulting function $p(k)$ in (2.171) *close* to zero for all values of $\boldsymbol{\gamma}\in\boldsymbol{\Gamma}$ when no failure has occurred. When considering the use of such a function for the detection of a particular failure one would also want to guarantee that $p(k)$ deviates significantly from zero for all $\boldsymbol{\gamma}\in\boldsymbol{\Gamma}$ when no failure has occurred. Such a parity structure–coefficient combination approximates the true parity function.

The problem then becomes one of finding coefficients $\boldsymbol{\alpha}$ and $\boldsymbol{\beta}$ for the parity function,

$$p(k) = \boldsymbol{\alpha}\left[\boldsymbol{C}(\boldsymbol{\gamma})\boldsymbol{x}(k-\rho) + \boldsymbol{\Phi}(\boldsymbol{\gamma})\tilde{\boldsymbol{w}}(k) + \boldsymbol{B}(\boldsymbol{\gamma})\boldsymbol{U}(k) + \tilde{\boldsymbol{v}}(k)\right] - \boldsymbol{\beta}\boldsymbol{U}(k)$$

Note the dependence of $p(k)$ on $\boldsymbol{\alpha}$, $\boldsymbol{\beta}$, $\boldsymbol{\gamma}$, $\boldsymbol{x}(k-\rho)$, and $\boldsymbol{U}(k)$. As $p(k)$ is a random variable, a convenient measure of the magnitude (squared) of $p(k)$ is its variance, $\mathrm{E}\{p^2(k)\}$, where the expectation is taken with respect to the joint probability density of $\boldsymbol{x}(k-\rho)$, $\tilde{\boldsymbol{w}}(k)$ and $\tilde{\boldsymbol{v}}(k)$ with the mean $\boldsymbol{x}_0(k-\rho)$ and the value of $\boldsymbol{U}(k)$ assumed known. This can be thought of as specifying a particular operating condition for the system. Note also that the statistics of $\boldsymbol{x}(k-\rho)$ depend on $\boldsymbol{\gamma}$. Define,

$$e(\boldsymbol{\alpha},\boldsymbol{\beta}) = \max_{\boldsymbol{\gamma}\in\boldsymbol{\Gamma}} \mathrm{E}\left\{p^2(k)\right\} \tag{2.173}$$

The quantity $e(\boldsymbol{\alpha},\ \boldsymbol{\beta})$ represents the worst case effect of noise and model uncertainty on the parity function $p(k)$ and is called the *parity error* for $p(k)$ with the coefficients $\boldsymbol{\alpha}$ and $\boldsymbol{\beta}$. A concervative choice of the parity coefficients is obtained by solving,

$$\min_{\boldsymbol{\alpha},\boldsymbol{\beta}}(\boldsymbol{\alpha},\boldsymbol{\beta})$$

Since this has a trivial solution ($\boldsymbol{\alpha}$=**0**, $\boldsymbol{\beta}$=**0**), this optimization problem has to be modified in order to give a meaningful solution. Recall that a parity equation primarily relates the sensor outputs, i.e., a parity equation always includes output terms but not necessarily input terms. Therefore, $\boldsymbol{\alpha}$ must be nonzero. Without loss of generality, $\boldsymbol{\alpha}$ can be restricted to have unit magnitude. The actuator input terms in a parity relation may be regarded as serving to make the parity function zero so that $\boldsymbol{\beta}$ is nominally free. In fact, $\boldsymbol{\beta}$ has only a single degree of freedom. Any $\boldsymbol{\beta}$ can be written as $\boldsymbol{\beta} = \lambda \boldsymbol{U}^{\mathrm{T}}(k) + \boldsymbol{z}^{\mathrm{T}}$, where $\boldsymbol{z}$ is a (column) vector orthogonal to $\boldsymbol{U}(k)$. The component $\boldsymbol{z}^{\mathrm{T}}$ in $\boldsymbol{\beta}$ will not produce any effect on $p(k)$. This implies for each $\boldsymbol{U}(k)$, only $\boldsymbol{\beta}$ of the form $\boldsymbol{\beta}=\lambda \boldsymbol{U}^{\mathrm{T}}(k)$ has to be considered, leading to the following problem:

$$\min_{\substack{\boldsymbol{\alpha},\lambda \\ s.t.\boldsymbol{\alpha}\boldsymbol{\alpha}^T=1}} \max_{\boldsymbol{\gamma}\in\Gamma} \mathrm{E}\left\{p^2(k)\right\} \tag{2.174}$$

where,

$$E\left\{p^2(k)\right\} = [\boldsymbol{\alpha},\lambda]\boldsymbol{S}[\boldsymbol{\alpha},\lambda]^{\mathrm{T}}$$

and $\boldsymbol{S}$ is the symmetric positive definite matrix,

$$\boldsymbol{S} = \begin{bmatrix} \boldsymbol{S}_{11} & \boldsymbol{S}_{12} \\ \boldsymbol{S}_{21} & \boldsymbol{S}_{22} \end{bmatrix}$$

$$\begin{aligned} \boldsymbol{S}_{11} &= \boldsymbol{C}(\boldsymbol{\gamma})\left[\boldsymbol{x}_0(k-\rho)\boldsymbol{x}_0^{\mathrm{T}}(k-\rho) + \boldsymbol{\Sigma}(\boldsymbol{\gamma})\right]\boldsymbol{C}^{\mathrm{T}}(\boldsymbol{\gamma}) \\ &\quad + \boldsymbol{\Phi}(\boldsymbol{\gamma})\boldsymbol{Q}\boldsymbol{\Phi}^{\mathrm{T}}(\boldsymbol{\gamma}) + \tilde{\boldsymbol{R}} + \boldsymbol{B}(\boldsymbol{\gamma})\boldsymbol{U}(k)\boldsymbol{U}^{\mathrm{T}}(k)\boldsymbol{B}^{\mathrm{T}}(\boldsymbol{\gamma}) \\ &\quad + \boldsymbol{C}(\boldsymbol{\gamma})\boldsymbol{x}_0(k-\rho)\boldsymbol{U}^{\mathrm{T}}(k)\boldsymbol{B}(\boldsymbol{\gamma}) + \boldsymbol{B}(\boldsymbol{\gamma})\boldsymbol{U}(k)\boldsymbol{x}_0^{\mathrm{T}}(k-\rho)\boldsymbol{C}^{\mathrm{T}}(\boldsymbol{\gamma}) \\ S_{12} &= S_{21}^{\mathrm{T}} = -S_{22}^{1/2}\left[\boldsymbol{B}(\boldsymbol{\gamma})\boldsymbol{U}(k) + \boldsymbol{C}(\boldsymbol{\gamma})\boldsymbol{x}_0(k-\rho)\right] \\ S_{22} &= \left[\boldsymbol{U}^{\mathrm{T}}(k)\boldsymbol{U}(k)\right]^2 \end{aligned}$$

Let $\boldsymbol{\alpha}^*$ and $\boldsymbol{\lambda}^*$ denote the values of $\boldsymbol{\alpha}$ and $\boldsymbol{\lambda}$ that solve (2.174), with $\boldsymbol{R}^*=\lambda^*\boldsymbol{U}(k)$. Then $\boldsymbol{e}^*$ is the parity error corresponding to the parity function $p^*(k)=\boldsymbol{\alpha}^*\boldsymbol{Y}(k)+\boldsymbol{\beta}^*\boldsymbol{U}(k)$. The quantity $\boldsymbol{e}^*$ measures the usefulness of $p^*(k)$ as a parity function around the operating point specified by $\boldsymbol{x}_0(k-\rho)$ and $\boldsymbol{U}(k)$.

Although the objective function of (2.174) is quadratic in $\boldsymbol{\alpha}$ and λ, (2.174) is generally very difficult to solve, because $\boldsymbol{S}$ may depend on $\boldsymbol{\gamma}$ arbitrarily.

With the coefficients and the associated parity errors determined for the candidate parity structures, the parity functions for residual generation using the parity function method can be chosen. As the squared magnitude of the coefficients $[\boldsymbol{\alpha}, \boldsymbol{\beta}]$ scales the parity error, the parity errors of different parity functions can be compared if they are normalized.

The *normalized parity error* $\bar{e}*$, the *normalized parity coefficients*, and the *normalized parity function* $\bar{p}*(k)$ are defined as follows:

$$\bar{e}* = e*/\theta$$
$$\bar{\boldsymbol{\alpha}}* = \boldsymbol{\alpha}*\theta$$
$$\bar{\boldsymbol{\beta}}* = \boldsymbol{\beta}*\theta$$
$$\bar{p}*(k) = \bar{\boldsymbol{\alpha}}*\boldsymbol{Y}(k) - \bar{\boldsymbol{\beta}}*\boldsymbol{U}(k)$$

where,

$$\theta^2 = [\boldsymbol{\alpha}*, \boldsymbol{\beta}*][\boldsymbol{\alpha}*, \boldsymbol{\beta}*]^{\mathrm{T}} = 1 + \boldsymbol{\beta}*\boldsymbol{\beta}*^{\mathrm{T}}$$

The parity functions with the smallest normalized parity errors are preferred as they are closer to being true parity functions under noise and model uncertainty, i.e., they are least sensitive to these adverse effects.

An additive consideration required for choosing parity functions for residual generation is that the chosen parity functions should provide the largest failure signatures in the residuals relative to the inherent parity errors resulting from noise and parameter uncertainty. A useful index for comparing parity functions for this purpose is the *signature–to–parity error ratio* π, which is the ratio between the magnitudes of the failure signature and the parity error. Using g to denote the effect of a failure on the parity function, π can be defined as,

$$\pi = |g| / \bar{e}*$$

For the detection and identification of a particular failure, the parity function that produces the largest π should be used for residual generation.

2.8 Applications

2.8.1 Fault detection in a jet engine system

The correct operation of a gas turbine is very critical for an aircraft and, if faults occur the consequences can be extremely serious. There is therefore a great need for simple and yet highly reliable methods for detecting and isolating faults in the jet engine. Patton *et al.* (1992), presented an example for the detection of jet engine sensor faults using the procedure described in Section 2.7.1. The jet engine model used, illustrated in fig. 2.9, has the measurement variables: N_L, N_H, T_7, P_6, T_{29} (N denotes compressor shaft speeds, the P variables denote pressures, whilst T represents temperature). The control inputs are: the main engine fuel flow rate and the exhaust nozzle area.

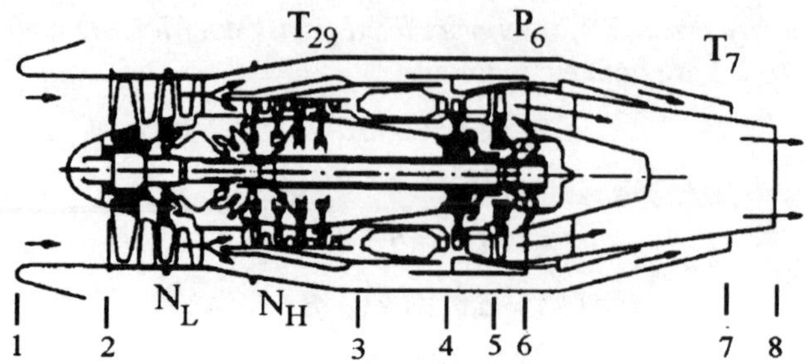

Figure 2.9 Jet engine

A thermodynamic simulation model of a jet engine is utilised as a test rig to assess the robustness of the FDI scheme. This model has 17 state variables; these include pressures, air and gas mass flow rates, shaft speeds, absolute temperatures and static pressure. The linearized 17*th* order model is used here to simulate the jet engine system. The nominal operating point is set at 70% of the demanded high spool speed (N_H). For practical reasons and convenience of design, a 5*th* order model is used to approximate the 17*th* order model. The model reduction and other errors are represented by the disturbance term $\boldsymbol{Ed}(t)$ of Eq. (2.106a). The 5*th* order model matrices are:

$$\boldsymbol{A} = \begin{bmatrix} -78 & 294 & -22 & 21 & -29 \\ 7 & -28 & 2 & -2 & 3 \\ -1325 & 5326 & -526 & 221 & -477 \\ 1081 & -4445 & 377 & -463 & 403 \\ 2152 & -8639 & 781 & -575 & 782 \end{bmatrix}$$

$$\boldsymbol{B} = \begin{bmatrix} -0.0072 & 0.0030 \\ 0.0035 & 0.0003 \\ 1.2185 & -0.0329 \\ 1.3225 & 0.0201 \\ -0.0823 & 0.0244 \end{bmatrix}, \quad \boldsymbol{C} = \boldsymbol{I}_{5\times5}, \boldsymbol{D} = \boldsymbol{0}_{5\times2}$$

As shown in Section 2.7.1, a necessary step for the robust residual generation design procedure is to find a matrix $\boldsymbol{H}$ to satisfy Eq. (2.118) (i.e. $\boldsymbol{HE}=\boldsymbol{0}$). The matrix E models structured uncertainty arising from the application of the 5*th* order observer to the 17*th* order plant, and is given by:

$$\boldsymbol{E} = \left[\boldsymbol{E}_1 \;\vdots\; \boldsymbol{E}_2 \;\vdots\; \boldsymbol{E}_3 \;\vdots\; \boldsymbol{E}_4\right] \; (\times 10^3)$$

where numerical values for the $\boldsymbol{E}_i$'s are defined in Patton *et al.* (1992). From these values, rank ($\boldsymbol{E}$)=5=n, and hence Eq. (2.118) has no solution. The singular values of $\boldsymbol{E}$ are

{1.5, 60, 198, 11268}, and the matrices S and T are omitted for brevity. The optimal low rank approximation of the distribution matrix $\boldsymbol{E}$ is,

$$\boldsymbol{E}^* = \boldsymbol{S}\,[\ \mathrm{diag}\ (\ 90,\ 5,\ 60,\ 198\ ,11268\ )\ \boldsymbol{0}_{5\times14}\ ]^{\mathrm{T}}.$$

Based on this matrix, an observer–based robust residual generator can be designed. The observer design is simplified by choosing all eigenvalues at −100. In this case, the gain matrix $\boldsymbol{K}$=$-(100\boldsymbol{I}_{5\times5}+\boldsymbol{A})$ as C is an identity matrix.

In fig. 2.10, the output estimation error norm is shown. This is very large, and *cannot* be used to detect the fault reliably. This represents the non–robust design situation. fig. 2.11 shows the fault–free residual. Compared with the output estimation error, the residual is very small, i.e., disturbance decoupling is achieved. This robust design can be used to detect incipient faults. In order to evaluate the power of the robust FDI design, a small fault is added to the exhaust gas temperature (T_7); this simulates the effect of an *incipient fault*, the effect of which is too small to be noticed in the measurements. fig. 2.12 shows the faulty output of the temperature measurement (T_7) and the corresponding residual. The fault is very small compared with the output, and consequently, is not detectable in the measurement. It can be seen that the residual has a very significant increase when a fault has occurred in the system. A threshold can easily be placed on the residual signal to declare the occurrence of faults. A fault signal is now added to the pressure measurement signal for P_6. The result is shown in fig. 2.13 which also demonstrates the efficiency of the robust residual in the role of robust FDI.

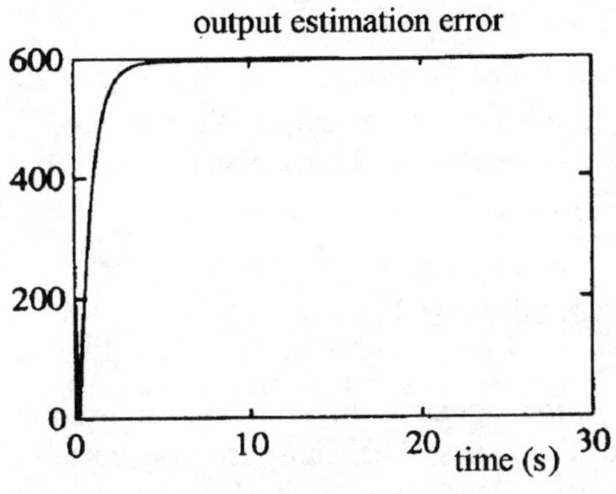

Figure 2.10 Norm of the output estimation error

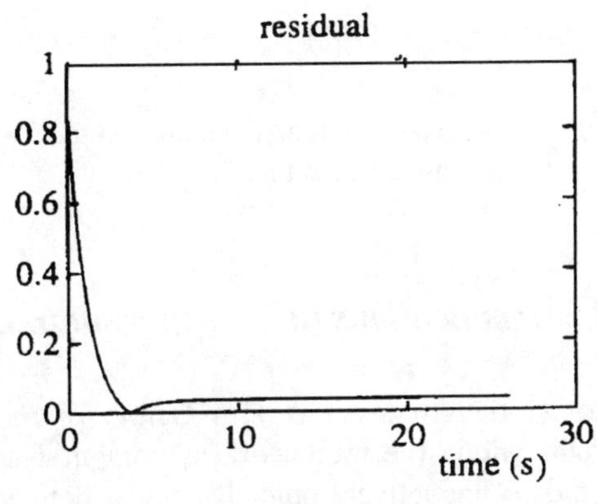

Figure 2.11 Absolute value of the fault–free residual

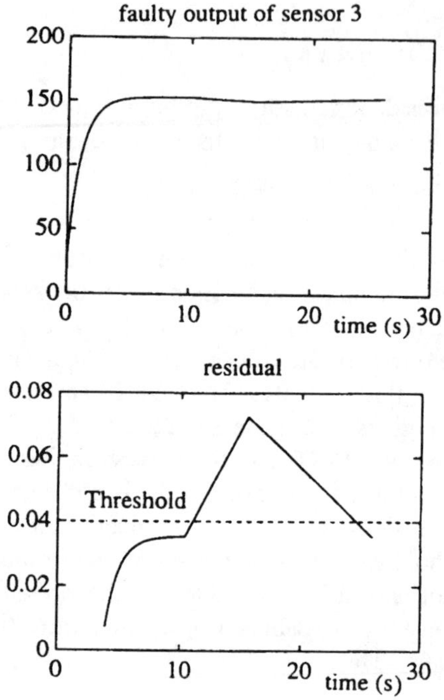

Figure 2.12 Faulty output and rsidual in the case of a fault in T_7.

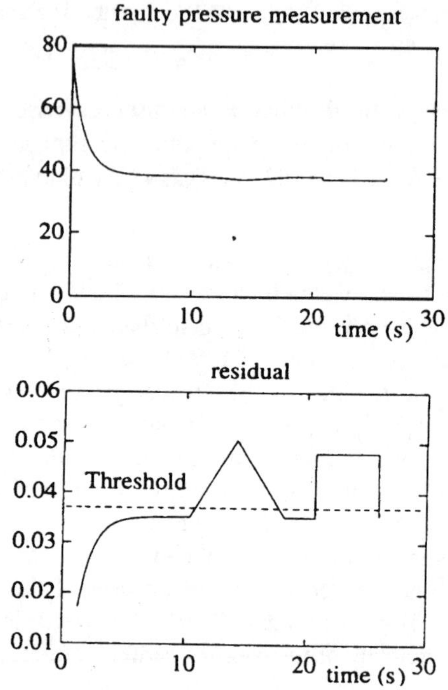

Figure 2.13 Faulty output of the pressure measurement P_6 and corresponding residual.

2.8.2 Applications in Transportation Engineering

Among other things, the term *factory of the future* contains driverless transportation systems within the framework of computer-aided logistics. The transportation vehicles are mostly inductively guided along a defined path. The principle of electronic track-guidance is applied in airports, container terminals at railway stations or harbours, within the service tunnel transportation system of the euro-tunnel and within modern public short-haul traffic systems. An example of this class of track-bounded transportation systems with automatic track-guidance, is the standard city bus O-305 of Mercedes-Benz.

The supervision of measuring instruments in such vehicles is of utmost importance, since they are self-driven. Thus, automatic sensor failure detection techniques have a potential field of application in this area. In a recent paper, van Schrick (1993) presents such an example. The main points of this work follow.

The bus follows a nominal track marked by an electro-magnetic field of a cable that is narrowly running under the road surface. The alternating current flowing through the cable generates an electromagnetic field, that induces a voltage into the measuring instrument located concentrically in front of the bus. The voltage induced is a measure for the deviation of track, $d(t)$, used as the only controller input. The digital controller calculates a steering signal, that with the aid of an active steering system, directly acts onto the front wheels to minimize the distance between the bus and the nominal track.

Additionally, a second measuring instrument was introduced to enhance the riding comfort of the bus by disturbance rejection control, (van Schrick, 1991). This instrument gives information on the directly measured steering angle, $\beta(t)$, of the front wheels. Both measuring instruments, the one for the deviation of track and the one for the steering angle have to be supervised. This is due to the very high requirements of safety on such transportation systems.

A linearised model of fifth order in sensor coordinates for the lateral motion of the city bus is given as follows:

$$\dot{\boldsymbol{x}}(t) = \boldsymbol{A}(\boldsymbol{p})\boldsymbol{x}(t) + \boldsymbol{b}u(t) + \boldsymbol{E}(\boldsymbol{p})\boldsymbol{d}(t)$$

$$y(t) = \boldsymbol{c}^{\mathrm{T}}\boldsymbol{x}(t)$$

where the states $x_1(t)$ to $x_5(t)$ are the displacement $d(t)$ between sensor and nominal track, its velocity $\dot{d}(t)$, the yaw angle rate $\varepsilon(t)$, the side slip angle $\alpha(t)$ and the steering angle $\beta(t)$. The control input $u(t)$ is the steering angle rate $\dot{\beta}(t)$ and the controlled variable $y(t)$=$x_1(t)$ is the measured deviation of track $d(t)$. Additionally, the disturbance vector $\boldsymbol{d}(t)$ consisting of bending $K(t)$, side wind moment $M(t)$ and side wind force $F(t)$ acts on the system. For the investigations described in the following, only the bending K is regarded. If necessary, the effects of the disturbances $M(t)$ and $F(t)$ can be treated in the same maner. The parameter vector $\boldsymbol{p} = [m \;\; \upsilon]$ contains the relative mass m corresponding to the friction coefficient μ ranging from 9950 kg to 27000 kg and the velocity v ranging from 0.6ms^{-1} to 14 ms^{-1}. The input vector $\boldsymbol{b}$ and the output vector $\mathbf{c}^{\mathrm{T}}$ are $\boldsymbol{b} = \boldsymbol{e}_5$ and $\boldsymbol{c}^{\mathrm{T}} = \boldsymbol{e}_1^{\mathrm{T}}$ The system matrix $\boldsymbol{A}(\boldsymbol{p})$ and the disturbance input vector $\boldsymbol{g}_K(\boldsymbol{p})$ of the disturbance input matrix $\boldsymbol{E}(\boldsymbol{p})$ are:

$$\boldsymbol{A}(\boldsymbol{p}) = \begin{bmatrix} 0 & 1 & 0 & 0 & 0 \\ 0 & 0 & \alpha_{23} & \alpha_{24} & \alpha_{25} \\ 0 & 0 & \alpha_{33} & \alpha_{34} & \alpha_{35} \\ 0 & 0 & \alpha_{43} & \alpha_{44} & \alpha_{45} \\ 0 & 0 & 0 & 0 & 0 \end{bmatrix}, \quad \boldsymbol{g}_K(\boldsymbol{p}) = \begin{bmatrix} 0 \\ g_{21} \\ 0 \\ 0 \\ 0 \end{bmatrix}$$

where the elements α_{ij} and g_{21} depend on the parameter vector $\boldsymbol{p}$ (cf. Darenberg (1987), for their evaluation).

The 5th order dynamic controller used for track control can be written as,

$$\dot{x}(t) = A_R x(t) + b_R y(t)$$
$$u_R(t) = c_R^T x(t) + d_R y(t)$$

where A_R, b_R, c_R^T and d_R consist of constant elements calculated by simulation, optimisation and riding tests along a test-track. This design result guarantees that the steering angle $|\beta| \leq 45^o$ and its derivative $|\dot{\beta}| \leq 23^o/s$ with an accuracy of track of 2cm and a maximum deviation in curves of 15cm.

In the work of van Schrick (1991), an additional term to the above control law was added, resulting in,

$$u(t) = u_R(t) + u_\upsilon(t) = u_R(t) - k_\upsilon^T(p)\hat{v}(t)$$

where $k_\upsilon^T(p)$ is the disturbance rejection gain vector. Investigations have shown that the gain vector can approximately be chosen k_υ^T=constant. The vector $\hat{v}(t)$ reflects the reconstructed unknown input $\hat{K}(t)$ and its derivative $\dot{\hat{K}}(t)$ supplied by a robust disturbance observer. For disturbance rejection control, the conditions for the control and compensation are fulfilled, but the condition for reconstructing the unknown input is fulfilled if a second measurement is introduced. This is the reason for using an additional sensor for measuring the steering angle β.

The development of a sensor supervision for the track-guided city bus with disturbance rejection control led to an extended system structure. fig. 2.14 illustrates the overall structure comprising the bus, the dynamic controller, the disturbance observer and the two IFD-parts.

The design task for a sensor supervision of the track-guided city bus is to determine a scalar decision function $f(t)$ that is robust against variations of the parameter vector p and the unknown disturbance $K(t)$ but sensitive to instrument faults.

For this task, it is not necessary to seperately investigate the residuals $e_d(t)$ and $e_\beta(t)$ defined as output estimation errors of the measurements $d(t)$ and $\beta(t)$. Therefore, the decision function,

$$f(t) = \left| w^T r(t) \right|$$

is the absolute value of a weighted linear combination of the residuals, where,

$$w^T = \begin{bmatrix} w_d & w_\beta \end{bmatrix}$$

is a weighting vector and,

$$r(t) = \begin{bmatrix} r_d(t) & r_\beta(t) \end{bmatrix}$$

the residual vector. In spite of parameter variations Δp and the unknown input $K(t)$, the fact that the decision function should be minimal in the fault-free case and maximal in the case of sensor faults simplifies the observer design. It is not required to minimise the estimation errors but to minimise the decision function. For this reason, the optimization ot the weighting vector w^T is included into the design procedure.

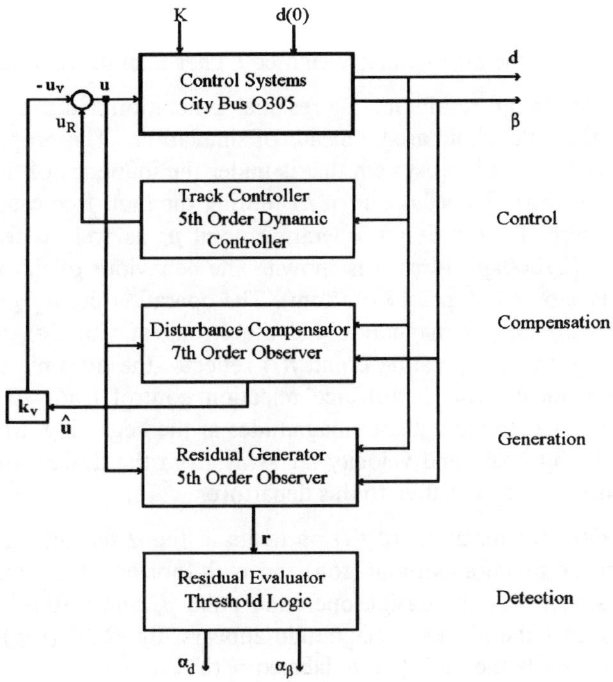

Figure 2.14 Overall structure of the IFD-scheme

The designed estimator is a fifth order observer,

$$\dot{\hat{x}}(t) = A_0\hat{x}(t) + bu(t) + L(y(t) - C\hat{x}(t))$$

that is based on the measurements $d(t)$ and $\beta(t)$. The matrix A_0 denotes that for the robust observed design a nominal bus model has to be taken into account, i.e. a design operating point p_e has to be chosen. The constant observer gain L and weighting w^T have then to be determined for $p \in P$ shown in fig.2.15, where p_i describe the admissible corner operating points of the track-guided city bus.

Unfortunately, a perfect decoupling of the decision function $f(t)$ from the unkown inputs (parameter variations included) is not possible because some rank conditions are not fulfilled. Consequently, the proposed design methods for unkonwn input observers as well as the parity space approach or the eigen-structure approach cannot be applied and an approximative solution has to be found. The adopted approach uses the multi-objective parameter optimization porcedure proposed by Kasper *et al.* (1990).

For a design operating point,

$$p_e = \begin{bmatrix} 13750\,kg & 4.5m/s \end{bmatrix}$$

the design results in a decision function that is robust against variations of $\boldsymbol{p}$ and the effect of the unknown bending $K(t)$. The optimized values of $\boldsymbol{w}^T$ are w_d = -0.1329 and w_β = 0.7661 which show that the β-residual is weighted higher than the d-residual.

For one channel, the design results for the residual generation and the resulting residual evaluation with a fixed threshold are by means of simulation. The proposed IFD-procedure is operated as a closed-loop system that is under the influence of the extended control law. In fig.2.16 (top), the behaviour of $f(t)$ during the fault-free case without disturbance rejection control for the design operating point $\boldsymbol{p}_e$ as well as for $\boldsymbol{p}_1$ = [9950kg 0.6ms^{-1}] and $\boldsymbol{p}_2$ = [27000kg 14ms^{-1}] is shown. The behaviour of $f(t)$ with disturbance rejection control is shown in fig. 2.15 (bottom). The operating point $\boldsymbol{p}_2$ results in a decision function with the largest magnitude and the strongest vibration for all operating points $\boldsymbol{p}$. Depending on the operating point, $f(t)$ reflects the dynamic bahaviour of the controlled bus. Obviously, the distrubance rejection control leads to vibrations with higher frequencies in general and higher magnitudes at the beginning. Moreover, for operating points with high mass and velocity far away from the design operating point $\boldsymbol{p}_e$ the decision function is very sensitive to this departure.

To illustrate the different reactions of $f(t)$ on faults in the d-sensor, fig.2.16 shows the course of the decision function without (top) and with (bottom) the influence of the disturbance compensation. For the design operating point $\boldsymbol{p}_e$ and at the instant of fault t_f =5s, a 5% decrease of the d-sensor magnitude appears. fig. 2.16 (top) shows that the controller nearly conceals the fault (curve labeled with f_mR) while a controller not influenced, gives an unmistakable course of function (curve labeled with f_oR). The bottom part reflects the same situation but now with disturbance compensation that results in a higher frequency of $f(t)$ and a complete concealing of the fault.

For all parameter and riding situations as well us different types of faults, further investigations have shown that this porcedure gives a very sensitive decision function to allow for the detection of incipient faults in the sensors of the track-guided city bus.

As a final comment, note that a fixed threshold was used. The threshold value depends on the input signal, operating point, riding situation and effects on unknown inputs as well as the quality of sensors. A worst-case estimation of the threshold resulted in T=5$\times$10^{-4}.

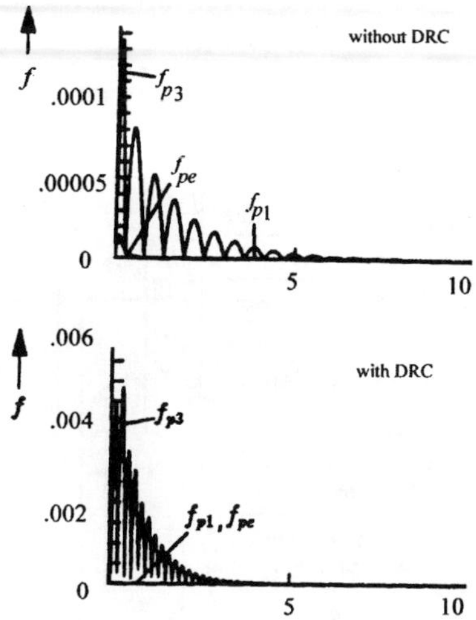

Figure 2.15 $f(t)$, no-fault case.

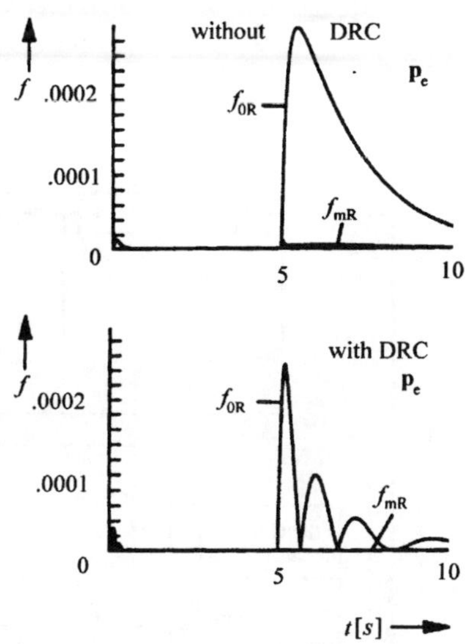

Figure 2.16 $f(t)$, 5% *d*-sensor fault.

2.8.3 Applications in aerospace engineering

Hydromechanical implementations of turbine engine control systems have matured into highly reliable units, resulting in increased engine complexity in order to meet ever increasing engine performance requirements. Consequently, the engine control has become increasingly complex. Because of this complexity trend and the revolution in digital electronics, the control has evolved from a hydromechanical to a full authority digital electronic control (FADEC) implementation. These FADEC type controls must demonstrate the same or improved levels of reliability as their hydromechanical predecessors.

DeLaat and Merrill (1990) describe such an implementation, termed Advanced Detection, Isolation and Accommodation (ADIA) concept, whose objective is to improve the overall reliability of digital electronic control systems for turbine engines.

The ADIA algorithm detects, isolates, and accommodates sensor failures in an F100 turbofan engine control system. The algorithm incorporates advanced filtering and detection logic and is general enough to be applied to different engines or other types of control systems. The algorithm detects two classes of sensor failures, hard and soft. Hard failures are defined as out-of-range or large bias errors that occur instantaneously in the sensed values. Soft failures are defined as small bias errors of drift errors that increase relatively slowly with time. The ADIA algorithm (fig.2.17) consists of four elements: (1) hard sensor failure detection and isolation logic; (2) soft sensor failure detection and isolation logic; (3) an accommodation filter; and (4) the interface switch matrix.

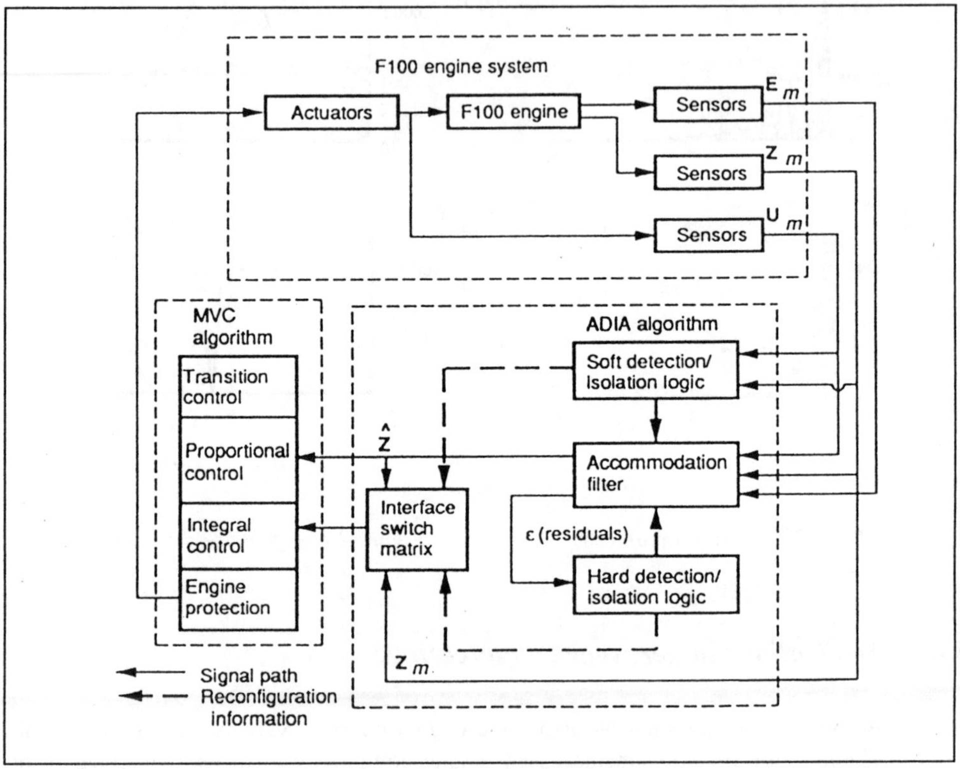

Figure 2.17 The ADIA block diagram

In the normal or unfailed mode of operation, the accommodation filter uses the full set of engine measurements to generate a set of optimal estimates, $\hat{y}(t)$, of the measurements. These estimates are used by the control law. When a sensor failure occurs, the detection logic determines that a failure has occurred. The isolation logic then determines which sensor is faulty. This structural information is passed to the accommodation filter. The accommodation filter then removes the faulty measurement from further consideration. The accommodation filter, however, continues to generate the full set of optimal estimates of the control. Thus the control mode does not have to be restructured for any sensor failure. The ADIA algorithm inputs as shown in fig. 2.17, are the sensed engine output variables $y_m(t)$, the sensed engine environmental variables $e_m(t)$, and the sensed engine input variables $u_m(t)$. The outputs of the algorithm, the estimates $\hat{y}(t)$ of the measured engine outputs $y_m(t)$, are used as input to the proportional part of the control. During normal mode operation, engine measurements are used in the integral control to ensure accurate steady-state operation. When a sensor failure is accommodated, the measurement in the integral control is replaced with the corresponding accommodation filter estimate by reconfiguring the interface switch matrix. Estimates are always used in the proportional control to avoid propagating a transient when switching from measurement to estimate.

The accommodation filter incorporates an engine model along with a Kalman gain update to generate estimates of the engine states $\boldsymbol{x}$ and the engine outputs $\boldsymbol{y}$ as follows:

$$\dot{\hat{\boldsymbol{x}}}(t) = \boldsymbol{A}(\boldsymbol{x}(t) - \hat{\boldsymbol{x}}_b) + \boldsymbol{G}(\boldsymbol{u}_m(t) - \boldsymbol{u}_b) + \boldsymbol{K}\boldsymbol{\gamma}(t)$$
$$\hat{\boldsymbol{y}}(t) = \boldsymbol{C}(\boldsymbol{x}(t) - \hat{\boldsymbol{x}}_b) + \boldsymbol{D}(\boldsymbol{u}_m(t) - \boldsymbol{u}_b) + \boldsymbol{y}_b$$
$$\boldsymbol{\gamma}(t) = \boldsymbol{y}_m(t) - \hat{\boldsymbol{y}}$$

where the subscript b represents the base point (steady-state engine operating point) and $\boldsymbol{x}$ is the 4×1 model state vector, $\boldsymbol{u_m}$ the 4×1 sensed control vector and $\boldsymbol{y_m}$ the 5×1 sensed output vector. The matrix $\boldsymbol{K}$ is the Kalman gain matrix and $\boldsymbol{\gamma}$ is the residual vector. The $\boldsymbol{A}$, $\boldsymbol{G}$, $\boldsymbol{C}$, and $\boldsymbol{D}$ matrices are appropriately dimensioned system matrices.

The basepoints are scheduled as nonlinear functions of the engine operating condition indicated by $\boldsymbol{e_m}$. The individual system matrix elements along with those of $\boldsymbol{K}$, are corrected by the engine inlet conditions $\boldsymbol{e_m}$ and scheduled as nonlinear functions of $\boldsymbol{y_b}$ (Beattie *et al.*, 1983). Reconfiguration of the accommodation filter after the detection and isolation of a sensor failure is accomplished by forcing the appropriate residual element to zero. For example, if a compressor speed sensor failure (*N*2) has been isolated, the effect of reconfiguration is to force γ_2=0. This is equivalent to setting sensed *N*2 equal to the estimate of *N*2 generated by the filter. The residuals generated by the accommodation filter are used in the hard failure detection logic.

The hard sensor failure detection and isolation logic is straightforward. To accomplish hard failure detection and isolation the absolute value of each component of the residual vector is compared to its own threshold. If the residual absolute value is greater than the threshold, then a failure is detected and isolated for the sensor corresponding to the residual element. Threshold sizes are initially determined from the standard deviation of the noise on the sensors. These standard deviation magnitudes are then increased to account for modeling errors in the accommodation filter. The hard detection threshold values are twice the magnitude of these adjusted standard deviations.

The soft failure detection and isolation logic consists of a multiple-hypothesis-based testing (fig.2.18). Each hypothesis is implemented using a Kalman filter. A total of six hypotheses filters are used, one for normal mode operation (H_0) and five for the failure modes (one for each engine output sensor H_1 to H_5). The structure for each hypothesis filter is identical to the accommodation filter. However, each hypothesis filter uses a different set of measurements. For example the first hypothesis filter (H_1) uses all of the sensed engine outputs except the first, *N*1. Thus, each hypothesis filter generates a unique residual vector $\boldsymbol{\gamma}_i$. From this residual each hypothesis filter generates a statistic or likelihood based upon a weighted sum of squared residuals (WSSR). Assuming Gaussian sensor noise, each sample of $\boldsymbol{\gamma}_i$ has a certain likelihood of probability L_i

$$L_i = p_i(\boldsymbol{\gamma}_i) = k\exp(-WSSR_i))$$

$$WSSR_i = \boldsymbol{\gamma}_i^{\mathrm{T}} \boldsymbol{\Sigma} \boldsymbol{\gamma}_i$$

$$\boldsymbol{\Sigma} = diag(\sigma_i^2)$$

where k is a constant and the σ_i are the adjusted standard deviations. These standard deviation values scale the residuals to dimensionless quantities that can be summed to form a WSSR. The WSSR statistic is smoothed to remove gross noise effects by a first order lag with a time constant of 0.1s. The log of the ratio of each hypothesis likelihood to the normal mode likelihood is calculated. If the maximum log-likelihood ratio exceeds the soft failure detection and isolation threshold, then a failure is detected and isolated and accommodation occurs. If a sensor failure has occurred in $N1$ for example, all of the hypothesis filters except H_1 will be corrupted by the faulty information. Thus each of the corresponding likelihoods will be small except for LR_1. Thus, LR_1 will be the maximum and it will be compared to the threshold to detect the failure.

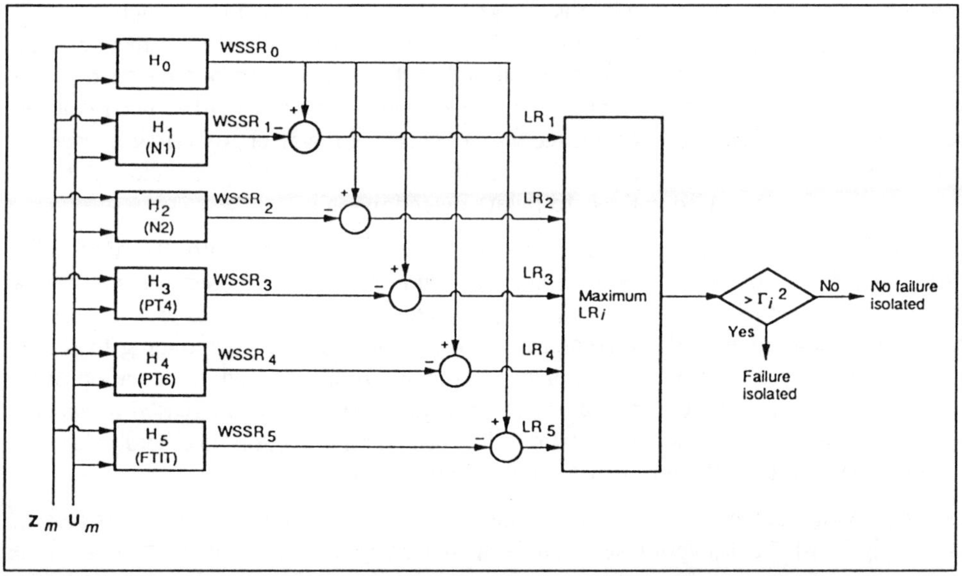

Figure 2.18 Soft failure detection and isolation logic

Initially, the soft failure detection/isolation threshold was determined by standard statistical analysis of the residuals to set the confidence level of false alarms and missed detections. The threshold was then modified to account for modeling error. It was soon apparent from initial evaluation studies that transient modeling error was dominant in determining the fixed threshold level. It was also clear that threshold was too large for desirable steady state operation. Thus, an adaptive threshold was incorporated, to make the algorithm more robust to transient modeling error while maintaining steady-state performance. The adaptive threshold d_i was heuristically determined and consisted of two parts. One part, d_{iss} is the steady-state detection/isolation threshold which accounts for steady-state, or low frequency modeling error. The second part, d_{EXP}, accounts for the transient, or high frequency modeling error. The adaptive threshold is triggered by an internal control system variable m_{tran} which is indicative of transient operation:

$$d_i = d_{iSS}(d_{EXP} + 1)$$

$$\tau(\dot{d}_{EXP}) + d_{EXP} = m_{tran}$$

The values of d_{iss}, τ, and M_{tran} were found by experimentation to minimize false alarms during transients. When the engine experiences a transient, M_{tran} is set to 4.5; otherwise is 0. The threshold time constant is τ =2s. The adaptive threshold expansion logic enabled d_{iss} to be reduced to 40% of its original value, which results in an 80% reduction in the detection/isolation threshold $d_i^2 d_{iss}$. The adaptive threshold logic is illustrated in fig. 2.19 for a power lever angle (PLA) pulse transient.

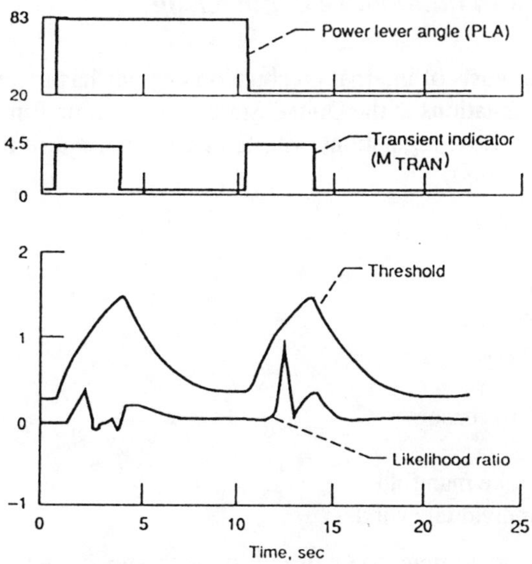

Figure 2.19 Adaptive threshold logic.

For failure accommodation, two separate steps were taken. First, all seven of the filters (the accommodation filter and the six hypothesis filters) were reconfigured (the appropriate residual in each filter is forced to zero) to account for the detected failure mode. Second, if a soft failure was detected, the states and estimates of all seven filters were updated to the values of the hypothesis filter which corresponds to the failed sensor. Stability of each of the filters after reconfiguration was verified during algorithm evaluation.

The algorithm was implemented in real-time to demonstrate its capabilities with a full scale engine using hardware and software typical of that to be used in next generation turbofan engine controls. The Multivariable Control (MVC)-ADIA implementation has several distinct hardware and software features. Three CPUs were used, operating in parallel. The schedules used to generate the engine model basepoints and table look-up routines within the ADIA algorithm were coded in assembly language.

The real-time microcomputer implementation of the combined MVC-ADIA algorithm performed extremely well. Sensor failure detection and accommodation were demonstrated at eleven different operating points which included subsonic and supersonic conditions and medium and high power operation of the engine.

2.8.4 Applications in automotive engineering

Interest in on-line diagnosis of internal combustion engines has recently increased due to new environmental regulations in the United States and in some European countries (e.g. the EFTA partners). These regulations will, for example, require the following components/faults to be diagnosed:

- individual fuel injectors
- O_2 sensor(s)
- mass air flow sensor
- manifold pressure sensor
- throttle position sensor
- A/F controller
- inlet air temperature sensor
- misfire detection
- vacuum leak in intake manifold
- loss of power in individual cylinder

In order for a diagnostic strategy to be useful for such applications, it should permit real time implementation in order to allow for continuous monitoring of system performance on-board the vehicle. Within this context, many of the currently employed service diagnostic strategies, for example those based on expert system methods, are not suitable.

Rizzoni *et al.* (1993) discuss simulation and experimental results of a study aimed at diagnosing faults associated with an automotive engine exhaust emissions control system, using the parity vector approach.

The engine model used was based on the works of Dobner (1982), Coats and Fruechte (1982), Moskwa and Hedrick (1987), Cho and Hedrick (1989) and Grizzle *et al.* (1990). The structure of the model is shown in fig. 2.20. The essential elements of the model are: intake manifold, rotating dynamics, volumetric efficiency, oxygen (lambda) sensor model, combustion and exhaust dynamics and fuelling controller and metering system.

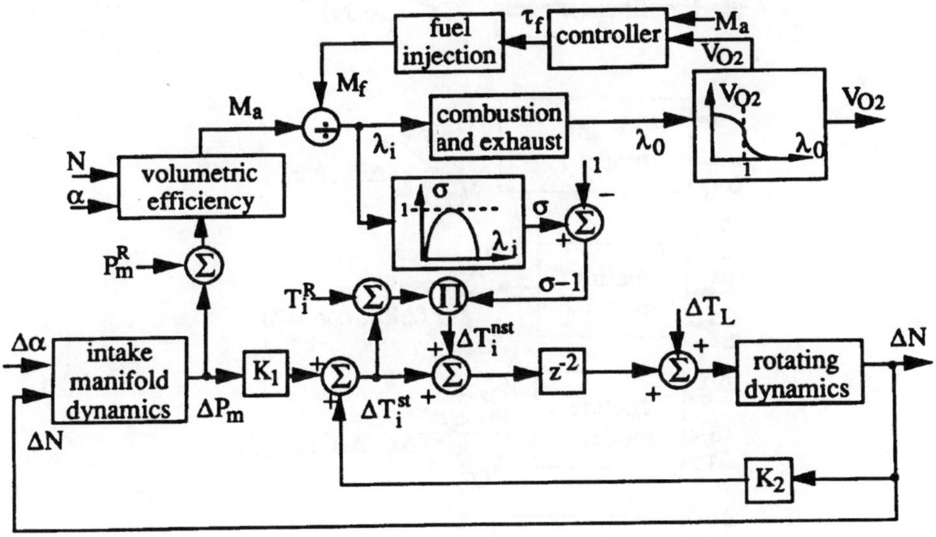

Figure 2.20 Engine model.

Details of the various subsystems can be found in Rizzoni *et al.* (1993). The whole system depicted in fig. 2.20, after suitable identification procedure, is put in the form:

$$\boldsymbol{x}(k+1) = \boldsymbol{A}\boldsymbol{x}(k) + \boldsymbol{B}\boldsymbol{u}(k) + \boldsymbol{E}\boldsymbol{d}(k)$$

$$\boldsymbol{y}(k) = \boldsymbol{C}\boldsymbol{x}(k) + \boldsymbol{D}\boldsymbol{u}(k) + \Delta\boldsymbol{y}(k)$$

where $\boldsymbol{d}(k)$ is a vector of plant or input faults, or unmeasured disturbance inputs, and $\Delta\boldsymbol{y}(k)$ is a sensor fault vector. In this case, residuals can simply be defined as follows:

$$\boldsymbol{r}(k) = \boldsymbol{y}(k) - \boldsymbol{G}(z)\boldsymbol{u}(k) = \boldsymbol{H}(z)\boldsymbol{d}(k) - \boldsymbol{G}(z)\Delta\boldsymbol{u}(k) + \Delta\boldsymbol{y}(k)$$

This equation is referred to as the primary parity equation. The residual vector can then be designed to have some desired properties by multiplying the primary parity equation by a suitable polynomial matrix $\boldsymbol{W}(z)$.

The model of fig. 2.20 was used to generate a set of residual generators, shown in fig. 2.21. Apart from the intake manifold model, which is modeled by a nonlinear static relation, each of the residual generators was obtained by employing system identification methods to identify the relevant dynamics. Thus, the range of operation of the residual generation strategy for these four subsystems is limited to the neighborhood of an operating point.

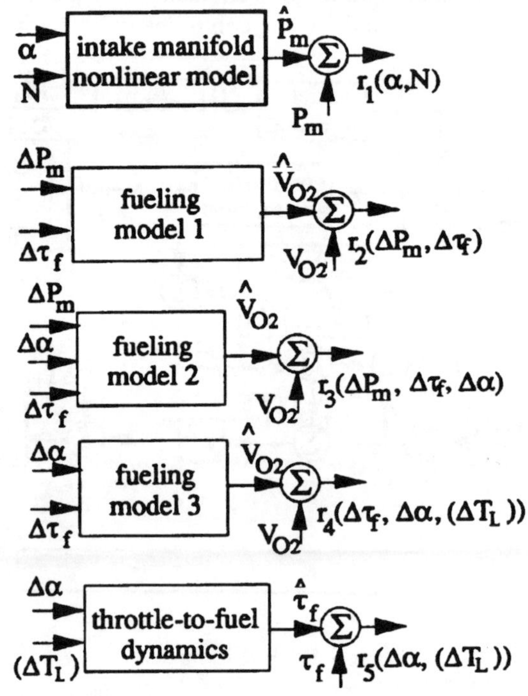

Figure 2.21 Residual generation strategy.

Another limitation of this design is that, because of lack of suitable experimental arrangements, the fueling dynamics of the engine have been simulated off-line; the engine simulator reflects each of the subsections except for the dependence of the fueling time, τ_f, on the estimated air mass. Thus, the fueling model may be thought of as a perturbation model.

Note that perturbations in the load torque, ΔT_L, appear as an input (or output) in two of the residual generators. This quantity is shown in parentheses to indicate that this disturbance input is not measured; part of the residual generation strategy consists of designing a decoupling matrix $\boldsymbol{W}(z)$ that can remove the effect of the load torque from the residual vector $\boldsymbol{r}(k)$. Details of this strategy can be found in Gertler (1991).

The residual generation strategy was tested using a mix of experiment and simulation. As an example one experiment is discussed, which included a throttle transient corresponding to a change in trottle position of 7.5° causing a change in engine speed from 3,000 to 2,700 rev/min, as shown in fig. 2.22. The sensor outputs were sampled in the crank angle domain, at a rate corresponding to one sample every 180° of crankshaft rotation. Various faults were injected, in correspondence of the 1800th sample. The faults consisted of 10% calibration errors in each of the sensors (throttle position, speed, manifold pressure), and in one of the fuel injectors. Figs 2.23 and 2.24 depict the residual vector created by the residual generators of fig. 2.22 for the no-fault case, and for a typical fault condition of 10% change in throttle calibration. As it was onserved, the transient caused some difficulties in discriminating among faults. However, once the effects of the transient had subsided, the residuals behaved according to the structure illustrated in Table 2.1.

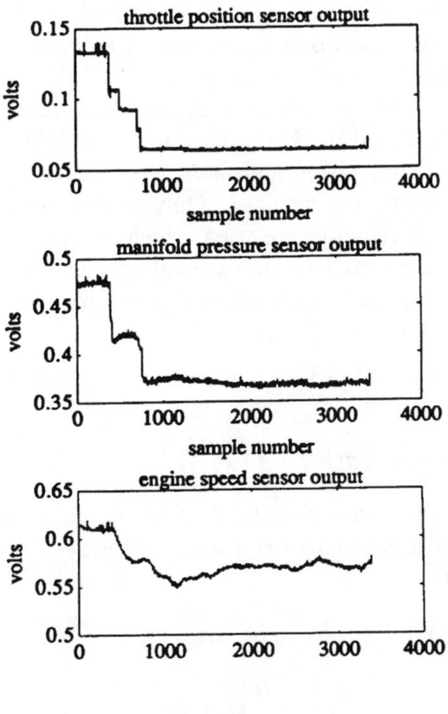

Figure 2.22 Experimental conditions for residual generation validation.

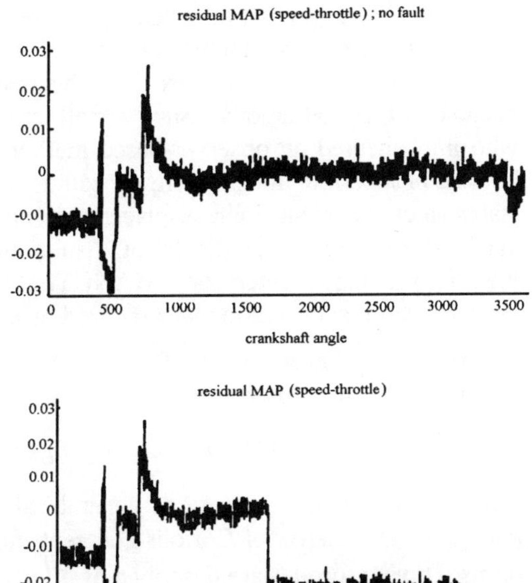

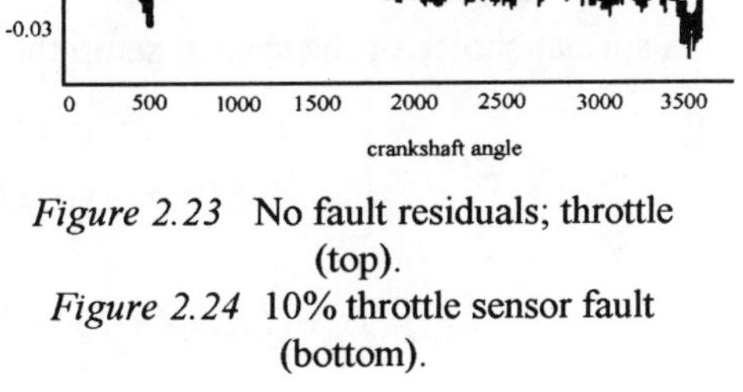

Figure 2.23 No fault residuals; throttle (top).

Figure 2.24 10% throttle sensor fault (bottom).

Table 2.1 Residual structure

no fault	throttle sensor	speed sensor	pressure sensor	fuel injec-tor
0	1	1	1	0
0	0	1	1	1
0	0	1	0	1
0	0	0	1	1
0	1	1	1	0

2.8.5 Applications in robotics

In flexible automation, unpredictable behavior can result from abnormal deviations. Therefore, a high demand in supervision, fault detection, fault isolation and fault recovery exists. The approach of installing as much sensors as possible to monitor automation and access all possible failure modes has the drawback of increasing costs and making automation even more complex. For this reason, software redundancy methods are applicable in these situations. Such a method is proposed by Schneider and Frank (1992) who implemented an observer-based method for supervision and fault detection of robotized productioin units, where no additional sensors are required.. Only internal robot states (motor current, joint displacement and joint velocity) are used which are already available for controlling the robot. Faults are considered to be external torques. The method was implemented for a MANUTEC r3 industrial robot where a major improvement for residual evaluation was realized by employing an adaptive threshold.

A dynamic, nonlinear, model of a robot can be derived by the Euler-Lagrange equations and presented in a closed form:

$$J(q(t))\ddot{q}(t) = X^d(q(t),\dot{q}(t)) + x^g(q(t)) + f(u(t))$$

Here, the vector $q(t)$ represents generalized position coordinates, J in an inertia mass matrix, X^d is a matrix of Coriolis and centrifugal terms and x^g is a vector of gravitional terms. Driving torques are described by $f(u(t))$.

With a special choice of the system setup the robot equation can be written:

$$\dot{x}(t) = \begin{bmatrix} \dot{q}(t) \\ J^{-1}\left(X^d(q,\dot{q}) + x^g(q) + f(u)\right) \end{bmatrix} + \begin{bmatrix} 0 \\ I \end{bmatrix} f_d(t)$$

$$= B(u(t), C_1(x(t)) + \begin{bmatrix} 0 \\ I \end{bmatrix} f_d(t)$$

$$y(t) = \begin{bmatrix} I & 0 \\ 0 & I \end{bmatrix} x(t), \quad \text{with } x(t) = \begin{bmatrix} q(t) \\ \dot{q}(t) \end{bmatrix}$$

This is of the form of Eqs. (2.146) with $\boldsymbol{A}_0=\boldsymbol{0}$, $\boldsymbol{K}=\boldsymbol{0}$, $\boldsymbol{F}=\boldsymbol{0}$, $\boldsymbol{G}=\boldsymbol{0}$. In order to detect a fault $f_d(t)$ in the i-th axis of an n-degree of freedom robot, applying conditions (2.147) yields the transformation matrix $\boldsymbol{T}$:

$$\boldsymbol{T} = \left[\boldsymbol{0}_{n\times 1} \;\vdots\; \boldsymbol{I}^*_{n\times 1}\right]$$

$$\boldsymbol{I}^* = \left[\begin{matrix}\boldsymbol{0}_{n\times 1} & \underset{i-th\ column}{\underline{1}} & \boldsymbol{0}_{n\times 1}\end{matrix}\right]$$

and

$$\boldsymbol{G} = \left[\boldsymbol{0}_{n\times 1} \;\vdots\; -\alpha_i \boldsymbol{I}^*_{n\times 1}\right]$$

$$\boldsymbol{L}_1 = \left[\boldsymbol{0}_{n\times 1} \;\vdots\; -\alpha_i \boldsymbol{I}^*_{n\times 1}\right]$$

$$\boldsymbol{L}_2 = \left[\boldsymbol{0}_{n\times 1} \;\vdots\; \alpha_i \boldsymbol{I}^*_{n\times 1}\right]$$

where α_i is a positive number determining the observer poles.

A matrix multiplication, $\boldsymbol{f}_d^*(t) = \boldsymbol{J}(\boldsymbol{q}(t))\boldsymbol{f}_d(t)$ decouples the torque on each generalized axis, representing unmodeled friction torques and any other additional torque disturbances.

Extensive investigations on this robot showed that friction cannot be modeled in a simple form, e.g. Coulomb and viscous friction. Instead, a displacement dependent friction characteristic had to be introduced (see e.g. Schneider and Frank, 1992, Armstrong, 1988). Typical friction characteristics of the r3 robot are displayed in fig. 2.25 where friction is obtained by the observer-based approach while constant velocity movements are performed. The residual, representing a friction torque, is not plotted as a time signal but as a correlation with the corresponding positional displacement, such that different measurements can easily be compared.

In experiments, external torques to the robot were calculated with the observer-based method and also measured with a force/torque sensor. An external torque was applied to the robot's tool center point while the robot performed a normal operation. As seen from fig. 2.26, the calculated residual matched the measured one within a ±3Nm range. Note that a dead-band of about 30Nm was needed when a fixed threshold was used. Moreover, only a small torque was applied in this experiment, degrading the results since the measurements were corrupted by noise.

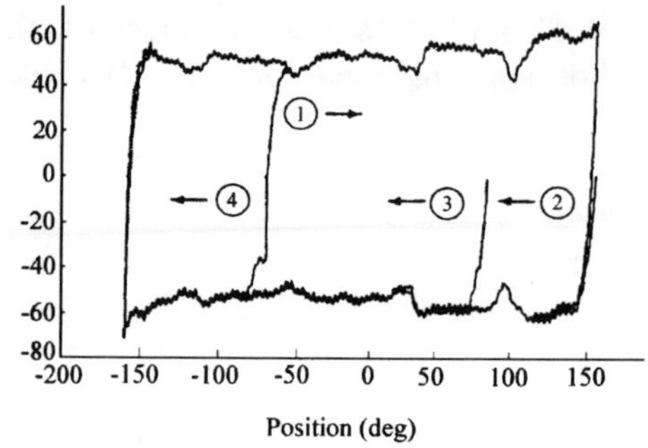

Figure 2.25 Friction characteristics of the MANUTEC r3 robot.

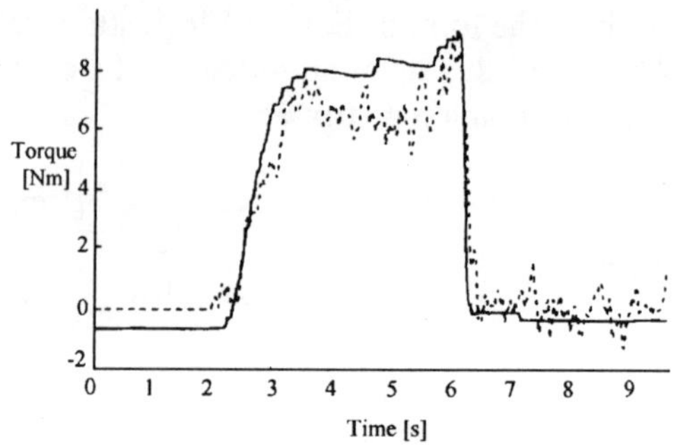

Figure 2.26 External torque estimation. Solid line: measured; dotted line: calculated.

References

Anderson T.W. (1958). An introduction to multivariate statistical analysis. John Wiley, N.Y.

Antsaklis P.J. (1980). Maximal order reduction and supremal *A, B* invariant and controllability subspaces. *IEEE Transactions on Automatic Control*, **AC–25**, 44–49.

Armstrong B. (1988). Friction: Experimental determination, modeling and compensation. Proceedings, *IEEE International Conference on Robotics and Automation*, Philadelphia.

Athans M. and D. Willner (1973). A practical scheme for adaptive aircraft control systems. *Symposium on Parameter Estimation Techniques and Applications in Aircraft Flight Testing*, NASA Flight Research Center, Edwards AFB, April 24–25.

Basseville M. (1981). Edge detection using sequential methods for change in level, Part I: A sequential edge detection algorithm. *IEEE Transactions on Acoustics, Speech and Signal Processing*, **ASSP-29**, 1, 32-50.

Basseville M. and A. Benveniste (1983). Design and comparative study of some sequential jump detection algorihms for digital signals. *IEEE Transactions on Acoustics, Speech and Signal Processing*, **ASSP-31**, 32-50.

Basseville M. and A. Benveniste, Eds. (1986). Detection of abrupt changes in signals and dynamical systems. Springer-Verlag.

Basseville M. and I. Nikiforov (1993). Detection of abrupt changes. Theory and application. Prentice Hall, Englewood Cliffs, N.J.

Baram Y. (1976). Information, consistent estimation and dynamic system identification. Phd. Thesis, M.I.T.

Beard R.V. (1971). Failure accomodation in linear systems through reorganisation. *Rep. MVT–71–1*, Man Vehicle Lab., M.I.T.

Beattie E.C. *et al.* (1981). Sensor failure detection systems for the F100 turbofan engine. *NASA CR-165515.*

Bickel, P.J. (1965). On some asymptotic competitors to Hotelling's T^2. *The annals of Mathematical Statistics*, **36**, 16–173.

Bonivento C. and A. Tonielli (1984). A detection–estimation multifilter approach with nuclear application. *Proceedings, 9th IFAC World Congress*, Budapest, Hungary, 1771–1776.

Cho D.and J.K. Hedrick(1989). Automotive Power Train Modeling for Control. *ASME Journal of Dynamic Systems, Measurement, and Control*, **111**, 568-576.

Chow E.Y. (1980). A failure detection system design methodology. Phd. Thesis, M.I.T.

Chow E.Y. and A.S. Willsky (1984). Analytical redundancy and the design of robust failure detection systems. *IEEE Transactions on Automatic Control*, **AC–29**, 603–614.

Chien T.T. and M.B. Adams (1976). A sequential failure detection technique and its application. *IEEE Transactions on Automatic Control*, **AC–21**, 750–757.

Clark R.N. (1978). A simplified instrument failure detection scheme. *IEEE Transactions on Aerospace and Electronic Systems*, **AES-14**, 4.

Coats F.E. and R.D. Fruechte (1982). Dynamic Engine Models for Control Development. Part II: Application to Idle Speed Control. *General Motors Research Laboratories Publication GMR-3789*, January.

Daley K.C., Gai E. and J.V. Harrison (1979). Generalised likelihood test for FDI in redundant sensor configurations. *Journal of Guidance and Control*, **2**, 9–17.

Darenberg W. (1987). Automatische Spurführung von Kraftfahrseugen - Ein Problem der robusten Regelung. *Automobil Industrie*, **2**, 155-159.

Deckert J.C., Desai M.N., Deyst, J.J. and A.S. Willsky (1977). F8 DFBW sensor failure identification using analytical redundancy, *IEEE Transactions on Automatic Control*, **AC–22**, 5, 795–803.

DeLaat J.C. and W.C. Merrill (1990). A real time microcomputer implementation of sensor failure detection for turbofan engines. *IEEE CS Magazine*, **10**, 4, 29-36.

Desai M. and A. Ray (1981). A fault detection and isolation methodology. *Proceedings, 20th Conference on Decision and Control*, 1363–1369.

Ding X. and P.M. Frank (1989). Komponentenfehler-detektion mittels auf Empfindlichkeitsanalyse basierender robuster detektionsfilter. *Automatisierungstechnik*, Oldenburg, Munich.

Dobner D.J. (1982), Dynamic Engine Models for Control Development. Part I: Nonlinear and Linear Model Formulation. *General Motors Research Laboratories Publication GMR-3783*, January.

Dowdle J.R., Willsky A.S. and S.W. Gully (1983). Nonlinear generalised likelihood ratio algorithms for manoeuver detection and estimation. *Proceedings, 1982 American Control Conference*, Arlington, Virginia, June 1983.

Frank P.M. (1987). Fault diagnosis in dynamic systems via state estimation-A survey. In *S. Tzafestas, M. Singh, G. Schmidt (Eds.), System fault diagnostics, reliability and related knowledge-based approaches*, D. Reidel, Dordrecht, 35-89.

Frank P.M. (1990). Fault diagnosis in dynamic systems using analytical and knowledge-based redundancy-A survey and some new results. *Automatica*, **26**, 3, 459-474.

Gantmacher R. (1974). The theory of matrices, Volume II. Chelsea Publishing Company, N.Y.

Gertler, J.J. (1991). Analytical redundancy methods in failure detection and isolation in complex plants. *Proceedings, IFAC/IMACS Symposium SAFEPROCESS '91*, Baden–Baden, September 10–13, pp.9–21.

Gertler, J.J., Costin, J., Kowalczuk, Z., Fang, X., Hira, R. and Q. Luo (1991). Model–based on–board fault detection and diagnosis for automotive engines. *Proceedings, IFAC/IMACS Symposium SAFEPROCESS '91*, Baden–Baden, 503–508.

Green C.S. (1978). An analysis of the multiple model adaptive control algorithm. Phd. Thesis, Report no. ESL-TH-843, Electronic Systems Laboratory, M.I.T.

Grizzle J.W., Cook J.A. and K.L. Dobbins (1990). Individual Cylinder Air to Fuel Ratio Control with a Single EGO Sensor. *Proceedings, 1990 American Control Conference*, San Diego, CA, May 1990, 2881-2886.

Gustafsson D.E., Willsky A.S., Wang J.-Y., Lancaster M.C. and J.H. Triebwasser (1978a). ECG/VCG Rhythm diagnosis using statistical signal analysis, I: Identification of persistent rhythms. *IEEE Transactions on Biomedical Engineering*, **BME-25**, 4, 344-353.

Gustafsson D.E., Willsky A.S., Wang J.-Y., Lancaster M.C. and J.H. Triebwasser (1978b). ECG/VCG Rhythm diagnosis using statistical signal analysis, I: Identification of transient rhythms. *IEEE Transactions on Biomedical Engineering*, **BME-25**, 4, 353-361.

Janssen K. and P.M. Frank (1984). Component failure detection via state estimation. *Proceedings, 9th IFAC World Congress*, Budapest, Hungary.

Jenkin G.M. and D.G. Watts (1968). Spectral analysis and its applications. Holden Day, San Francisco.

Jones H.L. (1973). Failure detection in linear systems. Phd. dissertation, Dept. Aeronautics and Astronautics, M.I.T.

Karkanias N. and B. Kouvaritakis (1979). The output zeroing problem and its relationship to the invariant zero structure. *International Journal of Control,* **30**, 395–415.

Kasper R., Lückel J., Jaker K.P. and J. Schroer (1990). CACE Tool for Multi-Input, . Multi-Output Systems using a New Vector Optimisation Method. *International Journal of Control*, **51**, 963-993.

Konik D. and S. Engell (1986). Sequential design of decentralised controllers using decoupling techniques. Part 2:Numerical aspects and application. *Proceedings, 4th IFAC/IFORS Symposium on Large Scale Systems: Theory and Applications*, Zurich, Switzerland.

Kumamaru K. (1984). Statistical failure diagnosis for dynamical systems. *Systems and Control*, **28**, 77–86.

Lainiotis D.G. (1971). Joint detection, estimation and system identification, *Information and Control*, **19**, 75–92.

Lou X.C., Willsky A.S. and G.C. Verghese (1986). Optimal robust redundancy relations for failure detection in uncertain systems. *Automatica*, **22**, 333–344.

Massoumnia, M.–A. (1986). A geometric approach to failure detection and identification in linear systems. Phd. Thesis, Dept. of Aeronautics and Astronautics, MIT.

Massoumnia M.–A. and W.E. Vanderwelde (1988). Generating parity relations for detecting and identifying control system component failures. *AIAA Journal of Guidance, Control and Dynamics,* **11**, 60–65.

Mehra, R.K. and I. Peschon (1971). An innovations approach to fault detection and diagnosis in dynamical systems. *Automatica*, **7**, 637–640.

Mironovskii L.A. (1980). Functional diagnosis of dynamic system. *Automation and remote control*, **41**, 1122–1143.

Moore B.C. (1976). On flexibility offered by state feedback in multivariable systems beyond closed loop eigenvalue assignment. *IEEE Transactions on Automatic Control,* **AC–21**, 689–692.

Moskwa J.J. and J.K. Hedrick (1987). Automotive Engine Modeling for Real Time Control Application. *Proceedings of the 1987 American Control Conference*, Vol. 1, pp. 341-346.

Ono T., Kumamaru T. and K. Kumamaru (1984). Fault diagnosis of sensors using a gradient method. *Transactions, Society of Instrument and Control Engineers*, **20**, 22–27.

Patton R.J. and J. Chen (1991). Robust fault detection using eigenstructure assignment: A tutorial consideration and some new results. *Proceedings, 30th IEEE Conference on Decision and Control*, Brighton, U.K., December 11–13, 2242–2247.

Patton R.J. and J. Chen (1992). Robust fault detection of jet engine sensor systems by using eigenstructure assignment. *Proceedings, AIIA Guidance, Navigation and Control Conference*, New Orleans, August.

Patton R.J., Chen J. and H.Y. Zhang (1992). Modelling methods for improving robustness in fault diagnosis of jet engine system. *Proceedings, IEEE 31st Conference on Decision and Control*, Tucson, Arizona, December 1992, 2330–2335.

Patton R.J., Frank P.M. and R.N. Clark, Eds. (1989). Fault diagnosis in dynamic systems, Theory and Applications. Prentice-Hall, Englewood Cliffs, NJ.

Patton R.J. and S.M. Kangethe (1989). Robust fault diagnosis using eigenstructure assignment of observers. In Patton R.J., Frank P.M. and R.N. Clark (Eds.), Fault diagnosis in dynamic systems: Theory and applications, Prentice Hall.

Patton R.J. and S.M. Willcox (1986). Fault diagnosis in dynamic systems using a robust output zeroing design method. *Proceedings, First European Workshop on Failure Diagnosis*, Rhodes, Greece, August 31–September 3.

Patton R.J., Willcox S. and S.J. Winter (1987). A parameter insensitive technique for aircraft sensor fault analysis, *AIAA Journal of Guidance, Control and Dynamics*, **10**, 359–367.

Patton R.J., Zhang H.Y. and J. Chen (1992b). Modeling of uncertainties for robust fault diagnosis. *Proceedings, IEEE 31st Conference on Decision and Control*, Tucson, Arizona, December 16-18.

Potter I.E. and M.C. Suman (1977). Thresholdless redundancy management with arrays of skewed instruments. *Integrity in electronic flight control systems*, **AGARDOGRAPH–224**, 15–11 to 15–25.

Pouliezos A. (1980a). Fault monitoring schemes for linear stochastic systems. Phd. Thesis, Brunel University, England.

Pouliezos A. (1980b). An iterative method for calculating the sample serial correlation coefficient. *IEEE Transactions on Automatic Control*, **AC-25**, 834-836.

Pouliezos A. and G. Stavrakakis (1987). Linear estimation in the presence of sudden system changes: An expert system. In P. Borne, S. Tzafestas (Eds.), *Applied modeling and simulation of technological systems*, North–Holland, 41-48.

Pouliezos A. and G. Stavrakakis (1991). A two-stage real-time fault monitoring system. *Proceedings, European Robotics and Intelligent Systems Conference EURISCON '91*, Corfu, Greece, June 23-28.

Pouliezos A., Stavrakakis G. and G. Tselentis (1993). Detection of multivariable system noise degradation. *Proceedings, IEEE Mediteranean Symposium on New Directions in Control Theory and Applications*, Hania, Greece, June 21-23.

Rizzoni G., Azzoni P.M. and G. Minelli (1993). On-board diagnosis of emission control malfunctions in electronically controlled spark igmition engines. *Proceedings, International Conference on Fault Diagnosis TOOLDIAG '93*, Toulouse, France, April 5-7, 238-248.

van Schrick D. (1991). Robust Observers for Track Control and Instrument Fault Detection of a City Bus. In Isermann R. (ed.), *Preprints, IFAC/IMACS Symposium on Fault Detection, Supervision and Safety for Technical Processes-SAFEPROCESS '91*, Baden-Baden, Germany, September 10-13, Vol. **I**, 227-231.

Schneider H. and P.M. Frank (1992). Observer-based fault detection of robots enhanced by characteristic curves. *Proceedings, IMACS/SICE International Symposium on Robotics, Mechatronics and Manufacturing Systems*, Kobe, Japan.

van Schrick D. (1993). Instrument fault detection scheme for inductive track-guided transportation vehicles. *Proceedings, International Conference on Fault Diagnosis TOOLDIAG '93*, Toulouse, France, April 5-7, 261-266.

Tanaka S.and P.C. Müller (1990). Fault detection in linear discrete dynamic systems by a pattern recognition of a generalised likelihood ratio. *Journal of Dynamic Systems, Measurement and Control, Transactions of the ASME*, **112**, 276–282.

Uosaki K. (1985). Failure detection using backward SPRT. *Proceedings, IFAC Symposium on Identification and System Parameter Estimation*, York, U.K., 1619–1624.

Uosaki K. and M. Kawagoe (1988). Backward SPRT failure detection system for detection of innovation variance change. *Proceedings, IFAC Identification and System Parameter Estimation*, Beijing, PRC.

Van Trees H.L. (1968). Detection, estimation and modulation theory. John Wiley and Sons.

Wald A. (1947). Sequential analysis. Wiley, New York.

Watanabe, K. and D.M. Himmelblau (1982). Instrument fault detection in systems with uncertainties. *International Journal of Systems Science*, **13**, 137–158.

Watanabe K., Yoshimura T. and T. Soeda (1979). A diagnosis design for a parametric failure. *Transactions, Society of Instument and Control Engineers*, **15**, 901-906.

Watanabe K., Yoshimura T. and T. Soeda (1981). A discrete-time adaptive filter for stochastic distributed parameter systems. *Transactions, ASME Journal of Dynamic Systems, Measurement and Control*, **103**, 266-278.

Willsky A.S. (1986). Detection of abrupt changes in dynamic systems. In M. Basseville and A. Benveniste (Eds.), *Detection of abrupt changes in signals and dynamical systems,* Springer-Verlag, 27-49.

Willsky A.S., Chow E.Y., Gershwin S.B., Greene C.S., Houpt P.K. and A.L. Kurkjian (1980). Dynamic model-based techniques for the detection of incidents on freeways. *IEEE Transactions on Automatic Control*, **AC-25**.

Willsky A.S., Deyst J.J. and B.S. Crawford (1974a). Adaptive filtering and self–test methods for failure detection and compensation. *Proceedings of the Joint Automatic Control Conference*, Austin, Texas, 637–645.

Willsky A.S., Deyst J.J. and B.S. Crawford (1974b). Two self–test methods applied to an inertial system problem. *Journal of Spacecraft and Rockets*, **12**, 434–437.

Willsky A.S. and H.L. Jones (1976). A generalised likelihood ratio approach to the detection and estimation of jumps in linear systems. *IEEE Transactions on Automatic Control*, **AC–21**, 108–112.

Wünnenberg J. and P.M. Frank (1987). Sensor fault detection via robust observers. In S. Tzafestas, M. Singh, G. Schmidt (Eds.), *System fault diagnostics, reliability and related knowledge–based approaches*, D. Reidel, 147–160.

Yoshimura T., Watanabe K., Konishi K. and T.A. Soeda (1979). Sequential failure detection approach and the identification of failure parameters, *International Journal of Systems Science*, **10**, 827–836.

CHAPTER 3

PARAMETER ESTIMATION METHODS FOR FAULT MONITORING

3.1 Introduction

Fault detection via parameter estimation relies on the principle that possible faults in the monitored process can be associated with specific parameters and states of a mathematical model of a process given in general by an input–output relation,

$$y(t) = f(\boldsymbol{u}, \boldsymbol{e}, \boldsymbol{\theta}, \boldsymbol{x}) \tag{3.1}$$

where $\boldsymbol{y}(t)$ represents the vector output of the process, $\boldsymbol{u}(t)$ the vector input, $\boldsymbol{x}(t)$ the partially measurable state variables, $\boldsymbol{\theta}$ the nonmeasurable process parameters likely to change and $\boldsymbol{e}(t)$ unmodeled or noise terms affecting the process. It is obvious therefore, that it is necessary to have an accurate theoretical dynamic model of the process in order to apply parameter estimation methods. This is usually derived from the basic balance equations for mass, energy, and momentum, the physico–chemical state equations and the phenomenological laws for any irreversible phenomena. The models will then appear in the continuous or discrete time domain, in the form of ordinary or partial differential or difference equations. Their parameters θ_i are expressed in dependence on process coefficients p_j, like storage or resistance quantities, whose changes indicate a process fault. Hence, the parameters θ_i of continuous time models have to be estimated. In this case there is a minimum number of independently measurable quantities which permit the estimation of various states and parameters. As an example consider a simple dynamic process model with lumped parameters, linearized about an operating point, which may be described by the differential equation,

$$y(t) + \dots + a_n y^{(n)}(t) = b_0 u(t) + b_1 u^{(1)}(t) + \dots + b_m u^{(m)}(t) \tag{3.2}$$

The process model parameters,

$$\theta^{T} = [a_1, \ldots, a_n \vdots b_0, \ldots, b_m] \tag{3.3}$$

are defined as relationships of several physical process coefficients, e.g. length, mass, speed, drag coefficient, viscosity, resistances, capacities. Faults which become noticeable in these physical process constants are therefore also expressed in the process model parameters. If the physical process coefficients, indicative of process faults, are not directly measurable, an attempt can be made to detect their changes via the changes in the process model parameters $\boldsymbol{\theta}$. The following procedure is therefore applicable in general:

(1) Establishment of the mathematical model of the normal process,

$$\boldsymbol{y}(t) = \boldsymbol{f}(\boldsymbol{u}(t), \boldsymbol{\theta}) \tag{3.4}$$

mainly from theoretical considerations. At this stage allowable tolerances for process coefficient values are also defined.

(2) Determination of the relationship between the model parameters θ_i and the physical process coefficients p_j,

$$\boldsymbol{\theta} = \boldsymbol{f}(\boldsymbol{p}) \tag{3.5}$$

(3) Estimation of the model parameters θ_i from measurements of $\boldsymbol{y}(t)$, $\boldsymbol{u}(t)$, by a suitable estimation procedure,

$$\hat{\theta}(t) = g(y(1), \ldots, y(t), u(1), \ldots, u(t)) \tag{3.5a}$$

(4) Calculation of process coefficients, via the inverse relationship,

$$\hat{\boldsymbol{p}}(t) = \boldsymbol{f}^{-1}(\hat{\theta}(t)) \tag{3.6}$$

(5) Decision on whether a fault has occurred, based either on the changes Δp_j calculated in step 4 or on the changes $\Delta\theta_i$ and tolerance limits from step 3. If decisions are made based on the $\Delta\theta_i$ the affected p_i's can be easily determined from 3.5. This may be achieved with the aid of a fault catalogue in which the relationship between process faults and changes in the coefficients Δp_j has been established. Decisions can be made either by simply checking against the predetermined threshold levels, or by using more sophisticated methods from the fields of statistical decision theory. A fault decision should include the fault location, fault size and time of occurrence. System reorganization should follow a positive fault decision. Such an action is essential, since usually controller design depends on the correctness of parameters. Robust schemes also benefit from this procedure.

The basis of this class of methods is the combination of theoretical modeling and parameter estimation of continuous time models. A block diagram is given in Figure 3.1. Since, however a requirement of this procedure is the existence of the inverse relationship (3.6) it may be restricted to well–defined process.

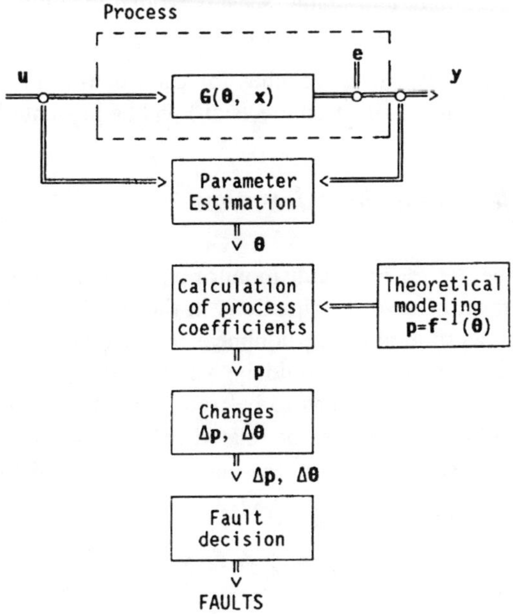

Figure 3.1 Fault detection based on parameter estimation and theoretical modeling.

Having in mind these limiting requirements it must be emphasized that parameter–estimation based fault detection performs all the tasks that are expected from a well–designed fault monitoring system:

- fault detection
- fault isolation
- fault estimation and
- enables systems reorganization following a fault.

Iserman (1987), points to some other advantages as well:

- Process faults can be detected earlier and localised more precisely since process parameters are in many cases closer to the process faults in terms of signal flow.
- Fault selectivity is improved since the monitored process parameter is changed directly by a fault and is not coupled to other parameters.
- Faults in closed loops can also be detected since the procedure considers the selection between inputs and outputs.
- Multiple fault detection is possible, since the procedure is checking for all values simultaneously.

The implementation of the full procedure requires however, more effort in modeling the process, more sophisticated and fault–sensitive identification methods and fast processing hardware suitable for on–line operation.

In the following sections different approaches for each of the three stages of the fault monitoring process (modeling, estimation, decision) will be presented in detail.

3.2 *Process modeling for fault detection*

The starting point in the design of a fault monitoring system is the development of the process model in the form of a set of equations in the continuous or discrete time domain. Usually these process models are nonlinear and should not be linearised since a process model for failure diagnosis should be valid over a large range of operating conditions. This process description has to be translated into equations which allow to estimate $\boldsymbol{\theta}$. Since usually least–squares or related methods are used for the estimation phase, these equations must be linear–in–the–parameters, i.e. of the following general form,

$$f(y_1,\ldots,y_n)=\sum_{i=1}^{r}\theta_i f_i(y_1,\ldots,y_n)$$

where y_i are the measured quantities. The functions f_i may be nonlinear; time–derivatives or integrals are also allowed since the derivatives can be obtained from the original signals by state variable filtering (SVF, Young, 1981) or standard difference techniques (Pouliezos and Stavrakakis, 1989). More valuable than the estimated parameters $\boldsymbol{\theta}$ are the physical process coefficients $\boldsymbol{p}$, since they are directly related to the monitored process. Thus deviations of the process coefficients from their "normal" values can be attributed to a significant change in the process itself and allow a precise fault diagnosis (localisation of fault) if there is a unique correspondence between $\boldsymbol{\theta}$, $\boldsymbol{p}$. A simple example will illustrate the importance of choosing the correct process model (Iserman, 1987).

Example 3.1 First order electrical circuit.

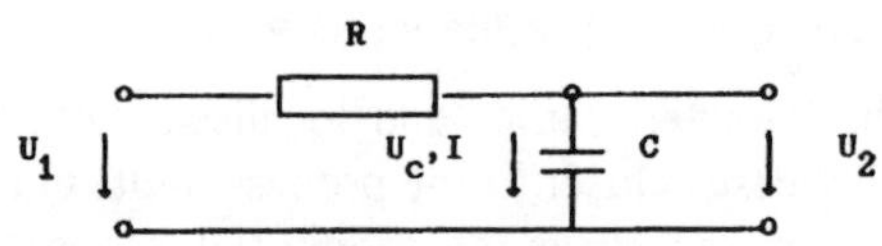

First approach: Defining voltage v_1 as input, voltage v_2 as output, and $a_1=RC$, $b_0=1$, the dynamic process model is:

$$a_1\dot{v}_2(t)+v_2(t)=b_0 v_1(t)$$

$$\dot{v}_2(t) = [-v_2 \quad v_1]\begin{bmatrix} 1/a_1 \\ b_0/a_1 \end{bmatrix}$$

This model has only one parameter a_1 (effectively) which contains two process coefficients R, C. Hence, one of them must be assumed as non–likely to fail. However, if $b_0 \neq 1$ a fault of the condensator (leakage current) is indicated.

Second approach: Defining voltage v_1 as input, current i as output, and $a_1=RC$, $b_1=C$, the dynamic process model is:

$$a_1 \frac{\mathrm{d}i(t)}{\mathrm{d}t} + i(t) = b_1 \frac{\mathrm{d}v_1(t)}{\mathrm{d}t}$$

$$\frac{\mathrm{d}i(t)}{\mathrm{d}t} = \begin{bmatrix} -i(t) & \frac{\mathrm{d}v_1(t)}{\mathrm{d}t} & v_1(t) \end{bmatrix}\begin{bmatrix} 1/\alpha_1 \\ \beta_1/\alpha_1 \\ 0 \end{bmatrix} = \boldsymbol{U\theta}$$

Hence, $\theta_1 = 1/RC$, $\theta_2 = 1/R$ and $R = 1/\theta_2$, $C = \theta_2/\theta_1$ and the two process coefficients can be determined uniquely.

This simple low order example shows that it depends on the selected input and output variables whether a change (fault) in one of the process coefficients can be detected. This capability is closely related to identifiability conditions for the process coefficients. If not all of the process coefficients can be determined uniquely then some of them must be assumed as not likely to fail.

Nold and Iserman (1986) give the following algebraic identifiability condition relating $\boldsymbol{\theta}$ and $\boldsymbol{p}$:

Theorem 3.1 A system of equations,

$$\theta_i = \sum_{\mu=1}^{n} c_{\mu i}\left[\hat{K} + \prod_{\nu=1}^{m} p_\nu^{k_{\mu\nu}}\right]; \; i = 1, \; \ldots, \; r$$

provides unique solutions for the process coefficients $p_1, \ldots, p_m$ in $\boldsymbol{R}^m$ with the exclusion of an open neighbourhood of $\mathbf{0}$ if the determinant,

$$\begin{vmatrix} u_{11} & u_{1,n-m-1} & 1 & k_{11} & k_{1m} \\ \cdot & & \cdot & \cdot & \\ \cdot & & \cdot & \cdot & \\ u_{n1} & u_{n,n-m-1} & 1 & k_{n1} & k_{nm} \end{vmatrix} \neq 0$$

where $\boldsymbol{u}_j = [\, u_{1j}, \ldots, u_{nj}]^{\mathrm{T}}$ are the base vectors of the equation system written in the form:

$$\boldsymbol{\theta} = \{c_{ii}\}\boldsymbol{z}$$

The number of base vectors has to be $n-m-1 \geq 0$. This result is illustrated by considering a second order electrical network as shown in Figure 3.2.

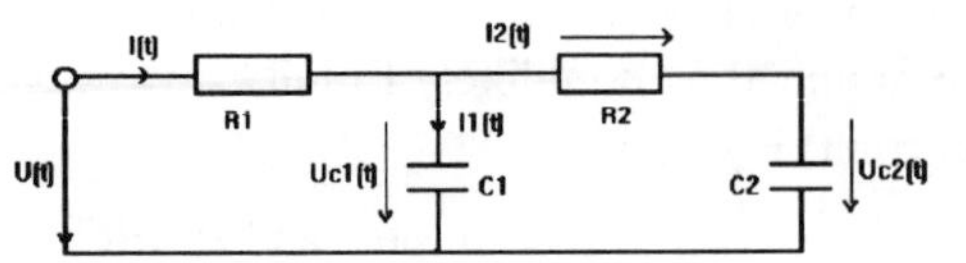

Figure 3.2 Second order electrical network

The variables which can be measured are only *u* and *i*. Here the circuit diagram is already a sufficient process description so that in this case, no block diagram is needed. One can directly write down the equation in matrix form:

$$\begin{bmatrix} 1 & -R_1 \\ 0 & 0 \\ 0 & 1 \\ 0 & 0 \\ 0 & 0 \\ 0 & 0 \end{bmatrix} \begin{bmatrix} U \\ I \end{bmatrix} = \underbrace{\begin{bmatrix} 1 & 0 & 0 & 0 & 0 \\ -1 & 1 & 0 & 1 & 0 \\ 0 & 0 & 1 & 0 & +1 \\ 0 & 0 & -1 & C_2 s & 0 \\ -C_1 s & 0 & 0 & 0 & 1 \\ 0 & 1 & -R_2 & 0 & 0 \end{bmatrix}}_{H} \begin{bmatrix} U_{C1} \\ U_{R2} \\ I_2 \\ U_{C2} \\ \mathrm{I}_1 \end{bmatrix}$$

The H–matrix is now rowwise tranformed in upper triangular form. After tranforming the first five rows one obtains:

$$\begin{bmatrix} 1 & -R_1 \\ 1 & -R_1 \\ 0 & 1 \\ 0 & 1 \\ C_1 s & -R_1 C_1 s \\ 0 & 0 \end{bmatrix} \begin{bmatrix} U \\ I \end{bmatrix} = \begin{bmatrix} 1 & 0 & 0 & 0 & 0 \\ 0 & 1 & 0 & 1 & 0 \\ 0 & 0 & 1 & 0 & 1 \\ 0 & 0 & 0 & C_2 s & 1 \\ 0 & 0 & 0 & 0 & 1 \\ 0 & 1 & -R_2 & 0 & 0 \end{bmatrix} \begin{bmatrix} U_{C1} \\ U_{R2} \\ I_2 \\ U_{C2} \\ I_1 \end{bmatrix}$$

Now the last row is stepwise tranformed:

$$\begin{bmatrix} -1 & R_1 \end{bmatrix} \boldsymbol{y} = \begin{bmatrix} 0 & 0 & -R_2 & -1 & 0 \end{bmatrix} \boldsymbol{x}$$

$$\begin{bmatrix} -1 & R_1 + R_2 \end{bmatrix} \boldsymbol{y} = \begin{bmatrix} 0 & 0 & 0 & -1 & R_2 \end{bmatrix} \boldsymbol{x}$$

$$\begin{bmatrix} -1 & R_1 + R_2 + \dfrac{1}{C_2 s} \end{bmatrix} \boldsymbol{y} = \begin{bmatrix} 0 & 0 & 0 & 0 & R_2 + \dfrac{1}{C_2 s} \end{bmatrix} \boldsymbol{x}$$

$$\begin{bmatrix} -1 - C_1 s\left(R_2 + \dfrac{1}{C_2 s}\right) & R_1 + R_2 + \dfrac{1}{C_2 s} + R_1 C_1 s\left(R_2 + \dfrac{1}{C_2 s}\right) \end{bmatrix} \boldsymbol{y} = 0$$

The last equation represents the estimation equation which can now be written in the form,

$$\left(\frac{C_1}{C_2} + 1\right) u(t) + R_2 C_1 \frac{\mathrm{d}u(t)}{\mathrm{d}t} = \left(R_1\left(1 + \frac{C_1}{C_2}\right) + R_2\right) i(t) + R_1 C_1 R_2 \frac{\mathrm{d}i(t)}{\mathrm{d}t} + \frac{1}{C_2} \int i(t)\mathrm{d}t$$

Therefore one can derive the equation system:

$$\begin{bmatrix} 1 & 1 & 0 & 0 & 0 & 0 & 0 & 0 \\ 0 & 0 & 1 & 0 & 0 & 0 & 0 & 0 \\ 0 & 0 & 0 & 1 & 1 & 1 & 0 & 0 \\ 0 & 0 & 0 & 0 & 0 & 0 & 1 & 0 \\ 0 & 0 & 0 & 0 & 0 & 0 & 0 & 1 \end{bmatrix} \begin{bmatrix} \hat{K} R_1^0 R_2^0 C_1^1 C_2^{-1} \\ \hat{K} R_1^0 R_2^0 C_1^0 C_2^0 \\ \hat{K} R_1^0 R_2^1 C_1^1 C_2^0 \\ \hat{K} R_1^1 R_2^0 C_1^0 C_2^0 \\ \hat{K} R_1^1 R_2^0 C_1^1 C_2^{-1} \\ \hat{K} R_1^0 R_2^1 C_1^0 C_2^0 \\ \hat{K} R_1^1 R_2^1 C_1^1 C_2^0 \\ \hat{K} R_1^0 R_2^0 C_1^0 C_2^{-1} \end{bmatrix} = \begin{bmatrix} \theta_1 \\ \theta_2 \\ -\theta_3 \\ -\theta_4 \\ -\theta_5 \\ -\theta_6 \end{bmatrix}$$

The base vectors are obtained by evaluating the homogeneous solution

$$z_1 + z_2 = \mathbf{0}$$

$$z_4 + z_5 + z_6 = \mathbf{0}$$

$$z_3 = z_7 = z_8 = \mathbf{0}$$

The solutions for this set of equations can be expressed in the form,

$$z = \lambda_1 \boldsymbol{u}_1 + \lambda_2 \boldsymbol{u}_2 + \lambda_3 \boldsymbol{u}_3; \quad \lambda_1, \lambda_2, \lambda_3 \in \boldsymbol{R}$$

A set of base vectors $\boldsymbol{u}_i$ for this equation system is,

$$\boldsymbol{u}_1^{\mathrm{T}} = [\,1 \;\; -1 \;\; 0 \;\; 0 \;\; 0 \;\; 0 \;\; 0 \;\; 0\,]$$

$$\boldsymbol{u}_2^{\mathrm{T}} = [\,0 \;\; 0 \;\; 0 \;\; 0 \;\; 0 \;\; -1 \;\; 0 \;\; 0\,]$$

$$\boldsymbol{u}_3^{\mathrm{T}} = [\,0 \;\; 0 \;\; 0 \;\; 0 \;\; 1 \;\; -1 \;\; 0 \;\; 0\,]$$

The determinant thus becomes,

$$\begin{vmatrix} 1 & 0 & 0 & 1 & 0 & 0 & 1 & -1 \\ -1 & 0 & 0 & 1 & 0 & 0 & 0 & 0 \\ 0 & 0 & 0 & 1 & 0 & 1 & 1 & 0 \\ 0 & 1 & 0 & 1 & 1 & 0 & 0 & 0 \\ 0 & 0 & 1 & 1 & 1 & 0 & 1 & -1 \\ 0 & -1 & -1 & 1 & 0 & 1 & 0 & 0 \\ 0 & 0 & 0 & 1 & 1 & 1 & 1 & 0 \\ 0 & 0 & 0 & 1 & 0 & 0 & 0 & -1 \end{vmatrix} = 1$$

The determinant is unequal to zero, the number of base vectors is 8–4–1=3, r=5<n=8 hence the system is structurally identifiable.

However, the theorem does not give the solutions for the process coefficients, it just says whether there is a unique solution or not. In this example, the process parameters can be easily determined from the estimated parameters. Using,

$$\theta_1 \frac{du(t)}{dt} + \theta_2 u(t) = i(t) + \theta_3 \frac{di(t)}{dt} + \theta_4 (t)$$

with,

$$\theta_1 = C_1 + C_2$$
$$\theta_2 = R_2\, C_1\, C_2$$
$$\theta_3 = R_1(C_1 + C_2) + R_2 C_2$$
$$\theta_4 = R_1\, R_2\, C_1\, C_2$$

one can immediately see that,

$$R_1 = \theta_4/\theta_2$$
$$C_1 = \theta_2^2 / (\theta_3\theta_2 - \theta_4\theta_1)$$
$$C_2 = \theta_1 - \theta_2^2 / (\theta_3\theta_2 - \theta_4\theta_1)$$
$$R_2 = (\theta_3 - \theta_4\theta_1 / \theta_2) / \left(\theta_1 - \theta_2^2 / (\theta_3\theta_2 - \theta_4\theta_1)\right)$$

3.3 Parameter estimation for fault detection

Parameter identification methods exist in a variety of approaches. However, not all of them are suitable if they are to be used for on–line fault detection schemes. Proper fault detection procedures should be oblivious to old data in order to be able to detect parameter changes soon after they occur and should perform necessary calculations fast enough for real–time operation. For the first requirement some method of overweighting old data must be used, but not at the expense of accuracy. A standard way to do that is to use sliding windows of data values. The windows may be either rectangular with equal weighting or exponential with some form of decreasing weights with distance from present. Appropriate techniques will be discussed in the next sections. The second requirement is fulfilled by using recursive methods and perhaps employing some modifications for reducing calculations. Memory requirement is not a problem with today's shrinking components.

In the following sections, several estimation techniques that have been used for fault detection will be presented. At this point the close relationship between identification algorithms for slowly or fast–varying parameters and for fault detection purposes must be pointed out. However, a subtle difference exists which should not be overlooked: a fault detection mechanism assumes that a parameter value is known until a change occurs. This means that any control loops utilizing these parameters should use the known and not the estimated value of the parameters.

3.3.1 *Recursive least squares algorithms*

Most schemes for fault detection employing L.S. parameter estimation methods, model the observed system (3.1) as,

$$y(t) = G(q^{-1};\ \boldsymbol{\theta})\ \boldsymbol{u}(t) + \boldsymbol{H}(q^{-1};\ \boldsymbol{\theta})\ \boldsymbol{e}(t) \qquad (3.7)$$

where $\boldsymbol{y}(t)$ is the p–dimensional output, $\boldsymbol{u}(t)$ the r–dimensional input, $\boldsymbol{e}(t)$ is a zero–mean white noise sequence with covariance $\boldsymbol{\Lambda}(\boldsymbol{\theta})$ and $\boldsymbol{G}(q^{-1};\ \boldsymbol{\theta})$, $\boldsymbol{H}(q^{-1};\ \boldsymbol{\theta})$ are filters of appropriate dimensions. Here q^{-1} denotes the backward shift operator $q^{-1}\{\boldsymbol{u}(t)\} = \boldsymbol{u}(t-1)$ etc. Finally, $\boldsymbol{\theta}$ denotes the n–dimensional parameter vector.

Equation (3.7) describes a quite general linear model which by proper choice of the matrices $\boldsymbol{G}$, $\boldsymbol{H}$, $\boldsymbol{\Lambda}$ and parameters θ_i can be put into more familiar forms. An ARMAX (autoregressive moving average with exogenous signals) model is obtained if,

$$G(q^{-1};\boldsymbol{\theta}) = \frac{B(q^{-1})}{A(q^{-1})}, \quad H(q^{-1};\boldsymbol{\theta}) = \frac{C(q^{-1})}{A(q^{-1})}, \quad \boldsymbol{\Lambda}(\boldsymbol{\theta}) = \lambda^2$$

where,

$$A(q^{-1}) = 1 + a_1 q^{-1} + \ \dots \ + a_{n_a} q^{n_a}$$
$$B(q^{-1}) = b_1 q^{-1} + \ \dots \ + b_{n_b} q^{n_b}$$
$$C(q^{-1}) = 1 + c_1 q^{-1} + \ \dots \ + c_{n_c} q^{n_c}$$

The parameter vector is taken as,

$$\boldsymbol{\theta} = \begin{bmatrix} a_1 & \dots & a_{n_a} & b_1 & \dots & b_{n_b} & c_1 & \dots & c_{n_c} & \lambda^2 \end{bmatrix}^{\mathrm{T}}$$

The ARMAX model is then,

$$A(q^{-1})y(t) = B(q^{-1})u(t) + C(q^{-1})e(t) \tag{3.8}$$

or in difference equation form,

$$y(t) + a_1 y(t-1) + \dots + a_{n_a} y(t-n_a) = b_1 u(t-1) + \dots + b_{n_b} u(t-n_b) + e(t) + c_1 e(t-1) + \dots + c_{n_c} e(t-n_c)$$

Here $y(k)$ and $u(k)$ are scalar signals.

If further n_c=0, an ARX (autoregressive with exogenous input) model is obtained given by,

$$A(q^{-1})y(t) = B(q^{-1})u(t) + e(t) \tag{3.9}$$

$$\boldsymbol{\theta} = [\ a_1 \dots a_{n_a}\ \ b_1 \dots b_{n_b}\ ]^{\mathrm{T}} \tag{3.9a}$$

This structure can in some sense be viewed as a linear regression. To see this, rewrite the model (3.9) as:

$$y(t) = \boldsymbol{\varphi}^{\mathrm{T}}(t)\boldsymbol{\theta} + e(t) \tag{3.9b}$$

where,

$$\boldsymbol{\varphi}(t) = [\ -y(t-1) \dots -y(t-n_a)\ \ u(t-1) \dots u(t-n_b)\ ]^{\mathrm{T}} \tag{3.9c}$$

Here though, the regressors (the elements of $\boldsymbol{\varphi}(t)$) are not deterministic functions.

Finally, note that a linear stochastic model in state space form,

$$\boldsymbol{x}(t+1) = \boldsymbol{A}(\boldsymbol{\theta})\boldsymbol{x}(t) + \boldsymbol{B}(\boldsymbol{\theta})\boldsymbol{u}(t) + \boldsymbol{w}(t) \tag{3.10}$$

$$\boldsymbol{y}(t) = \boldsymbol{C}(\boldsymbol{\theta})\boldsymbol{x}(t) + \boldsymbol{v}(t)$$

where $\boldsymbol{w}(t)$ and $\boldsymbol{v}(t)$ are multivariate white noise sequences with zero means and co-variances,

$$E\{\boldsymbol{w}(t)\boldsymbol{w}^{\mathrm{T}}(s)\} = \boldsymbol{R}_1(\boldsymbol{\theta})\delta_{t,s}$$

$$E\{\boldsymbol{w}(t)\boldsymbol{v}^{\mathrm{T}}(s)\} = \boldsymbol{R}_{12}(\boldsymbol{\theta})\delta_{t,s}$$

$$E\{\boldsymbol{v}(t)\boldsymbol{v}^{\mathrm{T}}(s)\} = \boldsymbol{R}_2(\boldsymbol{\theta})\delta_{t,s}$$

can be transformed into the general form (3.7) if the following definitions are made:

$$\boldsymbol{G}(q^{-1};\ \boldsymbol{\theta}) = \boldsymbol{C}(\boldsymbol{\theta})[q\boldsymbol{I} - \boldsymbol{A}(\boldsymbol{\theta})]^{-1}\boldsymbol{B}(\boldsymbol{\theta})$$

$$\boldsymbol{H}(\mathrm{q}^{-1};\ \boldsymbol{\theta}) = \boldsymbol{I} + \boldsymbol{C}(\boldsymbol{\theta})[q\boldsymbol{I} - \boldsymbol{A}(\boldsymbol{\theta})]^{-1}\boldsymbol{K}(\boldsymbol{\theta})$$

$$\boldsymbol{\Lambda}(\boldsymbol{\theta}) = \boldsymbol{C}(\boldsymbol{\theta})\boldsymbol{P}(\boldsymbol{\theta})\boldsymbol{C}^{T}(\boldsymbol{\theta}) + \boldsymbol{R}_2(\boldsymbol{\theta})$$

where $\boldsymbol{K}(\boldsymbol{\theta})$ is the Kalman gain,

$$\boldsymbol{K}(\boldsymbol{\theta}) = \left(\boldsymbol{A}(\boldsymbol{\theta})\boldsymbol{P}(\boldsymbol{\theta})\boldsymbol{C}^{\mathrm{T}}(\boldsymbol{\theta}) + \boldsymbol{R}_{12}(\boldsymbol{\theta})\right)\left(\boldsymbol{C}(\boldsymbol{\theta})\boldsymbol{P}(\boldsymbol{\theta})\boldsymbol{C}^{\mathrm{T}}(\boldsymbol{\theta}) + \boldsymbol{R}_2(\boldsymbol{\theta})\right)^{-1}$$

and $\boldsymbol{P}(\boldsymbol{\theta})$ is the symmetric positive definite solution of the Ricatti equation,

$$\boldsymbol{P}(\theta) = \boldsymbol{A}(\theta)\boldsymbol{P}(\boldsymbol{\theta})\boldsymbol{A}^{\mathrm{T}}(\theta) + \boldsymbol{R}_1(\theta) - \left(\boldsymbol{A}(\theta)\boldsymbol{P}(\theta)\boldsymbol{C}^{\mathrm{T}}(\theta) + \boldsymbol{R}_{12}(\theta)\right)$$
$$\times \left(\boldsymbol{C}(\theta)\boldsymbol{P}(\theta)\boldsymbol{C}^{\mathrm{T}}(\boldsymbol{\theta}) + \boldsymbol{R}_2(\theta)\right)^{-1}\left(\boldsymbol{C}(\theta)\boldsymbol{P}(\theta)\boldsymbol{A}^{\mathrm{T}}(\theta) + \boldsymbol{R}_{12}^{\mathrm{T}}(\theta)\right)$$

Spectral factorization theory can be used for the derivation of the above expressions (Söderström and Stoica, 1988).

Now let us consider the ARX model given by equations (3.9). This is the form most real–world applications use because of the existence of well behaved estimation methods. Then the one–shot least squares parameter estimate is given by:

$$\hat{\theta}(t) = \left[\sum_{s=1}^{t}\boldsymbol{\varphi}(s)\boldsymbol{\varphi}^{\mathrm{T}}(s)\right]^{-1}\left[\sum_{s=1}^{t}\boldsymbol{\varphi}(s)y(s)\right] \tag{3.11}$$

The argument t has been used to stress the dependence of $\hat{\theta}$ on time. The expression (3.11) can be computed in a recursive fashion. Introduce the notation:

$$\boldsymbol{P}(t) = \left[\sum_{s=1}^{t}\boldsymbol{\varphi}(s)\boldsymbol{\varphi}^{\mathrm{T}}(s)\right]^{-1} \tag{3.12}$$

Since trivially,

$$\boldsymbol{P}^{-1}(t) = \boldsymbol{P}^{-1}(t-1) + \boldsymbol{\varphi}(t)\boldsymbol{\varphi}^{\mathrm{T}}(t) \tag{3.13}$$

it follows that,

$$\hat{\boldsymbol{\theta}}(t) = \boldsymbol{P}(t)\left[\left(\sum_{s=1}^{t-1}\boldsymbol{\varphi}(s)y(s)\right) + \boldsymbol{\varphi}(t)y(t)\right]$$
$$= \boldsymbol{P}(t)\left(\boldsymbol{P}^{-1}(t-1)\hat{\boldsymbol{\theta}}(t-1) + \boldsymbol{\varphi}(t)y(t)\right)$$
$$= \hat{\boldsymbol{\theta}}(t-1) + \boldsymbol{P}(t)\boldsymbol{\varphi}(t)\left(y(t) - \boldsymbol{\varphi}^{\mathrm{T}}(t)\hat{\boldsymbol{\theta}}(t-1)\right)$$

Thus,

$$\hat{\boldsymbol{\theta}}(t) = \hat{\boldsymbol{\theta}}(t-1) + k(t)\varepsilon(t) \tag{3.14a}$$

$$\boldsymbol{k}(t) = \boldsymbol{P}(t)\boldsymbol{\varphi}(t) \tag{3.14b}$$

$$\varepsilon(t) = y(t) - \boldsymbol{\varphi}^{\mathrm{T}}(t)\hat{\boldsymbol{\theta}}(t-1) \tag{3.14c}$$

Here the term $\varepsilon(t)$ should be interpreted as a prediction error. It is the difference between the measured output $y(t)$ and the one–step–ahead prediction

$$\hat{y}\big(t|t-1;\hat{\theta}(t-1)\big)=\boldsymbol{\varphi}^{\mathrm{T}}(t)\hat{\boldsymbol{\theta}}(t-1)$$

of $y(t)$ made at time t–1 based on the model corresponding to the estimate $\hat{\theta}(t-1)$. If $\varepsilon(t)$ is small, the estimate $\hat{\theta}(t-1)$ is "good" and should not be modified very much. The vector $\boldsymbol{k}(t)$ in (3.14b) should be interpreted as a weighting or gain factor showing how much the value of $\varepsilon(t)$ will modify the different elements of the parameter vector.

To complete the algorithm, (3.13) must be used to compute $P(t)$ which is needed in (3.14b). However, the use of (3.13) needs a matrix inversion at each time step. This would be a time–consuming procedure. Using the matrix inversion lemma (Gantmacher, 1977) however, (3.13) can be rewritten in a more useful form. Then an updating equation for P(t) is obtained, namely,

$$\boldsymbol{P}(t)=\boldsymbol{P}(t-1)=\frac{\boldsymbol{P}(t-1)\boldsymbol{\varphi}(t)\boldsymbol{\varphi}^{T}(t)\boldsymbol{P}(t)}{1+\boldsymbol{\varphi}^{T}(t)\boldsymbol{P}(t-1)\boldsymbol{\varphi}(t)} \tag{3.15}$$

Note that in (3.15) there is now a scalar inversion instead of a matrix inversion. The algorithm consisting of (3.14)–(3.15) can be simplified further. From (3.14b),

$$\boldsymbol{k}(t)=\boldsymbol{P}(t-1)\boldsymbol{\varphi}(t)-\boldsymbol{P}(t-1)\boldsymbol{\varphi}(t)\boldsymbol{\varphi}^{\mathrm{T}}(t)\boldsymbol{P}(\mathrm{t}-1)\boldsymbol{\varphi}(t)/\big(1+\boldsymbol{\varphi}^{\mathrm{T}}(t)\boldsymbol{P}(t-1)\boldsymbol{\varphi}(t)\big)$$

$$=\frac{\boldsymbol{P}(t-1)\boldsymbol{\varphi}(t)}{1+\boldsymbol{\varphi}^{\mathrm{T}}(t)\boldsymbol{P}(t-1)\boldsymbol{\varphi}(t)} \tag{3.16}$$

This form for $\boldsymbol{k}(t)$ is more convenient to use for implementation than (3.14b). The reason is that the right–hand side of (3.16) must anyway be computed in the updating of $\boldsymbol{P}(t)$ (see 3.15). Further improvements for numerical implementations will be discussed later in this chapter.

The derivation of the ***Recursive Least Squares*** (*RLS*) method is now complete. The RLS algorithm consists of (3.16), (3.15), (3.14c) and (3.14a), which are now given in algorithmic order as:

$$k(t)=\frac{\boldsymbol{P}(t-1)\boldsymbol{\varphi}(t)}{1+\boldsymbol{\varphi}^{\mathrm{T}}(t)\boldsymbol{P}(t-1)\boldsymbol{\varphi}(t)} \tag{3.16}$$

$$\boldsymbol{P}(t)=\boldsymbol{P}(t-1)-\boldsymbol{k}(t)\boldsymbol{\varphi}^{\mathrm{T}}(t)\boldsymbol{P}(t-1) \tag{3.15}$$

$$\varepsilon(t)=y(t)-\boldsymbol{\varphi}^{\mathrm{T}}(t)\hat{\boldsymbol{\theta}}(t-1) \tag{3.14c}$$

$$\hat{\boldsymbol{\theta}}(t)=\hat{\boldsymbol{\theta}}(t-1)+\boldsymbol{k}(t)\varepsilon(t) \tag{3.14a}$$

The algorithm also needs initial values $\hat{\theta}(0)$ and $\boldsymbol{P}(0)$. These can be either provided from knowledge of system characteristics or calculated from an initial sample using

(3.11) and starting the recursion later.

As is well known the resulting estimate minimises,

$$V_N(\boldsymbol{\theta}) = \sum_{t=1}^{N} \varepsilon^2(t)$$

Also, under mild conditions the LS estimate is consistent, i.e. $\hat{\theta}$ tends to $\boldsymbol{\theta}$ as N tends to infinity, if,

$$E\{\boldsymbol{\varphi}(t)\boldsymbol{\varphi}^T(t)\} \text{ is non singular} \tag{3.17}$$

$$E\{\boldsymbol{\varphi}(t)\varepsilon(t)\} = \mathbf{0} \tag{3.18}$$

Condition (3.17) is usually satisfied. A common cause of singularity is the condition of *non–persistent excitation* of order n_b of the input. Remedies for this irregularity will be discussed later. Condition (3.18) is usually not satisfied unless $\varepsilon(t)$ is white, an assumption usually made for most systems. However it must be stressed that violation of (3.18) will render the whole fault detection scheme based on LS parameter estimation invalid. In such cases the designer should resort to methods circumventing this problem as for example *instrumental variable* methods (see Söderström and Stoica, 1988). Even small biases should not be tolerated since these may trigger false alarms in sensitive detectors.

There are several approaches for modifying the recursive LS algorithm to make it suitable as a real–time fault detection method:

- Use of a forgetting factor.
- Use of a Kalman filter as a parameter estimator.
- Use of sliding windows of data.

3.3.2 *Forgetting factors*

The approach in this case is to change the loss function to be minimized. Let the modified loss function be,

$$V_t(\boldsymbol{\theta}) = \sum_{s=1}^{t} \lambda^{t-s} \varepsilon^2(s) \tag{3.19}$$

The loss funtion used earlier had λ=1 but now the forgetting factor λ is a number somewhat less than 1 (for example 0.99 or 0.95). This means that with increasing t, the measurements obtained previously are discounted. The smaller the value of λ, the quicker the information in previous data will be forgotten. One can rederive the RLS method for the modified criterion (3.19). The calculations are straightforward. The recursive LS method with a forgetting factor is:

$$
\begin{aligned}
\boldsymbol{k}(t) &= \frac{\boldsymbol{P}(t-1)\boldsymbol{\varphi}(t)}{\lambda + \boldsymbol{\varphi}^{\mathrm{T}}(t)\boldsymbol{P}(t-1)\boldsymbol{\varphi}(t)} \\
\hat{\boldsymbol{\theta}}(t) &= \hat{\boldsymbol{\theta}}(t-1) + k(t)\left(y(t) - \boldsymbol{\varphi}^{\mathrm{T}}(t)\hat{\boldsymbol{\theta}}(t-1)\right) \\
\boldsymbol{P}(t) &= \frac{1}{\lambda}\left(\boldsymbol{P}(t-1) - \boldsymbol{k}(t)\boldsymbol{\varphi}^{\mathrm{T}}(t)\boldsymbol{P}(t-1)\right)
\end{aligned} \tag{3.20}
$$

Equations (3.20) are often referred to as the Recursive Weighted Least Squares (RWLS) identification method or Decreasing Gain Least squares (DGLS) method. Experiences with this simple rule for setting λ show that a decrease in the value of the forgetting factor leads to two effects:

(1) The parameter estimates converge to their true values quicker, thus decreasing the fault alarm delay time, t_d.
(2) But at the expense of increased sensitivity to noise. If λ is much less than 1 the estimates may even oscillate around their true values.

There are various ways around this problem:

Time–varying forgetting factor: In this method the constant λ in (3.20) is replaced by $\lambda(t)$. A typical choice is an exponential given by,

$$\lambda(t) = 1 - \lambda_0^t + \left(1 - \lambda(0)\right)$$

or recursively,

$$\lambda(t) = \lambda_0\lambda(t-1) + \left(1 - \lambda(0)\right) \tag{3.21}$$

Typical design values for λ_0 and $\lambda(0)$ are 0.99 and 0.95 respectively.

Equations (3.20) with (3.21) in place of λ minimise the quadratic cost function:

$$V_t(\boldsymbol{\theta}) = \sum_{s=1}^{t} \lambda(s)\varepsilon^2(s)$$

A more general class of variable weights is described by the relation,

$$w(\tau) = \mu(\tau)\prod_{i=\tau}^{t} \lambda(i) \geq 0 \tag{3.22}$$

where $\mu(\tau)$ is a multiplicative factor reflecting the quality of measurement $y(\tau)$ and $\lambda(\tau)$ is a time–varying forgetting factor given by (3.21). Minimising,

$$V_t(\boldsymbol{\theta}) = \sum_{s=1}^{t} \lambda(s)\varepsilon^2(s) \tag{3.23}$$

results in the following equations,

$$\sigma(t) = \frac{1}{\mu(t)} + \boldsymbol{\varphi}^{\mathrm{T}}(t)\boldsymbol{P}(t-1)\boldsymbol{\varphi}(t)$$
$$\boldsymbol{k}(t) = \boldsymbol{P}(t-1)\boldsymbol{\varphi}(t)\sigma^{-1}(t)$$
$$\boldsymbol{P}(t) = \frac{1}{\lambda(t)}\left(\boldsymbol{P}(t-1) - \boldsymbol{k}(t)\sigma(t)\boldsymbol{k}^{\mathrm{T}}(t)\right)$$
$$\hat{\boldsymbol{\theta}}(t) = \hat{\boldsymbol{\theta}}(t-1) + \boldsymbol{k}(t)\left(y(t) - \boldsymbol{\varphi}^{\mathrm{T}}(t)\hat{\boldsymbol{\theta}}(t-1)\right) \tag{3.24}$$

Constant Trace. In the case of abruptly changing systems, the tracking capability and consequently the fast response to parameter changes can be maintained by using the forgetting factor to keep the trace of $\boldsymbol{P}$ constant. This idea results in the recursive *Constant Trace Least Squares* (*CTLS*) algorithm (Shibata *et al*., 1988) implemented by the following set of equations:

$$\hat{\boldsymbol{\theta}}(i) = \hat{\boldsymbol{\theta}}(i-1) - \boldsymbol{P}(i-1)\boldsymbol{\varphi}(i-1)\varepsilon(i)$$
$$\varepsilon(i) = \left(\hat{\boldsymbol{\theta}}^{\mathrm{T}}(i-1)\boldsymbol{\varphi}(i) - y(i)\right)\alpha(i)$$
$$\boldsymbol{P}(i) = \lambda^{-1}(i)\left(\boldsymbol{P}(i-1) - \boldsymbol{P}(i-1)\boldsymbol{\varphi}(i)\boldsymbol{\varphi}^{\mathrm{T}}(i)\boldsymbol{P}(i-1)\alpha(i)\right)$$
$$\alpha(i) = \left(1 + \boldsymbol{\varphi}^{\mathrm{T}}(i)\boldsymbol{P}(i-1)\boldsymbol{\varphi}(i)\right)^{-1}$$
$$\lambda(i) = 1 - \alpha(i)\left(\boldsymbol{P}(i-1)\boldsymbol{\varphi}(i)\right)^{2} / \operatorname{tr}\boldsymbol{P}(0) \tag{3.25}$$

Here $\hat{\boldsymbol{\theta}}(0)$ and $\boldsymbol{P}(0)$ must be defined. This method eliminates the estimator *wind–up* problem which occurs when a constant forgetting factor is used and provides rapid convergence after the onset of a parameter change.

Kalman filters: Assuming that the parameters are constant, the underlying model

$$y(t) = \boldsymbol{\varphi}^{\mathrm{T}}\boldsymbol{\theta}(t) + e(t)$$

can be described as a state space equation,

$$\boldsymbol{x}(t+1) = \boldsymbol{x}(t) \tag{3.26}$$
$$y(t) = \boldsymbol{\varphi}^{\mathrm{T}}(t)\boldsymbol{x}(t) + e(t) \tag{3.27}$$

where the "state vector" $\boldsymbol{x}(t)$ is given by,

$$\boldsymbol{x}(t) = [\ a_1 \ \ldots \ a_{n_a} \ \ b_1 \ \ldots \ b_{n_b}\ ]^{\mathrm{T}} = \boldsymbol{\theta} \tag{3.28}$$

The optimal state estimate $\hat{x}(t+1)$ can be computed as a function of the measurements till time t using the Kalman filter. Note that usually the Kalman filter is presented for state space equations whose matrices may be time varying but do not depend on the data. The latter condition fails in the case of (3.27) since $\boldsymbol{\varphi}(t)$ depends on data up to (and inclusive of) time (t–1). However, it can be shown that also in such cases the Kalman filter provides the optimal (mean square) estimate of the system state vector (Åström,

1971).

Applying the Kalman filter to the state model (3.26) will give precisely the basic recursive LS algorithm. One way of modifying the algorithm so that time–varying parameters can be tracked better is to change the state equation (3.26) to

$$\boldsymbol{x}(t+1) = \boldsymbol{x}(t) + \boldsymbol{v}(t);\ E\{\boldsymbol{v}(t)\boldsymbol{v}^T(s)\} = \boldsymbol{R}_1\delta_{t,s} \tag{3.29}$$

This means that the parameter vector is modeled as a random walk or a drift. The covariance matrix $\boldsymbol{R}_1$ can be used to describe how fast the different components of $\boldsymbol{\theta}$ are expected to vary. Applying the Kalman filter to the model (3.29), (3.27) gives the following recursive algorithm:

$$\boldsymbol{k}(t) = \frac{\boldsymbol{P}(t-1)\boldsymbol{\varphi}(t)}{1+\boldsymbol{\varphi}^{\mathrm{T}}(t)\boldsymbol{P}(t-1)\boldsymbol{\varphi}(t)}$$

$$\boldsymbol{P}(t) = \left\{\boldsymbol{P}(t-1) - \boldsymbol{k}(t)\boldsymbol{\varphi}^{\mathrm{T}}(t)\boldsymbol{P}(t-1)\right\} + \boldsymbol{R}_1$$

$$\hat{\boldsymbol{\theta}}(t) = \hat{\boldsymbol{\theta}}(t-1) + \boldsymbol{k}(t)\left(y(t) - \boldsymbol{\varphi}^{\mathrm{T}}(t)\hat{\boldsymbol{\theta}}(t-1)\right) \tag{3.30}$$

Observe that for both algorithms (3.20) and (3.30) the basic method has been modified so that $\boldsymbol{P}(t)$ will no longer tend to zero. In this way $\boldsymbol{k}(t)$ also is prevented from decreasing to zero. The parameter estimates will therefore change continually.

In the algorithm (3.20) $\boldsymbol{R}_1$ has a role similar to that of λ in (3.20). These design variables should be chosen by a trade–off between fast detection (which requires λ "small" or $\boldsymbol{R}_1$ "large") on the one hand and reliability on the other (which requires λ close to 1 or $\boldsymbol{R}_1$ "small"). This trade–off may be resolved by fault simulation.

The Kalman filter interpretation of the RLS algorithm is also useful in another respect. It provides suggestions for the choice of the initial values $\hat{\boldsymbol{\theta}}(0)$ and $\boldsymbol{P}(0)$. These values are necessary to start the algorithm. Since $\boldsymbol{P}(t)$ (times λ^2) is the covariance matrix of $\hat{\boldsymbol{\theta}}(t)$ it is reasonable to take for $\hat{\boldsymbol{\theta}}(0)$ an a priori estimate of $\boldsymbol{\theta}$ and to let $\boldsymbol{P}(0)$ reflect the confidence in this initial estimate $\hat{\boldsymbol{\theta}}(0)$. If $\boldsymbol{P}(0)$ is small then $\boldsymbol{k}(t)$ will be small for all t and the parameter estimates will therefore not change too much from $\hat{\boldsymbol{\theta}}(0)$. On the other hand, if $\boldsymbol{P}(0)$ is large, the parameter estimates will quickly jump away from $\hat{\boldsymbol{\theta}}(0)$. Without any a priori information it is common practive to take,

$$\hat{\boldsymbol{\theta}}(0) = \boldsymbol{0};\quad P(0) = \sigma\boldsymbol{I}$$

where σ is a "large" number.

Increased flexibility in the choice of design parameters can be achieved if additionally to (3.29) one assumes,

$$E\{e(t)e(s)\} = r_2(t)\delta_{ts}$$

which results in the modified set of updating equations,

$$\hat{\boldsymbol{\theta}}(t) = \hat{\boldsymbol{\theta}}(t-1) + \boldsymbol{k}(t)\left(y(t) - \boldsymbol{\varphi}^{\mathrm{T}}(t)\hat{\boldsymbol{\theta}}(t-1)\right)$$

$$\boldsymbol{k}(t) = \frac{\boldsymbol{P}(t-1)\boldsymbol{\varphi}(t)}{r(t) + \boldsymbol{\varphi}^{\mathrm{T}}(t)\boldsymbol{P}(t-1)\boldsymbol{\varphi}(t)}$$

$$\boldsymbol{P}(t) = \boldsymbol{P}(t-1) - \frac{\boldsymbol{P}(t-1)\boldsymbol{\varphi}(t)\boldsymbol{\varphi}^{\mathrm{T}}(t)\boldsymbol{P}(t-1)}{r_2(t) + \boldsymbol{\varphi}^{\mathrm{T}}(t)\boldsymbol{P}(t-1)\boldsymbol{\varphi}(t)} + \boldsymbol{R}_1(t) \qquad (3.30a)$$

In this context $r_2(t)$ describes the confidence of the incoming measurements.

Recursive sliding window estimators. This approach utilizes rectangular sliding windows of length n_w, thus using only information contained in the last n_w samples. In this way similar behaviour to that of the forgetting factor approach is obtained. A recursive version of a least–squares sliding window algorithm is described by Stavrakakis and Pouliezos (1991):

Consider as usual the process described by,

$$y(k) = \boldsymbol{\varphi}^{\mathrm{T}}(k)\boldsymbol{\theta} + e(k);\ \dim(\boldsymbol{\theta})=n \qquad (3.31)$$

and define,

$$\boldsymbol{\Phi} = \begin{bmatrix} \varphi_1(1) & \dots & \varphi_n(1) \\ \cdot & & \cdot \\ \cdot & & \cdot \\ \varphi_1(k) & \dots & \varphi_n(k) \end{bmatrix} = \begin{bmatrix} \boldsymbol{\varphi}^{\mathrm{T}}(1) \\ \cdot \\ \cdot \\ \boldsymbol{\varphi}^{\mathrm{T}}(k) \end{bmatrix} \qquad (3.32)$$

and,

$$\boldsymbol{y} = [\, y(1) \dots y(\mathrm{k}) \,]^{\mathrm{T}}$$

Furthermore, for a moving window of length n_w, define,

$$\boldsymbol{\Phi}_k = \begin{bmatrix} \boldsymbol{\varphi}^{\mathrm{T}}(k - n_w + 1 \\ \cdot \\ \cdot \\ \cdot \\ \boldsymbol{\varphi}^{\mathrm{T}}(k) \end{bmatrix} = \begin{bmatrix} \boldsymbol{\varphi}^{\mathrm{T}}(k - n_w + 1) \\ \dots\dots\dots\dots \\ \boldsymbol{\Phi}(k, k - n_w + 2) \end{bmatrix} \qquad (3.33)$$

Then as shown in Appendix 3.A,

$$\hat{\boldsymbol{\theta}}(k+1) = \hat{\boldsymbol{\theta}}(k) - \boldsymbol{P}(k+1)\left[\boldsymbol{\Gamma}(k+1)\hat{\boldsymbol{\theta}}(k) - \boldsymbol{\delta}(k+1)\right]$$

$$\boldsymbol{P}^{-1}(k+1) = \boldsymbol{P}^{-1}(k) + \boldsymbol{\Gamma}(k+1)$$

where,

$$\boldsymbol{\Gamma}(k+1) = \boldsymbol{\varphi}(k+1)\boldsymbol{\varphi}^{\mathrm{T}}(k+1) - \boldsymbol{\varphi}(k - n_w + 1)\boldsymbol{\varphi}^{\mathrm{T}}(k - n_w + 1)$$

$$\boldsymbol{\delta}(k+1) = \boldsymbol{\varphi}(k+1)y(k+1) - \boldsymbol{\varphi}(k-n_w+1)y(k-n_w+1) \quad (3.34)$$

It should be remembered that $\hat{\theta}(k)$ is estimated using information from the last n_w samples. Equations (3.34) form the sliding window least squares estimator (SWLSE). Note that in this simple case a further reduction of $\boldsymbol{P}^{-1}$ is not needed since only one inversion is required. The reduction in speed is proportional to the length of the window since the dimensions of $\boldsymbol{P}$, $\boldsymbol{\Gamma}$ and $\boldsymbol{\delta}$ are independent of the window size.

The improvement in speed over the classical batch sliding window LSE is shown in the operations count table for the scalar case in Table 3.1. No special methods for better performance of individual operations (matrix inversion) are taken into account, since these would apply equally well to both cases. It should be noted however that memory requirements are not reduced, since at any one time all the window values must be accessible. The scalar case considered may serve as a guideline for speed improvement in the vector versions.

Table 3.1 Operations count for window size n_w (scalar output case).

Recursive version			Batch		
	Additions	Multiplies		Additions	Multiplies
Estimate updating	$3n^2$	$4n^2+2n$	$\boldsymbol{\Phi}^T\boldsymbol{\Phi}$	$(n_w-1)n$	$n_w n^2$
Covariance updatimg	n^2	0	$[\boldsymbol{\Phi}^T\boldsymbol{\Phi}]^{-1}\boldsymbol{\Phi}^T\boldsymbol{y}$	$n^2+(n_w-2)n$	$n^2+n_w n$
Total	$4n^2$	$4n^2+2n$		$n^2(n_w+1)+nn_w$	$n^2+2n_w n-3n$

3.3.3 Implementation issues

3.3.3.1 Covariance instability

In all the preceding algorithms it is necessary to recursively calculate the estimate's covariance matrix. Let us consider the updating formula (3.20),

$$\boldsymbol{P}(t) = \frac{1}{\lambda(t)}\left(\boldsymbol{P}(t-1) + \frac{\boldsymbol{P}(t-1)\boldsymbol{\varphi}(t)\boldsymbol{\varphi}^T(t)\boldsymbol{P}(t-1)}{\lambda(t)+\boldsymbol{\varphi}^T(t)\boldsymbol{P}(t-1)\boldsymbol{\varphi}(t)}\right)$$

where $\lambda(t)$ is generated by any of the previous methods.

Since $\boldsymbol{P}(t)$ is interpreted as a covariance matrix it should be positive definite. However computational rounding errors may accumulate and make the computed $\boldsymbol{P}(t)$ non–positive definite leading to numerical instability problems which manifest themselves as divergence of the parameter estimates. To avoid this problem, factorization algorithms can be used to update factors of $\boldsymbol{P}(t)$ in square–root or U–D form.

Square root algorithms. A square root algorithm is based on the following decomposition:

$$\boldsymbol{P}(t) = \boldsymbol{Q}(t)\boldsymbol{Q}^{\mathrm{T}}(t) \tag{3.35}$$

where $\boldsymbol{Q}(t)$ is a non–singular matrix. Potter, (1963), proposes the following square root algorithm for $\boldsymbol{P}(t)$:

$$\begin{aligned}
\boldsymbol{f}(t) &= \boldsymbol{Q}^{\mathrm{T}}(t-1)\boldsymbol{\varphi}(t) \\
\beta(t) &= \lambda(t) + \boldsymbol{f}^{\mathrm{T}}(t)\boldsymbol{f}(t) \\
\alpha(t) &= \frac{1}{\beta(t) + \sqrt{(\beta(t)\lambda(t))}} \\
\bar{\boldsymbol{k}}(t) &= \boldsymbol{Q}(t-1)\boldsymbol{f}(t) \\
\boldsymbol{Q}(t) &= \left(\boldsymbol{Q}(t-1) - \alpha(t)\bar{\boldsymbol{k}}(t)\boldsymbol{f}^{\mathrm{T}}(t)\right)\sqrt{\lambda(t)}
\end{aligned} \tag{3.36}$$

The algorithm is initialised by,

$$\boldsymbol{Q}(0)\boldsymbol{Q}^{\mathrm{T}}(0) = \boldsymbol{P}(0)$$

The quantity $\bar{\boldsymbol{k}}(t)$ is a normalized form of the gain vector since,

$$\boldsymbol{k}(t) = \bar{\boldsymbol{k}}(t) / \beta(t)$$

As an added bonus $\boldsymbol{k}(t)$ is not necessary to be computed since,

$$\hat{\boldsymbol{\theta}}(t) = \hat{\boldsymbol{\theta}}(t-1) + \bar{\boldsymbol{k}}(t)\left(\varepsilon(t)/\beta(t)\right)$$

i.e. the single division $\varepsilon(t)/\beta(t)$ is computed first. If the Kalman filter implementation (3.30) is used then,

$$\boldsymbol{P}(t) = \bar{\boldsymbol{P}}(t) + \boldsymbol{R}_1(t) \tag{3.37}$$

where $\bar{\boldsymbol{P}}(t)$ is given by (3.20) with $\lambda(t)=1$.

If a square root approach is taken, the algorithm (3.36) can be used for finding $\bar{\boldsymbol{P}}(t) = \bar{\boldsymbol{Q}}(t)\bar{\boldsymbol{Q}}(t)$ from $\boldsymbol{P}(t-1)=\boldsymbol{Q}(t-1)\boldsymbol{Q}^{\mathrm{T}}(t-1)$. It then remains to find $\boldsymbol{Q}(t)$ using (3.37). One way to do this is as follows: let $\boldsymbol{R}_1(t)$ be factored as,

$$\boldsymbol{R}_1(t) = \boldsymbol{V}(t)\boldsymbol{V}^{\mathrm{T}}(t)$$

where $\boldsymbol{V}(t)$ is a $(n \times s)$ matrix of full rank (recall that $n=\dim\boldsymbol{\theta}$). In most cases $\boldsymbol{R}_1(t)$ is a diagonal matrix with some diagonal elements equal to zero. In such cases it is easy to find $V(t)$. Then orthogonal transformations are applied to the rectangular matrix ($Q(t)$ $V(t)$). The problem is to find an orthogonal matrix $T(t)$ and a triangular matrix $Q(t)$ such that,

$$[\, Q(t) \mid V(t)\,]\, T(t) = [\, Q(t) \mid \mathbf{0}\,] \tag{3.38}$$

Then one has,

$$\begin{aligned}\bar{P}(t) + R_1(t) &= \bar{Q}(t)\bar{Q}^{\mathrm{T}}(t) + V(t)V^{\mathrm{T}}(t) \\ &= \left[\bar{Q}(t) \mid V(t)\right]T(t)T^{\mathrm{T}}(t)\begin{bmatrix}\bar{Q}^{\mathrm{T}}(t) \\ C^{\mathrm{T}}(t)\end{bmatrix} \\ &= \left[\bar{Q}(t) \mid \mathbf{0}\right]\begin{bmatrix}\bar{Q}^{\mathrm{T}}(t) \\ \mathbf{0}\end{bmatrix} = P(t),\end{aligned}$$

as required by (3.37). The matrices $T(t)$ and $Q(t)$ in (3.38) can be found using a *QR factorization* or a *Gram–Schmidt orthogonalization*. Such factorizations are common in numerical linear algebra for solving certain eigenvalue and least squares problems, and have appeared in many applications. An efficient procedure for a Gram–Schmidt orthogonalization is given in Appendix 3.B.

U–D factorisation algorithms: A U–D factorisation algorithm is based on the decomposition,

$$P(t) = U(t)D(t)U^{\mathrm{T}}(t) \tag{3.39}$$

where $U(t)$ is unit upper triangular and $D(t)$ diagonal. Following Bierman (1977), consider equation (3.20) again. Then the following algorithm produces $P(t)$ in U–D form:

- At time t, compute $k(t)$ and update $U(t-1)$ and $D(t-1)$ by performing steps 1–6.

1. Compute $f = U^{\mathrm{T}}(t-1)\varphi(t)$, $g = D(t-1)f$, $\beta_0 = \lambda(t)$.
2. For $j=1, \ldots, n$ go through steps 3–5 (subscripts denote matrix/vector elements).

   3. Compute:

$$\begin{aligned}\beta_j &= \beta_{j-1} + f_j g_j \\ D_{jj}(t) &= \beta_{j-1} D_{jj}(t-1) / \beta_j \lambda(t) \\ v_j &= g_j \\ \mu_j &= -f_j / \beta_{j-1}\end{aligned}$$

   4. For $i = 1, \ldots, j-1$, go through step 5. (If $j=1$, skip step 5).

5. Compute,

$$U_{ij}(t) = U_{ij}(t-1) + v_i \mu_j$$
$$v_i = v_i + U_{ij}(t-1) v_j$$

6. Compute,

$$\bar{\boldsymbol{k}}(t) = \begin{bmatrix} u_1 \\ \cdot \\ \cdot \\ u_d \end{bmatrix} \; ; \; \boldsymbol{k}(t) = \bar{\boldsymbol{k}}(t) / \beta_d$$

The scalar β_d obtained after the dth cycle of steps 3–5 is the *innovations variance*,

$$\beta_d = \lambda(t) + \boldsymbol{\varphi}^{\mathrm{T}}(t)\boldsymbol{P}(t-1)\boldsymbol{\varphi}(t).$$

The algorithm is initialised by $\boldsymbol{U}(0)\boldsymbol{D}(0)\boldsymbol{U}^{\mathrm{T}}(0) = \boldsymbol{P}(0)$

The U–D analogue of the Kalman filter updating given by (3.30) is discussed in Thornton and Bierman (1977) and consists of the following equations:

At time $t-1$, $\boldsymbol{U}(t-1)$ and $\boldsymbol{D}(t-1)$ are given, as well as the factorisation $\boldsymbol{R}_1(t)=\boldsymbol{V}(t)\boldsymbol{V}^{\mathrm{T}}(t)$ with $\boldsymbol{V}(t)$ a full rank $(n{\times}s)$ matrix.

1. Compute $\bar{\boldsymbol{k}}(t), \bar{\boldsymbol{U}}(t)$ and $\bar{\boldsymbol{D}}(t)$ by performing steps 1–6 of (3.40) ($\bar{\boldsymbol{U}}(t)$ and $\bar{\boldsymbol{D}}(t)$ are the matrices called $\boldsymbol{U}(t)$ and $\boldsymbol{D}(t)$ in (3.40)).
2. Define the $(n+s)$–column vector $\boldsymbol{W}_k^{(0)}$ as the kth column of $\boldsymbol{U}^{\mathrm{T}}(t)$ stacked on top of the kth column of $\boldsymbol{V}^{T}(t)$; $k=1, ..., d$.
3. Define the $(n+s)\times(n+s)$ diagonal matrix $\boldsymbol{D}$ as the block diagonal matrix formed from $\bar{\boldsymbol{D}}(t)$ and the $s{\times}s$ identity matrix.
4. For $j = n, n-1, ..., 2$ go through steps 5–8.
   5. Compute,

   $$\boldsymbol{U}_{ij}(t) = [\, w_j^{(d-j)} \,]^{\mathrm{T}} \boldsymbol{D}\, w_j^{(d-j)}$$

   6. For $i = 1, 2, ..., j-1$ go through step 7.
      7. Compute,

      $$U(t)_{ij} = [\, w_i^{(d-j)} \,]^{\mathrm{T}} \boldsymbol{D}\, w_j^{(d-j)} / D_{jj}(t)$$
      $$w_i^{(d-j+1)} = w_i^{(d-j)} - U_{ij}(t) w_j^{(d-j)}$$

   8. Compute,

   $$D_{11}(t) = [\, w_1^{(d-1)} \,]^{\mathrm{T}} \boldsymbol{D}\, w_1^{(d-1)}.$$

3.3.3.2 *Covariance singularity*

Identifiability conditions dictate that the monitored system be persistently excited in order for the covariance matrix to be nonsingular. However, there are cases when $\boldsymbol{\varphi}(t)$ becomes constant and in such cases a procedure of regularization must be employed. This happens if essentially the prediction is unaffected by changing certain linear combinations of the model parameters. This in turn implies that either the model contains too many parameters or the input signal is not general enough. The latter cause may be present at some time in every process, therefore remedies must be applied. Pot *et al.* (1984), proposed the following modification to the covariance updating equation (3.24):

1. Compute $\boldsymbol{P}'(t)$ defined as,

$$\boldsymbol{P}'(t) = \boldsymbol{P}(t-1) - \boldsymbol{k}(t)\sigma(t)\boldsymbol{k}^{\mathrm{T}}(t) \tag{3.40}$$

2. Choose the forgetting factor such that,

   - If $\mathrm{tr}\boldsymbol{P}'(t) \geqq tr_0$ then $\lambda(t)=1$, $\boldsymbol{P}(t)=\boldsymbol{P}'(t)$
   - If $\mathrm{tr}\boldsymbol{P}'(t) < tr_0$ then $\lambda(t) = \dfrac{\mathrm{tr}P''(t)}{tr_0}$, $\boldsymbol{P}(t) = \dfrac{1}{\lambda(t)}\boldsymbol{P}''(t)$

with,

$$\boldsymbol{P}''(t) = \boldsymbol{P}'(t) + \sigma(\,\mathrm{Diag}\,\boldsymbol{P}'(t) + \mathrm{tr}\boldsymbol{P}'(t)\boldsymbol{I}\,) \tag{3.41}$$

This algorithm possesses the additional property of keeping the trace of $\boldsymbol{P}(.)$ equal to tr_0. Note that the two last terms of the righ–hand side of (3.41) are introduced to prevent the eigenvalues of $\boldsymbol{P}(t)$ from becoming too small. Favier *et al.* (1988) computed the $\boldsymbol{P}'(t)$ matrix defined by (3.41) in U–D factorized form.

A similar problem might arise if a Kalman filter is used based on equations (3.30) or (3.30a). One way around this problem is to model the parameter variations as:

$$\boldsymbol{\theta}(t+1) = (1-\alpha)\,\boldsymbol{\theta}(t) + \alpha\boldsymbol{\theta}_p(t) + \boldsymbol{n}(t),\ 0\le\alpha<1 \tag{3.42}$$

where $\boldsymbol{\theta}_p(t)$ represents an a priori information on $\boldsymbol{\theta}(t)$. The Kalman filter built on the model (3.27) and (3.42) is given by,

$$\sigma(t) = \frac{1}{\mu(t)} + \boldsymbol{\varphi}^{\mathrm{T}}(t)\boldsymbol{P}(t-1)\boldsymbol{\varphi}(t)$$

$$\boldsymbol{k}(t) = (1-\alpha)\boldsymbol{P}(t-1)\boldsymbol{\varphi}(t)\sigma^{-1}(t)$$

$$\boldsymbol{P}(t) = (1-\alpha)^2\boldsymbol{P}(t-1) - \boldsymbol{k}(t)\sigma(t)\boldsymbol{k}^{\mathrm{T}}(t) + \boldsymbol{R}_1(t)$$

$$\hat{\boldsymbol{\theta}}(t) = (1-\alpha)\hat{\boldsymbol{\theta}}(t-1) + \alpha\boldsymbol{\theta}_p(t-1) + \boldsymbol{k}(t)\big(y(t) - \boldsymbol{\varphi}^{\mathrm{T}}(t)\hat{\boldsymbol{\theta}}(t-1)\big) \tag{3.43}$$

When $\boldsymbol{\varphi}(t)$ tends to zero, $\hat{\theta}(t)$ tends to $\boldsymbol{\theta}_p(t)$ and $\boldsymbol{P}(t)$ tends to $(2\alpha-\alpha^2)^{-1}\boldsymbol{R}_1(t)$. Favier *et al.* (1988), proposed a *U–D* factorised version of (3.43) consisting of the following steps:

Let $\boldsymbol{P}(t)=\boldsymbol{P}=\boldsymbol{UDU}^{\mathrm{T}}$, $\overline{\boldsymbol{P}} = \boldsymbol{P}(t-1) = \overline{\boldsymbol{UDU}}^{\mathrm{T}}$.

1. Form the matrices,

$$\boldsymbol{U}_a = \begin{bmatrix} 1 & \boldsymbol{\varphi}^{\mathrm{T}}\overline{\boldsymbol{U}} & \boldsymbol{0} \\ 0 & (1-\alpha)\boldsymbol{U} & \boldsymbol{I} \end{bmatrix}, \boldsymbol{D}_a = \begin{bmatrix} 1/\mu & 0 & 0 \\ 0 & \overline{\boldsymbol{D}} & 0 \\ 0 & 0 & \boldsymbol{R}_1 \end{bmatrix}$$

where $\boldsymbol{U}_a$ is $(n+1)\times(2n+1)$, $\boldsymbol{D}_a$ is $(2n+1)\times(2n+1)$.

2. Apply the modified Gram–Schmidt orthogonalization procedure,

$$\boldsymbol{U}_a = \overline{\boldsymbol{U}}_a \boldsymbol{T}$$

where $\overline{\boldsymbol{U}}_a$ is a $(n+1)\times(2n+1)$ unit lower triangular matrix, $\boldsymbol{T}$ is $(2n+1)\times(2n+1)$ such that $\boldsymbol{T}\boldsymbol{D}_a\boldsymbol{T}^{\mathrm{T}} = \boldsymbol{D}_a$ *and* $\overline{\boldsymbol{D}}_a$ is $(2n+1)\times(2n+1)$ diagonal.

3. Write $\overline{\boldsymbol{U}}_a$ and $\boldsymbol{T}$ as,

$$\overline{\boldsymbol{U}}_a = \begin{bmatrix} 1 & 0 \\ \boldsymbol{u} & \boldsymbol{U}_0 \end{bmatrix}, \boldsymbol{T} = \begin{bmatrix} \boldsymbol{v}_i^{\mathrm{T}} \\ \cdot \\ \cdot \\ \boldsymbol{v}_{n+1}^{\mathrm{T}} \end{bmatrix}$$

where $\boldsymbol{v}_i^T$ is the i–th row vector of $\boldsymbol{T}$. The gain $\boldsymbol{k}$ and the factors $\boldsymbol{U}$ and $\boldsymbol{D}$ are given by,

$$\boldsymbol{k} = \boldsymbol{u}, \;\; \boldsymbol{U} = \boldsymbol{U}_0$$

$$\boldsymbol{D} = \begin{bmatrix} \|\boldsymbol{v}_2\|^2_{\boldsymbol{D}_a} & & & \\ & \cdot & & \\ & & \cdot & \\ & & & \cdot \\ & & & & \|\boldsymbol{v}_{n+1}\|^2_{\boldsymbol{D}_a} \end{bmatrix}$$

with,

$$\|\boldsymbol{v}_i\|^2_{\boldsymbol{D}_a} = \boldsymbol{v}_i^{\mathrm{T}}\boldsymbol{D}_a\boldsymbol{v}_i \tag{3.44}$$

4. Use (3.43) to compute the parameters.

3.3.3.3 Speed – Fast algorithms

Computational speed is a fundamental requirement in every algorithm designed for on–line fault detection. Speed improvement can be achieved in two ways:

- By employing special algorithms (software improvement).
- By employing special instrumentation (hardware improvement).

In what follows we focus on the Kalman filter equations. Since LS algorithms are a special form of a Kalman filter, the discussion applies equally well to these cases.

Software improvement relies on fast algorithms which, in general, reduce the computational load required to calculate the covariance matrix by an order of magnitude. By looking at Table 3.2, adapted from Bierman (1977), one can see that these algorithms will compute $\boldsymbol{P}(t)$ in O(dim$\boldsymbol{\theta}$) operations. It should be noted however that the updating of $\boldsymbol{P}(t)$ is only a part of the total algorithm, so the total number of operations needed per time update is higher than that given in Table 3.2.

Table 3.2 Number of arithmetic operations used for updating $\boldsymbol{P}(t)$ once. The number of parameters is n.

Method	Additions	Multiplications	Divisions	Square roots
Conventional Kalman equation (3.30a)	$1.5n^2+3.5n$	$1.5n^2+4.5n$	1	0
Stabilized Kalman equation	$4.5n^2 + 5.5n$	$4n^2 + 7.5n$	1	0
Potter's square root	$3n^2+3n$	$3n^2+4n$	2	1
U-D factorization	$1.5n^2+1.5n$	$1.5n^2+5.5n$	n	0

Fast algorithms rely on the exploitation of the so called *shift structure*. To illustrate this notion consider (3.9) again,

$$y(t) + a_1y(t-1) + \dots + a_ky(t-k) = b_1u(t-1) + \dots + b_ku(t-k) + e(t)$$

Note that the orders of the polynomials are the same and equal to k. Now define,

$$\boldsymbol{x}(t) = \begin{bmatrix} -y(t) \\ u(t) \end{bmatrix}, \boldsymbol{\varphi}(t) = \begin{bmatrix} \boldsymbol{x}(t-1) \\ \dots \\ \boldsymbol{x}(t-k) \end{bmatrix} \tag{3.45}$$

These definitions lead to the following shift structure for $\boldsymbol{\varphi}$: introduce the vector,

$$\boldsymbol{\varphi}^*(t)=\begin{bmatrix}\boldsymbol{x}(t)\\ \boldsymbol{\varphi}(t)\end{bmatrix} \tag{3.46}$$

Then,

$$\boldsymbol{\varphi}^*(t)=\begin{bmatrix}\boldsymbol{x}(t)\\ \boldsymbol{\varphi}(t)\end{bmatrix}=\begin{bmatrix}\boldsymbol{\varphi}(t+1)\\ \boldsymbol{x}(t-k)\end{bmatrix}$$

Here the dimension of $\boldsymbol{x}$ is 2 since scalar signals are considered and the dimension of $\boldsymbol{\varphi}$ is $n=2k$.

Since the relevant calculations for the derivation of fast formulae are rather involved the necessary recursive equations are only cited and the interested reader is referred to the work of Kalouptsidis and co–workers and Ljung and co–workers.

Fast Calculation of the Gain Matrix

1. Initialize:

$$\boldsymbol{A}(0)=\boldsymbol{0},\ \boldsymbol{B}(0)=\boldsymbol{0},\ \boldsymbol{R}^e(0)=\delta\boldsymbol{I}\ \text{(for some } \delta>0),\ \boldsymbol{k}(1)=\boldsymbol{0}.$$

2. Given $\boldsymbol{A}(t-1)$, $\boldsymbol{B}(t-1)$, $\boldsymbol{R}^e(t-1)$, and $\boldsymbol{k}(t)$, update:

$$\boldsymbol{e}(t)=\boldsymbol{x}(t)+\boldsymbol{A}^{\mathrm{T}}(t-1)\boldsymbol{\varphi}(t) \tag{3.47}$$

$$\boldsymbol{A}(t)=\boldsymbol{A}(t-1)-\boldsymbol{k}(t)\boldsymbol{e}^{\mathrm{T}}(t) \tag{3.48}$$

$$\beta(t)=\boldsymbol{k}^{\mathrm{T}}(t)\boldsymbol{\varphi}(t) \tag{3.49}$$

$$\boldsymbol{e}_1(t)=(1-\beta(t))\boldsymbol{e}(t) \tag{3.50}$$

$$\boldsymbol{R}^e(t)=\lambda\boldsymbol{R}^e(t-1)+\boldsymbol{e}_1(t)\mathbf{e}^{\mathrm{T}}(t) \tag{3.51}$$

$$\boldsymbol{k}^*(t)=\begin{bmatrix}\left[\boldsymbol{R}^e(t)\right]^{-1}\boldsymbol{e}_1(t)\\ \boldsymbol{k}(t)+\boldsymbol{A}(t)\left[\boldsymbol{R}^e(t)\right]^{-1}\boldsymbol{e}_1(t)\end{bmatrix} \tag{3.52}$$

3. Partition $\boldsymbol{k}^*(t)$ as,

$$\boldsymbol{k}^*(t)=\begin{bmatrix}\boldsymbol{M}(t)\\ \cdots\cdots\cdots\\ \boldsymbol{\mu}(t)\end{bmatrix}\begin{matrix}\}\ 2k\ \text{rows}\\ \\ \}\ 2\ \text{rows}\end{matrix} \tag{3.53}$$

4. Compute,

$$\boldsymbol{r}(t)=\boldsymbol{x}(t-k)+\boldsymbol{B}^{\mathrm{T}}(t-1)\boldsymbol{\varphi}(t+1)$$

$$\boldsymbol{B}(t)=\left(\boldsymbol{B}(t-1)-\boldsymbol{M}(t)\boldsymbol{r}^{\mathrm{T}}(t)\right)\left(\boldsymbol{I}-\boldsymbol{\mu}(t)\boldsymbol{r}^{\mathrm{T}}(t)\right]^{-1}$$

$$\boldsymbol{k}(t+1) = \boldsymbol{M}(t) - \boldsymbol{B}(t)\boldsymbol{\mu}(t) \tag{3.54}$$

The algorithm computes the gain vector $\boldsymbol{k}(t)$ which is subsequently used to update the parameter estimate:

$$\hat{\boldsymbol{\theta}}(t) = \hat{\boldsymbol{\theta}}(t-1) + \boldsymbol{k}(t)\left(y(t) - \hat{\boldsymbol{\theta}}^{\mathrm{T}}(t-1)\boldsymbol{\varphi}(t)\right]^{\mathrm{T}} \tag{3.55}$$

Now the vector $\boldsymbol{x}(t)$ contains $y(t)$ since the first row of $\boldsymbol{x}(t)$ equals $-y(t)$ and therefore, by comparison of (3.55) with (3.47), (3.48), it is found that the first column of $A(t)$ will be equal to $\hat{\boldsymbol{\theta}}(t)$. Hence, when $A(q^{-1})\neq 1$ the updating of $\hat{\boldsymbol{\theta}}(t)$ is obtained as a by–product of the algorithm given by equations (3.47)–(3.54) to determine the gain.

A count of the number of operations involved gives,

$68n + 76/3$ multiplications, $80n + 17$ additions,

which is an order of magnitude less (in n) than previous algorithms (Table 3.2).

This fast algorithm has therefore a distinct advantage over conventional methods, when the model order n is high: it is much faster and requires much less memory.

Speed improvement using special hardware is usually based on implementations on parallel processors (transputers). These implementations require rewriting of the original algorithms so that full advantage of parallelism is taken. There are three important considerations involved in the implementation of an algorithm on a parallel architecture:

- Subdivision of the algorithm into parallel processes.
- Interprocess communication.
- Efficient execution of each process.

Maguire and Irwin, (1991), consider transputer implementations of three forms of the Kalman filter algorithm: the conventional filter, a mixed covariance/information filter and a square root filter. Two approaches were compared, one involving heuristic partitioning of the conventional algorithm and another employing a new strategy for mapping the systolic array descriptions onto parallel processors. Simulations on the T414 and T800 transputers showed a marked superiority of the systolic array approach. Rhodes, (1990), adopting a control systems viewpoint, proposes a Kalman filter decomposition based on a system representation which is a direct sum of observability subspaces. Kalouptsidis and Theodoridis, (1987), developed a highly concurrent implementation for LS algorithms requiring O(dim$\boldsymbol{\theta}$) computing time and O(dim$\boldsymbol{\theta}$) processors. The basis of the formulation are Levinson–type recursions, while dot products are computed via a scheme based on Schur's algorithm. Finally, Stavrakakis *et al.* (1990) have implemented the recursive sliding window estimator on an INMOS transputer, and applied the whole scheme to fault detection in robotic systems. This example is elaborated in the applications section.

A different strategy would be to code the Kalman filter algorithm directly onto hardware.

Yeh, (1991), implemented a conventional and a U–D filter on a DSP32C processor using assembly and C languages.

3.3.3.4 Data weights selection

As explained in previous sections the selection of the quantities that affect the contribution of past data to the calculation of the estimate, plays a crucial role in the fault detection process. In fact, it may be argued that the tuning of these parameters is the weak link in the detection chain since usually they can be assigned only by simulation. In a parameter identification context, especially for slow–or rapidly–varying systems, the role of the weighting parameters is well understood. Consider for example the process model,

$$y(t+1) = \alpha(t)\, y(t) + e(t+1)$$

where,

$$\alpha(t) = -0.9 \text{ for } t < 100.$$

$$= -0.3 \text{ for } t \geq 100.$$

This model is a typical fault situation, where the estimated parameter jumps in a step–like fashion. Less rapid changes can usually be well approximated by a series of step changes. In Figure 3.3 the outcome of the estimation procedure using (3.20) for three different values of λ is shown. A clear conclusion is therefore drawn: the larger the value of λ the better the quality of the estimate but at the expense of slow response. On the other hand smaller values of λ result in poorer estimates but which are reached faster. A useful relation of λ and effective window length is given by,

$$T_0 = \frac{1}{1-\lambda}$$

This means that data older than T_0 time units have a weight less than $e^{-1} \approx 36\%$ of that of the most recent data. Analogous conclusions can be reached for the algorithm of (3.30) for parameter $\boldsymbol{R}_1(t)$. However in this case $\boldsymbol{R}_1$ lies on the opposite size of the scale, close to 0. Window length for sliding window algorithms can also be determined by similar arguments since λ is directly related to effective data length.

To develop appropriate weight sequences for fault monitoring schemes one has to observe that fault monitoring differs from time–varying system identification in that the quality of the parameter estimates has not the same significance at every time instance. In fault monitoring, parameters are assumed known until a change occurs, therefore until then the most important operation is change detection. Following a positive change decision, the part of parameter identification becomes important, while change detection may be even suspended. This is acceptable if subsequent faults can be assumed to happen in longer intervals than these needed for parameter estimation.

This reasoning points to the use of two different weighting sequences: a pre– fault and a post–fault one. The whole procedure is shown in Figure 3.4. Two problems remain: how to decide on a fault occurrence and how to change the weighting sequence. The first question will be answered in subsequent chapters. Let us concentrate on the second.

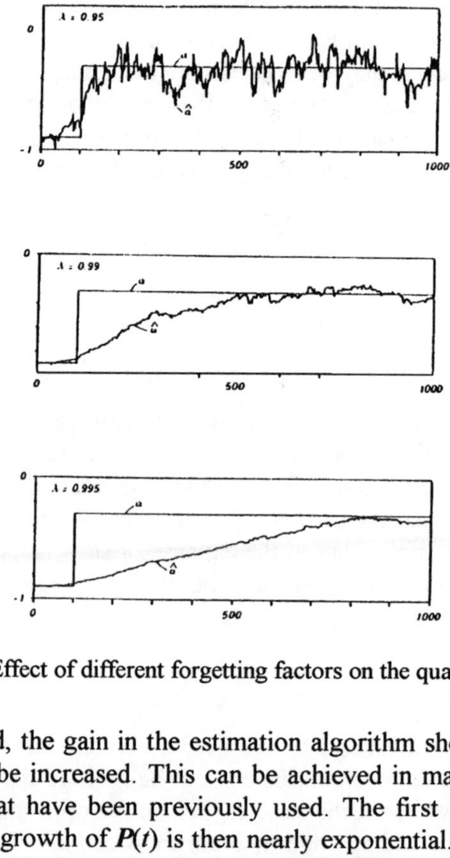

Figure 3.3 Effect of different forgetting factors on the quality of estimates.

When a fault is detected, the gain in the estimation algorithm should be increased. This means that $\boldsymbol{P}(t)$ should be increased. This can be achieved in many ways, but there are mainly two methods that have been previously used. The first one is to decrease the forgetting factor λ. The growth of $\boldsymbol{P}(t)$ is then nearly exponential. This approach can be implemented by the variable forgetting factor method of Fortescue *et al.*., (1981). The necessary recursions are:

1. Prediction $\hat{y}(t) = \boldsymbol{\varphi}^{\mathrm{T}}(t-1)\hat{\boldsymbol{\theta}}(t-1) + u(t-1)$
2. Error $\varepsilon(t) = y(t) - \hat{y}(t)$
3. Gain $\boldsymbol{k}(t) = \dfrac{\boldsymbol{P}(t-1)\boldsymbol{\varphi}(t-1)}{1+\boldsymbol{\varphi}^{\mathrm{T}}(t-1)\boldsymbol{P}(t-1)\boldsymbol{\varphi}(t-1)}$
4. Estimate $\hat{\boldsymbol{\theta}}(t) = \hat{\boldsymbol{\theta}}(t-1) + \boldsymbol{k}(t)\varepsilon(t)$

5. Forgetting $\lambda(t) = 1 - \left(1 - \boldsymbol{\varphi}^{\mathrm{T}}(t-1)\boldsymbol{k}(t)\right)\varepsilon^2(t) \big/ \sigma_0$

note: if $\lambda(t)<\lambda_{\min}$ then set $\lambda(t)=\lambda_{\min}$

6. Covariance $\boldsymbol{P}(t) = \dfrac{1}{\lambda(t)}\left(1 - \boldsymbol{k}(t)\boldsymbol{\varphi}^{\mathrm{T}}(t-1)\boldsymbol{P}(t-1)\right)$ (3.56)

This strategy is based on the requirement that the imformation content of the filter expressed recursively as,

$$\sigma(t) = \lambda(t)\sigma(t-1) + \left(1 - \boldsymbol{\varphi}^{\mathrm{T}}(t-1)\boldsymbol{k}(t)\right)\boldsymbol{\varepsilon}^2(t) \tag{3.57}$$

is kept constant and equal to σ_0. In other words, the amount of forgetting will at each step correspond to the amount of new information in the latest measurement, thereby ensuring that the estimation is always based on the same amount of information. Thus from (3.57),

$$\lambda(t) = 1 - 1/N(t)$$

where,

$$N(t) = \sigma_0 \,/\, (1 - \boldsymbol{\varphi}^{\mathrm{T}}(t-1)\boldsymbol{k}(t))\varepsilon^2(t)$$

is the equivalent asymtotic memory length if $\lambda=\lambda(t)$ were to be used throughout the estimation. Since σ_0 is related to the sum of the squares of the errors, one possible guideline on how to choose it is to express σ_0 as,

$$\sigma_0 = \sigma^2 N_0$$

where σ^2 is the expected measurement noise variance based on real knowledge of the process. Then N will control the speed of adaptation as it corresponds to a nominal asymptotic memory length. Simulations have shown that if σ_0 is chosen in this manner, then for a stationary process,

$$E\{N(t)\} \to N_0 \text{ as } t \to \infty$$

and,

$$\mathrm{E}\{\boldsymbol{\Phi}\boldsymbol{\Phi}^{\mathrm{T}}\} \to \boldsymbol{\Phi}\boldsymbol{\Phi}^{\mathrm{T}}$$

where $\boldsymbol{\Phi}\boldsymbol{\Phi}^{\mathrm{T}}$ is a steady–state information matrix related to system dynamics and choice of σ_0. Further details of this are given by Ydstie (1981). The sensitivity of the system is governed by the choice of N_0. A small value of N_0 will give a large covariance matrix and a sensitive system; a larger value will give a less sensitive estimator and slower adaptation.

An exact solution to the problem of keeping $\sigma(t)=\sigma_0$ would give an algorithm where the forgetting factor is updated before the gain calculation in step 3, and would involve solving a quadratic and hence more complex relationship for $\lambda(t)$. The practical difference between this and the algorithm above is in most cases small, but when the simpler algorithm is used, a test on $\lambda(t)$ must be introduced to prevent this from becoming too small or even negative. A discussion of some of the convergence characteristics of the algorithm is presented by Cordero and Mayne (1981).

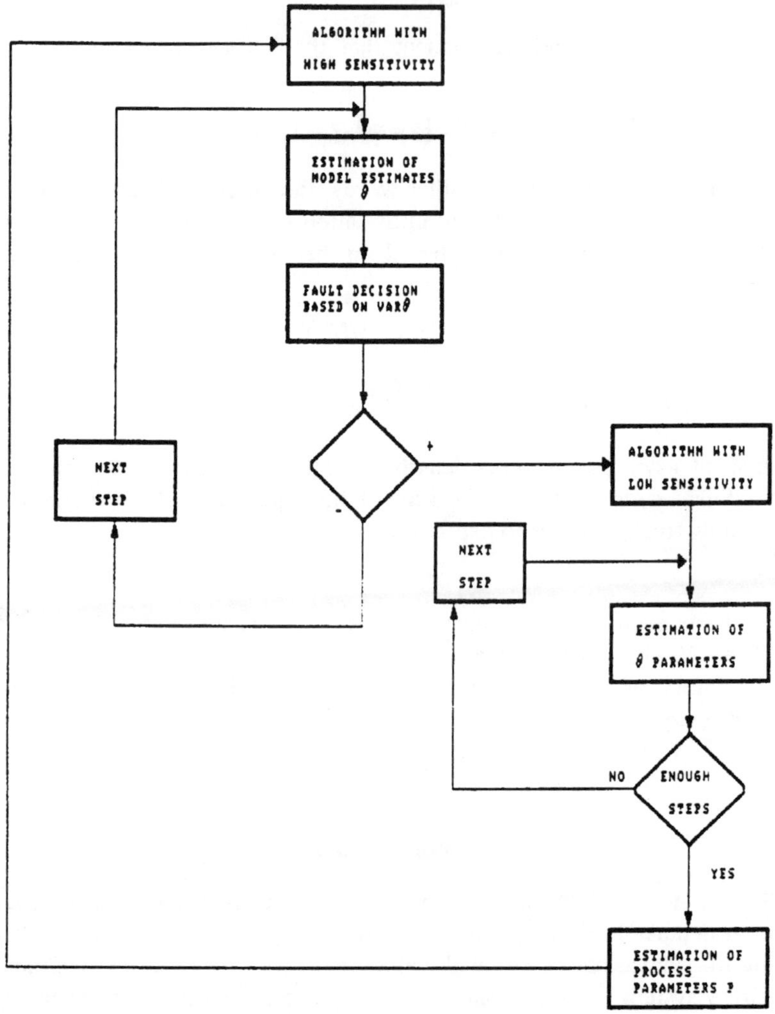

Figure 3.4. A unified strategy for fault detection based on parameter estimation

A typical simulation is illustrated below. The system to be identified is described by the equation,

$$y(t)-1.2y(t-1)+0.2y(t-2)+0.02y(t-3) = u(t-1)-0.62u(t-2)-0.03u(t-3)+e(t)$$

where $e(t)$ is a sequence of pseudo–random numbers with zero mean and standard deviation 0.005. To simulate one aspect of a nonlinear plant, after a period of operation the setpoint is changed and the system model is modified simultaneously to:

$$y(t)-1.4y(t-1)+0.3y(t-2)+0.05y(t-3) = u(t-1)-0.5u(t-2)-0.08u(t-3)+e(t).$$

The results shown in Figure 3.5 indicate the behaviour of the forgetting factor and estimates. Under steady–state operations $\lambda(t)$ is close to one and the estimator behaves very much like an unweighted filter. During the setpoint change the system description is changed; this results in a poor fit between model and process, and a smaller forgetting factor is produced. After a while, the parameters are "retuned" and the forgetting factor returns to its former value close to 1.

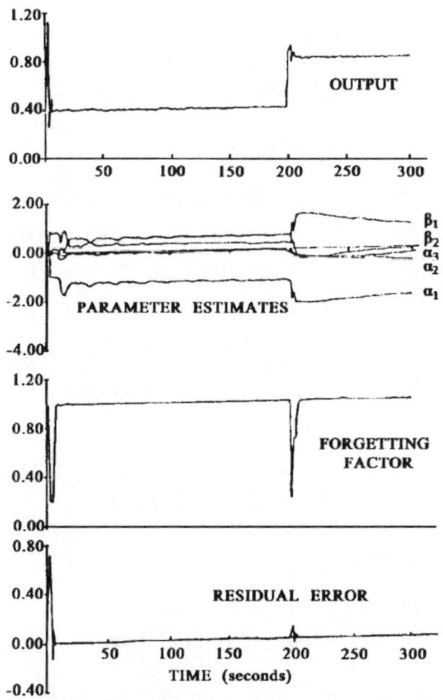

Figure 3.5. Simulation of self–tuning estimator with variable forgetting factor σ_0=0.0125.

This feature is extremely usuful in a fault monitoring situation: if a fault occurs all information prior to the fault is really useless. The small value of $\lambda(t)$ at this instant actually automatically "restarts" the algorithm at the time of the fault. A drawback of this approach is that "restarting" is introduced to faulty and non–faulty parameters simultaneously.

The second method is to add a constant times the unit matrix to $\boldsymbol{P}(t)$ in which case it is increased instantaneously. Equation (3.20) will then be substituted by,

$$\boldsymbol{P}(t)=\frac{1}{\lambda}\left[\boldsymbol{P}(t-1)-\frac{\boldsymbol{P}(t-1)\boldsymbol{\varphi}(t)\boldsymbol{\varphi}(t)^{\mathrm{T}}\boldsymbol{P}(t-1)}{\lambda+\boldsymbol{\varphi}(t)^{\mathrm{T}}\boldsymbol{P}(t-1)\boldsymbol{\varphi}(t)}\right]+\beta(t)\boldsymbol{I} \tag{3.58}$$

where $\beta(t)$ is a nonnegetive scalar variable which is zero except when a fault is detected. When a fault is detected, a positive $\beta(t)$ has the effect of increasing $\boldsymbol{P}(t)$. The final problem is to choose a suitable $\beta(t)$. When no fault is detected, $\beta(t)$ is zero. When a fault is detected, it is reasonable to let $\beta(t)$ depend on the actual value of $\boldsymbol{P}(t)$ and on how significant the alarm is. This may be done in many ways, and the following proposal is just one possibility.

In the noise–free case, the progress of the estimation error, when $\boldsymbol{\theta}(t)$ is constant, is given by,

$$\begin{aligned}\boldsymbol{\theta}_{\mathrm{e}}(t) &= \boldsymbol{\theta}_{\mathrm{e}}(t-1)-\boldsymbol{P}(t)\boldsymbol{\varphi}(t)\varepsilon(t)\\ &= (\boldsymbol{I}-\boldsymbol{P}(t)\boldsymbol{\varphi}(t)\boldsymbol{\varphi}^{\mathrm{T}}(t))\boldsymbol{\theta}_{\mathrm{e}}(t-1)\\ &= \boldsymbol{U}(t)\boldsymbol{\theta}_{\mathrm{e}}(t-1)\end{aligned}$$

All eigenvalues of $\boldsymbol{U}(t)$ are one, except the one corresponding to the eigenvector $\boldsymbol{P}(t)\boldsymbol{\varphi}(t)$. This eigenvalue determines the step length in the algorithm. A small eigenvalue causes large steps, while an eigenvalue close to one means that the step length in the algorithm is small. Using (3.58) the eigenvalue can be written as:

$$\begin{aligned}1-\boldsymbol{\varphi}^{\mathrm{T}}(t)\boldsymbol{P}(t)\boldsymbol{\varphi}(t) &= 1-\frac{\boldsymbol{\varphi}^{\mathrm{T}}(t)\boldsymbol{P}(t-1)\boldsymbol{\varphi}(t)}{\lambda+\boldsymbol{\varphi}^{\mathrm{T}}(t)\boldsymbol{P}(t-1)\boldsymbol{\varphi}(t)}-\beta(t)\boldsymbol{\varphi}^{\mathrm{T}}(t)\boldsymbol{\varphi}(t)\\ &= \frac{\lambda}{\lambda+\boldsymbol{\varphi}^{\mathrm{T}}(t)\boldsymbol{P}(t-1)\boldsymbol{\varphi}(t)}-\beta(t)\boldsymbol{\varphi}^{\mathrm{T}}(t)\boldsymbol{\varphi}(t)\end{aligned}$$

When $\beta(t)=0$, the elgenvalue is thus,

$$v_0(t)=\frac{\lambda}{\lambda+\boldsymbol{\varphi}^{\mathrm{T}}(t)\boldsymbol{P}(t-1)\boldsymbol{\varphi}(t)}$$

The eigenvalue is obviously between zero and one as long as $\boldsymbol{P}>0$. Suppose now, that an eigenvalue equal to $v(t)$ is desired when a fault is detected. Then $\beta(t)$ has to be chosen as,

$$\beta(t)=\frac{1}{\boldsymbol{\varphi}^{\mathrm{T}}(t)\boldsymbol{\varphi}(t)}\left(v_0(t)-v(t)\right)$$

The eigenvalue $v(t)$ should lie in the interval,

$$0<v(t)\leq v_0(t)$$

in order to keep $\boldsymbol{P}(t)$ positive definite. In practice, this choice of $\beta(t)$ must also be combined with a test for nonsingularity of $\boldsymbol{\varphi}^{\mathrm{T}}(t)\boldsymbol{\varphi}(t)$.

It remains to determine a suitable $v(t)$. One choice is to model $v(t)$ as a piecewise linear function of the significance of the fault alarm (Figure 3.6).

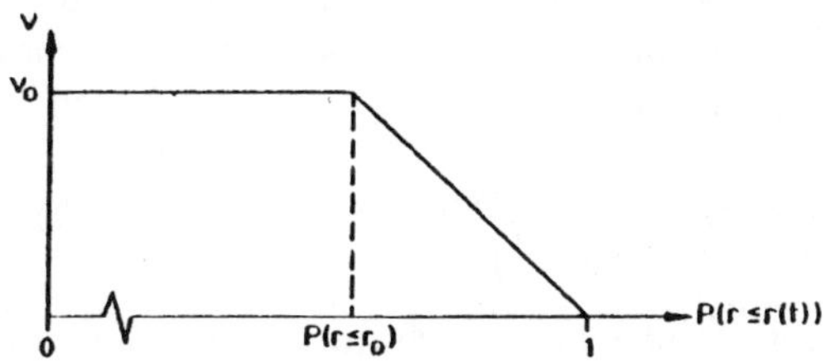

Figure 3.6. Choice of $v(t)$

3.3.5 Robustness issues

The preceding sections indicate that most, if not all, traditional approaches to estimation problems assume that the model is perfect and hence the only source of modeling error that is considered is due to process or measurement noise. However, in practical situations, it is frequently the case that the major source of modeling errors is the approximate description of the system's response. This gives rise to the need for robust (in the presence of modeling errors) parameter estimation techniques. The importance of this problem has already been discussed in Chapter II, with reference to residual–based fault detection methods.

Efforts have also been made to produce robust parameter–estimation methods that could be used in fault detection applications. This is a very important development, since it is well known that if model mismatch exists, it is the source of false alarms, a condition that will invalidate even the most sophisticated algorithm. Weiss, (1988), computed an uncertainty bound in the frequency domain which accounts for modeling errors. This bound is then used to construct a test variable for fault detection. Carlsson *et al.* (1988), developed a related strategy in which the unmodeled response is embedded in a stochastic process, so that bounds can be computed on the unmodeled errors.

In this section two approaches to robust parameter estimation–based fault detection are briefly summarized: the "*black box*" approach suggested by Wahlberg (1990) and the "*embedding*" approach of Carlsson, extended by Kwon and Goodwin (1990). It must be noted that both approaches are in fact off–line and have been mainly used for model validation.

The "stochastic embedding" approach.

This approach provides robust fault detection when unmodeled dynamics, linearization errors and noisy inputs are present. It functions by calculating upper bounds for parameter estimates in the face of the previous anomalies.

The model mismatch is described by incorporating into the system description additive unmodeled dynamics. System dynamics are described by the general linear model of Equation (3.7). Two models are needed: the *nominal model,* with transfer function,

$$G(q^{-1};\boldsymbol{\theta}) = \frac{B(z^{-1};\boldsymbol{\theta})}{A(z^{-1})} \tag{3.59}$$

and the *unmodeled dynamics* model with tranfer function $G_\Delta(q^{-1})$. The system output then satisfies,

$$\begin{aligned} y(k) &= [G(q^{-1}) + G_\Delta(q^{-1})]u(k) + v(k) \\ &= \boldsymbol{B}(q^{-1};\, \boldsymbol{\theta})u_F(k) + \eta(k) \end{aligned} \tag{3.60}$$

where $v(k)$ models measurement noise of zero mean and spectral density $\Phi(\omega)$ and,

$$u_F(k) = \frac{1}{A(q^{-1})}u(k)$$

$$\eta(k) = G_\Delta(q^{-1})u(k) + v(k)$$

Equation (3.60) can be represented in standard linear regression form as,

$$y(k) = \boldsymbol{\varphi}^T(k)\boldsymbol{\theta} + \eta(k) \tag{3.61}$$

where,

$$\boldsymbol{\varphi}(k) = [\ u_F(k-1)\ \ u_F(k-2)\ \ \dots\ \ u_F(k-n_B)\ ] \tag{3.62}$$

The parameters are estimated using ordinary least squares as,

$$\hat{\boldsymbol{\theta}} = \left[\boldsymbol{\Phi}^{\mathrm{T}}\boldsymbol{\Phi}\right]^{-1}\boldsymbol{\Phi}^{\mathrm{T}}\boldsymbol{Y} \tag{3.63}$$

where,

$$\boldsymbol{\Phi} = [\ \varphi(1)\ \ \varphi(2)\ \ \dots\ \ \varphi(N)\ ]^{\mathrm{T}} \tag{3.64}$$

$$\boldsymbol{Y} = [\ y(1)\ \ y(2)\ \ \dots\ \ \eta(N)\ ]^{\mathrm{T}} \tag{3.65}$$

From (3.61) and (3.63), the following expression for the estimation error is derived:

$$\boldsymbol{\theta}_e = \hat{\boldsymbol{\theta}} - \boldsymbol{\theta} = \left[\boldsymbol{\Phi}^{\mathrm{T}}\boldsymbol{\Phi}\right]^{-1}\boldsymbol{\Phi}^{\mathrm{T}}\boldsymbol{S}$$

where,

$$\boldsymbol{S} = [\ \eta(1)\ \eta(2)\ \dots\ \eta(N)\ ]^{\mathrm{T}} \tag{3.66}$$

Denoting the impulse response of G_Δ as $\{h_0(.)\}$, $\eta(k)$ can be expressed as,

$$\eta(k) = \sum_{i=0}^{k} h_0(i)u(k-1) + v(k)$$

Then the following relationship is obtained from (3.66):

$$\boldsymbol{S} = \boldsymbol{\Psi H} + \boldsymbol{V}$$

where,

$$\boldsymbol{V} = [\ v(1)\ v(2)\ ....\ v(N)\ ]^{\mathrm{T}}$$

$$\boldsymbol{\Psi} = \begin{bmatrix} u(1) & 0 & \dots & 0 \\ u(2) & u(1) & \dots & 0 \\ \cdot & \cdot & \dots & \cdot \\ u(N) & u(N-1) & \dots & u(1) \end{bmatrix}$$

$$\boldsymbol{H} = [\ h(0)\ \ h(1)\ \dots\ h(N\!-\!1)\ ]^{\mathrm{T}} \tag{3.67}$$

A priori knowledge is assumed to be available which allows to give a prior distribution to $\{h(.)\}$. This is known as *stochastic embedding*, ie. the unmodeled dynamics are embedded in a stochastic process (Goodwin *et al.*, 1989). In the present context, knowledge of the mean and covariance function for these distributions is adequate. Given these, the expected value of the estimation error can be calculated. This will be the basis of the fault detection method.

Assume that two sets of data, I_n and I_f, corresponding to normal and suspected faulty operation are available. In this situation, estimated parameters $\hat{\boldsymbol{\theta}}$ and corresponding nominal transfer function $G(z)$ may differ for the two cases:

$$\hat{\boldsymbol{\theta}} = \begin{cases} \hat{\boldsymbol{\theta}}_n, \text{ for data set } I_n \\ \hat{\boldsymbol{\theta}}_f, \text{ for data set } I_f \end{cases} \tag{3.68}$$

$$G(q^{-1}, \hat{\boldsymbol{\theta}}) = \begin{cases} \hat{G}_n(q^{-1}) = G_n\left(q^{-1}, \hat{\boldsymbol{\theta}}_n\right) \text{ for } I_n \\ \hat{G}_f(q^{-1}) = G_f\left(q^{-1}, \hat{\boldsymbol{\theta}}_f\right) \text{ for } I_f \end{cases} \tag{3.69}$$

The fault detection procedure now amounts to comparing $\hat{\boldsymbol{\theta}}_n$ and $\hat{\boldsymbol{\theta}}_f$ (or $\hat{G}_n$ and $\hat{G}_f$) and deciding if the observed changes can be explained satisfactorily in terms of the effects of noise or undermodeling. If not, then it may be concluded that a system fault has occurred. The covariance functions of $(\hat{\boldsymbol{\theta}}_n - \hat{\boldsymbol{\theta}}_f)$ and $(\hat{G}_n - \hat{G}_f)$ under nonfaulty conditions will be used as measures of the uncertainty due to noise and undermodeling. An upper bound for the covariance of $(\hat{\boldsymbol{\theta}}_n - \hat{\boldsymbol{\theta}}_f)$ is given by Kwon and Goodwin, (1990), as:

$$\boldsymbol{C} = \mathrm{Cov}\left\{\hat{\boldsymbol{\theta}}_n - \hat{\boldsymbol{\theta}}_f\right\} = \mathrm{E}\left\{\left[\hat{\boldsymbol{\theta}}_n - \hat{\boldsymbol{\theta}}_f\right]\left[\hat{\boldsymbol{\theta}}_n - \hat{\boldsymbol{\theta}}_f\right]^{\mathrm{T}}\right\}$$

$$\leq [\boldsymbol{Q}_n - \boldsymbol{Q}_f]\boldsymbol{R}\,[\boldsymbol{Q}_n - \boldsymbol{Q}_f]^{\mathrm{T}} + [\boldsymbol{P}_n + \boldsymbol{P}_f]\sigma_v^2 \tag{3.70}$$

where,

$$\left.\begin{aligned} \boldsymbol{Q}_i &= \boldsymbol{P}_i \boldsymbol{\Phi}_i^{\mathrm{T}} \boldsymbol{\Psi}_i \\ \boldsymbol{P}_i &= \left[\boldsymbol{\Phi}_i^{\mathrm{T}} \boldsymbol{\Phi}_i\right] \\ \boldsymbol{R}_i &= \mathrm{E}\left\{HH^{\mathrm{T}}\right\} \end{aligned}\right\} i = n, f$$

and σ_v^2 is the equivalent noise variance corresponding to the upper bound $\overline{S}$ of the power spectrum of v.

The first term on the right side of (3.70) accounts for the effects of undermodeling and the difference in input signals for the two experiments. Note that if there is no undermodeling or if the inputs are identical, these terms vanish. The second term corresponds to measurement noise. The higher the SNR (signal–to–noise ratio) is, the smaller the norm of this term. The matrix $\boldsymbol{C}$ can now be used to formulate appropriate test variables for fault detection. For example, one may use,

$$T_1 = \left[\hat{\boldsymbol{\theta}}_n - \hat{\boldsymbol{\theta}}_f\right]^{\mathrm{T}} \boldsymbol{C}^{-1} \left[\hat{\boldsymbol{\theta}}_n - \hat{\boldsymbol{\theta}}_f\right] \tag{3.71}$$

$$T_2 = \left[\hat{\boldsymbol{\theta}}_n - \hat{\boldsymbol{\theta}}_f\right]^{\mathrm{T}} \left[\mathrm{diag}(\boldsymbol{C})\right]^{-1} \left[\hat{\boldsymbol{\theta}}_n - \hat{\boldsymbol{\theta}}_f\right] \tag{3.72}$$

These test variables are based on a comparison between the observed value and the expected value of $[\hat{\boldsymbol{\theta}}_n - \hat{\boldsymbol{\theta}}_f][\hat{\boldsymbol{\theta}}_n - \hat{\boldsymbol{\theta}}_f]^{\mathrm{T}}$. If the test variable is larger than a fixed threshold, then this is evidence that a fault has occurred.

Uncertainty bounds in the frequency domain can easily be obtained by extending the previous results. Kwon and Goodwin, (1990), have shown that the expected value of the difference between the estimated transfer functions in two nonfaulty experiments is given by,

$$D(\boldsymbol{\omega}, n) = \mathrm{E}\left\{\Delta\hat{\boldsymbol{G}}(\boldsymbol{\omega}, n)\Delta\hat{\boldsymbol{G}}^*(\boldsymbol{\omega}, n)\right\}$$

$$= V(\boldsymbol{\omega}, n)\, C\, V^*(\boldsymbol{\omega}, n) \tag{3.73}$$

where,

$$\Delta\hat{\boldsymbol{G}}(\boldsymbol{\omega}, n) = \left[\hat{G}_n(e^{j\omega_1}) - \hat{G}_f(e^{j\omega_1}) \;\; \ldots \;\; \hat{G}_n(e^{j\omega_n}) - \hat{G}_f(e^{j\omega_n})\right]^{\mathrm{T}}$$

$$V(\boldsymbol{\omega}, n) = \left[\frac{\partial}{\partial\theta}G(e^{j\omega_1},\theta) \;\; \frac{\partial}{\partial\theta}G(e^{j\omega_2},\theta) \;\; \ldots \;\; \frac{\partial}{\partial\theta}G(e^{j\omega_n},\theta)\right]^{\mathrm{T}}$$

$$\boldsymbol{\omega} = [\omega_1 \;\; \omega_2 \;\; \ldots \;\; \omega_n]^{\mathrm{T}}$$

and G is the nominal transfer function, C is given by (3.70), n is the number of frequencies and * denotes conjugate transpose. Based on this result, test variables analogous to the ones of equations (3.71), (3.72) can be devised.

Evaluation of test variables require knowledge of σ_v^2 and $\boldsymbol{R}$. This knowledge can be acquired from experiments performed on non–faulty operation. For example, σ_v^2 can be estimated by supplying a constant input and recording the output variations around the mean value. For $\boldsymbol{R}$, simplification may be necessary. For example, it may be assumed that,

$$\mathrm{E}\{h(k)h(j)\} = r(k)\delta_{kj}$$

where

$$r(k) = \sigma_0^2 e^{-\beta k}; \quad k, j = 0,\ 1,\ \ldots \tag{3.74}$$

In (3.74), $2/\beta$ can be considered an average time constant for the class of unmodeled dynamics. With this simple description, the parameters of (3.74) can now be estimated from experiments of non–faulty operation.

This idea can also be extended to nonlinear situations by including terms which will account for linearization errors around the operating point. In this situation, the system output, given by equation (3.60), will be,

$$y(k) = G(q^{-1})u(k) + G_\Delta(q^{-1})u(k) + G_{n\Delta}(q^{-1})u^2(k)\mathrm{sign}(u(k)) + v(k) \tag{3.75}$$

where $G_{n\Delta}$ accounts for model mismatch due to input nonlinearity. Proceeding similarly to the linear case, bounds in the parameter space are:

$$\boldsymbol{C} \le [\boldsymbol{Q}_n - \boldsymbol{Q}_f]\boldsymbol{R}[\boldsymbol{Q}_n - \boldsymbol{Q}_f]^{\mathrm{T}} + [\boldsymbol{Q}_{nn} - \boldsymbol{Q}_{nf}]\boldsymbol{R}_n[\boldsymbol{Q}_{nn} - \boldsymbol{Q}_{nf}]^{\mathrm{T}} + [\boldsymbol{P}_n + \boldsymbol{P}_f]\sigma_v^2 \tag{3.76}$$

where,

$$Q_{ni} = P_i \Phi_i^{\mathrm{T}} \Psi_{ni}; \; i = n, f$$

$$\boldsymbol{\Psi}_{ni} = \begin{bmatrix} u^2(1)\mathrm{sign}(u(1)) & 0 & \ldots & 0 \\ u^2(2)\mathrm{sign}(u(2)) & u^2(1)\mathrm{sign}(u(1)) & \ldots & 0 \\ \cdot & \cdot & \ldots & \cdot \\ u^2(N)\mathrm{sign}(u(N)) & u^2(N-1)\mathrm{sign}(u(N-1)) & \ldots & u^2(1)\mathrm{sign}(u(1)) \end{bmatrix}$$

$$\boldsymbol{R}_n = \mathrm{E}\{\boldsymbol{H}_n \boldsymbol{H}_n^{\mathrm{T}}\}$$

$$\boldsymbol{H}_n = [h_n(0) \quad h_n(1) \quad \ldots \quad h_n(N-1)]^{\mathrm{T}}$$

In the above, $h_n(k)$ denotes the impulse response of $G_{n\Delta}$ and the remaining variables are as in (3.70). An example of this procedure will be given in the applications section.

Black box identification approach.

This method, proposed by Wahlberg, (1990), aims at providing robust fault detection when undermodeling exists, ie. when a finite order structure is used to model systems of infinite order. This is accomplished by using flexible models of a finite order in the frequency domain, which approximate the process in the frequency range of interest. The approach uses the *high order black–box* identification theory of Ljung, (1987).

The situation is again viewed as a standard hypothesis testing problem: given two data sets, I_n and I_f, decide if they relate to the same underlying system. Decision is made based on the estimates of the system parameter vector $\boldsymbol{\theta}$ obtained from the two data sets. The general model described by (3.7) and parameterized by $\boldsymbol{\theta} \in \boldsymbol{R}^n$ is used:

$$y(k) = G(q^{-1}; \boldsymbol{\theta})u(k) + H(q^{-1}; \boldsymbol{\theta})e(k) \tag{3.77}$$

Equation (3.77) is supposed to model time–invariant exponentially stable linear system with additive noise of the form,

$$y(k) = G_{\mathrm{T}}(q^{-1})u(k) + v(k)$$

where $v(k)$ is the part of the output that cannot be explained from the input signal in open–loop experiments. It is a zero mean, stationary stochastic process with spectral representation,

$$v(k) = \sum_{\tau=0}^{\infty} h_T(\tau)e(k-\tau); \quad h_T(0) = 1$$

where $\{e(k)\}$ is a sequence of independent random variables with zero mean and variances λ_T.

The two hypotheses are then given by,

$$H_0\text{: } \hat{\theta}_n \text{ and } \hat{\theta}_f \text{ relate to the same system}$$

$$H_1\text{: } H_0 \text{ is not true}$$

To decide in a robust way between these two hypotheses, firstly a robust estimate of $\boldsymbol{\theta}$ must be obtained and then a suitable test statistic must be devised. Let (3.77) be rewritten in transfer function form as,

$$y(k) = \frac{B(q^{-1})}{A(q^{-1})}u(k) + \frac{C(q^{-1})}{D(q^{-1})}e(k)$$

where $A(q^{-1})$, $C(q^{-1})$ are given stable polynomials of fixed degree and

$$B(q^{-1}) = b_1 q^{-1} + \ \ldots \ + b_n q^{-n}$$
$$D(q^{-1}) = 1 + d_1 q^{-1} + \ \ldots \ + d_n q^{-n}$$

Then the parameter vector is,

$$\boldsymbol{\theta} = \begin{bmatrix} \boldsymbol{b} & \boldsymbol{d} & \lambda \end{bmatrix}$$

Based on N system measurements, the system model is estimated by the two–step procedure:

1. Estimate $B(q^{-1})$ by minimising the criterion,

$$\frac{1}{N}\sum_{k=1}^{N}\left[L(q^{-1})\left(y(k)-\frac{B(q^{-1})}{A(q^{-1})}u(k)\right)\right]^2$$

where $L(q^{-1})$ is prefilter used to modify the bias distribution.

2. Determine $D(q^{-1})$ by minimising the criterion,

$$\frac{1}{N}\sum_{k=1}^{N}\left[D(q^{-1})\bar{y}(k)\right]^2$$

where,

$$\bar{y}(k)=\frac{1}{C(q^{-1})}\left(y(k)-\frac{\hat{B}(q^{-1})}{A(q^{-1})}u(k)\right)$$

The spectral density estimate is then given by,

$$\hat{\Phi}_v(e^{i\omega})=\hat{\lambda}\frac{\left|A(e^{i\omega})\right|^2}{\left|\hat{D}(e^{i\omega})\right|^2}$$

where,

$$\hat{\lambda}=\frac{1}{N}\sum_{k=1}^{N}\left[\hat{D}(q)\bar{y}(k)\right]^2$$

A suitable test statistic is then the following:

$$T=\sum_{m=1}^{l}\left[\left|E(\omega_m)\right|^2+\left|F(\omega m)\right|^2\right]$$

$$E(\omega_m)=G(e^{i\omega_m},\hat{\theta}_n)-G(e^{i\omega_m},\hat{\theta}_f)\times\left(\frac{n_1}{N_1}\frac{\hat{\Phi}_v^1(e^{i\omega_m})}{\hat{\Phi}_u^1(e^{i\omega_m})}+\frac{n_2}{N_2}\frac{\hat{\Phi}_v^2(e^{i\omega_m})}{\hat{\Phi}_u^2(e^{i\omega_m})}\right)^{-1/2}$$

$$F(\omega_k)=H(e^{i\omega_m},\hat{\theta}_n)-H(e^{i\omega_m},\hat{\theta}_f)\times\left(\frac{n_1}{N_1}\left|H(e^{i\omega_m},\hat{\theta}_n)\right|^2+\frac{n_2}{N_2}\left|H(e^{i\omega_m},\hat{\theta}_f)\right|^2\right)^{-1/2} \tag{3.78}$$

where N_1, N_2 are the sample sizes for the two experiments, n_1, n_2 are the estimated orders of $\hat{B}(q)$ and $2\ell \le \min(n_1, n_2)$. The input spectral density estimates may be given by a–priori information or estimated using a smoothed DFT. By noting that under H_0, $E(\omega_m)$ and $F(\omega_m)$; $m=1, \ldots, \ell$, are asymptotically jointly complex normally distributed with zero mean and identity variance, the test variable T is χ^2 distributed with 4ℓ

degrees of freedom. Hence for a given confidence level α, a confidence interval can be calculated. If T falls outside this interval, a fault is declared.

3.4 Decision rules

Following the parameter estimation phase of the fault monitoring algorithm a decision has to be made, based on the estimated values, on whether a change/fault has occurred. There are two basic approaches to this problem:

- decide based on the estimated values of the model parameters $\hat{\theta}_i$
- decide based on the calculated values of the process parameters $\hat{p}_i$ using the inverse relationship (3.6).

It should be stressed that decisions made using the first approach do not prohibit inferences about the state of the process parameters. As an example consider the following set of relations amongst $\boldsymbol{\theta}$ and $\boldsymbol{p}$. (Pouliezos and Stavrakakis, 1989):

$$\theta_1 = \frac{R}{L}, \theta_2 = \frac{K_m N}{L}, \theta_3 = \frac{1}{L}, \theta_4 = -\frac{K_m}{J_m}, \theta_5 = \frac{\rho}{J_m}, \theta_6 = -\frac{1}{J_m N}$$

resulting in,

$$p_1 = R = \frac{\theta_2}{\theta_3}, p_2 = L = \frac{1}{\theta_3}, \quad p_3 = K_m = \frac{\theta_2}{\theta_3 N}, p_4 = J_m = -\frac{\theta_2}{\theta_3 \theta_4 N} = -\frac{1}{\theta_6 N}$$

$$p_5 = \rho = -\frac{\theta_2 \theta_5}{\theta_3 \theta_4 N}$$

Reasoning backwards, a change in p_3=K_m would appear in θ_2, θ_4 and a change in p_1=R in θ_1. It is obvious therefore that by constructing a table of interrelations, inference on change of p_i can be made by simple observations of the θ_j.

Let us examine these two approaches in detail.

Decision based on $\hat{\theta}$. Hägglund, (1984), proposes the following procedure for detecting a change in one or more of the estimated parameters:

Define the variables,

$$\Delta\hat{\boldsymbol{\theta}}(t) = -\hat{\boldsymbol{\theta}}(t) + \hat{\boldsymbol{\theta}}(t-1) \tag{3.79}$$

$$\boldsymbol{w}(t) = \gamma_1 \boldsymbol{w}(t-1) + \Delta\hat{\boldsymbol{\theta}}(t) \quad 0 \le \gamma_1 < 1 \tag{3.80}$$

In the case a fault has occurred, $\boldsymbol{w}(t)$ can be viewed as an estimate of the direction of the parameter change. The test sequence that will be studied is $(s(t))$, where $s(t)$ is defined as,

$$s(t) = \text{sign}\left[\Delta\hat{\boldsymbol{\theta}}(t)^{\mathrm{T}} \boldsymbol{w}(t-1)\right] \tag{3.81}$$

The sign function makes the test sequence insensitive to the noise variance. It is now clear in principle how to carry out the fault detection:

- Inspect the latest values of $s(t)$: if $s(t)$ is +1 unlikely many times, conclude that a fault has occurred.

Under normal operation, i.e. when the parameter estimates are close to their true values, $s(t)$ has approximately a symmetric two point distribution with mass 0.5 each at +1. When a fault has occurred, the distribution is no longer symmetric, but the mass at +1 is larger than the mass at −1. To add the most recent values of $s(t)$, the stochastic variable $r(t)$ defined as

$$r(t) = \gamma_2 r(t-1) + (1-\gamma_2)s(t);\quad 0 \le \gamma_2 < 1$$

is introduced. The sum of the most recent values of $s(t)$ is replaced by an exponential smoothing in order to obtain a simple algorithm. When the parameter estimates are close to the true ones, $r(t)$ has a mean value close to zero. When a fault has occurred, a positive mean is expected. The parameter γ_2 determines, roughly speaking, how may $s(t)$ values should be included. For example γ_2=0.95 corresponds to about 20 values, which is a reasonable choice in many applications. A small γ_2 allows a fast fault detection, although at the price of reduced security against false alarms. When the signal to noise ratio is small, it is not possible to detect the faults as fast as otherwise. It is then necessary to have more information available to decide whether a fault is present. This can be achieved by increasing γ_2.

For values of γ_2 close to one, $r(t)$ will have an approximately Gaussian distribution with variance,

$$\sigma_2 = \frac{1-\gamma_2}{1+\gamma_2}$$

Since γ_2 is generally chosen in this region, it will be assumed that $r(t)$ has a Gaussian distribution. If $r(t)$ exceeds a certain threshold r_0, a fault may be concluded with a confidence determined from the value of the threshold. In the present algorithm, the threshold can be computed directly as a function of the rate of false alarms f_f. If a false alarm frequency equal to f_f is acceptable, a fault should be declared every time $r(t)$ is greater than the threshold r_0, defined by,

$$\mathrm{P}\left[r(t) \ge r_0\right] = \frac{1}{\sqrt{(2\pi)}\sigma}\int_{r_0}^{\infty} \exp\left[-\frac{x^2}{2\sigma^2}\right]\mathrm{d}x = f_f \tag{3.82}$$

If a small value of the threshold is chosen to make it possible to detect faults quickly, the false detection rate will be high. This is seen in (3.82) where there is an inverse relation between r_0 and f_f. The determination of r_0 by this method has the advantage that it is formulated in terms of the expected frequency of false detections, which may be chosen to suit any particular application.

This procedure is not sensitive to changes in noise level. This characteristic is not found in other fault detection methods. If this property is not needed, study of the quantity $w(t)^T w(t)$ can result in more powerful tests. However, it does suffer from the fact that it cannot locate the faulty parameter. To overcome this shortcoming, the following test based on the time dependancy of the estimated parameter can be performed:

Given the time–dependent quantity $a(k)$ define the function $f(a, N_1, N_2)$ as:

$$f(a,N_1,N_2) = \frac{1}{N_2 - N_1 + 1}\sum_{k=N_1}^{N_2} a(k)$$

Use the following detection criterion for each estimated parameter $\hat{\theta}_i$, $i\in[1,n]$,

$$J_i(t) = \frac{\left|\bar{\theta}_{S,i}(t) - \bar{\theta}_{L,i}(t)\right|}{\left|\bar{\theta}_{L,i}(t)\right|}$$

where,

$$\bar{\theta}_{s,i}(t) = \mathrm{f}\left(\hat{\theta}_i, t - N_S + 1, t\right)$$

$$\bar{\theta}_{L,i}(t) = f\left(\hat{\theta}_i, t - N_L - N_S + 1, t - N_S\right)$$

and N_S and N_L satisfying $0<N_S<N_L$ denote the short– and long–term estimator computation windows respectively. Then a decision on fault occurrence can be made according to the value of $J_i(t)$ as follows:

If	$J_i(t) \le J_{i\min}$ $\forall i\in[1,n]$:	no change
If	$J_{i\min} < J_i(t) < J_{i\max}$ for some $i\in[1,n]$:	alarm for ith parameter
If	$J_i(t) > J_{i\max}$ for some $i\in[1,n]$:	change in ith parameter

This procedure suffers however from the difficulty of choosing the following parameters: N_S, N_L, $J_{i\max}$, $J_{i\min}$; $i=1, \ldots, n$. These may be chosen through simulation studies only since their probability distributions are not known. Furthermore the conducted tests showed that this method is difficult to implement because the detection criterium value fluctuations are much greater after model changes, making the choice of thresholds even more difficult.

Decision based on $\hat{\boldsymbol{p}}$. After calculation of the physical process coefficients consider $\hat{\boldsymbol{p}}(k)$ as a gaussian vector with its components statistically independent and its realizations $\hat{\boldsymbol{p}}(i)$ and $\hat{\boldsymbol{p}}(j)$ in different sample instants $i<>j$ statistically independent. It is assumed that the mean vector $\boldsymbol{\mu}(k) = \mathrm{E}\{\hat{\boldsymbol{p}}(k)\}$ and the covariance matrix,

$$\boldsymbol{C}(k) = \mathrm{E}\left\{\left(\hat{\boldsymbol{p}}(k) - \boldsymbol{\mu}(k)\right)\left(\left(\hat{\boldsymbol{p}}(k) - \boldsymbol{\mu}(k)\right)^{\mathrm{T}}\right)\right\}$$

are invariant for the non–error case, i.e.

$$\boldsymbol{\mu}(k) = \boldsymbol{\mu}, C(k) = \text{diag}\left[\sigma_1^2, \ \ldots, \ \sigma_m^2\right]; \text{ all } k$$

Under these conditions the joint probability density function over N samples is defined as,

$$f(\hat{p}(1), \ldots, \hat{p}(n)) = \prod_{i=1}^{N} f(\hat{p}(i))$$

A fault will be declared by a significant deviation of the mean $\hat{\mu}_i$ and/or variance $\hat{\sigma}_i^2$ of $\hat{p}_i(k)$, the ith component of $\hat{p}(k)$, from the non–error values μ_i, σ_i^2.

This is a classical hypothesis test problem and can be handled by the formulation of $(m+1)$ hypotheses H_i, $0 \leq i \leq m$:

$$H_i = \begin{cases} \text{no fault in the mean and / or the variance of } \ \hat{\boldsymbol{p}}(i=0) \\ \text{fault of type } i \text{ (significant deviation of mean and / or variance of } \ \hat{p}_i \end{cases}$$

Each hypothesis H_i can be associated with a gaussian conditional density function, where $\mu(H_i)$ and $\sigma^2(H_i)$ denote conditional mean and variance for hypothesis H_i. Therefore, the non–error case is described by $\mu(H_0)$, $\sigma^2(H_0)$, whereas $\mu(H_i)$, $\sigma^2(H_i)$, $1<i<m$, describe a fault of parameter i.

Fault detection and localization is possible by computing appropriate logarithmic likelihood ratios. The algorithm used here for estimating the non–error statistics and the fault detection and localization is described by Geiger (1986), where a common window technique is used in order to calculate the relevant statistics of the fault occurrence. However, the procedure employed is non–iterative and this greatly increases computational time. An iterative procedure described by Pouliezos and Stavrakakis (1989) is:

(i) For the training/non–error case, H_0:

a) $$\mu_i(k) = \frac{1}{k}\left[(k-1)\mu_i(k-1) + \hat{p}_i(k)\right], \ i = 1, \ldots, m, \ \ k = 1, \ldots, N_s$$

The algorithm is initialized by $\mu_i(1) = \hat{p}_i(1)$.

b) $$\sigma_i^2(k) = \frac{k-2}{k-1}\sigma_i^2(k-1) + \frac{1}{k}\left(\hat{p}_i(k) - \mu_i(k-1)\right)^2 \quad i = 1, \ldots, m, k = 2, \ldots, N_s.$$

The algorithm is self–initialized by,

$$\sigma_i^2(2) = \frac{1}{2}\left(\hat{p}_i(1) - \hat{p}_i(2)\right)^2.$$

The sample size, N_s, is chosen so that an accurate estimate for the above statistics is obtained.

(ii) For the error–case, H_i, i=1, ..., m, three quantities are needed:

a) Window sample mean

$$\hat{\mu}_i(k) = \frac{1}{N_W} \sum_{j=k-N_w+1}^{k} \hat{p}_i(j)$$

$$= \hat{\mu}_i(k-1) - \frac{1}{n_W}\left(\hat{p}_i(k-N_w) - \hat{p}_i(k)\right)$$

$$= \hat{\mu}_i(k-1) - \frac{1}{N_w}\gamma_i(k); \;\; i = 1, \ldots, m$$

where,

$$\gamma_i(k) = \hat{p}_i(k-N_w) - \hat{p}_i(k); \;\; i = 1, \ldots, m$$

b) Window sample variance

$$\hat{\sigma}_i^2(k) = \frac{1}{N_W} \sum_{j=k-N_w+1}^{k} \left(\hat{p}_i(j) - \hat{\mu}_i(k)\right)^2$$

$$= \hat{\sigma}_i^2(k-1) + \frac{1}{N_W}\left[2\gamma_i(k)\hat{\mu}_i(k-1) - \frac{1}{N_W}\gamma_i^2(k) - \hat{p}_i^2(k-N_w) + \hat{p}_i^2(k)\right]$$

c) Window sample variance based on non–error mean

$$\tilde{\sigma}_i^2(k) = \frac{1}{N_W} \sum_{j=k-N_w+1}^{k} \left[\hat{p}_i(j) - \mu_i\right]^2$$

These three iterative schemes need a starting window of N_w sample values $\hat{p}_i$ in order to be initialized. The window size, N_w, is chosen so that reasonable rates for missed alarms and false detections are achieved. The advantage of using a small moving window of sample parameter values $\hat{p}_i(k)$ is in the improved speed of detection. This is offset, however, by an increasing false detection rate.

A fault in the ith parameter is declared at time k, if the quantity,

$$\Lambda_i(k) = \frac{N_W}{2}\left[\frac{\tilde{\sigma}_i(k)}{\sigma_i(k)} - 2\ln\left(\frac{\hat{\sigma}_i(k)}{\sigma_i(k)}\right) - 1\right]$$

exceeds a predetermined threshold in M consecutive time instants.

Alternatively the comparable quantity may be:

$$\tilde{\Lambda}_i(k) = \frac{\tilde{\sigma}_i(k)}{\sigma_i(k)} - 2\ln\left(\frac{\hat{\sigma}_i(k)}{\sigma_i(k)}\right)$$

resulting in less operations per time update. The threshold may be set by defining an a–priori probability of no–fault P_{i0}. Then a suitable test is,

$$\Lambda_i(k) \underset{<}{>} \ln \frac{P_{i0}}{1-P_{i0}}$$

An alternative discrimination measure is the *Kullback* discrimination index (KDI) proposed for fault detection purposes by Kumamaru *et al.* (1988). The KDI can be seen as distortion measure to compare two probability density functions. In fault detection situations one pdf refers to an interval of no–fault, while the other to subsequent intervals where fault monitoring is in effect. If these intervals are denoted I_1, I_2, their estimates $\hat{\boldsymbol{\theta}}_1, \hat{\boldsymbol{\theta}}_2(t)$ their points numbered by {1, 2, ..., N}, {1, 2, ..., t} respectively and parametric models of the form given by equation (3.7) are considered, the KDI can be written as,

$$I_t^N[1,2] = \int \rho(\boldsymbol{Y}_N / \hat{\boldsymbol{\theta}}_1, \boldsymbol{U}_{N-1}) \ln \frac{\rho(\boldsymbol{Y}_N / \hat{\boldsymbol{\theta}}_1, \boldsymbol{U}_{N-1})}{\rho(\boldsymbol{Y}_N / \hat{\boldsymbol{\theta}}_2(t), \boldsymbol{U}_{N-1})} \mathrm{d}\boldsymbol{Y}_N$$

where $\boldsymbol{Y}_i$,$\boldsymbol{U}_i$ are data collections up to point i for interval I_1, and ρ is the likelihood function. The index can be calculated iteratively using,

$$I_t^N[1,2] = I_t^N[1,2]^{(2)} + I_t^N[1,2]^{(3)}$$

where,

$$I_t^N[1,2]^{(2)} = \frac{1}{2}\sum_{k=1}^{N} \left\| H_2^{-1}(G_1 - G_2)u(k) \right\|_{\Lambda^{-1}}^2$$

$$I_t^N[1,2]^{(3)} = \frac{N}{2}\frac{1}{2\pi i}\oint \left[H_2^{-1}(z)H_1(z) - 1\right] \times \left[H_2^{-1}(z^{-1})H_1(z^{-1}) - 1\right]\frac{\mathrm{d}z}{z}$$

the index i denotes identified models using $\hat{\boldsymbol{\theta}}_i$, and $\Lambda=\sigma^2$ is the noise variance.

Once the index has been calculated, it leads to a decision problem using a threshold η:

$$I_N^t \underset{<}{>} \eta \Rightarrow \begin{cases} \hat{\theta}_1 \neq \hat{\theta}_2: \text{ fault} \\ \hat{\theta}_1 = \hat{\theta}_2: \text{ no fault} \end{cases}$$

The value of the threshold can be determined according to the statistical properties of the KDI. Under the normal situation Söderström and Kumamaru (1985), have shown that all the terms involved in the iteration have asymptotic χ^2 distributions with degrees equal to the dimension of the parameters included in their expressions. This index has performed quite well in adaptive control schemes, where a sensitive fault index is required because of the system's adaptation properties. In those cases monitoring of the estimation history alone is not enough to trigger fault alarms.

3.5 Practical examples

In this section several examples of fault detection methods using parameter estimation applied to real technological problems will be presented. These examples use different mixtures of models/estimation/decision approaches and thus present an interesting framework for comparison.

3.5.1 Evaporator fault detection

Dalla Molle and Himmelblau (1987), have applied real time parameter estimation techniques for fault detection in an evaporator. The complexities of a real evaporator have been simplified, so that the model reduces to,

$$\frac{\mathrm{d}x_1}{\mathrm{d}t} = F - (wx_1 + E_c) - V \tag{3.83a}$$

$$\frac{\mathrm{d}x_2}{\mathrm{d}t}\left[\beta F x_F + (V - F)(x_2 - TB)\right]/x_1 \tag{3.83b}$$

where,

$$[x_1\ x_2] = [\ W\ T\ ]$$

$$V = [UA(T_s - T) - FC_p(T - T_F) - Q_L]/\Delta H_v$$

and,

- UA : (heat transfer coefficient)×(area of heat transfer),
- T_S : steam temperature in steam chest,
- T_B : normal boiling point of solvent,
- C_p : heat capacity of solution,
- T_F : temperature of feed system,
- Q_L : rate of heat loss to the surrounding,
- ΔH_V : heat of vaporization of solvent,
- Q_s : total rate of heat transfer from steam,
- w : constant (0.6),
- β : boiling point elevation per mass fraction of solute,
- E_c : constant (0.1).

Figure 3.7 shows the rest of the notation.

Here the states of the model are the hold–up (w) and temperature (T) and two parameters of interest for process degradation are the heat transfer coefficient UA and the composition of feed x_F.

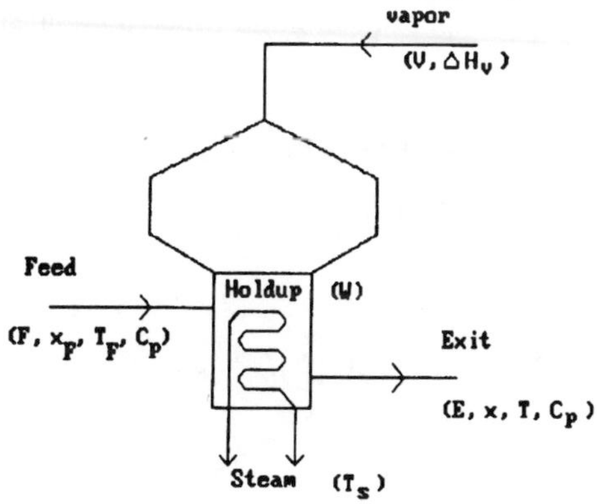

Figure 3.7. Evaporator configuration and notation

As the heat transfer surface becomes fouled or scaled, the heat transfer rate is decreased and the efficiency of the process is reduced. On the other hand composition at the input of the evaporator could be useful in determining if the previous unit was operating properly.

To illustrate the types of trajectories that occur for the two parameters, the following faults were simulated:

	UA	x_F
% change in value	−10.0 ramp	−20.0 square
Starting time of change (min)	75.0	165.0
Stopping time of change (min)	375.0	258.0

Noise was added to the process measurements to represent randomness, and was also introduced into the inputs. For the simulations all process parameters were assumed to remain constant (except for the fault parameters). The standard deviations of the noise factors are listed in Dalla Molle, (1985).

Two fault detection methods were used:

a. *Least squares with forgetting factor.* As they are, equations (3.83) are not suitable for applying the standard L.S. procedure (3.20). However, as shown by Dalla Molle, (3.83) can be put into the form:

$$\boldsymbol{S}(k)\boldsymbol{p} = \boldsymbol{b}(k) + \boldsymbol{\varepsilon}(k)$$

where,

$$\boldsymbol{b}(k) = \left[\boldsymbol{x}(k) - \boldsymbol{x}(k-1)\right]/\tau - \boldsymbol{A}\boldsymbol{x}(k) - \boldsymbol{B}\boldsymbol{u}(k) - \boldsymbol{r}(k)$$

and $s(k)$, $r(k)$ contain the coefficient terms of the fault parameters and other non–linear terms respectively and $\hat{o}$ is the discretization time constant. Then the parameters $p(k)$ can be estimated from the following L.S. equations:

$$p(k+1) = \left[I_s - U(k)V(k)S(k+1)\right]\left[p(k) + U(k)b(k+1)\right]$$

$$U(k) = (1/\lambda)\left[R^{\mathrm{T}}(k)R(k)\right]^{-1} S^{\mathrm{T}}(k+1)$$

$$V(k) = \left[I_n + S(k+1)U(k)\right]^{-1}$$

$$\left[R^{\mathrm{T}}(k+1)R(k+1)\right]^{-1} = (1/\lambda)R^{\mathrm{T}}R(k)\Big]^{-1} U(k)V(k)U^{\mathrm{T}}(k)$$

The initial values for the algorithm are given by,

$$p(0) = \left[R^{\mathrm{T}}(0)R^{-1}(0)b(0)\right];\ \left[R^{\mathrm{T}}(0)R(0)\right]^{-1} = \left[S^{\mathrm{T}}(0)S(0)\right]^{-1}$$

The results of the simulations are shown in Figures 3.8, 3.9.

These results demonstrate that the least squares estimation scheme is valid even when two faults occur simultaneously. However, it can be seen in Figure 3.9 that for abrupt changes such as that applied to x_F, the estimate responds slowly and convergence to the new value takes nearly 60 minutes. The rate of convergence of the estiamte could be increased by decreasing the weighting factor, but the variance of the faster response would be larger. The effect of making process measurements at discrete intervals rather than continuously is a slower response in the estimates. A larger sampling interval reduces the effect of measurement noise on the approximation of the derivative of the states because the approximations are averaged over a longer time.

Hence, the variance of the estimates does not necessarily increase with the slower response due to sampling. The speed of the response for the discrete case can also be increased by decreasing λ, but only at the expense of the variance at a constant sampling interval.

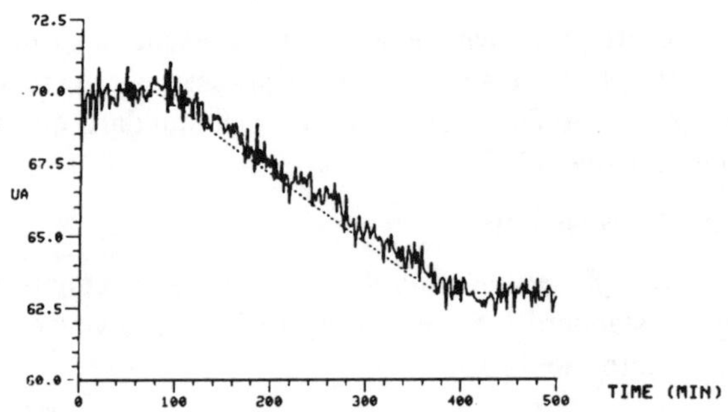

Figure 3.8. Estimate of UA for λ=0.95; (——) Estimate; (– – –) Measured value.

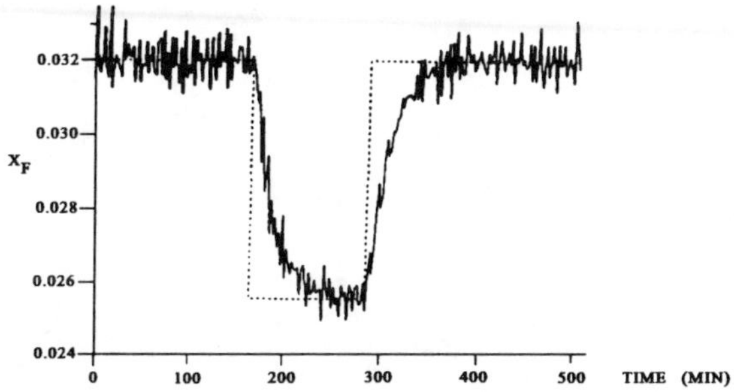

Figure 3.9. Estimate of x_F for λ=0.95; (——) Estimate; (- - -) Measured value.

b. *Kalman Filter:* By modeling the parameter evolution as,

$$\mathrm{d}\boldsymbol{p}(t)/\mathrm{d}t = \boldsymbol{p}(t) + \boldsymbol{w}_p(t)$$

where $\boldsymbol{w}_p(t)$ is $N(\mathbf{0}, \boldsymbol{Q}_p(t))$, and augmenting this differential equation to the state equation, the following system is obtained:

$$\dot{\boldsymbol{x}}_{\mathrm{a}} = \begin{bmatrix} \dot{\boldsymbol{x}} \\ \dot{\boldsymbol{p}} \end{bmatrix} = \begin{bmatrix} \boldsymbol{f}(\boldsymbol{x},\boldsymbol{u},\boldsymbol{p},t \\ 0 \end{bmatrix} + \begin{bmatrix} \boldsymbol{G}(\boldsymbol{x},\boldsymbol{p},t) & \mathbf{0} \\ \mathbf{0} & \boldsymbol{I}_p \end{bmatrix} \begin{bmatrix} \boldsymbol{w}(t) \\ \boldsymbol{w}_p(t) \end{bmatrix} \tag{3.84}$$

To implement an extended Kalman filter for (3.83) using the representation (3.84) initial conditions must be supplied for the states, parameters, and the error covariance matrix of the augmented state vector. For the states and parameters the design or normal operating values can be used as the initial conditions. The initial error covariance matrix is assumed to be diagonal with large values of the elements to express uncertainty in the initial values of the states and parameters. In addition to the initial values, the noise covariance matrices $\boldsymbol{R}(t)$ and $\boldsymbol{Q}(t)$ must also be supplied. Normally, the measurement noise covariance matrix, $\boldsymbol{R}(t)$, is assumed to be diagonal. The variances of each measurement can be guessed or estimated from sample output values. The input noise covariance matrix, $\boldsymbol{Q}(t)$, is also assumed to be diagonal. Although the values of the elements in $\boldsymbol{R}(t)$ and $\boldsymbol{Q}(t)$ might be obtained from process measurements, the filter is usually "tuned" to dynamics of the process so that the response for the parameter trajectories is reasonably fast and their covariance matrix elements are of reasonable size. Figures 3.10a and 3.10b illustrate the trajectories of x_F and UA for the Kalman filter in which confidence limits (P=0.95) have been put in place, based on the period of normal operation.

The simulation runs demonstrated that there was a neeed for some heuristics in the analysis of the estimates to avoid misdiagnosing nonexistent faults when more than one fault occurs at a time. Factors such as decision rules, confidence coefficients and so on were obtained based on the selected filter parameters and the dynamics of the process.

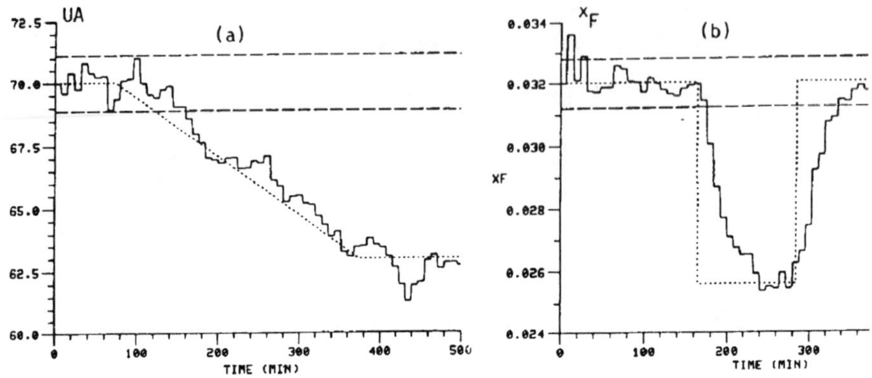

Figure 3.10 EKF estimate of *UA* and x_F with confidence intervals (—) Estimate; (- -) True value; (— —) Confidence limits.

3.5.2 Gas turbine fault detection and diagnosis

Gas turbine performance degrades over time due to the influence of many effects including tip clearance changes in the rotating components, seal wear, blade fouling, blade erosion, blade warping, foreign object damage, actuator wear, blocked fuel nozzles and sensor problems. In some applications, such as in the commercial transport field, the availabily of reliable cruise data facilitates the use of performance trending techniques for alerting maintenance personnel to emerging problems. However, the successful implementation of any trending techniques to gas turbine performance data still depends very largely on the skill and experience of the operator, especially when trying to diagnose some faults to module or line replaceable unit level. This situation is futher exacerbated in the military area because combat aircraft, in particular, seldom operate with their engines in a steady–state condition for extended periods. Thus, the selection of a suitable data capture window to provide maintenance personnel with reliable steady–state data is often difficult without resorting to dedicated tests, either on the ground or in– flight. In view of this, it would be convenient if operational transient engine data could be used for assessing engine condition and for diagnosing some of the more difficult engine faults.

Current generation military aircraft are often equipped with an Engine Monitoring System (EMS) which can be configured to capture selected engine data under certain conditions. These conditions include, during each take–off and in–flight if one or more of the measured parameters exceed predetermined limit values. The take–off data, in particular, has the potential to provide a consistent data base for assessing engine condition provided the analytical means are available for extracting the fault information.

Because these data comprise engine accelerations from part–power positions, the current steady– state methods for assessing engine condition are not suitable. Methods have been developed for extracting fault information from gas turbine transient data (Baskiotis *et al*., (1979), Carlsson *et al*., (1988), Henry (1988), Smed *et al*. (1988)). However, these methods suffer in their ability to detect small changes which usually accompany the presence of degraded engine components.

A L.S. method of analysing transient engine data based on the *stochastic embedding* principle has been implemented by Merrington *et al*., (1991). This method, which has the potential to detect the presence of degraded engine components from the actual EMS take–off measurements, follows.

Exact models of aircraft engines are highly nonlinear (Merrill, 1984) and thus simplified linearized models are usually employed (Dehoff *et al*., 1978). For example, taking the engine fuel flow W_F as the input and the fan–spool speed N_L as the output, an appropriate linearized nominal model is given as follows:

$$\Delta N_L(t) = \frac{b_{1c}p + b_{0c}}{p^2 + f_{1c}p + f_{0c}} \Delta W_F(t) \qquad (3.85)$$

where p denotes the differential operator.

Taking noise and linearization errors into consideration, the underlying system can be described by the following discretized model:

$$\Delta \acute{I}_L(k) = G(q^{-1})\Delta W_F(k) + G_{n\Delta}(q^{-1})[\Delta W_F(k)]^2 + v(k) \qquad (3.86)$$

where,

$$G(q^{-1}) = \frac{B(q^{-1},\theta)}{A(q^{-1})} = \frac{b_1 q^{-1} + b_2 q^{-2}}{1 + a_1 q^{-1} + a_2 q^{-2}}$$

The denominator $A(q^{-1})$ is determined from a priori information about the system, e.g., approximate values of dominant poles or by some prior estimation experiments.

Using this system description the system ouput has the form,

$$\Delta N_L(k) = B(q^{-1},\boldsymbol{\theta})\Delta W_F' + \eta(k) \qquad (3.87)$$

where,

$$\Delta W_F' = \frac{1}{A(q^{-1})}\Delta W_F(k)$$

$$\eta(k) = G_{n\Delta}\left[\Delta W_F'(k)\right]^2 + v(k)$$

Equation (3.87) can be put in the standard regression form of (3.61) if,

$$\varphi(k) = \left[\Delta W_F'(k-1) \quad \Delta W_F'(k-2)\right]$$
$$\theta = \left[b_1 \quad b_2\right]$$

Two noise–free non–faulty data sets (CLF6 and CLF61) and a faulty data set (LTEF) with a –2% change in the low pressure turbine efficiency were chosen for the study (Figure 3.11). Note that LTEF has the same operating point as that of CLF6 but that CLF61 has a different operating point with a very similar output as LTEF.

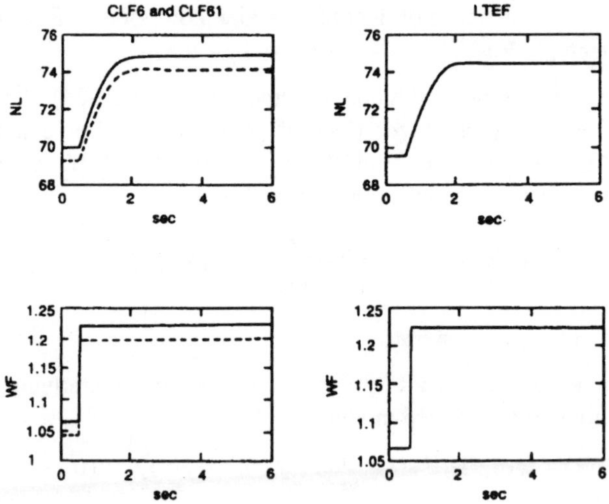

Figure 3.11 Non–faulty data sets (—CLF6; – –CLF61) and faulty data set (LTEF) in aircraft engines.

Using the data sets and the theory of section 3.3.5 appropriate test variables for fault detection can be formulated. For example, equations (3.71), (3.72) may be used:

$$T_1 = \left[\hat{\theta}_n - \hat{\theta}_f\right]^{\mathrm{T}} C^{-1}\left[\hat{\theta}_n - \hat{\theta}_f\right] \tag{3.71}$$

$$T_2 = \left[\hat{\theta}_n - \hat{\theta}_f\right]^{\mathrm{T}}\left[\mathrm{diag}(C)\right]^{-1}\left[\hat{\theta}_n - \hat{\theta}_f\right] \tag{3.72}$$

The following constants were chosen: sampling period T_s=0.02, number of data points N=350, $\sigma_v^2 = 0.15^2$ (with a reference value of 100%) and the input ΔW_F was assumed to be corrupted by white noise with variance $\sigma_u^2 = 0.003^2$ (with a fuel range of 0 to 1). The fixed denominator was taken by prior experiments as a_1=–1.8238 and a_2=0.8294 and the values of β_n and σ_n^2 as $\beta_n = 0.0837$ and $\sigma_n^2 = 0.0818$.

Simulation results for test T_1 are shown in Figure 3.12 and summarized in Table 3.3.

Note that 100 trials were conducted with different noise realizations. These results show that this fault detection method works very well even under the effect of linearization errors.

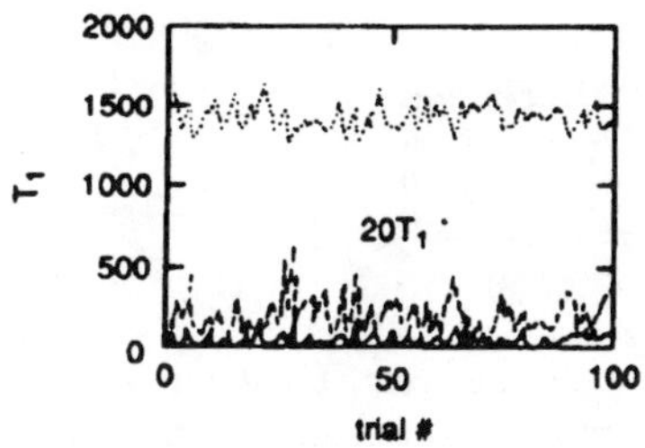

Figure 3.12 Simulation results
(_NF1: no–fault; —NF2: no– fault; ...F3: fault)

Table 3.3 Cases in aircraft engine fault detection simulation

Case no.	Experiment n	Experiment f	T_1
NF1	CLF6	CLF6	2.45±2.42
NF2	CLF6	CLF61	9.67±6.18
F3	CLF6	LTEF	1459.88±85.07

3.5.3 Fault detection for electromotor driven centrifugal pumps

The early detection of process faults is especially atttractive for engines. In this example a centrifugal pump with a water circulation system , driven by a speed–controlled direct current motor is considered (Figure 3.13, after Iserman, 1984). The goal is to detect changes (faults) in the d.c. motor, the pump and the circulation system based on theoretically derived process models and parameter estimation.

The dynamic models of the d.c. motor, the centrifugal pump and the pipe system are gained by stating the balance equations for energy and momentum and by using special physical relationships. In order not to obtain too many parameters, appropriate simplifications have to be made, as lumping of more than one process coefficients together, e.g. the friction coefficients of the motor c_{FM1} and the pump c_{FPw} and the torque coefficient g_w of the pump.

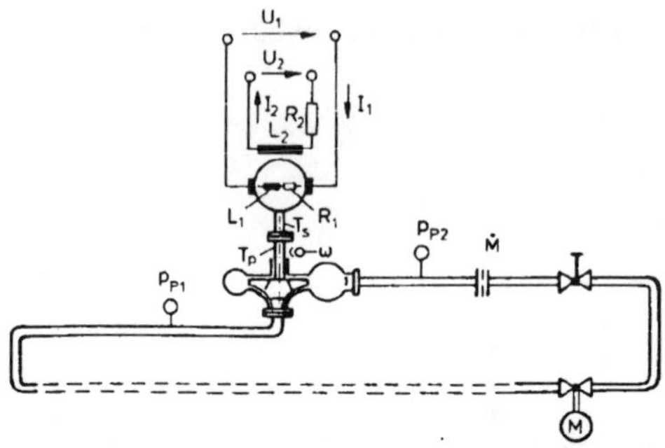

Figure 3.13 Scheme of a speed controlled d.c. motor and centrifugal pump.
M mass flow, angular velocity, T torque, U voltage, I current, R resistance, L inductance.

The resulting four basic equations will be used for parameter estimation in the following form:

(a) Armature circuit:

$$\frac{dI_1(t)}{dt} = \alpha_{11}\Delta I_1(t) + \alpha_{12}\Delta w(t) + b_1\Delta U_1(t)$$

(b) Mechanics of motor and pump:

$$\frac{dw(t)}{dt} = \alpha_{21}\Delta I_1(t) + \alpha_{22}\Delta w(t) + \alpha_{23}\Delta M(t) \tag{3.88}$$

(c) Pipe system:

$$\frac{d\dot{M}}{dt} = \alpha'_{33}\Delta\dot{M}(t) + d_3\Delta Y(t) \tag{3.89}$$

(d) Pump (specific energy Y):

$$\Delta(Y) = h_w(t) + h_M\Delta M(t) \tag{3.90}$$

The parameters are,

$$\alpha_{11} = -\frac{R_1}{L_1},\ b_1 = \frac{1}{L_1},\ \alpha_{12} = -\frac{\Psi}{L_1},\ \alpha_{21} = \frac{\Psi}{\theta} = \frac{\Psi}{\theta_M + \theta_p}$$

$$\alpha_{22} = -\frac{c_{F1} + g_w}{\theta},\ \alpha_{23} = -\frac{g_M}{\theta},\ a'_{33} = -\frac{\alpha_R}{\alpha_{ac}}\ \ d_3 = \frac{1}{\alpha_{ac}} \tag{3.91}$$

A state variable representation,

$$\dot{\boldsymbol{x}}(t) = \boldsymbol{A}\boldsymbol{x}(t) + \boldsymbol{b}u(t)$$
$$\boldsymbol{y}(t) = \boldsymbol{C}\boldsymbol{x}(t)$$

can be given with the following definitions:

$$\boldsymbol{x}(t) = \begin{bmatrix} \Delta I_1(t) \\ \Delta w(t) \\ \Delta \dot{M}(t) \end{bmatrix}, \quad \boldsymbol{A} = \begin{bmatrix} \alpha_{11} & \alpha_{12} & 0 \\ \alpha_{21} & \alpha_{22} & \alpha_{23} \\ 0 & \alpha_{32} & \alpha_{33} \end{bmatrix}$$

$$\boldsymbol{y}(t) = \begin{bmatrix} \Delta I_1(t) \\ \Delta w(t) \\ \Delta \dot{M}(t) \\ \Delta Y(t) \end{bmatrix} \quad \boldsymbol{C} = \begin{bmatrix} 1 & 0 & 0 \\ 0 & 1 & 0 \\ 0 & 0 & 1 \\ 0 & h_w & h_M \end{bmatrix}$$

$$\boldsymbol{b} = \begin{bmatrix} b_1 \\ 0 \\ 0 \end{bmatrix}, \; u(t) = \Delta U_1(t), \; \alpha_{32} = \frac{h_w}{\alpha_{ac}}, \; \alpha_{33} = \alpha'_{33} + \frac{h_w}{\alpha_{ac}}$$

A block diagram of the modeled system is given in Figure 3.14. The parameters of (3.87)–(3.90) can be estimated by bringing them into the form of (3.9) and applying the least–squares method. The simple case of the d.c. motor and pump with closed valve and measured signals ΔU_1, ΔI_1 and Δw will be considered.

In this case $M(t)$=0, so that only (3.87) and (3.88) are to be used. Both equations are written due to (3.9),

$$y_1(t) = \boldsymbol{\psi}_1^{\mathrm{T}}(t)\hat{\boldsymbol{\theta}}_1, \; y_2(t) = \boldsymbol{\psi}_2^{\mathrm{T}}(t)\hat{\boldsymbol{\theta}}_2$$

where,

$$y_1(t) = \mathrm{d}I_1(t)/\mathrm{d}t; \; y_2(t) = \mathrm{d}w(t)/\mathrm{d}t$$

$$\psi_1^{\mathrm{T}}(t) = \begin{bmatrix} \Delta I_1(t) & \Delta w(t) & \Delta U_1(t) \end{bmatrix}$$

$$\hat{\theta}_1^{\mathrm{T}}(t) = \begin{bmatrix} \hat{\alpha}_{11} & \hat{\alpha}_{12} & \hat{b}_1 \end{bmatrix}$$

$$\boldsymbol{\psi}_2^{\mathrm{T}}(t) = \begin{bmatrix} \Delta I_1(t) & \Delta w(t) \end{bmatrix}$$

$$\hat{\boldsymbol{\theta}}_2^{T}(t) = \begin{bmatrix} \hat{\alpha}_{21} & \hat{\alpha}_{22} \end{bmatrix}$$

Using (3.91), the following five process coefficients can be calculated based on the five parameter estimates $\hat{\boldsymbol{\theta}}_1$ and $\hat{\boldsymbol{\theta}}_2$:

$$\hat{L}_1 = 1/\hat{b}_1, \; \hat{R}_1 = -\hat{\alpha}_{12}\hat{L}_1 = -\hat{\alpha}_{11}/\hat{b}_1$$

$$\hat{\Psi} = -\hat{\alpha}_{12}\hat{L}_1 = -\hat{\alpha}_{12}/\hat{b}_1, \; \hat{\theta} = \hat{\Psi}/\hat{\alpha}_{21} = -\hat{\alpha}_{12}/\hat{b}_1\hat{\alpha}_{21}$$

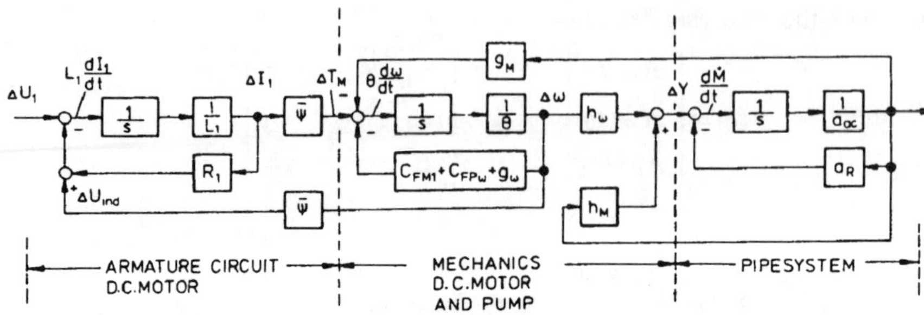

Figure 3.14: Block diagram of the linearized d.c. motor–pump–pipe system

$$\hat{c}_{F1} = c_{FM1} + c_{FMw} = \hat{\alpha}_{22}\hat{\theta} = \hat{\alpha}_{22}\hat{\alpha}_{12} / \hat{b}_1\hat{\alpha}_{21} \tag{3.92}$$

Hence, all process coefficients which describe the linearized dynamic behaviour can be calculated. However, the friction coefficients of the motor c_{FM1} and the pump c_{FMw} and the moments of inertia θ_M and θ_P are lumped together so that only their sum can be gained. If not the dynamic behaviour, but only the static behaviour could be identified, L_i and θ could be not obtained. This shows that by identifying the dynamics, more parameters can be estimated and therefore more process coefficients can be monitored.

A disadvantage of the linearized dynamic relationships is that the coefficients c_{FM0} for the adhesive friction do not appear. However, it can be shown that with the assumption that the friction torque only depends linearly on the speed,

$$T_{FM}(t) = c_{FM0} + c_{FM1}w(t)$$

$$T_{FP}(t) = c_{FP0} + c_{FP1}w(t)$$

$$(\theta_P + \theta_M)(\,\mathrm{d}w(t)/\mathrm{d}t\,) = \Psi I_1(t) - c_{F0} - c_{F1}w(t) \tag{3.93}$$

$$c_{F0} = c_{FM0} + c_{Fp0}$$

$$c_{F1} = c_{FM1} + c_{FP1}$$

Then the absolute values $w(t)$ and $I_1(t)$, and not their deviations, are used and the estimation of c_{F0} also becomes possible (Geiger, 1982).

Experiments were made with a centrifugal pump driven by a speed controlled d.c. motor. The technical data are,

(a) D.C. motor:
maximum power $P_{max} = 4kW$
maximum rotation speed $N_{max} = 3000$ rev/min

(b) Centrifugal pump, one stage:
maximum total head $H_{\max} = 39m$
for $N_{\max} = 3000$ rev/min

(c) Pipe system:

length = 10*m*
diameter, d_i = 50*mm*

The d.c. motor is controlled by an a.c./d.c. converter with cascade control of the speed and the armature current as auxiliary control variable. The manipulated variable is the armature current U_1. For the experiments the reference value *W*(*t*) of the speed control has been changed stewise with a magnitude of 2% of N_{max}, i.e. 60 rev/min every 60*s*. The measured signals were sampled with sampling time T_0=2*ms* over a period of 2*s*, so that 1000 samples were obtained.

After 2*s* measurements the parameters were estimated off–line, using the recursive least–squares method with state variable filters for the determination of the time derivatives. As the noise is negligibly small, the parameter estimates can be assumed to be unbiased. In order to obtain the adhesive friction coefficient, (3.93) has been used,

$$\theta\frac{\mathrm{d}w(t)}{\mathrm{d}t}=\Psi I_1(t)-c_{F0}-c_{F1}w(t)$$

together with,

$$L_1=\frac{\mathrm{d}I_1(t)}{\mathrm{d}t}=U_1(t)-R_1I_1(t)-\Psi w(t).$$

for the armature circuit. Therefore the deviations of the signals have to be replaced in (3.91) by their absolute values. The process coefficients are obtained by (3.92) with $\hat{c}_{F0}=-\hat{\alpha}_{23}\hat{\theta}$ in addition.

In Figures 3.15–3.18 results of the parameter monitoring are presented. Figure 3.15 shows the step responses after a speed setpoint change. The resulting process coefficients after a start of the cold engine (Figure 3.16), indicate that the armature resistance increases during the first 10min, the flux linkage decreases during 20 min and the friction torque coefficient decreases during the first hour. Hence, small changes of the process coefficients can be detected. Figures 3.17 and 3.18 show the reaction on artificial changes (faults). A significant change of the armature resistance estimate is detectable after a 7% change (Figure 3.17). The effect of tightening and loosening the screws of the pump packing box cap is clearly seen in Fig. 3.18. More details are given in Geiger (1982). Results of more experiments including multiple hypothesis testing for the fault decision are described in Geiger (1984).

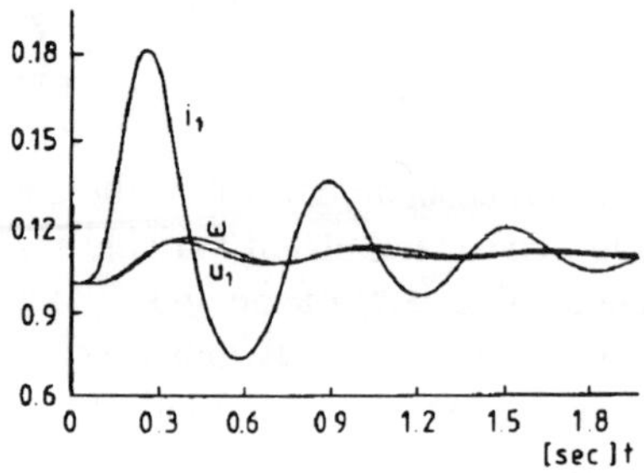

Figure 3.15 Step responses for a change of the speed setpoint.
$u_1=U_1/\overline{U}_1$, armature voltage $\overline{U}_1=60V$
$i_1=I_1/\overline{I}_1$, armature current, $\overline{I}_1=0.5A$
$w=w_1/\overline{w}_1$, angular velocity, $\overline{w}_1=62.83s^{-1}$ ($\approx$600 rev min^{-1})

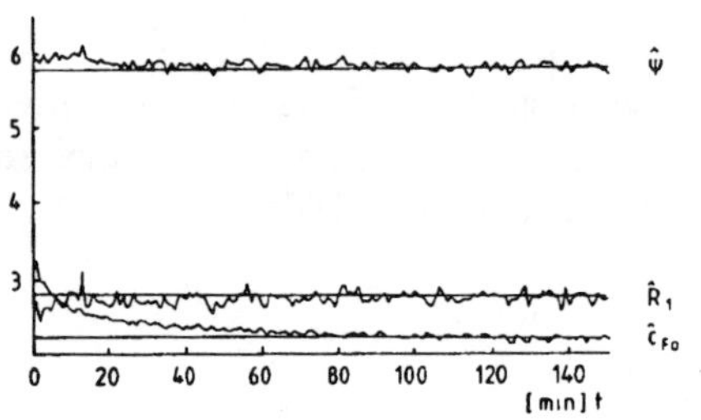

Figure 3.16 Process coefficient estimates after start of the cold engine.
R_1 armature resistance, Ψ flux linkage, c_{F0} friction coefficient.

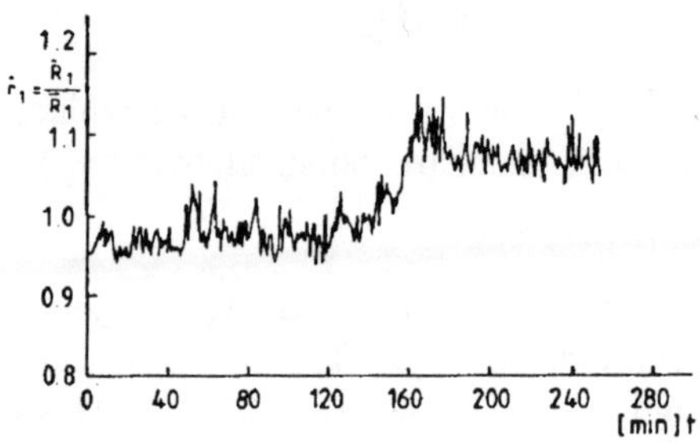

Figure 3.17: Change of armature circuit resistance.

$\Delta R_1 \approx 0.2$, $\overline{R}_1 \approx 3\Omega$

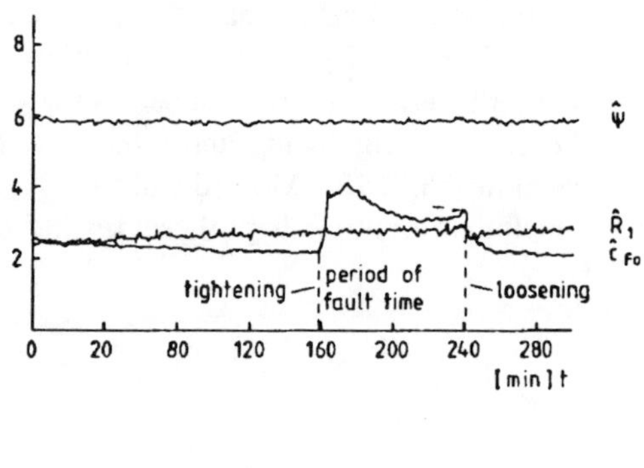

Figure 3.18: Change of pump packing box friction by tightening and loosening of the cap screws.

3.5.4 Fault detection in power substations

Stavrakakis and Dialynas, (1991), have used recursive least squares estimation with forgetting factor and hypothesis testing techniques on the process parameter values, for improving the reliability performance of power substations. Following a positive fault decision, the substation is reconfigured according to a detailed fault tree. The fault detection methodology adopted was applied to the following power substation components:

A. Power transformers, modeled by their one–phase equivalent circuit, described by,

$$V_i = R_1 I_i + L_{1e}\frac{dL_i}{dt} - M\frac{dI_0}{dt} \tag{3.94}$$

$$V_0 = M\frac{dI_i}{dt} - R_2 I_0 - L_{2e}\frac{dI_0}{dt} \tag{3.95}$$

where,

V_i, V_0 : actual input (primary) and output (secondary) voltages,
I_i, I_0 : actual input (primary) and output (secondary) currents,
R_1, R_2 : primary and secondary winding resistances,
L_1, L_2 : primary and secondary winding self–inductances,
L_m : mutual inductance between windings on the same core, and,

$$M = \frac{L_m}{a},\ L_{1e} = L_1 + L_m\ L_{2e} = L_e + \frac{L_m}{a_2}.$$

The faults that most frequently arise in practice in the power transformers, were classified as follows:

1. Failures in the magnetic circuits (cores, yokes and clamping structure).
2. Failures in the windings (coils and minor insulation and terminal gear).
3. Failures in the dielectric circuit (oil and major insulation).
4. Structural failures.

By monitoring the estimated values of R_1, R_2, L_{1e}, L_{2e}, M and performing a hypothesis testing using the likelihood ratio test, a change in these parameters can be detected, leading to a decision regarding one of the failures 1–4, described above.

B. Substation lines and cables, modeled by their equivalent one–phase circuit which neglects entirely the susceptance and leakance, and is described by the simple first order differential equation,

$$V_i = V_0 + RI_i + L\frac{dI_i}{dt} \tag{3.96}$$

The most important failures occurring on the lines or cables of power substations are the short circuits which are generally due to insulation breakdown. By applying the previously described method on the parameters R and L of this model, short circuits can be detected and localised early, in this way avoiding further degradation of the system.

C. *Synchronous generators.* The model used corresponds to an unsaturated cylindrical–rotor machine under balanced polyphase conditions, and is described by,

$$E_f = V_0 + r_a I_0 + L_s \frac{\mathrm{d}I_0}{\mathrm{d}t} \tag{3.97}$$

where,

- V_0 : actual value of terminal voltage,
- E_f : actual value of the excitation voltage,
- L_s : synchronous reactance (constant at constant frequency),
- r_a : armature resistance.

Here, deviations of L_s, r_a from their nominal values, will indicate a voltage failure of the ac synchronous generator, which is a result of an open in the field circuit, an open in the field rheostat or a failure of the exciter generator. The loss of field excitation to a generator operating in parallel with others, causes it to lose load and overspeed. High armature current caused by the high voltage differential between the armature and the bus, and the high currents induced in the field iron and field windings by the armature current, will cause rapid heating of the apparatus. This is avoided, in the case of failure, by the fast detection which the proposed method provides.

Substation configuration after the diagnosis of a failure condition. After the diagnosis of a failure condition on a substation component has been deduced by the previously described methodology, the circuit breakers which surround the component are instructed to open. System restoration follows with automatic or manual switching sequences which aim at minimising the effects of the outage by returning to service the healthy components and circuits as quickly as possible. The effect that the outage will have on the continuity of supply to each circuit load–point can then be assessed. Two types of load point failures can generally be recognised: (1) total loss of continuity which causes no load to be supplied; (2) partial loss of continuity which causes part of the load to be supplied. Load–point supply restoration can be achieved by closing components (breakers and/or isolators) which are in an open condition to provide alternative routes for power supply. These routes can be deduced from the list of available normally open paths leading to the load–point of interest from all the sources. An approach must therefore be developed to identify the suitable substation configurations and evaluate the supplied load. In most cases the outages being considered can be assumed to be of first order but a more complete assessment of substation operation would require the simulation of up to third order outages so that common mode faults can be also considered. The outages which may occur in power substations can be generally divided into the following six categories according to the type of the available restoration procedures:

Category A: Outage on components belonging to the incoming or outgoing circuits. The circuit node can not be reconnected to the substation because no alternative route is available. However, if the fault has occurred on an outgoing circuit component, alternative restoration procedures for other load– points may exist.

Category B: Outage on the isolators belonging to the interlocking scheme of the substation. More than one alternative restoration procedures may exist and can be deduced from the list of the isolators (branches) belonging to the same interlocking sequence with these taken out. In the case that the outage being considered is of a second or third order, the possible alternative restoration procedures are the respective second and third order combinations of all the isolators (branches) belonging to the corresponding interlocking sequences. If an outage occurs on isolators 16 and 24 in the substation of Figure 3.19 four possible alternative procedures are available by operating either on the pairs of isolators (17,25), (17,25), (18,26) as can be seen from Table 3.4. If a busbar configuration contains two or more bus–tie breaker schemes, one of the schemes is normally in operation while the breakers of the other schemes remain open. In the event of an outage on a component of the scheme being closed, one of the other schemes may become effective to provide additional routes of supply.

Table 3.4 Interlocking scheme of substation of Figure 3.19

No.	Isolator sequence	Branch sequence
1	7, 8	2, 3
2	9,10	4,5
3	33,34	15,16
4	35,36	17,18
5	62,63	28,29
6	64,65	30,31
7	87,88	41,42
8	89,90	43,44
9	16,17,18	7,8,9
10	24,25,26	11,12,13
11	42,43,44	20,21,22
12	50,51,52	24,25,26
13	70,71,72	33,34,35
14	78,79,80	37,38,39
15	95,96,97	46,47,48
16	103,104,105	50,51,52

Category C: Outage on the busbar section(s). All the incoming or outgoing circuits connected to these busbar sections are disconnected and each circuit can be transferred to any of the available busbars by closing the appropriate breakers and isolators.

Category D: Outage on the components (breaker, isolator) of the busbar sectionilising branches. After isolation of the outage, the respective busbars are divided into two or more parts not directly connected to each other. If this substation configuration is not operationally accepted, all the circuits connected to the affected busbar sections can be transferred to other busbars with the same restoration procedure followed after the occurence of an outage on busbar sections (Category C).

Category E: Outage on components belonging to a branch containing a transformer. Since the power supply from the superior to the inferior voltage level is decreased, alternative restoration procedures may exist and can be deduced from the list of the transformer branches being open.

Category F: Outage on the remaining substation components. Alternative restoration procedures may exist.

The basic steps of the developed algorithm for deducing the suitable substation configuration after the diagnosis of a substation abnormality are the following:

(i) Consider the detected faults and simulate the corresponding outage.
(ii) Depending on the outage category:
(a) For outage category B, detect the isolators (and their corresponding branches) belonging to the same interlocking sequence with these taken out. Deduce their second and third order combinations of the outage contains two and three isolators of such type respectively.
(b) For outage category E, detect the substation open branches containing transformers.
(c) For outage categories F and A on outgoing circuit components, detect the substation open branches not considered in steps (a) and (b) and their second and third order combinations.
(d) For outage categories C and D, detect the breakers and isolators which may close to transfer the disconnected circuits to healthy busbars. Deduce all the alternative restoration procedures by considering the substation interlocking shceme.
(iii) Deduce the list of possible alternative restoration procedures by combining the relevant switching actions obtained in step (ii).
(iv) For each circuit load–point to be considered:
(a) Read the paths from data base.
(b) Identify the closed and open paths.
(c) For each open path deduce the order of its discontinuity by counting the contained open components.
(d) For either total loss of continuity (no path in operation) or partial loss of continuity (one or more paths in operation, the supplied load less than required),

consider all the possible alternative restoration procedures and for each of them:

- Detect the paths which can be closed by considering only the paths with order of discontinuity less than or equal to the order of the procedure.
- If one or more paths can be closed, evaluate thc load supplied to the load–point being considered by performing a load–flow on the modified substation configuration. This configuration contains a limited number of nodes since all the substation busbars connected to each other by branches having zero impedance are linked together.

In order to illustrate the increased and more meaningful information for substation opearation that can be achieved using the described computational techniques, a typical 400/150 KV high voltage substation was analysed. The substation employs the triple busbar scheme for all system busbars and its detailed one line diagram is shown in Figure 3.19. It consists of 34 nodes, 58 branches and 109 components while its interlocking scheme is shown in Table 3.4. An opearational substation configuration was studied by assuming the breakers and isolators status shown in Figure 3.19. Source points are assumed to be the circuit busbar L8 and the generator busbar L17 while load–points are the nodes L25 and L33. The minimal paths leading to each load–point from all sources were deduced and retained in compact form in a data base. Finally, parameter estimation methods and hypothesis testing on the process parameters were used to deduce the alternative restoration procedures which are available after the diagnosis of faults on the substation components. The category of each fault and the components to close are shown in Table 3.5. For category E and F faults, it has also been assumed that breakers 40 and 93 are open and 32 and 86 are closed.

Table 3.5 Substation configuration after restoration of supply

Fault on component	Fault category	Alternative restoration switching operations	
		Components to close	Additional routes of suuply to L33
101	A	–	no
104	B	105	no
		103	yes
62	B	63	yes
		86	yes
64	B	65	no
		85	yes
24	B	25	yes
		26	no
82(L28)	C	103	yes
		105,65	yes
53	D	26,8,10	yes
67	E	40,93	yes
66	E	40,93	yes
14	F	40,93	yes

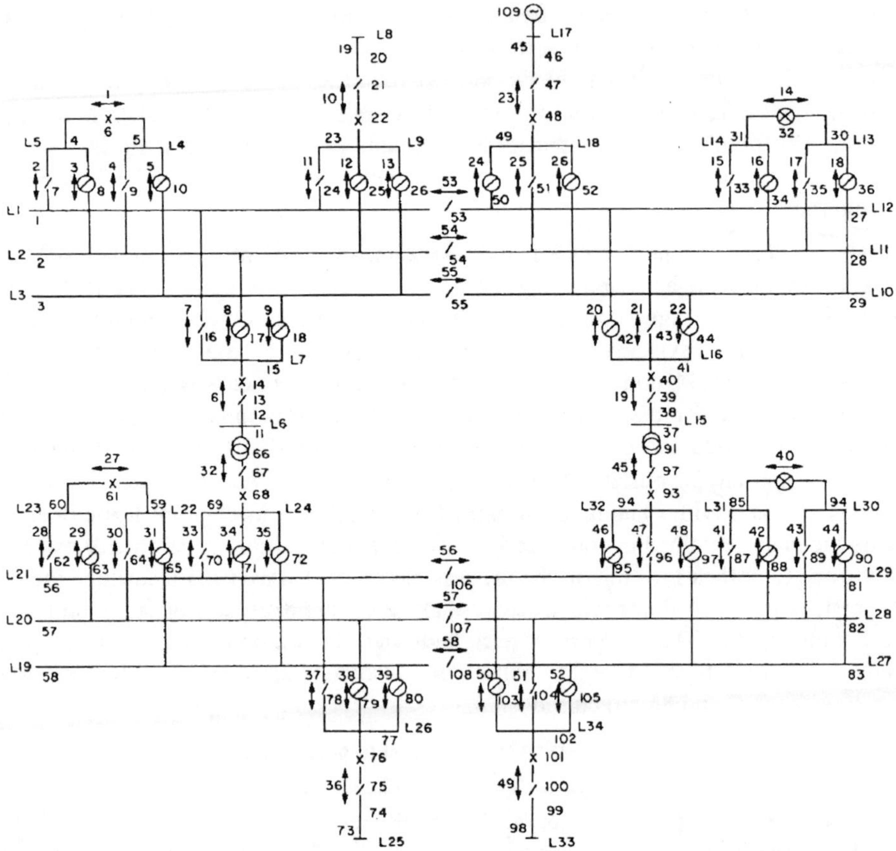

Figure 3.19: Detailed one line diagram of a typical high voltage substation. L5, node; x, circuit breaker (closed); ⊗, circuit breaker (open); /, isolator (closed); ⊘, isolator (open); ⦶⦶, transformer; –, line; ⊖, generator.

3.5.5 *Fault diagnosis in robotic systems*

Stavrakakis *et al.*, (1990), describe a fast fault detection system for robotic D.C. motor drives. The detection system is implemented on a commercially available parallel processing machine.

Using the global dynamic model of a 3 degrees of freedom robotic manipulator derived by Tzafestas and Stavrakakis (1986), the state–space representation for the actuator of the *i*'th link of the robot can be written as,

$$y^{(1)}(t) + A_1 y^{(1)}(t) + A_0 y(t) = B_0 u(t) \qquad (3.98)$$

where,

$$A_1 = 0 \in R^{4\text{x}4}$$

$$A_0 = \begin{bmatrix} \dfrac{R_i}{L_i} & \dfrac{K_{mi}N_i}{L_i} \\ -\dfrac{K_{mi}}{J_{mi}N_i} & \dfrac{\rho_i}{J_{mi}} \end{bmatrix}, \quad B_0 = \begin{bmatrix} \dfrac{1}{L_i} & 0 \\ 0 & -\dfrac{1}{J_{mi}N_i^2} \end{bmatrix}$$

$$u^{\mathrm{T}}(t) = \begin{bmatrix} V_i(t) & T_{Li}(t) \end{bmatrix}, \; y^{\mathrm{T}}(t) = \begin{bmatrix} i_i(t) & \omega_i(t) \end{bmatrix}$$

where,

V_i : applied armature voltage,
T_{Li} : disturbance torque referred to the link side of the drive shaft,
i_i : armature current,
ω_i : shaft angular velocity referred to link side of the drive shaft,
N_i : gear ratio,
J_{mi} : moment of inertia of drive rotor,
K_{mi} : electromechanical constant of the motor (the back–emf constant is equal to the torque constant),
R_I : armature resistence,
L_i : armature inductance,
ρ_i : viscous friction coefficient.

The subscript i denotes the ith joint of the robotic manipulator. Define,

$$\theta_1 = \frac{R_i}{L_i}, \; \theta_2 = \frac{K_{mi}N_i}{L_i}, \; \theta_3 = \frac{1}{L_i}, \; \theta_4 = -\frac{K_{mi}}{J_{mi}N_i},$$

$$\theta_5 = \frac{\rho_i}{J_{mi}}, \; \theta_6 = -\frac{1}{J_{mi}(N_i)^2} \qquad (3.99)$$

i.e.

$$\theta^T = \begin{bmatrix} \theta_1 & \theta_2 & \theta_3 & \theta_4 & \theta_5 & \theta_6 \end{bmatrix} \in R^6$$

The following variables are measured for each motor:

- armature current,
- angular velocity,
- armature voltage,
- shaft torque.

The former two are the system outputs, whereas the latter are the system inputs. Input and output signal measurements are available at discrete times $t = kT_0$, $k = 0, 1, \ldots, N, \ldots$, where T_0 is the sampling time, defined as $i_i(k)$, $\omega_i(k)$, $V_i(k)$, $T_{Li}(k)$. The following observation equations are therefore obtained:

$$y_1^{(1)} = \boldsymbol{\psi}_1^T(k)\boldsymbol{\theta}_a + e_1(t)$$
$$y_2^{(1)} = \boldsymbol{\psi}_2^T(k)\boldsymbol{\theta}_b + e_2(t)$$

where,

$$\boldsymbol{\psi}_1^T(k) = \left[-\boldsymbol{y}^T(k) \quad u_1(k)\right] \in \boldsymbol{R}^3$$
$$\boldsymbol{\psi}_2^T(k) = \left[-\boldsymbol{y}^T(k) \quad u_2(k)\right] \in \boldsymbol{R}^3$$

The fault detection algorithm for this case consists of the following steps (tasks) carried out at every sampling instant k:

Measurements: Measure $i_i(k)$, $\omega_i(k)$, $V_i(k)$, $T_{Li}(k)$ and compute the derivatives $i_i^{(1)}(k)$ and $\omega_i^{(1)}(k)$ by a third order backward formula:

$$i_i^{(1)}(k) = (1/2h)\{3i_i(k) - 4i_i(k-1) + i_i(k-2)\}$$
$$\omega_i^{(1)}(k) = (1/2h)\{3\omega_i(k) - 4\omega_i(k-1) + \omega_i(k-2)\}$$

where $h=T_0$ is the sampling interval.

Task 1: Perform one iteration of the recursive least squares (RLS) parameter estimation algorithm for parameters,

$$\boldsymbol{\theta}_a^{\mathrm{T}} = [\theta_1 \quad \theta_2 \quad \theta_3]$$

Task 2: Perform one iteration of the parameter estimation algorithm for parameters,

$$\boldsymbol{\theta}_b^{\mathrm{T}} = [\theta_4 \quad \theta_5 \quad \theta_6]$$

Task 3(a): Calculate the physical parameters $p_i(k)$, i = 1, 2, 3 from the previously computed estimates $\hat{\boldsymbol{\theta}}_a$ and $\hat{\boldsymbol{\theta}}_b$ using,

$$p_1(k) = R_i = \frac{\theta_1(k)}{\theta_3(k)}, \; p_2(k) = L_i = \frac{1}{\theta_3(k)}, \; p_3(k) = N_i K_{mi} = \frac{\theta_2(k)}{\theta_3(k)} \quad (3.100)$$

The case of a fault occurrence into the gearbox is considered as an event with probability 0.

Task 3(b): Redefine the data window by accepting the new estimates $\hat{p}_i(k)$, $i = 1, 2, 3$, dropping the oldest estimates $\hat{p}_i(k - N_w - 1)$ and recalculating the real time parameter mean and variance estimates (i.e. the parameter statistics are estimated over the N_w+1 most recent parameter estimates). The recursive ralations described in Section 3.4 are used.

Task 3(c): Compute the likelihood ratio for the hypothesis detection problem.

Task 3(d): Decide on whether a fault condition exists. The decision is taken by comparing the likelihood ratio obtained in Task 3(c), against a predefined threshold. To avoid false alarms, the fault condition is signalled if the threshold is exceeded in M consecutive instants. The optimal threshold value and M are best chosen by trial and error using simulation.

Task 4: Perform Tasks 3(a) to 3(d) for parameters $p_4(k)$, $p_5(k)$, using,

$$p_4(k) = N_i^2 J_{mi} = -\frac{\theta_2(k)}{\theta_3(k)\theta_4(k)}, \; p_5(k) = N_i^2 \rho_i = -\frac{\theta_2(k)\theta_5(k)}{\theta_3(k)\theta_4(k)}$$

The above procedure assumes that the algorithm is run initially on a fault free d.c. motor. From this run the non–error statistics are obtained and are used subsequently in Tasks 3(b), 3(c), 4(b) and 4(c).

The effectiveness of the method was verified using simulated data. For this purpose the d.c. motor robotic actuator parameters were chosen as,

$$R = 1.04\ \Omega\,,\ L = 0.00089\ H\,,\ K_m = 0.0224\ V\text{sec/rad},$$

$$J_m = 0.00005\ \text{kgm}^2\,,\ \rho = 0.005\ \text{kgm2/sec}\,,\ N = 64\,.$$

A 2KHz sampling frequency is considered. The non–error statistics are calculated using N_s = 300 samples, whereas the detection window was N_w = 50. The first parameter estimate to be used by the detection procedure was taken at time k = 70, giving a large initial sample. The likelihood ratio fault detection threshold value is 11.2 and M = 10. From sample time k = 1 to 130, the normal operating d.c. drive was simulated. A simulated fault occurred at k = 131, indicated by a 4.8% change in the armature resistance R_i (R_{if}= 1.09 Ω). A recursive least squares (RLS) estimator with a forgetting factor of λ = 0.95 for estimating $\boldsymbol{\theta}_a$, and λ = 0.99 for $\boldsymbol{\theta}_b$ is used. All estimates converge quickly to their respective true values. The exact estimated values are shown in Table 3.6.

A major factor for the success of the algorithm is the 2KHz sampling rate. This means that the algorithm must be implemented on a computer capable of performing all the above calculations in 0.5ms. The above procedure however is suitable for implementation in commercially available parallel processing machines e.g. the INMOS transputer system. This algorithm was implemented in a system employing four processors operaring as two–stage pipeline as shown in Figure 3.20. At the input a measurement unit M feeds the first two processors. At the output, a fault decision unit operates as a separate unit, having however a light computational load and it is therefore a low cost processor system. The numbers shown in Figure 3.20 correspond to the tasks performed by each machine according to the task partition described earlier. This implementation forms a 2–stage pipeline where its first stage consists of machines 1 and 2 and its second of machines 3 and 4. The FD machine which is underutilised by the algorithm, leaves power for suitable presentation of the results. The computational complexity (i.e. the multiplications and divisions per recursion, MADPR) is 30 for each estimator and 60 for the detection procedure.

Table 3.6: True and estimated values for test run

	R_1	L_1	$K_{m1}N_1$	$J_{m1}N_1^2$	$\rho_1 N_1^2$
True value	1.09	0.000890	1.4336	0.2048	20.48
Estimated values at k=300	1.10	0.000896	1.4476	0.2038	20.82

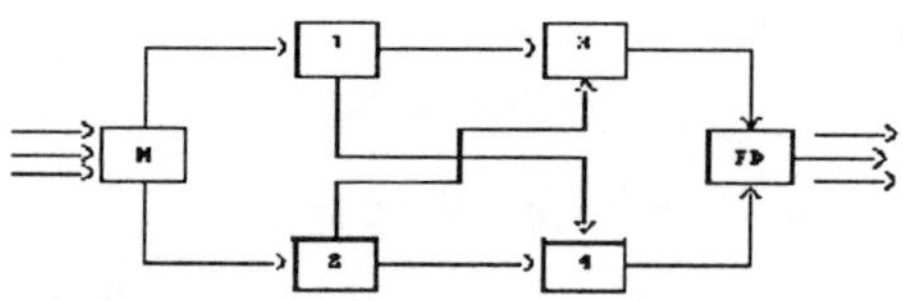

Figure 3.20: Four processor real–time computer implementation of DC–drive fault detection algorithm

3.6 *Additional references*

The field of fault detection based on parameter estimation techniques is vast. The preceding sections present only a small sample of what has been developed. The interested engineer can look at the relevant references for more information. Some additional work follows in summarised form.

The team around Iserman has published several reports of application of parameter estimation fault detection methods to industrial processes. Reiβ (1991), developed models for drilling processes and applied the LS algorithm to the detection of tool wear in two machining centers. Wanke and Reiβ (1991) and Reiβ *et al.* (1990), applied similar techniques to milling machine drives. Janik and Fuchs (1991), used a singular decomposition technique to enhance LS estimation performance in order to detect tool wear and grinding chatter of grinding processes. Neumann (1991), applied a two–step identification algorithm for the estimation of a parametric signal model with ARMAX structure. Signal spectra are then used for fault detection of machine tools. Freyermuth and Iserman (1991), have combined parameter estimation techniques with statistical feature classification methods. This idea was tested on detecting malfunctions of sensors, actuators and gears in industrial robots. Iserman (1991) and Iserman *et al.* (1990), proposed a general hybrid framework for machine fault detection using parameter estimation techniques with knowledge processing. Finally, Cho *et al.* (1992) studied the detection of broken rotor bars in induction motors. This was done by estimating the rotor resistance from measurements of stator voltage, stator current, stator excitation frequency and rotor velocity.

Appendix 3.A

Using the definitions (3.33), it follows,

$$\boldsymbol{\Phi}_{k+1} = \begin{bmatrix} \boldsymbol{\varphi}^{\mathrm{T}}(k-n_{\mathrm{w}}+2) \\ \hline \cdot \\ \cdot \\ \hline \boldsymbol{\varphi}^{T}(k+1) \end{bmatrix} = \begin{bmatrix} \boldsymbol{\Phi}(k,k-n_{\mathrm{w}}+2) \\ \hline \boldsymbol{\varphi}^{\mathrm{T}}(k+1) \end{bmatrix}$$

The iteration for $\boldsymbol{\Phi}_k^T\boldsymbol{\Phi}_k$ is considered first.

$$\boldsymbol{\Phi}_k^{\mathrm{T}}\boldsymbol{\Phi}_k = \left[\boldsymbol{\varphi}(k-n_w+1) \,\middle|\, \boldsymbol{\Phi}^{\mathrm{T}}(k,k-n_w+2)\right]\begin{bmatrix} \boldsymbol{\varphi}^{\mathrm{T}}(k-n_w+1) \\ \hline \boldsymbol{\Phi}(k,k-n_w+2) \end{bmatrix}$$

$$= \boldsymbol{\varphi}(k-n_w+1)\boldsymbol{\varphi}^{\mathrm{T}}(k-n_w+1) + \boldsymbol{\Phi}^{\mathrm{T}}(k,k-n_w+2)\boldsymbol{\Phi}(k,k-n_2+2)$$

and,

$$\boldsymbol{\Phi}_{k+1}^{\mathrm{T}}\boldsymbol{\Phi}_{k+1} = \left[\boldsymbol{\Phi}^{\mathrm{T}}(k,k-n_w+2) \;\middle|\, \boldsymbol{\varphi}(k+1)\right]\begin{bmatrix} \boldsymbol{\Phi}(k,k-n_w+2) \\ \hline \boldsymbol{\varphi}^{\mathrm{T}}(k+1) \end{bmatrix}$$

$$= \boldsymbol{\varphi}(k+1)\boldsymbol{\varphi}^{\mathrm{T}}(k+1) + \boldsymbol{\Phi}^{\mathrm{T}}(k,k-n_w+2)\boldsymbol{\Phi}(k,k-n_w+2)$$

$$= \boldsymbol{\Phi}_k^{\mathrm{T}}\boldsymbol{\Phi}_k + \boldsymbol{\varphi}(+1)\boldsymbol{\varphi}^{\mathrm{T}}(k+1) - \boldsymbol{\varphi}(k-n_w+1)\boldsymbol{\varphi}^{\mathrm{T}}(k-n_w+1)$$

$$= \boldsymbol{\Phi}_k^{\mathrm{T}} \boldsymbol{\Phi}_k + \boldsymbol{\Gamma}(k+1)$$

where,

$$\boldsymbol{\Gamma}(k+1) = \boldsymbol{\varphi}(k+1)\boldsymbol{\varphi}^{\mathrm{T}}(k+1) - \boldsymbol{\varphi}(k-n_w+1)\boldsymbol{\varphi}^{\mathrm{T}}(k-n_w+1)$$

Secondly, the iteration on $\boldsymbol{\Phi}_k^{\mathrm{T}} \boldsymbol{y}_k$ is considered. Define,

$$\boldsymbol{Y}_k = \begin{bmatrix} y(k-n_w+1) \\ \cdot \\ \cdot \\ \cdot \\ y(k) \end{bmatrix} = \begin{bmatrix} y(k-n_w+1) \\ \hline \boldsymbol{y}(k, k-n_w+2) \end{bmatrix}$$

then,

$$\boldsymbol{y}_{k+1} = \begin{bmatrix} y(k-n_w+2) \\ \cdot \\ \cdot \\ \cdot \\ y(k+1) \end{bmatrix} = \begin{bmatrix} \boldsymbol{y}(k, k-n_w+2) \\ \hline y(k+1) \end{bmatrix}$$

Hence,

$$\boldsymbol{\Phi}_k^{\mathrm{T}} \boldsymbol{y}_k = \left[\boldsymbol{\varphi}(k-n_w+1) \mid \boldsymbol{\Phi}^{\mathrm{T}}(k, k-n_w+2) \right] \begin{bmatrix} \boldsymbol{y}^T(k-n_w+1) \\ \hline \boldsymbol{y}(k, k-n_w+2) \end{bmatrix}$$

$$= \boldsymbol{\varphi}(k-n_w+1) y^{\mathrm{T}}(k-n_w+1) + \boldsymbol{\Phi}^{\mathrm{T}}(k, k-n_w+2) \boldsymbol{y}(k, k-n_2+2)$$

and,

$$\boldsymbol{\Phi}_{k+1}^{\mathrm{T}} \boldsymbol{y}_{k+1} = \left[\boldsymbol{\Phi}^{\mathrm{T}}(k, k-n_w+2) \mid \boldsymbol{\varphi}(k+1) \right] \begin{bmatrix} \boldsymbol{y}(k, k-n_w+2) \\ \hline y^{\mathrm{T}}(k+1) \end{bmatrix}$$

$$= \boldsymbol{\varphi}(k+1) y(k+1) + \boldsymbol{\Phi}^{\mathrm{T}}(k, k-n_w+2) \boldsymbol{y}(k, k-n_w+2)$$

$$= \boldsymbol{\Phi}_k^{\mathrm{T}} \boldsymbol{y}_k + \boldsymbol{\varphi}(k+1) y(k+1) - \boldsymbol{\varphi}(k-n_w+1) y(k-n_w+1)$$

$$= \boldsymbol{\Phi}_k^{\mathrm{T}} \boldsymbol{y}_k + \boldsymbol{\delta}(k+1)$$

where,

$$\boldsymbol{\delta}(k+1) = \boldsymbol{\varphi}(k+1)\, y(k+1) - \boldsymbol{\varphi}(k-n_w+1)\, y(k-n_w+1)$$

Now,

$$\hat{\theta}(k+1) = \left(\boldsymbol{\Phi}_{k+1}^{\mathrm{T}} \boldsymbol{\Phi}_{k+1} \right)^{-1} \boldsymbol{\Phi}_{k+1}^{\mathrm{T}} \boldsymbol{y}_{k+1}$$

Defining,

$$P(k+1) = \left(\boldsymbol{\Phi}_{k+1}^{\mathrm{T}} \boldsymbol{\Phi}_{k+1}\right)^{-1}$$

which is the covariance of the estimate $\hat{\boldsymbol{\theta}}(k+1)$, yields,

$$\begin{aligned}
\hat{\boldsymbol{\theta}}(k+1) &= \boldsymbol{P}(k+1)\left[\boldsymbol{\Phi}_k^{\mathrm{T}} \boldsymbol{y}_k + \boldsymbol{\delta}(k+1)\right] \\
&= \boldsymbol{P}(k+1)\left[\boldsymbol{P}^{-1}(k)\hat{\boldsymbol{\theta}}(k) + \boldsymbol{\delta}(k+1)\right] \\
&= \boldsymbol{P}(k+1)\left[\boldsymbol{P}^{-1}(k+1) - \boldsymbol{\Gamma}(k+1)\hat{\boldsymbol{\theta}}(k) + \boldsymbol{\delta}(k+1)\right] \\
&= \hat{\boldsymbol{\theta}}(k) - \boldsymbol{P}(k+1)\left[\boldsymbol{\Gamma}(k+1)\hat{\boldsymbol{\theta}}(k) - \boldsymbol{\delta}(k+1)\right]
\end{aligned}$$

and,

$$\boldsymbol{P}^{-1}(k+1) = \boldsymbol{P}^{-1}(k) + \boldsymbol{\Gamma}(k+1)$$

Appendix 3.B

Modified Gram–Schmidt orthogonalization algorithm.

The triangularization problem can be stated in the following general form:

Given:

$$\boldsymbol{A}\boldsymbol{x} = \boldsymbol{b}$$

where $\boldsymbol{A} \in \boldsymbol{R}^{nxr} \times \boldsymbol{R}^{n}$, $\boldsymbol{x} \in \boldsymbol{R}^{n}$, $\boldsymbol{b} \in \boldsymbol{R}^{n+r}$, find an orthogonal tranformation $\boldsymbol{T}$ such that

$$\boldsymbol{TA} = \begin{bmatrix} \boldsymbol{W} \\ \cdots\cdots \\ 0 \end{bmatrix}$$

where $\boldsymbol{W}$ is upper triangular (or equivalently, find $\boldsymbol{W}$ and $\boldsymbol{b}' = \boldsymbol{Tb}$ directly).

The following algorithm is a numerically improved adaptation of the classical Gram–Schmidt orthogonalization procedure. When computations are made exactly (no round–off) the result is equivalent to the classical Gram–Schmidt result. However, when round–off errors occur, Björck (1967) has shown that the MGS procedure is much more accurate. The algorithm can be derived from the classical Gram–Schmidt orthogonalizing procedure, and is essentially the classical Gram–Schmidt procedure in reverse order (Kaminski, 1971). The MGS algorithm is stated in a form which computes $\boldsymbol{W}$ and $\boldsymbol{b}'$ directly.

MGS Algorithm: For $k = 1, \ldots, n$ compute,

$$\sigma_k = \sqrt{A_k^{\mathrm{T}(k)} A_k^{(k)}}$$

$$W_{kj} = \begin{cases} 0, & j = 1, \ldots, k-1 \\ \sigma_{k,} & j = k \\ (1/\sigma_k) A_k^{\mathrm{T}(k)} A_j^{(k)}, & j = k+1, \ldots, n \end{cases}$$

$$b_k' = \frac{1}{\sigma_k} A_k^{\mathrm{T}(k)} b^{(k)}$$

$$A_j^{(k+1)} = A_j^{(k)} - \frac{W_{kj}}{\sigma_k} A_k^{(k)}, \quad j = k+1, \ldots, n$$

$$b^{(k+1)} = b^{(k)} - \frac{b_k'}{\sigma_k} A_k^{(k)}$$

(here single suffix denotes column of matrix and double suffix denotes element of matrix). If $\sigma_k = 0$ at any stage in the algorithm, then the rank of A is less than n.

References

Åström K.J. and P. Eykhoff (1971). Special Issue, *AUTOMATICA*, 7, 123– 162.

Åström K.J. and B. Wittenmark (1971). Problems of identification and control. *Journal of Mathematical Analysis and Applications*, **34**, 50–113.

Baskiotis C., Raymond J. and A. Rault (1979). Parameter identification and discriminant analysis of jet engine mechanical state diagnosis. Proceedings, *IEEE Conference on Decision and Control*, Fort Lauderdale.

Björck A. (1967). Solving linear least square problems by Gram–Schmidt orthogonalization. *BIT*, 7, 1–21.

Biermann G.J. (1977). Factorization methods for discrete sequential estimation. Academic Press, N.Y.

Carayannis G., Manolakis D. and N. Kalouptsidis (1983). A fast sequential algorithm for least squares filtering and prediction. *IEEE Transactions on Acoustics, Speech and Signal Processing*, **ASSP–31**, 1394–1402.

Carayannis G., Manolakis D. and N. Kalouptsidis (1986). A unified view of parametric processing algorithms for prewindowed signals. *Signal Processing*, **10**, 335–368.

Carlsson B., Salgado M. and G.C. Goodwin (1988). A new method for fault detection and diagnosis. *Technical Report EE8842*, Dept. of Electrical Eng. and Computer Science, University of Newcastle, Australia.

Cho K.R., Lang J.H. and S.D. Umans (1992). Detection of broken rotor bars in induction motors using state and parameter estimation. *IEEE Transactions on Industry Applications*, **28**, 3, 702–709.

Cordero A.O. and D.Q. Mayne (1981). Deterministic convergence of a self tuning regulator with variable forgetting factor. *Proceedings IEE, Part–D*, **128**, 1, 19–23.

Dehoff R.L. Hall W.E. Jr. Adams R.J. and N.K. Gupta (1977). F100 multivariable control synthesis program. *AFAPLTR–77–35*, Vol. **I** and **II**.

Dehoff R.L. and W.E. Hall Jr. (1978). Models for jet engine systems. Part II: state space techniques and modeling for control. *Control and Dynamic Systems*, **14**, 259–299.

Dalla Molle D.T. (1985). Fault detection via parameter estimation ia a single effect evaporator. MS Thesis, University of Texas, Austin.

Dalla Molle D.T. and M.D. Himmelblau (1987). Fault detection in an evaporator via parameter estimation in real time. *Fault Detection and Reliability: Knowledge–based and other approaches*, Pergamon Press, 131–138.

Favier G., Rougerie C., Bariani J.P., de Amaral W., Gimena L. and L.V.R. de Amanda (1988). A comparison of fault detection methods and adaptive identification algorithms. *Proceedings, IFAC Identification and System Parameter Estimation*, Beijing, PRC, 535–542.

Fortescue T.R., Kershenbaum L.S. and B.E. Ydstie, (1981). Implementation of self–tuning regulators with variable forgetting factors. *Automatica*, **17**, 6, 831–835.

Freyermuth B. (1991). An approach to model based fault diagnosis of industrial robots. *Proceedings, IEEE International Conference on Robotics and Automation*, April 7–12, 1991, Sacramento, USA.

Freyermuth B. and R. Iserman (1991). Model based incipient fault diagnosis of industrial robots via parameter estimation and feature classification. *Proceedings, European Control Conference ECC '91*, 2–5 July 1991, Grenoble, France.

Gantmacher F.R. (1977). The theory of matrices. Chelsea Publishing Company.

Geiger G. (1982). Monitoring of an electrical driven pump using continuous– time parameter estimation methods. *Proceedings, 6th IFAC Symposium on Identification and Parameter Estimation*, Washington.

Geiger G. (1984). Fault identification of a motor–pump system using parameter estimation and pattern classification. *Proceedings, 9th IFAC Congress*, Budapest.

Geiger G., (1986). Fault identification using a discrete square root method. *International Journal of Modeling and Simulation*, **6**, 1, 26–31.

Goodwin G.C. and M.E. Salgado (1989). Quantification of uncertainty in estimation using an embedding principle. *International Journal of Adaptive Control and Signal*

Processing, **8**, 232–345.

Hägglund T. (1984). Adaptive control of systems subject to large parameter changes. *Proceedings, IFAC 9th Triennial World Congress*, Budapest, Hungary, 993–998.

Henry J.R. (1988). CF–18F404 transient performance trending. *AGARD*, Paper No. 448, Quebec City.

Iserman R. (1984). Process fault detection based on modelling and estimation methods – A survey. *Automatica*, **20**, 387–404.

Iserman R. (1987). Experiences with process fault detection methods via parameter estimation. In *System Fault Diagnostics and Related Knowledge–Based Approaches*, S. Tzafestas *et al.*. (eds.), D. Reidel.

Iserman R. (1991). Fault diagnosis of machines via parameter estimation and knowledge processing. *Proceedings, IFAC/IMACS Symposium "SafeProcess '91"*, 10–13 September 1991, Baden–Baden, Germany.

Iserman R., Appel W., Freyermuth B., Fuchs A., Janik W., Neumann D., Reiss Th. and P. Wanke (1990). Model based fault diagnosis and supervision of machines and drives. *Proceedings, IFAC 11th Triennial World Congress*, Tallinn, Estonia.

Janik W. and A. Fuchs (1991). Process– and signal–model based fault detection of the grinding process. *Proceedings, IFAC/IMACS Symposium "SafeProcess '91"*, 10–13 September 1991, Baden–Baden, Germany.

Kaminski P.G., (1971). Square root filtering smoothing for discrete processes. Phd. Thesis, Dept. Aeronautics and Astronautics, Stanford University.

Kalouptsidis N. (1987). Efficient transversal and lattice algorithms for linear phase multichannel filters. *IEEE Transactions on Circuits and Systems*, **CAS–37**, 805–813.

Kalouptsidis N., Carayannis G. and D. Manolakis (1984). A fast covariance type algorithm for sequential least squares filtering and prediction. *IEEE Transactions on Automatic Control*, **AC–29**, 8, 752–755.

Kalouptsidis N., Manolakis D. and G. Carayannis (1983). A family of computationally efficient algorithms for multichannel signal processing. *Signal Processing*, **5**, 1, 5–19.

Kalouptsidis N. and S. Theodoridis (1987). Parallel implementation of efficient LS algorithms for filtering and prediction. *IEEE Transactions on Acoustics, Speech and Signal Processing*, **ASSP–35**, 11, 1565–1569.

Karaboyas S. and N. Kalouptsidis N. (1991). Efficient adaptive algorithms for ARX identification. *IEEE Transactions on Acoustics, Speech and Signal Processing.*

Kumamaru K., Söderström T., Sagara S. and K. Morita (1988). On–line fault detection in adaptive control systems by using Kullback discrimination index. *Proceedings, IFAC Identification and System Parameter Estimation,* 1135–1140.

Kwon O.–K. and G.C. Goodwin (1990). A fault detection method for uncertain systems with unmodeled dynamics, linearization errors and noisy inputs. *Proceedings, 11th IFAC Triennial World Congress*, Tallinn, Estonia, 367–372.

Liu J.S.H. (1977). Detection, isolation and identification techniques for noisy degradation in linear, discrete-time systems. *Proceedings, 1977 CDC*, 1132-1139.

Ljung L. (1987). System Identification: Theory for the User, Prentice Hall, Inc. Englewood Cliffs, N.J.

Ljung L., Morf M. and D. Falconer (1978). Fast calculations of gain matrices for recursive estimation schemes. *International Journal of Control*, **27**, 1–19.

Maguire L.P. and G.W. Irwin (1991). Transputer implementation of Kalman filters. *IEE Proceedings–D*, **138**, 4, 355–362.

Manolakis D., Carayannis G., Kalouptsidis N., (1980). Fast inversion of vector generated matrices for signal processing. Signal Processing: Theories and Applications, North–Holland, 525–532.

Merrill, W. (1984). Identification of multivariable high–performance turbofan engine dynamics from closed loop data. *Journal of Guidance*, **7**, 677–683.

Merrington G., Kwon O.K., Goodwin G. and B. Carlsson (1991). Fault detection and diagnosis in Gas Turbines. *Transactions of the ASME*, **113**, 276–282.

Neumann D. (1991). Fault diagnosis of machine–tools by estimation of signal spectra. *Proceedings, IFAC/IMACS Symposium "SafeProcess '91"*, 10–13 September 1991, Baden–Baden, Germany.

Nold S. (1987). Fault detection in AC–drives by process parameter estimation. *Proceedings, IFAC 10th Triennial World Congress*, Munich, Germany.

Nold S.and R. Iserman (1986). Identifiability of process coefficients for technical failure diagnosis. *Proceedings, 25th IEEE Conference of Decision and Control*, Athens, Greece, Dec. 1986, 1587–1592.

Pot J., Falinower, V.M.and E. Irving (1984). Regulation multivariable adaptative des fours. *Colloque CNRS "Commande Adaptative. Aspects Pratique et Theoriques"*, St. Martin d'Heres.

Potter J.E. (1963). New statistical formulas. Memo 40, Instrumentation Laboratory, MIT.

Pouliezos A., Stavrakakis G. and C. Lefas (1989). Fault detetcion using parameter estimation – A survey. *Quality and Reliability International*, **5**, 4, 283–290.

Pouliezos A. and G.S. Stavrakakis (1989). Fast fault diagnosis for industrial processes applied to the reliable operation of robotic systems. *International Journal of Systems Science*, **20**, 7, 1233–1258.

Reiβ T. (1991). Model based fault diagnosis and supervision of the drilling process. *Proceedings, IFAC/IMACS Symposium "SafeProcess '91"*, 10–13 September 1991, Baden–Baden, Germany.

Reiβ T., Wanke P. and R. Iserman (1990). Model based fault diagnosis of a flexible milling center. *Proceedings, IFAC Triennial World Congress*, Tallinn, Estonia.

Rhodes I.B. (1990). A parallel decomposition for Kalman filters. *IEEE Transactions on Automatic Control*, **AC–35**, 3, 322–326.

Shibata, H. Ikeda, Y., Maruoka, G., Aoki, S. and T. Ogawa (1988). Application of estimation techniques to failure detection for A.C. electric machines. *Proceedings, IFAC Identification and System Parameter Estimation*, Beijing, PRC, 1147–1152.

Smed, T., B. Carlsson, C.E. de Souza and G.C. Goodwin (1988). Fault detection and diagnosis applied to gas turbines. *Technical Report EE8815*, Dept. of Electr. Engr. and Computer Science, Univ. of Newcastle, Australia.

Söderström T. and K. Kumamaru (1985). On the use of Kullback discrimination index for model validation of fault detection. *Report UPTEC 8520R*, Uppsala University, Sweden.

Söderström T. and P. Stoica (1988). System Identification. Prentice Hall.

Stavrakakis G.S. and E.N. Dialynas (1991). Efficient computer based scheme for improving reliability performance of power substations. *International Journal of Systems Science*, **22**, 9, 1527–1539.

Stavrakakis G.S. and A. Pouliezos (1991). Fatigue life prediction using a new moving window regression method. *Mechanical Systems and Signal Processing*, **5**, 4, 327–340.

Stavrakakis G.S., Lefas Ch. and A. Pouliezos (1990). Parallel processing computer implementation of a real time DC motor drive fault detection algorithm. *IEE Proceedings, Part B*, **137**, 5, 309–313.

Thornton C.L. and G.J. Bierman (1977). Gram–Schmidt algorithms for covariance propagation. *International Journal of Control*, **25**, 243–260.

Tzafestas S.G. and G.S. Stavrakakis (1986). Model reference adaptive control of industrial robots with actuator dynamics. *IFAC/IFIP/IMACS International Symposium on Theory of Robots*, Vienna, Austria, December 3–5.

Wahlberg B. (1990). Robust frequency domain fault detection/diagnosis. *Proceedings, 11th IFAC Triennial World Congress*, Tallinn, Estonia, 373–378.

Wanke P. and T. Reiβ (1991). Model based fault diagnosis and supervision of the main and feed drives of a flexible milling center. *Proceedings, IFAC/IMACS Symposium "SafeProcess '91"*, 10–13 September 1991, Baden–Baden, Germany.

Watanabe K. and D.M. Himmelblau (1983). Fault diagnosis in nonlinear chemical processes; Part I. Theory. *AIChE Journal*, **29**, 2, 243–249.

Weiss J.L. (1988). Threshold computation for detection of failures in SISO systems with transfer function errors. *Proceedings, American Control Conference*, 2213–2218.

Ydstie B.E., (1981), Phd. Thesis, University of London.

Yeh H.G. (1991). Processing performance of two Kalman filter algorithms with a DSP32C by using assembly and C languages. *IEEE Transactions on Industrial Electronics*, **38**, 4, 298–302.

Young P.C. (1981). Parameter estimation for continuous time models – a survey", *Automatica*, **17**, 23–29.

CHAPTER 4

AUTOMATIC EXPERT PROCESS FAULT DIAGNOSIS AND SUPERVISION

4.1 Introduction

Automatic fault diagnosis, supervision and control of very complex systems are becoming extrememly important. This is the direct consequence of the occurrence of recent disasters because of unsatisfactory control or missed diagnosis of failures (Three Mile Island and Tchernobyl are but a few examples). Control and fault diagnosis cannot be realized without a good methodology of modeling, i.e. representing the structure and behavior of the systems under consideration in the significant states of their operation. The conventional methods of large scale modeling require comprehensive knowledge about the system consisting of conforming elements (e.g. a set of ordinary differential equations), and no gaps in the knowledge are allowed. Complex physical systems (e.g. a nuclear power plant, chemical processes) contain several types of elements and processes (e.g. nuclear, mechanical, electrical, electronic, etc.) with different types of description and eventually gaps in the available knowledge. The purely numerical-mathematical approach of large scale systems modeling could not offer adequate methodology to solve the problems arising in this field, therefore, symbolic and artificial intelligence methods have been tried to obtain an adequate solution.

Diagnosis is currently one of the largest application domains of expert systems. Strategies and capabilities for diagnosis have been evolving rapidly. Most of the past applications involving diagnosis have been rule-based. That is, they use simple production rules to provide a mapping between the possible causes and inputs of a system and the possible faults.

The most primitive approach to automation would be to store diagnostic procedures in a computer and activate them when symptoms arise. This approach is valid, however, only when the symptoms are anticipated and the corresponding procedures can be predetermined by the designer of the diagnostic system.

The leading wave of technology, however, provides powerful new techniques that are applicable in a broad range of situations. These techniques give the ability to build and reason about deep models and can operate with a wide range of information, such as learning from experience, probabilistic information, fuzzy reasoning and learning from examples. At the same time, a clearer picture has emerged about the range of strategies available and when they are most appropriate.

In this chapter these strategies will be examined and the nature of automatic expert diagnostic and supervision systems, with respect to them, will be revealed . A framework for coupling them together and their real-time implementation features will be provided. Examples from the recent expert diagnostic practice in industry will be presented to help the reader to delve into the matter.

4.2 Nature of automatic expert diagnostic and supervision systems

4.2.1 Expert systems for automatic process fault diagnosis

4.2.1.1 The terminology of knowledge engineering

An *industrial Expert System* (*iES*), in its most basic sense, is no more than a tool to organize and codify for the computer, the experience and thought processes of a human with expertise concerning the operation of a technological process or an industrial plant or a given piece of equipment.

Knowledge engineering (*KE*) is the process of building expert systems. Such systems are medium-to large-scale software products which are designed to solve problems of different kinds using a knowledge-based approach, where the knowledge is represented in an explicit manner. They have a wide area of applicability, particularly in industrial control.

Generic categories of knowledge engineering applications are interpretation, prediction, diagnosis, design, planning, monitoring, debugging, repair, instruction and control. Such systems normally contain two main components: the inference mechanism (the problem solving component) and the knowledge base (which may actually comprise a number of knowledge bases). Generally speaking, expert systems work best in narrow application domains.

The process of building an expert system consists of two main activities which usually overlap: acquiring the knowledge and implementing the system. The acquisition activity involves the collection of knowledge about facts and reasoning strategies from the domain experts. Usually, such knowledge is elicited from the experts by so-called knowledge engineers, using interviewing techniques or observational protocols. However, machine induction, which automatically generates more elaborate knowledge from an initial set of basic knowledge (usually in the form of examples), has also been extensively used. In the system construction process, the system builders (i.e. knowledge engineers), the domain experts and the users work together during all stages of the process, which traditionally has involved extensive prototyping.

To automate the problem solving process, the relevant task knowledge in the domain of interest needs to be understood in great detail. However, acquiring the knowledge for expert system building is generally regarded as a hard problem. This is not surprising, as acquiring knowledge from an expert entails answering some really fundamental questions such as:

- What is the relationship between knowledge and language?
- How can different domains be characterized?
- What constitutes a theory of problem solving?

The process of extracting knowledge from an expert is not the process of transferring a mental model lying in the brain of an expert into the mind of the system builder, but the formalization of a domain for the first time, and this is inherently a difficult process. Ideally, models of conceptual structures of problem solving behavior are required as a prerequisite to the knowledge transfer process. However, cognitive science approaches have not yet yielded sufficient information to enable a full understanding of the knowledge structures and problem solving strategies of experts to be applied, so that current approaches are incomplete and often ad hoc.

The situation is further complicated by the fact that experts often have faulty memories or provide inconsistencies. This means that separate validation of the expertise elicited from experts is essential. Furthermore, experts exhibit cognitive biases such as overconfidence, simplification, and a low preference for the abstract, the relative and conflicting evidence. It is therefore important to test and validate expert systems both by analyzing the expertise in the knowledge base and by examining failures in actual performance. As far as possible, cognitive biases should be filtered out during the elicitation process.

A great deal of experimental evidence exists about the limitations of human decision making and it has been suggested that the development of systems which mimic human problem solving should be approached with some degree of caution. In order to reduce the chances of bias, experts should be made aware of commonly found biases in judgment, the elicitation process should include probes to foster the consideration of alternatives and when experts run through sample problems in the elicitation process, it should be borne in mind that the way in which the problems are presented, will have an impact as to how far any derived rules will exhibit cognitive bias.

Appraising the knowledge engineering process from a cognitive engineering viewpoint, the following six stages, termed *mainstream development*, are suggested:

1. Knowledge elicitation.
2. Cognitive bias filtering.
3. Knowledge representation and control scheme selection.

4. Software development and integration.
5. System evaluation and validation.
6. Advanced prototype expert system.

Stages three and four ideally should only be carried out after the elicitation and cognitive bias stages have been completed. In reality, this is not possible and researchers' experience suggests that several iterations through the first five stages are required before stage six can be contemplated.

It is important to realize that experts change their solution strategies dependent upon the boundedness of the problem. In well-bounded problems an expert's approach differs dramatically from that of a novice. Conceptual models of experts reflect the physical processes that actually occur. By contrast, the models used by non experts do not account for all the process parameters driving the problem. When problems concerning the processes are less well understood, expert and novice models appear superficially to be similar, though the experts seem to recognize that the simple models are not accurate and that the use of a precise model is not practicable. So experts know what they do not know and can readily identify features of uncertainty that preclude the use of precise solution strategies; in such circumstances they will adopt simple and somewhat inaccurate process models. Knowledge acquisition techniques must be able to cope with this variation in expert strategy.

There are a number of terms used to describe the expert system building process which are not well defined and appear to overlap. Such terms include *knowledge elicitation*, *knowledge acquisition*, *system implementation*, *machine induction* and even the term knowledge engineering itself. Knowledge acquisition is defined as the transfer and transformation of problem-solving expertise from some knowledge source to a program. This definition covers the whole process including identification of the problem, its conceptualization, formalization, implementation, testing and prototype revision.

The process of building ***knowledge-based systems*** (*KBS*) is essentially one of knowledge engineering, thus the relationship between the different terms should be as in fig. 4.1.

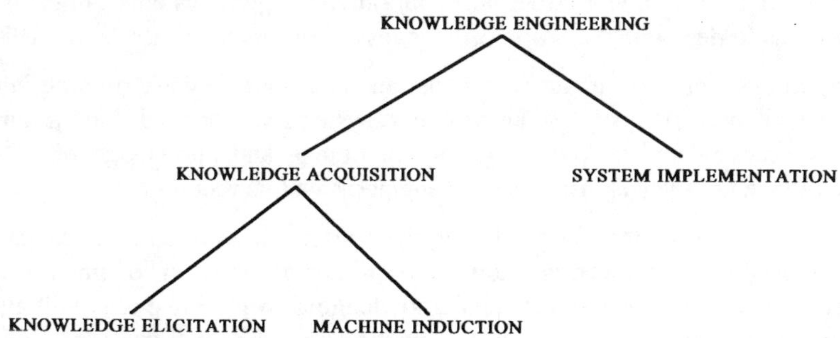

Figure 4.1 Relationship between terms in knowledge engineering.

Developing knowledge-based systems is a far from trivial process. Those who build knowledge-based systems for industrial systems supervision and diagnosis know that no significant systems development can sensibly take place without a structured approach. Systematic and structured approaches to KBS development available today can be found in Hickman *et al.* (1989) and Luger and Stubblefield (1989).

Expert systems can be introduced into industrial systems to provide support for different classes of people such as designers, operators and maintenance personnel. In general such systems will be off-line (for designers and maintenance personnel) and on-line (for operators). The knowledge engineering task will be different for each of these applications since the tasks involved will comprise different knowledge sources and structures. One difference is that between technological/scientific knowledge and experimental knowledge. This difference was described as a *knowledge of functioning* versus a *knowledge of utilization*. The former knowledge is used by designers and maintenance personnel whereas the latter characterizes the one used by operators.

Off-line knowledge-based systems are not time critical. They may utilize several knowledge sources including technical documents, reference literature, handbooks, ergonomic knowledge, and knowledge about operator personnel (for use in user modeling). Whilst their operation is not time critical, operator time constraints may still have to be taken into account.

The most critical and challenging industrial expert systems are those developed for system operation. They may encompass support for the automatic monitoring system as well as support for the operators, and may provide heuristic control, fault diagnosis, consequence prediction and procedural support. The latter is particularly suitable for consistency checking of input sequences or for operator intent recognition (Johannsen and Alty, 1991).

Expert systems will be more effective when linked to dynamic databases. Knowledge can then be applied to new situations by periodically executing rules and queries. Because of this linkage, knowledge acquisition costs can be amortized over many instances of reuse. Users will not lose their motivation to employ the system because of the the pain of having to enter data each time. Furthermore, data-driven expert systems can propose and rank suggestions to deal with the world as changes are observed (Kaiser *et al.* 1988).

Support expert systems work under time constraints because they are running in parallel with the dynamic industrial process. These expert systems will depend upon a number of knowledge sources related to knowledge of functioning and knowledge of utilization. Additional knowledge such as that of senior engineers will be required.

Whilst a support expert system for predicting the consequences of some technical failure will normally need only engineering knowledge, procedural support, diagnosis and heuristic control modules will need operational knowledge as well. Since they will also have to be integrated with the supervision and control system, they will need to support numerical as well as symbolic knowledge.

The importance of signal and symbol processing has been emphasized by Nawab *et al.* (1987) and Rouse *et al.* (1989). They point out that models of symbol processing are much harder to be identifiable than those of signal processing because semantics and pragmatics play a large role in symbol processing systems. In particular, they stress the need for symbolic representations in industrial process control applications.

In all cases of knowledge-based systems development, it will be necessary to define carefully the goals and functionalities of the various systems and their interdependencies at an early stage. It is also important to realize that in the industrial environment, not all applications are suitable for the application of knowledge-based techniques. For example, existing numerical supervision and control systems are based upon thorough engineering methodologies and replacement by knowledge-based techniques would, in most cases, lead to performance degradation.

Finally, it must be realized that most industrial applications are very complex and this makes the problem of acquiring and assembling the knowledge in the industrial environment much more severe than in traditional computing domains. The elicitation and conceptualization processes are liable to be far more complex and attempts to prove the consistency of the knowledge will be very time-consuming. The full process is likely to take years rather than months. In the absence of a powerful methodology, one is forced to work with inadequate tools for some time to come.

4.2.1.2. Techniques for knowledge acquisition

The most time consuming portion of constructing an expert system is the knowledge acquisition phase (Forsythe and Buchanan (1989), Adelman (1989)). Conceptually, knowledge engineering is a measurement problem This measurement problem is a complex one because there are five sources of variation: domain experts, knowledge engineers, knowledge representation schemes, elicitation methods and problem domains. Adelman, (1989), suggests the use of two or three distinctly different knowledge engineers, knowledge representation schemes and elicitation methods when working with two or more domain experts. The expert system development team will then be able to identify which, if any, of these sources of variability result in disagreement in the predictions of the knowledge base and, thereby, resolve them.

The techniques used in knowledge acquisition can be broadly divided into two categories: *elicitation* and *machine induction*. Strictly speaking, there is a continuum between human-human elicitation and automatic induction. Three general principles have been proposed for the acquisition process by Gruber and Cohen, (1987). They are concerned with primitives and generalizations.

The first principle prescribes that task-level primitives should be designed in order to capture important domain concepts defined by the expert. The knowledge engineer must use a language of task-level terms rather than imposing implementation-level primitives.

This principle stresses the importance of separating out acquisition from implementation. These task-level primitives must be natural constructs for describing information, hypotheses, relations, and actions, in the language of the domain expert. This would suggest that task analyses should be combined with knowledge analyses.

The second principle suggests that explicit declarative representational primitives are preferable to procedural descriptions. This principle is based upon the observation that most experts understand declarative representations more easily. Formulating procedural aspects in this way can facilitate acquisition, explanation, and maintenance. Gruber and Cohen, (1987), suggest that an expert should be asked "for the parameters of a domain that affect control decisions, and then to formulate control knowledge in terms of these parameters".

The third principle requires representations at the same level of generalization as the expert's knowledge. Experts should not be forced to generalize except when absolutely necessary and they should not be asked to specify information not available to them. An example of an oversimplified generalization would be the requirement to categorize a process variable as high, medium or low, when the expert needs to differentiate between many more steps or even a full range of numbers.

Knowledge elicitation.

A number of techniques for knowledge elicitation are now in use. They usually involve the collection of information from the domain expert(s) either explicitly or implicitly. Originally, reports written by the experts were used, but this technique is now out of favour since such reports tend to have a high degree of bias and reflective thought. Current techniques include interviews (both structured and unstructured), questionnaires or observational techniques such as protocol analyses and walkthroughs.

Knowledge elicitation methodologies have more in common with the field-work orientation of anthropology and qualitative sociology than with the experimental orientation of many cognitive sciences. It is suggested that knowledge engineers also use the large amount of literature and experience as well as the much longer tradition of the social sciences in field work, particularly data-gathering methods such as face-to-face interviewing. Some pitfalls of knowledge elicitation are described on the basis of this experience in the social sciences. In particular, some interviewing problems such as obtaining data versus relating to the expert as a person, fear of silence and failing to listen, difficulty in asking questions, interviewing without a record as well as conceptual problems such as treating interview methodology as unproblematic or blaming the expert, are explained.

Interviews. In a structured interview, the knowledge engineer is in control. Such interviews are useful for obtaining an overall sense of the domain. In an unstructured interview, the domain expert is usually in control; however, such interviews can, as the name implies, yield a somewhat incoherent collection of domain knowledge. The result can be a very unstructured set of raw data that needs to be analyzed and conceptualized. It is obviously important for the knowledge engineer to have some knowledge of the domain before wasting the valuable time of the expert. This might be obtained through textbooks, manuals and other well-documented sources. Group interviews can be useful particularly in the phase of cognitive bias filtering.

Far from coming naturally, interviewing is a difficult task that requires planning, stage-management, technique and a lot of self-control. Forsythe and Buchanan, (1989), present some ethnographic techniques that can be applied to the problem of identifying and mitigating difficulties of communication between knowledge engineers and experts during interviews.

Questionnaires and rating scales. Questionnaires can be used instead or in addition with interviews. The interviews can be standardized in question-answer categories or questionnaires can be applied in a more formal way. However, the latter should be handled in most cases in a relaxed manner for reasons of building up an atmosphere of confidence and not disturbing the expert too much when applied in actual work situations.

Rating scales are formal techniques for evaluating single items of interest by asking the expert to cross-mark a scale. Verbal descriptions along the scale such as from "very low" to "very high" or from "very simple" to "very difficult" are used as a reference for the expert. The construction, use and evaluation of rating scales is described very well in the psychological and social sciences literature. Rating scales can also be combined with interviews or questionnaires.

Observations. Observations are another technique for knowledge elicitation. They require little or no active participation of the expert. All actions and activities of the expert are observed as accurately as possible by the knowledge engineer who makes recordings of all the observed information. A special mixture of interview and observation techniques are the observation interviews. Sequences of activities are observed and questions about causes, reasons and consequences asked by the knowledge engineer during these observations. The combined technique is very powerful because the sequence of activities is observable whereas decision criteria, rules, plans etc. are elicited in addition through what-, how- and why- questions.

Protocol analysis. Protocol analyses are useful for obtaining detailed knowledge. They can involve verbal protocols in which the expert thinks aloud whilst carrying out the task, or motor protocols in which the physical performance of the expert is observed and recorded (often on videotape). Eye movement analysis is an example of a very specialized version of this technique. Motor protocols, however, are usually only useful when used in conjunction with verbal protocols.

In a verbal protocol, the expert thinks aloud and a time-stamped recording is made of his utterances. In such protocols, the expert should not be allowed to include retrospective utterances. He or she should avoid theorizing their behavior and should "only report information and intentions within the current sphere of conscious awareness". As a verbal protocol is transcribed, it is broken down into short lines corresponding roughly to meaningful phrases. This technique can collect the basic objects and relations in the domain and establish causal relationships. From these a domain model can be built. Experience with the use of verbal protocols for the analysis of trouble-shooting in maintenance work of technicians, is described by Rasmussen, (1984).

The **critical decision method** (*CDM*) as described by Klein *et al.*, (1989), is a special protocol analysis which elicits knowledge from experts and novices in a retrospective way. Non-routine cases such as critical incidents are selected in order to discriminate the true knowledge of the expert(s). Sources of bias are minimized by asking for uninterrupted incident descriptions. Subsequently, the history of the incident is reconstructed by means of time lines and decision points are identified and probed. It is claimed that, by using the critical decision method knowledge can be elicited with relatively little effort .

The **cognitive task analysis approach**. Roth and Woods, (1989), and Lancaster *et al.* (in SIGART newsletter, p. 152, 1989) suggest a multi-phase progression from initial informal interview techniques (to derive a preliminary mapping of the semantics of the domain), to more structured knowledge elicitation techniques (to refine the initial semantic structure), to controlled experiments designed to reveal the knowledge and processing strategies utilized by domain practitioners.

The first phase (*categorial knowledge structure*) gives preliminary cognitive description of the task as a guide for further analysis. It is important here not to home in on specific rules. One possibility is to get the experts to provide an overview presentation. Only when an overview of the semantics of the application has been developed, can more structured techniques be used.

The second phase (*temporal event organization*) concentrates on how practitioners perform their tasks; thus, there is an emphasis on observation and analysis of actual task performance. It involves techniques such as critical incident review, discussion of past challenges, or the construction of test cases on which to observe the experts at work. During this phase the use of *expert panels* it is also recommended in order to obtain a corpus of challenging cases for identifying critical elements and strategies for handling them.

The third phase (*causal structures in knowledge*) uses observational techniques under controlled conditions to observe expert problem solving strategies. The practitioner is observed and asked to provide a verbal commentary (i.e. the why and how of a particular domain). The task can be deliberately manipulated, for example, by forcing the expert to go beyond reasonable routine procedures. In some cases, the expert himself controls the information gathering. Alternatively, it is controlled by the observer. Each approach provides useful information; the former provides data on the diagnostic search process and the latter on the effect (or bias) of particular types of information on expert interpretations. Another useful technique is to compare the performance of experts with different levels of expertise, so as to isolate what factors really account for superior performance.

Teachback interviewing. In this technique, the expert first describes a procedure to the knowledge engineer, who then teaches it back to the expert in the expert's terms until the expert is completely satisfied with the explanation. Johnson and Johnson, (1987), describe this technique and illustrate its use in two case studies. Their approach is guided by *Conversation Theory*, in which interaction takes place at two levels: specific and general. The paper gives a useful set of guidelines on the strengths and weaknesses of this technique.

Walkthroughs. More detailed than protocol analysis and often better because they can be done in the actual environment, resulting in better memory cues. They need not, however, be carried out in real time. Indeed, such techniques are useful in a simulated environment where states of the system can be frozen and additional questions pursued.

Time lines. Tables in which several items of knowledge are contained in columns. The left column has to be filled with the time of occurrence of particularly interesting events such as failures or operator actions. Related information about the behavior of the technical process, the automatic system and the human operators at these times is recorded in separate columns with as much detail as is felt appropriate.

Formal techniques. These include multidimensional scaling, repertory grids and hierarchical clustering. Such techniques tend to elicit declarative knowledge. The most commonly used is the *repertory grid technique* based on the *personal construct theory*. It is used in *ETS*, (Boose, 1986), which assists in the elicitation of knowledge for classification type problems. In ETS, the expert is interviewed to obtain elements of the domain. Relationships are then established by presenting triads of elements and asking the expert to identify two traits which distinguish the elements. These are called *constructs*. They are then classified into larger groups called *constellations*. Various techniques such as statistical, clustering and multidimensional scaling are then used to establish classification rules which generate conclusion rules and intermediate rules together with certainty factors. The experts are interviewed again to refine the knowledge. ETS is said to save 2-5 months over conventional interviewing techniques. The system has been modified and improved and is now called *AQUINAS* (Boose and Bradshaw, 1988). To obtain procedural knowledge, techniques such as verbal protocols can be used.

Hypertext as a means of knowledge acquisition. Hypertext is an approach to information management in which information is organized as a network of *nodes* connected by *links*. Nodes may contain text, graphics, audio, video and generally software for operating on numeric and/or symbolic data. While other software paradigms are promising similar things, the essence of hypertext is that linking is machine-supported. At the development level most hypertext environments feature control buttons (*link icons*) which can be arbitrarily embedded within the content material by a user. Hypertext allows for easy and intuitive access to documents and programs by linking dispersed yet interrelated information throughout a document, a program or a series of documents/programs. Traditional document structure is *sequential*, in other words there is a single linear sequence defining the order in which the text is to be accessed. Hypertext is *nonsequential*, i.e. there is no single order that determines the sequence in which text is to be read.

Hypertext is simply the nonlinear presentation of any informational medium. Some communities prefer to reserve this term for textual information only and use *hypermedia* as a more general one. The term hypertext will be used here in the more general sense, as applied to the broad scope of informational media including graphics, video disc, and other such media. Hypertext systems have been used for information management and intelligent computer aided instruction systems, and they should prove equally useful in the area of expert systems technology.

A general knowledge acquisition tool designed around a hypertext concept could allow a knowledge engineer to list important concepts, create nodes attached to these concepts which explain their relevance, connect related concepts by linking their nodes, use graphics to explain difficult concepts, and even critique information entered into the system previously. In such a system, knowledge acquisition would not be confined to linear input of information. The knowledge engineer could use the hypertext system to compile knowledge gathered from an expert after interviewing, or (s)he could enter the knowledge into the system as the expert sits there saying what information to encode.

It is the nonsequential capabilities of hypertext systems that make them attractive as automated knowledge acquisition tools. The user of a hypertext system can dynamically create new nodes for information, make notes to her/himself in these nodes, and attach these nodes to the places in the hyperspace which have caused her/him to think of the nodes. The user could even make a node and leave it unattached to any reference, but displayed on the screen as a reminder of information which needs to be added to the hypertext database. A hypertext system facilitates linking together related information in hierarchical structures that resemble the relationships between the nodes of information, regardless of how complex the structure of the relationship may be.

Consider an example session with a knowledge engineer or an expert sitting down for a first pass at amassing the knowledge relevant to a project at hand. The user brings up the hypertext knowledge acquisition tool and enters a name for the knowledge base. The system confirms a new knowledge base and displays a node for relevant concepts. The user enters in list format short phrases to describe the topics which (s)he believes the expert system will have to know about in order to complete its function. Each concept will represent a link to a node in which that concept is described in further detail. There can be multiple links from concepts to other nodes. Also, concepts can be linked to one another to indicate their relationships (similarly to semantic network connections). To the nodes describing the concepts, other nodes can be attached. Such nodes can be definition nodes, primarily for use in the expert system for explanation and help facilities, graphic nodes, also for user help in the expert system, note nodes, as discussed above, and more informational nodes describing subconcepts within the main concepts. The key here is that the node attachments are created by the user as (s)he sees they would best be able to represent the knowledge required for implementing the expert system.

Another capability which the hypertext environment offers, is the contribution of knowledge to the knowledge-base from multiple experts. Once an initial attempt has been made at gathering the knowledge for the expert system, the experts can navigate through the hypertext system and judge the validity of knowledge represented. Each expert can attach comment nodes to currently existing nodes which indicate what (s)he thinks about the information contained within the original node. Experts can even critique the information input by other experts (this is a capability which may or may not be desirable, so there should be provisions for enabling and disabling such a feature).

There exists, however, a fundamental drawback in hypertext: *user disorientation* and even confusion. Increasing the number of connections, or links, increases the possibility that a user will get lost in irrelevant information. User disorientation may be particularly severe for large scale applications such as those involved in the utilization of power plant databases. The process of moving through a hypertext information base is referred to as navigation. Tsoukalas *et al*., (1991), outline navigational tools based on the theory of fuzzy graphs and fuzzy relations. These tools quantify context-dependent user preferences and application-specific constraints in such a manner that a user may direct (him)herself to an information island of interest. A numerical example and a prototype for monitoring special material in a nuclear power plant are included.

Bottom-up and top-down knowledge capturing.

There are two competing views about the knowledge acquisition task which might be described as bottom-up and top-down.

The bottom-up proponents aim to prise data and concepts out of the expert and then iteratively refine it. The implication is that deeper mining will reveal more relevant knowledge, but this assumes that there is a simple relationship between what is verbalized by experts and what is actually going on in their minds.

The basic assumption underlying the bottom-up approach is that an expert system is based upon a large body of domain specific knowledge, and that there are a few general principles underlying the organization of the domain knowledge in an expert's mind. However, the existence of underlying principles and causal relationships may be an indication that expert knowledge is somehow domain independent. So, expert behavior that is seemingly domain-specific, may originate from higher level problem solving methods which are well structured and have some degree of domain independence. Currently, the most popular heuristic rule generation procedures are based on schemes of ***top-down induction of decision trees*** (*TDIDT*) (Gray, 1990).

The essence of a TDIDT program for decision tree generation is depth-first search. One starts with some training set of example objects, each characterized by attribute values and a class designation. The program selects and tests a binary attribute, resulting in two recursive subproblems. Each subproblem involves a subset of the original example set. The first subproblem is then analyzed in the same manner leading to two further recursive subproblems. Search proceeds depth-first until either all objects associated with a general subproblem are in the same class (completing a decision tree branch), or all attributes have been utilized (demonstrating that either data are incorrect or the attribute set is inadequate). When a branch is completed, the program backtracks to a previous choice point and from there explores the subproblem(s) associated with alternative values for the tested attribute.

If the attribute set is sufficient, many decision trees may exist that correctly classify training set examples. The program uses inductive inference to construct a decision tree that correctly classifies other objects in addition to those in the training set. Consequently, a decision tree must capture some meaningful relation between an object class and its attribute values. Given several trees that correctly classify training set data, the "simplest" is usually chosen on the grounds that it is more likely to capture the inherent structure of the problem.

TDIDT methods restrict search by employing a heuristic measure: they considere combinations of attributes appearing to have a high information content. This useful restriction can make rule generation feasible, but other aspects of the TDIDT approach are unhelpful. Its depth-first recursive search with backtracking, causes it to impose inappropriate context restrictions on rule search. These restrictions lead to opaque knowledge representations and excessive sensitivity to noise.

Gray, (1990), proposed algorithmic modifications so that users can still measure heuristically the information content of attributes in order to guide the search. The program iteratively examines all positive instances remaining to be covered, along with negative training set instances. Moreover, the search does not take place with irrelevant context restrictions. This algorithm is no more complex than usual TDIDT, it is just as fast, it is less sensitive to noise, and it leads to clearer representations of the information present in trainingset data.

Machine induction.

Machine induction is a special case of machine learning which encompasses heuristics for generalizing data types, candidate elimination algorithms, methods for generating decision trees and rule sets, function induction and procedure synthesis. MacDonald and Witten, (1989), developed a framework for describing such techniques that allows an evaluation of the usefulness of any method in solving particular knowledge engineering problems. They have concentrated upon decision tree and rule set generation approaches because these techniques have been successfully used in a number of knowledge acquisition situations.

It is commonly observed that experts have great difficulty in explaining the procedures which they use to arrive at decisions. Indeed, experts often make use of assumptions and beliefs which they do not explicitly state, and are surprised when the consequences of these hidden assumptions are pointed out. The inductive approach relies on the fact that experts can usually supply examples of their expertise even if they do not understand their own reasoning mechanisms. This is because creating an example set does not require any understanding of how different evidence is assessed or what conflicts were resolved to reach a decision. Sets of such examples are then analyzed by an inductive algorithm (one of the most popular being the ID3 algorithm, see knowledge acquisition tools in the following) and rules are generated automatically from these examples.

The problem with inductive techniques is that the rules induced depend both upon the example set chosen and the inductive algorithm used. There is no guarantee that the rules induced will be valid knowledge. Therefore, the approach normally involves a check with the expert to validate the induced rules. It is not uncommon to cycle a number of times through the induction process, refining the knowledge base with the domain expert.

The most important guidelines in the appropriate use of inductive techniques are:

- The technique is useful if there are documented examples or if they can be obtained easily. It is not suitable where an unpredictable sequence of observations drives the system (e.g. as in some real-time situations).
- The technique is consistent and unbiased and is very suitable for domains where rules form a major part of the knowledge representation.
- Induction provides the knowledge engineer with questions, results and hypotheses which form a basis for consultation with the expert.
- There is no explanation for the rules produced. All output must be examined critically.
- The process assumes that the example set is complete and current.
- Results should not be sensitive to small changes in the training set.

The inductive technique has been used successfully for weather prediction, predicting the behavior of a new chemical compound, diagnosing plant disease, symbolic integration, improved debt collection, and designing gas-oil separators.

Knowledge acquisition tools.

A large number of tools for supporting the knowledge acquisition process have been developed in the academic environment and some of these have been mentioned already. The general aim of all these tools is to minimize the number of iterations needed for the whole knowledge engineering process by bridging the gap between the problem domain and the implementation. Boose and Gaines, (1988), and Johannsen and Alty, (1991), give a brief summary of the main tools under development and provide a summary. Some tools endeavour to make the process fully automatic. *KRITON* for example, has a set of procedures and pre-stored interviews, and caters for incremental text analysis and protocol analysis. Repertory grids are used to pull out declarative knowledge. An intermediate knowledge representation system is suggested for supporting the knowledge elicitation techniques. The knowledge representation scheme involves a propositional calculus for representing transformations during the problem solving process and a descriptive language for functional and physical objects. This is then translated semi-automatically into the run-time system but this commits the knowledge engineer to a particular representation.

Other tools, for example *KADS* and *ACQUIST*, merely provide a set of tools to aid in a more methodological approach. Thus, KADS aims only at producing a document describing the structure of the problem in the form of a documentation handbook.

The KADS methodology is based upon the following principles (Hickman *et al.*, 1989):

- Knowledge and expertise should be analyzed before the design and implementation starts, i.e. before an implementation formalism is chosen.
- The analysis should be model driven as early as possible.
- Expert problem solving should be expressed as epistemological knowledge.
- The analysis should include the functionality of the prospective system.
- The analysis should be breadth-first allowing incremental refinement.
- New data should only be elicited when previous data has been analyzed.
- All collected data and interpretations should be documented.

KRITON supports only bottom-up knowledge acquisition but KADS supports both top-down and bottom-up through a ***hypertext protocol editor*** (*PED*) and hierarchies are developed and manipulated by a ***context editor*** (*CE*). Top-down is supported by a set of interpretation models each describing the meta-level structure of a generic task.

KEATS-1 provides a ***cross reference editing facility*** (*CREF*) and a ***graphical interface system*** (*GIS*), to support data analysis and domain conceptualization. CREF organizes the verbal transcript text into segments and collections and GIS allows the knowledge engineer to draw and manipulate domain representations on a sketch pad. In KEATS-2, these have been replaced by ACQUIST, a hypertext application for structuring the knowledge from the raw text data. Fragments from the data are collected around concepts, concepts are factored into groups, and groups into meta-groups. Links can then be defined between any of these entities. The emerging structure is displayed graphically. ACQUIST provides support for both bottom-up approaches (fragments to concepts to groups to meta-groups) and top-down approaches (using what are called *coding sheets* on which a caricature of the observed behavior of the domain expert is captured). In this approach, the knowledge engineer uses a redefined abstract model to guide the knowledge acquisition process. Use of such models (even if incomplete or inadequate) can dramatically improve the knowledge acquisition process. The coding sheet is a set of hypertext cards.

A further knowledge acquisition tool is *ROGET*. It conducts a dialogue with a domain expert in order to acquire (his)her conceptual structure. ROGET gives advice on the basis of abstract categories and evidence. Initial conceptual structures are selected on this basis. Only a small set of example cases have been tested on this system.

The systematic acquisition of knowledge about the faulty behavior of a technical system was suggested by Narayanan and Viswanadham, (1987). A procedure involving the development of a hierarchical failure model with fault propagation digraphs and cause-consequence knowledge bases for a given system, is proposed. It uses the so called *augmented fault tree* as an intermediate knowledge representation. Fault propagation *digraphs* describe the hierarchical structure of the system with respect to faults in terms of propagation. The cause-consequence knowledge bases characterize failures of subsystems dependent on basic faults by means of production rules. The knowledge acquisition process can be reduced to defining parameters required by the knowledge representation scheme and transforming human expertise into these parameter values. The augmented fault tree is a conceptual structure, which describes causal aspects of failures as in conventional fault trees as well as probabilistic, temporal and heuristic information (see also Contini, 1987). The production rules of cause-consequence relations are derived from the augmented fault tree by decomposing it into mini fault trees. The proposed methodology has reached a relatively high level of formal description. However, it cannot yet deal with inexact knowledge by using ranges of parameters. An example of a failure event in a reactor system is given.

Strategic knowledge acquisition tools have been built especially for diagnosis purposes: *TEIRESIAS, ASK, CART, ODYS, NEOMYCIN, HERACLES, BDM-KAT, CATS, ID3, ELI, HELIOS, MILKAM, KSS0* assistants and their evolutions are some of the more recent knowledge elicitation systems which are in use in medical and industrial diagnosis practice. The reader is referred to the SIGART Newsletter (1989), Mussi and Morpurgo (1990) and Brule and Blount (1989) for details.

4.2.1.3 Expert system approaches for automatic process fault diagnosis

Having concluded that the human's cognitive limitations, biases, errors, and lack of knowledge are major obstacles in diagnostic performance, it seems reasonable to attempt full or partial automation of the diagnostic procedure in order to aid the human diagnostician in real-time. One approach in this category that has recently received widespread attention is that of *expert systems* (*ES*).

Some recent surveys of ES appropriate in fault detection/diagnosis of technological processes are provided by Milne (1987), Tzafestas (1987, 1989, 1991), Majstorovic (1990), Prasad and Davis (1993) and Dounias *et al.*(1993).

Several designs have been efficiently employed for fault diagnosis in various domains such as medical, electronic, computer H/W and S/W, and industrial process diagnosis. A brief description of the main characteristics of the most important current approaches is given in the following.

Shallow reasoning approach.

Since shallow reasoning is highly domain-specific, diagnosis is fast if the symptom has been experienced and thus has been included in the knowledge base. This reasoning typically uses (production) rules which consist of *antecedents* and *consequents*. An antecedent is a condition part and a consequent is an action part of a rule. If certain conditions are met, then some actions are performed. For this reason, such a rule is often called an IF-THEN rule. These rules can be classified by their behavior, i.e. self-managing rules or *meta-rules*. A self-managing rule is one in which actions are performed without referring to any other rules. A meta-rule is one in which its actions result from the triggering of other rules.

A disadvantage however, is that shallow reasoning is rigid in the sense that substantial changes in the rules may have to be made if a single component is added or deleted. Consequently, the number of rules becomes practically unmanageable as the number of components of the system being diagnosed increases. If multiple faults can occur simultaneously, then this approach becomes combinatorially explosive.

The classical example of a shallow expert diagnostic system is MYCIN and NEOMYCIN (see Section 4.2.2). The main shortcomings of this approach are:

- Difficult knowledge acquisition.
- Unstructured knowledge requirements.
- Diagnosability or knowledge-base completeness non guaranteed.
- Excessive number of rules.
- Knowledge-base highly specialized to the individual process.

These disadvantages can be overcome by decomposing the problem into smaller problems either in a hierarchical manner or according to unit operations.

Deep knowledge approach.

This approach is appropriate for man-made technological system diagnosis (causal oriented systems) and is based on a structural and functional model of the problem domain. A deep knowledge ES attempts to capture the underlying principles of a domain (or process) explicitly, and so the need to predict every possible fault scenario is eliminated. Obviously, this approach leads to expert system tools that are able to handle a wider range of problem types and larger problem domains (Yoon and Hammer, 1988).

Deep reasoning, compared to shallow reasoning, is more flexible and thorough, but slower. "More flexible" means that since deep reasoning is not domain-specific, it is easier to modify the model when a single component is added or deleted. Deep reasoning may not be sensitive to the change. "More thorough" means that deep reasoning may answer "what-if" type questions which may not be possible in shallow reasoning. This implies that there is no limitation of fault coverage in deep reasoning. "Slower" means that the speed of reasoning is slower than that of shallow reasoning because a deep knowledge base does not contain every detail of a symptom.

The principal deep-knowledge diagnosis methods are:

- The causal knowledge search method.
- The physical system mathematical model method.

- The hypothesis test method.

The *causal knowledge search* method is based on the tracing of process malfunctions to their source. Causality is usually represented by directed graphs (digraphs), the nodes of which represent state variables, alarm conditions or failure origins, and the branches (edges) represent the influences between nodes. The digraph can include (besides the positive (+) or negative (-) influences) the intensity of the effect, the time delays and probabilities of fault propagation.

Since the space of possible hypotheses can be very large if multiple disorders can be present simultaneously, some means are required to focus an expert system's attention on those hypotheses that are most likely to be valid. A domain-independent algorithm is proposed by Peng and Reggia, (1987), that uses symbolic causal knowledge and numeric probabilistic knowledge to generate and evaluate plausible hypotheses during diagnostic problem solving with multiple simultaneous disorders.

The use of causal knowledge for diagnosis has been more popular in applications such as electronics and mechanics because the problems and mechanisms there are better defined (Fink and Lusth, 1987).

Hudlicka and Lesser, (1987), integrated a number of known techniques (diagnostic paths, simulation, qualitative reasoning, constraint networks) and described two new ones (comparative reasoning and the use of under-constrained abstracted objects) in an attempt to solve, using a causal model of the system, the problems encountered in representing and reasoning about problem-solving system behaviour. The diagnosis system selects its own "correct behavior criteria" from objects within the problem-solving system which did achieve some desired situation.

The *physical system mathematical model* method relies on redundancy between process measurements and consists of an analytic problem solution, a process knowledge base, a knowledge acquisition component and a rule-based inference mechanism. The analytic problem solution uses a physico-mathematical model of the process through which process state or parameter estimation can be performed. Detection of changes in process variables or parameters are then considered to be symptoms of process faults. Classification of the changes which may occur in the process state can be performed. The process knowledge base is comprised of analytical knowledge in the form of process models, and heuristic knowledge in the form of fault and/or event trees and fault statistics (Contini, 1987). In the phase of knowledge acquisition, the process specific knowledge, eg. theoretical process models, classification of changes, normal behavior, event trees and fault trees, is compiled. The inference mechanism performs the fault diagnosis, based on the observed symptoms, the fault and/or event trees, fault probabilities and the process history (Pouliezos and Stavrakakis (1987), Frank (1990), van Soest *et al.* (1990), Isermann and Freyermuth (1991), Rhodes and Karakoulas (1991), Gertler and Anderson (1992)).

The *hypothesis formulation/hypothesis testing method* follows the usual human diagnostics path, i.e. a cause for a system malfunction (upset) is postulated, the symptoms of the postulated fault are determined, and the result is compared with the process observables. Of course the search for the location of a fault can be narrowed by using appropriate heuristics. Hypothesis testing requires qualitative simulation of the effects of the postulated malfunctions. Qualitative (non numerical) simulation requires prediction of the direction of deviation of measured variables of the process as a result of faults. Qualitative simulation models need to be enriched with suitable heuristics and precedence rules in order to be able to resolve competing causal influences on the same process variable.

A means of interpreting observations made of a physical system across time, in terms of qualitative physics theory, is described by Forbus, (1987), and Bandekar, (1989). The theory described is ontology-independent as well as domain-independent. This means that it only requires a qualitative description of the domain capable of supporting envisioning and domain-specific techniques for providing an initial qualitative description of numerical measurements, even when noisy data should be handled. For the diagnosis problem this theory provides a general method for testing fault hypotheses using an analogous AI model, to test if it actually explains the observed behavior. Trave-Massuyes *et al.*, (1990), present a qualitative calculus (qualitative equations setting, the orders of magnitude qualitative algebras, qualitative equations solving techniques) which is a key point in qualitative simulation for automatic intelligent fault diagnosis and supervision. Although qualitative calculus gives rise to numerous problems, this paper and its authors' previous and current work provide a quite complete resolution scheme.

Ontological analysis approach. Another general knowledge engineering methodology which is based on the deep knowledge approach is the so called *ontological structure analysis*. Ontological analysis proceeds in a step-by-step articulation of the knowledge structures needed to perform a task by following the objects and relationships that occur in the task domain itself. The application of ontological analysis to practical troubleshooting/diagnosis problems can be done very effectively by decomposing the notion of ontological structure into three levels, namely:

1. *Static ontology*: Definition of actual physical objects in the problem domain and their properties and relationships.
2. *Dynamic ontology*: Definition of the state space in which the problem solving must occur and the actions that transform the problem from one state to another.
3. *Epistemic ontology*: Definition of the form of constraints and heuristics that control navigation in the state space.

For example, in the case of electronic instrument troubleshooting, static ontology encompasses the components and knobs of the instrument which are connected electronically by nodes, and are grouped in blocks. Dynamic ontology defines the states of the problem which consist of belief states and instrument states. A *belief state* consists of diagnostic beliefs about the diagnostic condition of each module (component, knob or block), which have associated justifications. An *instrument state* consists of knob settings and signal inputs that are used to stimulate the instrument under test. Transformations are measurements of signals and electrical parameters. Measurements are grouped into tests, and tests are grouped into strategies. Each test and strategy has implications, i.e. diagnostic beliefs implied by the test results. Finally, epistemic ontology defines appropriate types of knowledge that make it possible to choose effective transformations and thus navigate the problem state space in a reasonable amount of time. Most experts use the hypothesis formulation/hypothesis testing method presented previously, i.e. they diagnose an instrument by having (or formulating) a set of hypotheses about which modules might be good and bad. In order to test such hypotheses, they have heuristic diagnostic strategies, which relate each module to the method by which it may be tested.

Formal tools exist for defining and communicating the ontological analysis. These tools are illustrated in Freiling *et al.*, (1986), with an example domain equation language called *SPOONS* (***SP**ecification **O**f **ON**tological **S**tructure*) and the presented results are based on the knowledge-based system *HIPE* (***H**ierarchical **I**nference **P**rocessing **E**ngine*).

Hybrid reasoning.

Shallow reasoning has been widely used but, because of the disadvantages mentioned above, deep reasoning has emerged. However, since it requires more search time and thus shows an undesirable speed of reasoning for some complex systems, deep reasoning alone is not satisfactory either. Hence, hybrid reasoning, combining these two approaches has been attempted in order to perform the diagnostic process efficiently (from deep reasoning) and effectively (from shallow reasoning). In other words, hybrid reasoning utilizes both deep and shallow reasoning methodologies in an attempt to take advantage of the strengths of each. Two directions exist:

1. Deep first, then shallow (D-S).
2. Shallow first, then deep (S-D).

Examples of an S-D approach are *CHECK* (***C**ombining **H**euristic and **C**ausal **K**nowledge*) by Torasso and Console, (1989), and *IDM* (***I**ntegrated **D**iagnostic **M**odel*) by Fink and Lusth, (1987). An example of a D-S approach is *ISA* (***I**ntegrated **S**tatus **A**ssessment*) developed by Marsh, (1988).

The current trend in diagnosis is toward hybrid reasoning. Yet, there has been no comparative study of the various types of reasoning. However, when the system to be diagnosed is relatively small (this also implies a small number of rules), an S-D approach seems to be preferred. For a large-scale system like a manufacturing plant, a D-S approach is often chosen. The D-S type hybrid reasoning diagnosis model is analytically described in Appendix 4.A for the interested reader.

Attribute grammar approach.

The discussion on ES-based fault diagnosis will be closed with a brief discussion of the attribute grammar approach to building process fault diagnosis tools. By employing the attribute grammar model for knowledge representation, both *declarative* (factual) knowledge and *inferential* (procedural) knowledge can be combined in a single tool. Since there exist many implementations of compilers and interpreters for processing attribute grammars, this approach promises a lot for knowledge engineers and fault diagnosis experts. In cases where knowledge can be expressed in the form of logic rules, an attribute grammar interpreter is written in the form of logic rules in a PROLOG-like way, i.e.,

$R_0(t_{01}, t_{02}, \ldots, t_{0k_0})$	is true if
$R_1(t_{11}, t_{12}, \ldots, t_{1k_1})$	is true and
....	
$R_{m_e}(t_{m_e1}, t_{m_e2}, \ldots, t_{m_ek_{m_e}})$	is true,

where t_{ij}; $0 \le i \le m_e$, $i \le j \le k_i$, $1 \le e \le n$, is a constant or a variable, and e is the rule number (assuming that the rules are numbered from 1 to n). If $m_e = 0$, then the rule reduces to a fact.

A syntax rule corresponding to the above logic rule for an *homologous attribute grammar* (i.e. a grammar which when processed by its interpreter will give the same results as those coming from the successful application of the above logic rules) has the form,

$$< R_0 > \; :: \; = \; < R_1 >< R_2 > \; \ldots \; < R_{m_e} > \; |\text{-}|$$

where the combination of the last two characters means the end of the syntax rule.

The parser of the homologous attribute grammar has the following features.

- No terminal symbols are used (i.e. it is degenerate).
- An extended stack is used for saving the attribute values as well.
- Calls to the attribute evaluator are included.
- A meta-variable, named FLAG, is used to show variable matching (when a value mismatch occurs, the FLAG takes the value false).

A false value of FLAG results in a backtracking of the parsing process. The semantic rules that perform the unification of variables can be written in a straightforward way. The interpreter includes a facility of providing "why" and "how" explanations.

To deal with inexact knowledge (i.e. uncertain facts and rules or imprecise items of evidence) each rule is assigned a certainty measure which can be the conditional probability of the validity of its conclusion, given the corresponding premises (de Kleer, 1990). Each of the premises is assigned a posterior probability evaluated from previous inference. The updating of these posterior probabilities can be done using Bayes inference rule. De Kleer, (1990), applied a similar procedure for the construction of a diagnostic engine in order to identify automatically the faulty components of a malfunctioning device in the fewest number of measurements. A minimum entropy technique is used (Pandelidis, 1990) to select the next best measurement to be used for the diagnosis procedure.

Alternatively, one can use an upper or lower bound of the validity probability of each item, i.e. the so called possibility and necessity measures, or the well known Shortliffe's certainty factors employed in the expert system MYCIN, mentioned previously.

4.2.1.4 High-speed implementations of rule-based diagnostic systems

In the previous sections of this chapter, rule-based knowledge systems for modeling intelligent behavior and building expert systems for automatic process fault monitoring, are described.

However, most rule-based programs are extremely computationally intensive and run quite slowly. The slow speed of execution has prohibited the use of rule-based knowledge systems in domains requiring high performance and real-time response such as in real-time process fault diagnosis. In this section various methods for speeding up the execution of rule-based knowledge systems are explored. In particular, the role of parallelism in high-speed execution of rule-based knowledge systems is examined and the architectural issues in the design of computers for rule-based systems are studied. It is shown that contrary to initial expectations, the speed-up that can be obtained from parallelism is quite limited, only about tenfold. The reasons for this small speed-up are:

1. The small number of rules relevant to each change in data memory.
2. The large variation in the processing requirements of relevant rules; and
3. The small number of changes made to data memory between synchronization steps.

Furthermore, in order to obtain this limited factor of tenfold speed-up, it is necessary to exploit parallelism at a very fine granularity. A suitable architecture to exploit such fine-grain parallelism is a shared-memory multiprocessor with 32-64 processors. Using such a multiprocessor, it is possible to obtain execution speeds of about 3800 rule-firings/sec (Gupta *et al.*, 1989).

A rule-based knowledge system is composed of a set of IF-THEN rules (also called *productions*) that make up the *rule memory*, and a database of assertions called the *working memory*. The assertions in the working memory are called *working memory elements*. Each rule consists of a conjunction of *condition elements* corresponding to the IF part of the rule (also called the *left-hand side* of the rule), and a set of actions corresponding to the THEN part of the rule (also called the *right-hand side* of the rule).

The actions associated with a rule can add, remove, or modify working memory elements, or perform input-output.

The *rule interpreter* is the underlying mechanism that determines the set of satisfied rules and controls the execution of the rule-based knowledge system. The interpreter executes a rule-based program by performing the following *recognize-act* cycle:

- *Match:* In this first phase, the left-hand sides of all rules are matched against the contents of working memory. As a result a *conflict set* is obtained, which consists of *instantiations* of all satisfied rules. An instantiation of a rule is an ordered list of working memory elements that satisfies the left-hand side of the rule.
- *Conflict resolution:* In this second phase, one of the rule instantiations in the conflict set is chosen for execution. If no rules are satisfied, the interpreter halts.
- *Act:* In this third phase, the actions of the rule selected in the conflict-resolution phase are executed. These actions may change the contents of working memory. At the end of this phase, the first phase is executed again.

The recognize-act cycle forms the basic control structure in rule-based programs. During the match phase, the knowledge of the program (represented by the rules) is tested for relevance against the existing problem state (represented by the working memory). During the conflict-resolution phase, the most promising piece of knowledge that is relevant is selected. During the act phase, the action recommended by the selected rule is applied to the existing problem state, resulting in a new problem state.

Parallelism is possible to be used while performing each of the above three phases. It is further possible to overlap the processing performed within the match phase and the conflict-resolution phase of the same recognize-act cycle, and that within the act phase of one cycle and the match phase of the next cycle. However, it is not possible to overlap the processing within the conflict-resolution phase and the subsequent act phase, because the conflict-resolution must finish completely before the next rule to fire can be determined and its right-hand side evaluated. Thus, the possible sources of speed-up are:

1. Parallelism within the match phase.
2. Parallelism within the conflict-resolution phase.
3. Parallelism within the act phase.
4. Overlap between the match phase and the conflict-resolution phase of the same cycle.
5. Overlap between the act phase of one cycle and the match phase of the next cycle.

Parallelism within the match phase.

In the following, several ways in which parallelism may be used to speed up the match phase, are discussed.

Rule-level parallelism. When using rule-level parallelism, the rules in a program are divided into several partitions and the match for each of the partitions is performed in parallel. In the extreme case, the number of partitions equals the number of rules in the program, so that the match for each rule in the program is performed in parallel. One of the main advantages of using rule-level parallelism is that no communication is required between the processes that perform a match for different rules or different partitions.

Contrary to all expectations, Gupta *et al.* (1989), show that the true speed-up expected from rule-level parallelism is really quite small, only about twofold. Some of the reasons for this are given below:

- Simulations show that the average number of rules affected per change in working memory is around 28. (A rule is said to be affected by a change in working memory, if the new working memory element matches at least one of the condition elements of that rule). In most matches, determining the set of affected rules is much faster than processing the state changes associated with the affected rules. Thus the number of affected rules bounds the amount of speed-up that can be achieved using rule-level parallelism.
- The speed-up obtainable from rule-level parallelism is further reduced by the variance in the processing time required by the affected rules. The maximum speed-up that can be obtained is proportional to the ratio t_{avg}/t_{max}, where t_{avg} is the average time taken by an affected rule to finish match and t_{max} is the maximum time taken by any affected rule to finish match. The parallelism is inversely proportional to t_{max} because the next recognize-act cycle cannot begin until all rules have finished match. Note that *nominal* speed-up (or *concurrency*) is defined to be the average number of processors that are kept busy in the parallel implementation. Nominal speed-up is to be contrasted against *true* speed-up, which refers to the speed-up with respect to the highest performance uniprocessor implementation, assuming that the uniprocessor is as powerful as the individual nodes of the parallel processor. True speed-up is usually less than the nominal speed-up because some of the resources in a parallel implementation are devoted to synchronizing the parallel processes, scheduling the parallel processes, recomputing some data that are too expensive to be communicated, etc.
- The third factor that influences the speed-up is the loss of sharing in the data flow network when rule-level parallelism is used. The loss of sharing happens because operations that would have been performed only once for similar rules are now performed independently for such rules, since the rules are evaluated on different processors.
- The fourth factor that influences the speed-up is the overhead of mapping the parallel algorithm on to a parallel hardware architecture. The overheads may take the form of memory-contention costs, synchronization costs or task-scheduling costs.

Some implementation issues associated with using rule-level parallelism are now discussed. The first point that emerges from the previous discussion is that it is not advisable to allocate one processor per rule for performing match. If this is done, most of the processors will be idle most of the time and the hardware utilization will be poor. When using only a small number of processors, two alternative mapping strategies can be considered. The first is to divide the rule-based program into several partitions so that the processing required by rules in each partition is almost the same, and then allocate one processor for each partition. The second strategy is to have a task queue shared by all processors in which entries for all rules requiring processing are placed. Whenever a processor finishes processing one rule, it gets the next rule that needs processing from the task queue. Some advantages and disadvantages of these two strategies are given below.

The first strategy is suitable for both shared memory multiprocessors and non-shared memory multicomputers, since little or no communication is required between processors. The main difficulty, however, is to find partitions of the rule-based system that require the same amount of processing. Note that even if one finds partitions with only one affected rule per partition, the variance in the cost of processing the affected rule still destroys most of the speed-up. The task of partitioning is also difficult because good models are not available for estimating the processing cost of rules, and also because the processing cost of rules varies over time. A discussion of the various issues involved in the partitioning task is presented in Carriero and Gelernter, (1989).

The second strategy is suitable only for shared memory architectures, because it requires that each processor has access to the code and state of all rules in the program (while it is possible to replicate the code in the local memories of all the processors, it is not possible to do so economically for the dynamically changing state associated with the rules). Since the tasks are allocated dynamically to the processors, this strategy has the advantage that the load distribution problem is not present. Another advantage of this strategy is that it extends very well to finer granularities of parallelism. However, this strategy loses some performance due to the synchronization, scheduling, and memory contention overheads.

Node parallelism. When node parallelism is used, activations of different multi-input nodes in the data-flow network are evaluated in parallel.

It is important to note that node parallelism subsumes rule-level parallelism, in that node parallelism has a finer grain than rule-level parallelism. Thus, using node parallelism, both activations of two-input nodes belonging to different rules (corresponding to rule-level parallelism), and activations of two-input nodes belonging to the same rule (resulting in the extra parallelism) are processed in parallel.

The main reason for going to this finer granularity of parallelism is to reduce the value of t_{max}, the maximum time taken by any affected rule to finish match. This decreased granularity of parallelism, however, leads to increased communication requirements between the processes evaluating the nodes. When using node parallelism, a process must communicate the results of a successful match to the successors of that two-input node. However, no communication is necessary if the match fails. To evaluate the effectiveness of exploiting node parallelism, it is necessary to weigh the advantages of reducing t_{max}

against the cost of increased communication and the associated limitation on feasible hardware architectures.

Another advantage of using node parallelism is that some of the sharing lost when using rule-level parallelism is recovered. If two rules need a node with the same functionality, it is possible to keep only one copy of the node and to evaluate it only once, since it is no longer necessary to have separate nodes for different rules. The gain due to the increased amount of sharing is a factor of 1.3, which is quite significant.

Action parallelism. Usually when a rule fires, it makes several changes in the working memory. Processing these changes concurrently, instead of sequentially, leads to increased speed-up from rule, node, and intranode parallelism. This source of parallelism is named *action parallelism*, since matches for multiple actions in the right-hand side of the rule are being processed in parallel.

Data parallelism. A still finer grain of parallelism may be exploited by performing the processing required by each individual node activation in parallel. This task can be speeded up using data parallelism (Carriero and Gelernter, 1989). Such parallelism is expected to reduce t_{max} even further, and thus help increase the overall speed-up. The disadvantage of exploiting data parallelism of conventional shared memory multiprocessors is that the overhead of scheduling and synchronizing these very fine grained tasks (a few instructions) nullifies the advantages. However, exploiting data parallelism is not as hard on highly parallel machines.

Parallelism in conflict resolution.

The conflict-resolution phase is not expected to be a bottleneck in the near future. The reasons for this are:

- Current rule-based interpreters spend only about 5 percent of their execution time on conflict-resolution. Thus the match phase has to be speeded up considerably before conflict-resolution becomes a bottleneck.
- In rule-level and node parallelism, the matches for the affected rules finish at different times because of the variation in the processing required by the affected rules. Thus many changes to the conflict set are available to the conflict-resolution process while some rules are still performing match. Thus much of the conflict-resolution time can be overlapped with the match time, reducing the chances of conflict-resolution becoming a bottleneck.

If the conflict-resolution does becomes a bottleneck in the future, there are several strategies for avoiding it. For example, to begin the next execution cycle, it is not necessary to perform conflict-resolution for the current changes to completion. It is only necessary to compare each current change to the highest priority rule instantiation so far. Once the highest priority instantiation is selected, the next execution cycle can begin. The complete sorting of the rule instantiations can be overlapped with the match phase for the next cycle. Hardware priority queues provide another strategy.

Parallelism in RHS evaluation.

The RHS-evaluation step, like the conflict-resolution phase, takes only about 5 percent of the total time for the current rule-based systems. When many rules are allowed to fire in parallel, it is quite straightforward to evaluate their right-hand sides in parallel. Even when the right-hand side of only a single rule is to be evaluated, it is possible to overlap some of the input/output with the match for the next execution cycle. Also when the right-hand side results in several changes in the working memory, the match phase can begin as soon as the first change to working memory is determined.

It was stated before that the conflict-resolution phase must finish completely before the right-hand side can be evaluated (until that time it is not sure which rule will fire next). However, if one takes a speculative approach, it is possible to overlap the conflict resolution and the RHS evaluation step. The solution is to make an intelligent guess about which rule is going to fire next. For example, one may guess that the second best rule from the previous conflict resolution phase is the rule that is going to fire next. After making the guess, one can go ahead and evaluate the RHS of that rule; that is, determine what changes are going to be made in the working memory. Actually, the working memory is not modified at this point. When the winning rule is found out at the end of the conflict resolution phase, if the guess was correct, the RHS evaluation step is already done. If the guess was wrong, then some processing resources have only been wasted, which is not too bad, especially if they were idle in any case.

Application parallelism.

There is extra speed-up to be gained from application parallelism, where a number of cooperating, but loosely coupled, rule-based tasks execute in parallel. The cooperating tasks may arise in the context of search, where there are a number of paths to be explored, and it is possible to explore each of the paths in parallel (similar to *OR-parallelism* in logic programs). Alternatively, the cooperating tasks may arise in the context where there are a number of semi-independent tasks, all of which have to be performed, and they can be performed in parallel (similar to *AND-parallelism* in logic programs). It is also possible to have cooperating tasks that have a producer-consumer relationship among them (similar to *stream-parallelism* in logic programs). The maximum speed-up that can be obtained form application parallelism is equal to the number of cooperating tasks, which can be significant. Unfortunately, most current rule-based systems do not exploit such parallelism, because,

1. The rule-based programs were expected to run on a uniprocessor, where no advantage is to be gained from having several parallel tasks, and
2. Current rule-based languages do not provide the features to write multiple cooperating rule-based tasks easily.

Gupta *et al.*, (1989), describe the architecture of the ***Production System Machine*** (*PSM*), a hardware structure suitable for executing in parallel rule-based systems of dataflow-like nature. The performance that would be obtained as a result of parallel implementation of rule-based knowledge systems was evaluated through a large number of simulations. They also compared this performance rating to that expected from other proposed architectures for parallel implementation of rule-based knowledge systems. Very useful information for eventual application problems concerning parallel implementation of rule-based diagnosis systems can be found there.

4.2.1.5 Validating expert systems

Like typical software development, expert system development has a life cycle. Validation is formally included in most expert system development frameworks, in the form of phased or task-stepwise decomposition of the complete development process. The term validation is used many times inconsistently and often confused with evaluation. Validation is defined here to be distinct form evaluation.

Validation is the process of determining that an expert system accurately represents an expert's knowledge in a particular problem domain. This definition of validation focuses on the expert system and the expert. In contrast, *evaluation* is defined as the process of examining an expert system's ability to solve real-world problems in a particular problem domain. Evaluation focuses on the expert system and the real world. Grogono *et al.*, (1991), outline some of the issues involved in evaluating expert systems and cite almost 200 significant papers on this topic.

Validation has two dimensions, verification and substantiation. *Verification* is the authentication that the formulated problem contains the actual problem in its entirety and is sufficiently well structured to permit the derivation of a sufficiently credible solution. *Substantiation* is defined as the demonstration that a computer model within its domain of applicability possesses a satisfactory range of accuracy consistent with the intended application of the model.

Among the many concerns expressed about developing and validating expert systems are the following:

- What should be validated?
- How is it validated?
- What are the procedures for validation?
- How is bias controlled?
- How is validation integrated into development?
- How are costs controlled?

These concerns are particularly relevant when developing demonstration prototypes, where costs and time resources are constrained. In these situations, it is easy to minimize or overlook validation. All too often validation becomes highly informalized and, as a result, does not become an integral part of development. O' Leary *et al.*, (1990), extending Buchanan's and previous testing tasks, presented a specific formal validation paradigm for prototype expert system development within time and cost constraints. It incorporates many of the descriptive elements addressed by others, and explicitly incorporates validation into the development life-cycle approach for prototype development.

The validation process involves verification that the model sufficiently addresses the real problem in its entirety, and substantiation that the model possesses a sufficient range of accuracy. Verification and substantiation are evaluated through a three stage procedure ensuring face validity, establishing subsystem validity and comparing input-output transformations.

These stages and processes are related by the interaction of the knowledge engineering term, the expert(s), the prototypical expert system and the real world. Central to the validation process are the expert(s) and the knowledge engineering team, consisting of at least two members. One member, the system designer, has primary responsibility for knowledge acquisition and encoding the prototypical expert system. The other member, the third-party validator, has primary responsibility for validation.

The development process begins as the system designer interacts with the expert to develop a view of the expert system. (S)he then creates a tangible representation of this view in the form of an initial prototype (Buchanan's identification, conceptualization, formalization, and implementation tasks).

During formal validation (Buchanan's testing task), the third-party validator, the system designer and the expert's work closely together. The validator examines the prototype to ensure that the system designer's view and the expert's view are consistently represented and that the prototype is able to respond to domain-specific real world situations. This examination iterates through three stages: face validity, subsystem validity, and input-output comparison. As the team members find inconsistencies or unacceptable limitations in the prototype, they make system reformulations, redesigns, and refinements, and revisit appropriate tasks. In this manner, validation becomes the driver as the initial prototype evolves into a demonstration prototype.

This paradigm is especially relevant to expert system endeavors where demonstrating feasibility and potential performance is necessary or appropriate before making a substantial resource investment. As organizations consider integrating expert system technology into their repertoire of computer-based applications, it is important that experience precede development work.

4.2.2 Event-based architecture for real-time fault diagnosis

A new class of diagnostic systems is emerging from recent programs directed toward vehicle operator aids for fighter aircraft, submarines and helicopters. These systems are neither static off-line aids nor real-time controllers. Instead they are expert control advisory systems which span the time scales of both regimes. These systems interface with controllers to interpret the error codes and to conduct tests and implement reconfigurations. On the other hand, these systems also interact with the vehicle operator to prioritize their activity consistent with the operator's goals and to recommend diagnostic/emergency procedures. The extension of the applicability of these methods to the industrial fault diagnosis practice is straightforward.

System status (*SS*) is the function responsible for in-flight diagnosis of aircraft equipment failures and SS examples will be used here to describe the requirements for diagnosis in expert control advisory systems (Pomeroy *et al.*, (1990), Passino and Antsaklis (1988)).

The diagnostic architecture developed for SS integrates a number of separate technologies to achieve coverage of all the requirements. This architecture is a fusion of statistical fault detection techniques like Kalman filters (see Chapters 2 and 3) with artificial intelligence techniques such as rule-based logic, blackboards, causal nets and model-based reasoning (see Section 4.2.1). This approach exploits the strengths of each technique and provides a mechanism for automated reasoning using both quantitative and qualitative information. Furthermore, the concept of an "event" has been introduced to track multiple faults and maintain diagnostic continuity through priority interrupts from the SS controller. A specific application of this approach to jet engine diagnosis is described by Pomeroy *et al.* (1990).

Levels of architecture. In the real-time environment of system status any diagnostic activity must be structured so that it can be interrupted and restarted as SS control reacts to new events and changing priorities. The diagnostic process must also provide answers with varying degrees of resolution depending upon the time available for processing. Both of these requirements are met by dividing the diagnostic process into four levels:

1. Monitor for abnormal data.
2. Generate hypotheses that might explain the abnormal data.
3. Evaluate the available data to confirm or rule-out the hypothesized faults; if more data are required request tests to be done.
4. Execute the tests, and monitor for the results. Tests may consist of running models of the systems, initiating non-intrusive built-in tests (BITs) in the systems, or requesting operator approval for intrusive or operator-initiated tests.

These levels communicate through messages as shown in fig. 4.2, and each level is a knowledge source within the SS blackboard control scheme. While these messages provide the internal/external communication functions of diagnosis, something more is needed to provide coordination of the multiple diagnostic processes which can occur with overlapping time frames. This problem is solved by linking the overall diagnostic procedure to the concept of an *event*.

Events. An event is triggered by a new abnormality appearing in the bus data stream. An event includes all of the subsequent diagnostic steps leading to isolation of the fault which caused the abnormality. A frame-based data-structure is used to track each event and keep it untangled from other events which may be proceeding through processing at the same time. This structure also provides a record of the event that may be useful for post-operation maintenance.

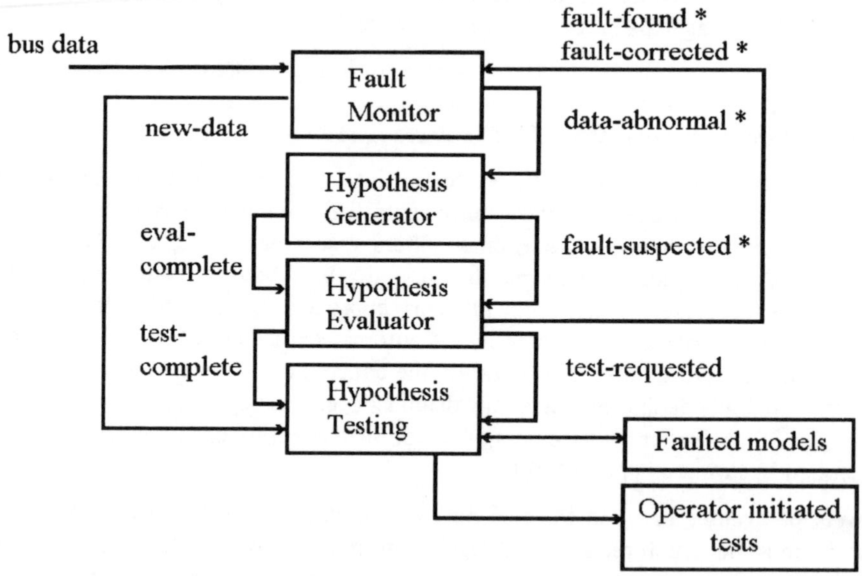

Figure 4.2 Event based diagnostic architecture and messages.

Each event is an instance of a general event class; event frames have the following slots:

BUS DATA: a list of data samples connected with the event; this is a "snapshot" of the situation near the event, and may include later samples collected during testing.
ANOMALIES: a list of abnormal data items which triggered this event. This list is used by the Fault Monitor to suppress further data-abnormal messages once an event has been spawned; it provides a "we know about that and we're working on it" sort of behavior.
HYPOTHESES: a list of possible faults.
TESTS PENDING: a list of tests that are to be performed.
TESTS COMPLETED: a list of tests and their results.
FAULTS CONFIRMED/RULED-OUT: the hypotheses are sorted into one of these two categories.
STATUS OF EVENT: is pending until it becomes resolved or unresolved.

Diagnosis stops when there are no new hypotheses.

Interaction with other functions. Communication between SS Diagnosis and the SS Limits Estimation and Corrective functions is provided by the activity of a causal network (see Section 4.2.1.3).

Communication with the outside world consists of the input and output streams discussed earlier in connection with fig. 4.2. All communication between the system subsystems is by means of the bus data stream, which implements the following division of labor between SS diagnosis and the local system diagnosis:

1. All fault detection is performed within the local systems. Detection requires continuous screening of sensor data at the sampling rate of the local controller, and detection processes are typically included in the control loop to protect against sensor failure. Transmitting the sensor data to a central detection process in most cases would require high bandwidth communication. Fault detection can be done more efficiently in the local systems.
2. Isolation of faults is shared between SS and the local systems. In general, fault isolation can be most efficiently done by the central diagnostic process (SS) which can bring multiple sources of information to bear on the problem, and which can execute tests beyond the scope of the local systems.
3. On the other hand, there are classes of faults which must be isolated by the local system in order to reconfigure quickly enough to avoid loss of control.

Thus the bus interface to SS normally reports only the results of continuously running built-in tests (BITs), i.e. error alerts; in the case of jet engines these BITs are generated by a Kalman filter that continuously compares the engine sensor data to outputs from an engine model. Only when a fault occurs and SS begins isolation, does SS request access to detailed data sampling streams.

Multiple faults that are related through a common mode can be addressed within the event-based architecture by adding a Fault Predictor to the four functions in fig. 4.2. Whenever a fault is found this predictor searches for common mode relations, e.g. functionally connected or physically connected, and posts the names of components which may be effected to act as a focus mechanism for the hypothesis generator.

4.2.3 Curve analysis techniques for real-time fault diagnosis

Process parameters and some process observables are gathered during the process execution, so they may be represented as discrete curves with time as the independent variable. Frequently, an *ideal curve* can be associated with each process. This is what is expected from a perfectly executed process.

Problems in operation are often identifiable when the input curve deviates from the ideal curve. The deviation may be a difference in slope, amplitude, or duration between the input and ideal curves. The difference in curves may be caused by malfunctioning equipment, processing an already damaged part, or processing problems (e.g., operator errors). In all of these cases, it is important to identify the problem in order to make appropriate corrections. Analysis of curves is therefore an important tool for diagnosis.

Diagnostic techniques have been developed to analyze process parameters and observables that change over time (Dolins and Reese, 1992). These techniques can use specific digital signal-processing algorithms to transform the input signal into symbolic data. Knowledge-based diagnosis is performed on the symbolic data to determine malfunctions. The monitoring system informs appropriate personnel of problems by sounding an alarm or printing a message.

Curve analysis involves detecting and identifying deviations of an input curve from an ideal curve. There are two alternative ways to perform analysis: one approach is to compare the input curve to a set of curves that result from unsuccessful processing. Another compares the input curve only to the ideal curve using qualitative analysis of the differences.

In the first approach, a knowledge base of *abnormal* curves is defined, where each curve is a characteristic representative of a particular problem. Associated with each characteristic abnormal curve is a diagnosis. If the input curve closely matches one of the abnormal curves, then the associated cause of the problem is reported. The advantage of this approach is implicit diagnosis; when the input curve matches successfully, it already has an associated diagnosis. However, this approach has two disadvantages. First, it may be difficult to build a complete knowledge base as the anomalous curves must be defined to match closely with actual erroneous measurements. The second disadvantage is that curves are hard wired, i.e., if the process changes, then the entire knowledge base must be changed to support the new data describing the correct and incorrect behavior of the process.

The second approach compares an input curve to the ideal curve only. Ideal and input curves are composed of regions. A region is a continuous group of data points where each point has approximately the same slope. Regions can be inclining, flat, or declining. If a process engineer is uninterested in several contiguous regions, then (s)he may elect to aggregate them into one region. In general, region divisions correspond to significant changes in the process, e.g., an abrupt change in the value of a parameter. This approach is possible if the user has some technique available to describe anomalous curves with respect to the ideal curve. Such a description should allow the user to express deviations using qualitative as well as quantitative criteria, and associate causes using symbolic processing. Suppose one uses a technique based on this approach, to interpret an input curve that has a flat region with a longer duration than the ideal curve. The technique should allow her(him) to describe the problem in terms of the flat region having a duration that lasts too long. Also, the user must be able to associate causes of problems with the different anomalous curves. Several diagnostic systems have been developed to diagnose manufacturing problems based on this second approach (Dolins and Reese, 1992).

Dolins and Reese (1992), developed a technique that allows manufacturing and process engineers to describe abnormal curves. The abnormal curves are described in terms of their differences from the ideal curve, which is the curve that best describes a process parameter or observable after a given industrial process successfully finishes processing.

Manufacturing engineers can describe the differences symbolically, e.g., "if the first region of the curve lasts too long then the machine must have a gas leak". The user can also input numeric values to set tolerances for determining unacceptable input curves. The technique is independent of any industrial process, and all domain specific information is input by the user, who is an expert in the process, to the program.

The technique has two operating modes: process definition and process monitoring/diagnosis.

In process definition, the human expert has to describe the ideal curve and anomalies. An ideal curve is initially input into the computer program, and the user manually selects regions. Each region is an interesting feature in the ideal curve which corresponds to a specific manifestation of the process.

After defining the ideal curve the human expert describes input curve anomalies by creating a knowledged-base of *process-specific rules*. Process-specific rules relate *generic tests* to input and ideal curve regions for a given process. Generic tests are built-in functions provided by the diagnostic technique that compare different symbolic attributes of input and ideal regions. For example, length is a symbolic attribute of a region, and the result of a comparison of the length of two regions can be described as either *too long*, *too short*, or *okay*.

In the process monitoring/diagnosis mode, the technique analyzes input curves in two steps: signal-symbol transformation and knowledge-based diagnosis, see fig. 4.3. The signal-symbol transformation step identifies regions of the input curve by matching all of the points of the input curve to the ideal curve. After all points are matched, the regions of the ideal curve are used to find the regions of the input curve. The second step applies the complete knowledge-base of process-specific rules to compare the regions of the input curve to the regions of the ideal curve.

An expert is required to select an ideal curve for a particular process and input the curve to the program. Some machines may have idiosyncracies that make their ideal curve differ in shape from the ideal curves generated by the other machines of the same type. In these cases, an ideal curve has to be defined for each machine.

Once the ideal curve is input, the expert divides the curve into meaningful regions, i.e., (s)he marks divisions where process-related changes occur. These regions are stored and used later in the analysis. The expert also defines a set of rules for testing input curves. Entering an ideal curve, dividing the ideal curve into regions, and defining rules are initialization tasks required of the human expert. These tasks constitute the process definition mode.

The diagnostic system can now run automatically without human intervention until an error is detected, i.e., the program can operate in a process monitoring/diagnosis mode.

The combination of signal-to-symbol transformations and rule-based reasoning has several advantages, but it is not a panacea for all diagnostic problems based on curve interpretation. One disadvantage of the diagnostic technique is that two potential processing problems may have identical input curves. In this case, a better diagnosis can only be provided if more data are available and more reasoning provided. A second disadvantage of this technique is that an abnormality in a curve may mask other problems. One approach is to explain only the first difference between the ideal and input curves.

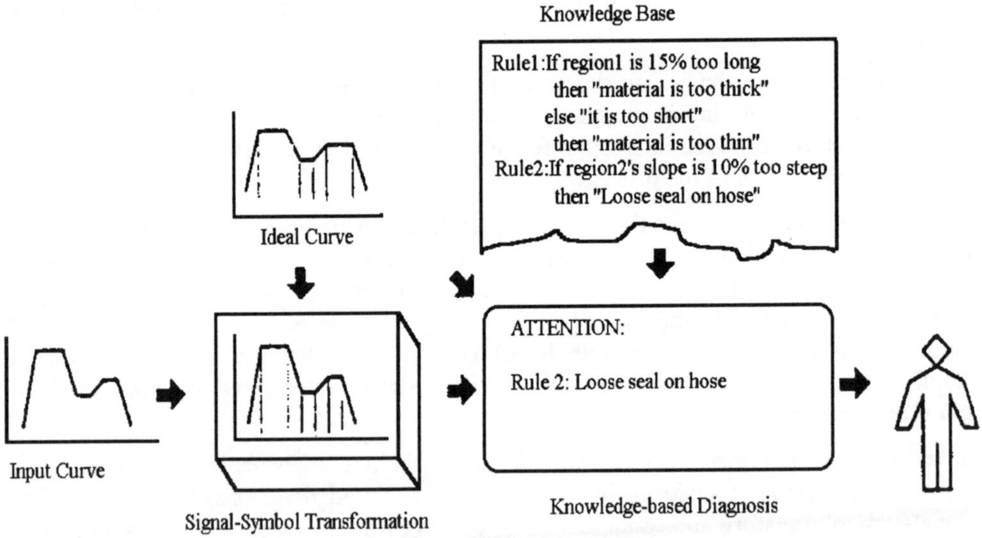

Figure 4.3 Curve analysis based diagnosis combining digital signal processing and rule-based reasoning.

One advantage of this method is that the signal processing algorithm used to transform the input signal into symbolic data allows the fast analysis of regions that vary with respect to time. This is important because the durations of regions may vary due to unsuccessful processing. Regions of the input curve, with varying durations, can match directly to corresponding regions of the ideal curve. This processing allows the user to examine regions symbolically.

A second advantage is that few false alarms are generated with this method. Problems are detected by the process-specific rules, and the process engineer has complete control over the criteria for judging acceptable and unacceptable traces. False alarms can only be caused when process engineers define rules that incorrectly diagnose problems or incorrectly set thresholds.

The system's ease of use is a third advantage. Only an ideal curve and process-specific rules have to be defined. Furthermore, few rules are needed for the system to be effective, which is unlike most knowledge-based systems. For these cases, a process engineer many only need to define a single rule to detect a commonly occurring error.

Several applications to detect manufacturing problems as soon as they occur are discussed by Dolins and Reese, (1992), to illustrate the general purpose use of this technique.

4.2.4 Real-time fault detection using Petri nets

Petri nets are a powerful tool for system description (Al-Jaar, 1990). Nevertheless up to the present they have mainly been used only for simulation purposes. The problem of process fault monitoring in an industrial plant can be stated as follows: The measurement signals come from the system with a constant scanning rate. When processing these data, a computer-based system should decide on-line in real time if an error has occurred or not. To perform this, the computer program needs some expert knowledge about the system (or the "total" process, which is composed of several partial processes, like bigger subsystems in a power plant or in a chemical factory) under consideration.

By modeling the system as a Petri net, failures with slow time constants are detectable in real-time. Sensor or process errors which are manifested in signals related to physical conservation quantities can be identified. After a fault is detected, a prognosis of the future system's behavior can be provided.

The original Petri net theory only describes the causal correlation between places and transitions within a system (an event is a consequence of another one). There were no statements about its temporal behavior. This, however, is absolutely necessary for describing events and processes in the manufacturing area. There are different theories how to link Petri nets with time. In the manufacturing techniques the processes (milling, drilling, assembling, ...) are responsible for the consumption of time. This is the reason, why one has to associate time with the transitions. Thus, in the case of firing a transition, the tokens of the places before a transition will be removed. If the firing time is over they will be at the place behind the transition. An example is the time which takes a slide from one limit switch to the following one.

The nets for diagnosis purposes represent the temporal progress of a plant or machine, which are to be controlled, as a model. This explains why the nets used for control form the basis for the construction of the nets for diagnosis (see fig.4.4). The places in both nets represent the inputs and the outputs of the PLC and, therefore, they are the interface between control and diagnosis. Thus, in both nets the count and the indication of the places must be identical.

The most important function of a diagnosis system is the monitoring component. Its capability defines the nature, the scope and the precision of the failure detection. Only after the detection of a failure a specific diagnosis can start. The power of monitoring is equivalent to the quality and quantity of information from the machines. This is especially the case when sensors and actuators do not have their own information processing and, therefore, they are not able to monitor themselves.

The range of methods for monitoring depends extremely on the support for the methods, which is provided directly from the model.

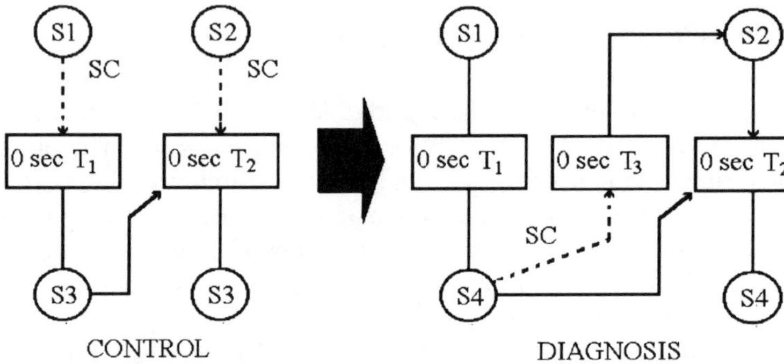

S1,S2:input signal from sensor
S3,S4:output signal to actuator
SC:secondary condition. This condition is necessary for firing of a transition.
If the transition fires, no tokens will be removed from the place before the transition.

Figure 4.4 Diagnosis of sensors.

Within the concept of monitoring, one can distinguish between a functional and a temporal comparison. The required state is determined by the interpretation of the Petri net data structure. This takes place on the facility level as well as on the station level. The actual state on the station level results from the inputs and the outputs of the PLC, which are assigned to the places of the Petri net. On the higher levels, the actual state results from the condensing of the state reports from the different PLCs which control plant components as single machines, conveyors, robots etc.

In order to show the different monitoring methods clearly, the following cases have to be distinguished (see fig. 4.6):

1. The real process has kept to the required time.
2. The real process has fallen short of the required time.
3. The real process has exceeded the required time.

In the case of time monitoring, the duration of performing a real action is recorded and compared to the required time. The required state of time is taken from the active transition in the Petri net. If more than one transition is simultaneously active, the time monitoring will be processed in a parallel way. If microcomputers are used on the facility level and PLC's on the station level, their operating system provides several timers. These timers can be used for monitoring.

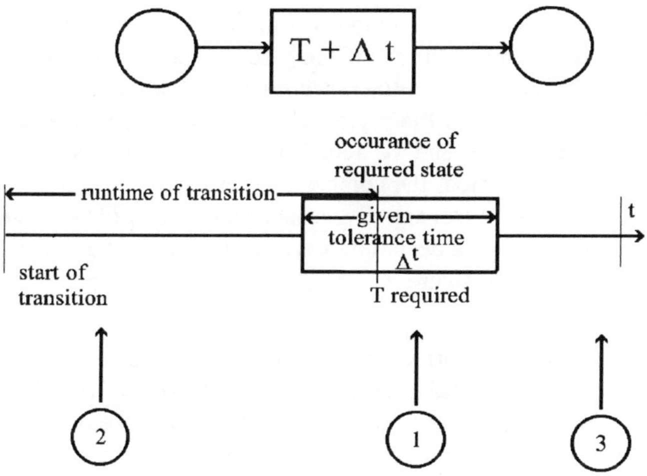

Figure 4.5 Diagnosis of sensors.

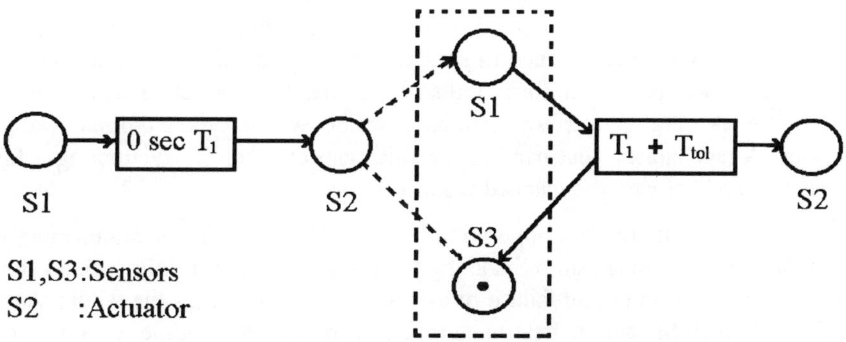

Figure 4.6 Different states in the Petri net based monitoring concept.

The interpreter of the Petri nets within the diagnosis system always determines the next required state and, by means of the time component, also the precise time of its occurrence one step in advance compared to the real plant. During the runtime of the system it is important that the diagnosis program and the control program work concurrently (Maâberg and Seifert, 1991).

In order to prevent the indication of a failure in cases of small deviations from the required time, a tolerance time is additionally implemented. A tolerance time can be clearly assigned to a transition. After the required time of an active transition within the diagnosis net model has passed, the component for monitoring of the tolerance time will be activated and the required conditions of the places behind that transition will be actualized. Within the monitoring of the tolerance time, a continuous comparison between the required and the actual state of the places which are directly connected to the transition, is performed. If the required and the actual state of those places are equal (case 1 in fig 4.6), a failure has not occurred, the comparison between the required and the actual state will be broken off, and control and diagnosis of the plant will be continued. If, even after finishing the tolerance time a deviation between the required and the actual state of the places can be determined, a failure will be detected from the time monitoring (case 3 in fig 4.6).

In order to select and define the correct reaction of the diagnosis system in case of a failure, a thorough analysis of all possible failures by the operator of the plant is necessary. The failures have to be classified according to their effects and the reactions correspondingly defined. In case of serious failures the diagnosis system must react with emergency shutdown or emergency stop. Deviations, which do not represent any failure can therefore be ignored. This means that the diagnosis system causes reactions which do not stop the plant, but make an operation possible in those individual manufacturing parameters (e.g. velocity of motion etc.) which are changed. Another possibility of the *limited operation* (*LO*) of the plant is the activation of alternative predefined control strategies, which, for example, transfer the plant into a secure condition. With this concept minor failures can be compensated or even corrected by control instructions. As soon as the classification of failures has finished and individual failures as well as combination failures have been assigned to the correct reactions, the results are made available to the diagnosis system in the form of a so called reaction model. This model consists of IF-THEN rules. The causal correlations describe which preconditions lead to which reactions. A mechanism, that handles the rules can choose, after detecting a failure, the correct rule and activate the planned reaction.

The essential module of the cooperation between the functions of monitoring, diagnosis and therapy is the mechanism of handling the rules. It does not only process the reaction model as the collection of all failure rules, but it also administers the failure vector as the interface between the above mentioned mechanism and the module, which compares the actual and the required state.

The failure vector is structurally identical with the vector of the required and actual value and consequently identical with the structure of the IF-part of the failure rules. Each column of that part of the rule represents a vector, which has as many elements as there are states in the Petri net. This sort of data compatibility guarantees a very fast processing in the mechanism of rule handling (see fig. 4.7).

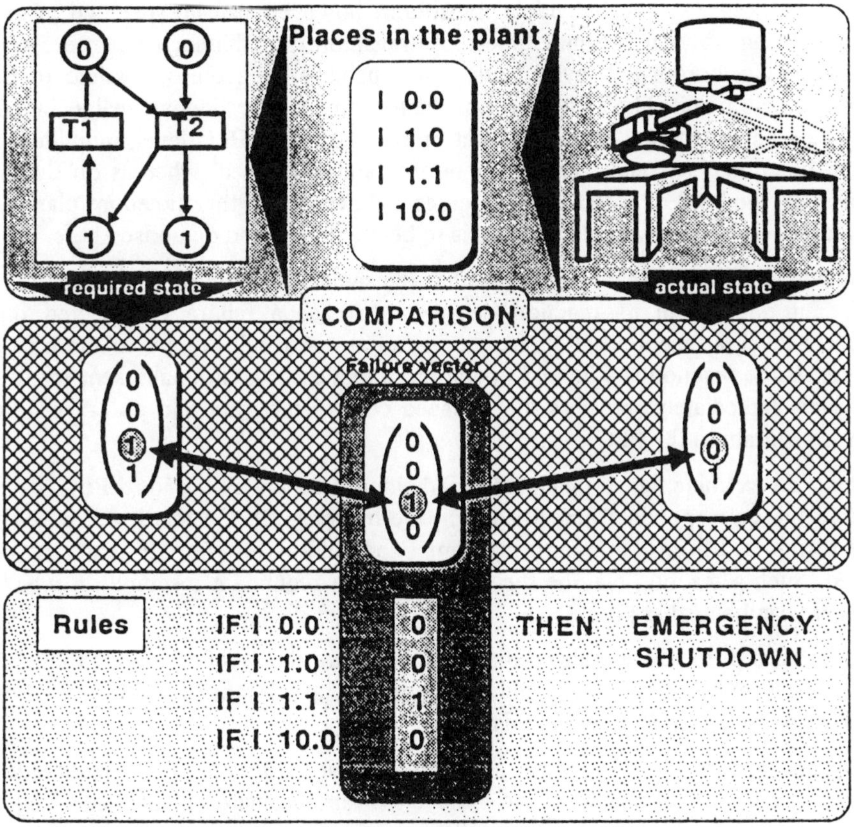

Figure 4.7 Concept of the mechanism, which handles the rules in Petri net based fault diagnosis

In the case of an emergency shutdown or emergency stop, control orders for stopping the operations are given out, the diagnosis is stopped and a report about this interruption is produced.

In case the reaction intends a limited operation, the control orders which the operator has defined in advance in the form of a program for a PLC, are activated. For this purpose, a message is sent from the station level to the area level which immediately selects the corresponding Petri net for control.

Such a diagnosis system must be designed in such a way, so as to be able to be implemented on all levels of a hierarchical control structure. This concept is supported by the capability of the Petri nets to decompose complex systems. By means of Petri nets it is possible to describe a manufacturing system as a rough net in which a transition for example, represents an individual machine, robot or conveyor belt. The places within the net represent, according to their definition, static components as a storage system or as a buffer for workpieces. The individual transitions, however, can be specified in greater detail depending on their meaningfulness. A transition on a higher level represents an entire net in a subordinated level. With that capability it becomes possible to model individual function units like an actuator or the movement of a slide, as well as individual places like a limit switch. This means, that on the station, or PLC level, individual units and their functional sequences can be monitored and diagnosed, whereas on the higher control levels the level of abstraction increases and because of this the entire plant or the cooperation between individual machines is to be monitored and diagnosed here.

The module which activates the reactions has the same allocation of tasks. Each control level is autonomous in its reaction behavior. In case a failure is detected on the component level, a reaction will be activated and a report will be sent to the area level computer. If the reaction is a limited operation, a request is additionally sent to the area level which immediately transfers the replaced control program and its corresponding diagnosis program to the PLC.

An event, for example the stoppage of an individual machine by the local PLC, causes a change of the corresponding actual state on the higher level. On this level, a deviation will be recognized from the monitoring component and an appropriate reaction will be activated. Such a reaction may be the activation of a redundant machine on which the workpieces can be further processed.

The advantages of the decentralized diagnosis tasks are the relief of the area level computer, the uniformity of failure recognition algorithms at all control levels and therefore the high response speed in case of deviations.

Maâberg and Seifert, (1991), present a *Computer Aided Automation Environment* supporting a user during all phases of the life cycle of an automated plant, beginning from the planning and projection phase up to the runtime of the plant. Petri nets are the integrating components. They are generated in the projection phase by translating the function charts. In the realization phase they are used for simulation and planning and finally in the running phase of the plant they are used for controlling, monitoring and diagnosis of the plant.

Prock, (1991), describes a new technique of on-line fault detection in real time, in a process independent formulation, using *place/transition nets*. Place/transition nets are a subclass of Petri nets. For readers who are unfamiliar with the Petri net theory, some basic definitions of place/transition nets (hereafter called *pt nets*) are given in Appendix 4.B. This formal presentation for pt nets theory will help the reader to understand related application examples as well as to deal with related diagnostic problems from engineering practice.

Prock, (1991), applied this method to the real time fault monitoring of a secondary cooling loop of a nuclear power plant. The detection of abnormal process behavior or measurement faults with low time constants was possible and a prognosis of the future system behavior was given in the error case. Due to the simplicity of the fault detection criterion no diagnosis of the failure localization could be provided. This is not a real drawback because fast transients, as a consequence of serious faults, are well managed by the automatic plant safety systems.

The Petri net fault monitoring methods are predestinated for the surveillance of complex technical systems like production lines or transport circuits (Wiele *et al.*, 1988). Because of the lack of the diagnosis feature, this method should be considered as part of an on-line process information system which is able to trigger a (possible off-line and thus more practical from the implementation point of view) diagnosis and interpretation unit.

4.2.5 Fuzzy logic theory in real-time process fault diagnosis

Rule-based approaches have been proposed as capable of realizing flexible diagnostic methods by paying attention to the rules describing the relationship between the causes and symptoms of failures (see the previous sections of this chapter).

It is important to recognize the fact that a large part of expertise consists of heuristic knowledge, which relies mostly upon subjective judgments and may include incomplete, ambiguous and imprecise information. As a consequence, the application of such uncertain knowledge results in inexact reasoning that the expert system has to deal with. There are numerous methods which show how the expert system copes with uncertain knowledge and inexact reasoning (de Kleeer (1990), Maruyama and Takahashi (1985), Rhodes and Karakoulas (1991)).

Generally it can be said that the theory of probability is employed to solve the problems of *plausible reasoning*, while fuzzy set theory is used to solve the problems of . Approximate reasoning, as opposed to plausible reasoning, means drawing conclusions by taking into account the linguistic consistency of the facts. In all expert systems based on symbolic manipulation and plausible reasoning, uncertainly resides in the state of one's knowledge. In expert systems based on semantic manipulation and approximate reasoning, the emphasis is on fuzziness viewed as an intrinsic property of the natural language.

The elicited knowledge, which allows interpretation and diagnostics, is organized in the knowledge base as a set of *fuzzy conditional statements* that relate test results to conclusions about process condition or possible failures. The natural, logical way of reasoning and data reduction is applied. At the first level, a statement which estimates the possible situation based on the analytical input data is selected. Then, at the second level, adequate additional input data are collected and the situation is specified more closely. The *fuzzy conditional statements* are of the form:

If $A_{1.1}$ and $A_{1.2}$ and ... and $A_{1.N}$ then B_1 or
If $A_{2.1}$ and $A_{2.2}$ and ... and $A_{2.N}$ then B_2 or
....
If $A_{M.1}$ and $A_{M.2}$ and ... and $A_{M.N}$ then B_M

where $A_{i.j}$ is a linguistic variable. A *linguistic variable* is a variable whose value can be presented by a linguistic term used by experts such as "high", "normal" or "low" (i.e. words or sentences in a synthetic language). A linguistic variable includes an adjective-like term and its antonym, a modifier and a connective. The *modifier* is a measure of intensity which is associated with a possibility distribution. This is often referred to as the *membership function* in the literature. The fuzzy logic connectives are the well known *conjunction*, *disjunction* and *negation* operations. The value of a linguistic variable can be presented by a fuzzy set which permits the definition of a membership function μ, reflecting the degree to which an element belongs to the set. The membership function for elicited expert knowledge about the fuzzy test limits can be represented by a piecewise linear function. Such a function is presented in fig. 4.8. The four values *a*, *b*, *c*, and *d* are numerical values stated by the experts in the process of knowledge acquisition B_i is a possible conclusion.

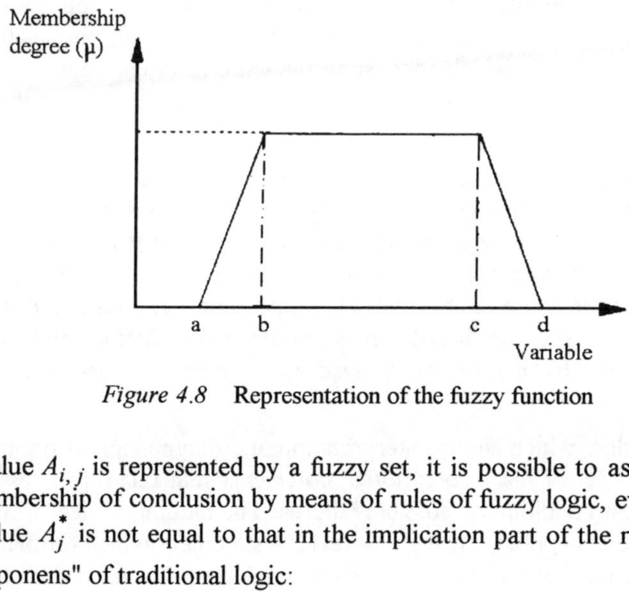

Figure 4.8 Representation of the fuzzy function

Since the value $A_{i,j}$ is represented by a fuzzy set, it is possible to associate it with the grade of membership of conclusion by means of rules of fuzzy logic, even in cases where the input value A_j^* is not equal to that in the implication part of the rule, as contrary to the "modus ponens" of traditional logic:

A^*	input value
$A \to B$	fuzzy statement
$A^* \circ (A \to B) = B^*$	fuzzy conclusion

The truthfulness designated as a grade of membership for this simple implication is evaluated through the operation called *Zadeh's Composition* and provided by the min-max operator:

$$m_{B^*}(y) = \max(\min(m_{A^*}(x), m_{A\times B}(x,y)))$$

Another way of writing the above equation is:

$$m_{B^*}(y) = \min(m_B(y), \max(\min(m_A(x), m_{A^*}(x))))$$

The practical solution of the above equation for m_B=1, is shown in fig. 4.9.

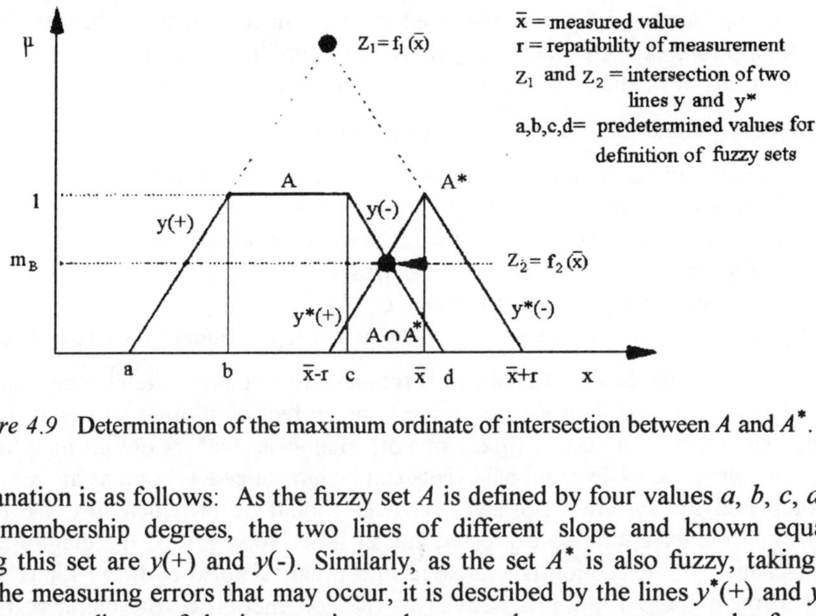

Figure 4.9 Determination of the maximum ordinate of intersection between A and A^*.

The explanation is as follows: As the fuzzy set A is defined by four values a, b, c, d and by their membership degrees, the two lines of different slope and known equation describing this set are $y(+)$ and $y(-)$. Similarly, as the set A^* is also fuzzy, taking into account the measuring errors that may occur, it is described by the lines $y^*(+)$ and $y^*(-)$. The maximum ordinate of the intersection z, between these two sets, can be found by means of the following three rules:

1.	If $z_{(1,2)} \leq 1$ and $z_{(2,1)} > 1$	then $m_{B^*}(y) = z \leq 1$
2.	If $z_{(1,2)} > 1$ and $z_{(2,1)} > 1$	then $m_{B^*}(y) = 1$
3.	If $z_{(1,2)} > 1$ and $z_{(2,1)} \leq 1$	then $m_{B^*}(y) = 0$

The quantitative analysis of the possibility of a certain situation in the system described by the fuzzy conditional statements, is made through the evaluation of its grade of membership according to Zadeh's compositional equation,

$$m_{B^*}(y) = \max_{1 \le i \le m} (\min(m_{B_i}(y), \min_{1 \le j \le n} (\max(\min(m_{A_j^*}(x_j), m_{i,j}(x_j))))))$$

taking into account the solution for the intersections mentioned above. Since each B_i; i=1, 2, ..., n, in the fuzzy conditional statements can be considered as a fuzzy singleton over a domain consisting of certain situations y, the starting value for m_{B_i} while evaluating $m_{B^*}(y)$ at the first level is 1, and at the second level the evaluated $m_{B^*}(y)$ becomes the starting value.

The uncertainty of the knowledge in the knowledge base is taken into consideration by giving different weight factors to fuzzy conditional statements. The choice of weight factors is rather subjective. Trained *artificial neural networks* (*ANN*s), as generators of membership functions and weight factors in fuzzy conditional statements, are potential tools for the purpose of fuzzy logic process fault monitoring. Details and ANN application examples are given in Chapter 5.

Zadeh's compositional inference rule is adopted as an inference mechanism. It accepts fuzzy descriptions of the process symptoms and infers fuzzy descriptions of the process faults by means of the fuzzy relationships described above.

The main characteristics of a fuzzy logic diagnosis system performance are:

1. Automatic interpretation of relations among the test (observation results) and possible situations, pointing out the process condition.
2. Detailed explanation of how the particular conclusion has been reached.
3. Indication of the possible causes of failures.
4. Description of the possible consequences.
5. Recommendation for process maintenance and repair under new circumstances.

There may be difficulties for the above techniques to comprise the elements of the failures and symptoms and their logical connections perfectly. In other words, it is a plausible criticism against rule-based (fuzzy or not) diagnosis, that its design may be beyond human knowledge since exceptional events can be introduced as soon as an improvement has completed the system. A practical answer should be provided for this argument. How should the exception be expressed and included to reinforce the diagnosis? In the present chapter the exception is expressed in a practical form of fuzzy logic. First, the logical form of the exception is derived as the conjunction of the dictative functions. Second, the cancellation law in binary logic is fuzzified in order to give an arithmetic for calculating the linguistic truth value for reasoning. The logical form of the exception is derived in Appendix 4.C and requirements for its use are clarified there as well. The cancellation law is extended to fuzzy logic in order to devise the diagnostic method with the exception (Maruyama and Takahashi, 1985).

The introduction of a practically defined exception may be a solution to a plausible criticism against rule-based diagnosis, which emphasized that its design was beyond human knowledge since exceptional events always exist. As will be shown in Section 4.4, where application examples will be presented, complementary utilization of the exception generates reinforcement of the existing expert fuzzy diagnostic method to identify the leaking location of a Boiling Water Reactor power plant (Takahashi and Maruyama, 1987).

4.3 Application examples

4.3.1 Automatic expert diagnostic systems for nuclear power plant (NPP) safety

4.3.1.1. Diagnostic expert systems for NPP safety

Research in applying expert systems software to nuclear power plants (NPP) has substantially increased in the last two decades. The dynamically complex system of a NPP is a most challenging topic for artificial intelligence (AI) specialists. Malfunction diagnosis of NPP's systems generally uses shallow knowledge incorporated in fault models. This derives from the fact that most NPP's systems receive signals from sensors and that their possible malfunction causes and effects on system variables are well known.

In recent years many important results have been obtained about representation and reasoning on structure and behavior of complex physical systems using qualitative causal models. The current AI trend in this aspect is qualitative reasoning using deep knowledge representation of physical behavior (Soumelidis and Edelmayer, 1991).

The suitability and the limits of a qualitative model based on deep knowledge for fault detection and diagnosis of the emergency feed water system (EFWS) in a NPP is presented here. The EFWS has been chosen because of its importance in the safe functioning of the NPP.

The EFWS is a standby system which is not operated during normal plant operation. The role of the EFWS is to provide full cooling of the Reactor Coolant System in emergency conditions. The EFWS is automatically activated in three cases of NPP malfunction:

1. Loss of offsite power (LOOP).
2. Low-low level in any steam generator.
3. Loss of alternative current (LOAC).

The possible malfunctions which can occur in the EFWS and their causes and effects on system variables are well known. They are associated with cracks in the pump or condensate storage tank (CST) casing, pipe or valve ruptures, and pump or valve operation failure.

As the EFWS is working only in emergency conditions, the occurrence of a malfunction in the EFWS will lead to catastrophic results. Safety insurance is an acute problem in NPP. It is expected that expert systems can contribute to the improvement of flexibility and man-machine communication in NPP.

The expert system diagnostic process is performed by a forward-chaining inference engine that operates on the knowledge base. The inference mechanisms adopted in deep modeling techniques are used.

The diagnostic process module consists of two modules: Fault Detection and Fault Diagnosis (see fig. 4.10).

The process starts with the Fault Detection module which detects a symptom of malfunction by observing any qualitative variation of the output parameters. Several information sets are instantiated in the initialization phase. The process then continues with the identification of the causes of malfunction by exploiting the information contained in the Model. The Model actually represents the Knowledge Base. It contains descriptions of the Physical System (generic components, initial measurements, connections, possible measurements, actual components). Note that only the correct systems behavior is described in the model (Obreja, 1990). The Fault Diagnosis module then propagates the observed qualitative variation through the system model using a constraint propagation method (de Kleer, 1987). Thus, all possible fault models are generated. This step ends when some input parameters (i.e. parameters in the LHS of the rules) are unknown, thus making further propagation impossible. The qualitative reasoning process can continue only if new measurements are taken.

The decision on the choice of optimum measurements is taken according to heuristic criteria, i.e. probabilities of component failures. From these probabilities one can compute candidate probabilities and Shannon's entropy function (de Kleer, 1990, 1987).

After the most appropriate point to measure next has been identified, the measurement is taken, and the qualitative propagation is continued for this measurement.

The Knowledge Base contains qualitative information derived from the EFWS model. This information is used by the Diagnostic Process as presented above. The EFWS model is described by components, connections, equations involving: process variables and design parameters. Components are manual isolation valves, pumps, tanks and "t-connections", i.e. pipes.

Components are connected together by process variables. The component's behavior involves process variables and design parameters. Design parameters have nominal values stated by design. Qualitative analysis considers design parameters as constants.

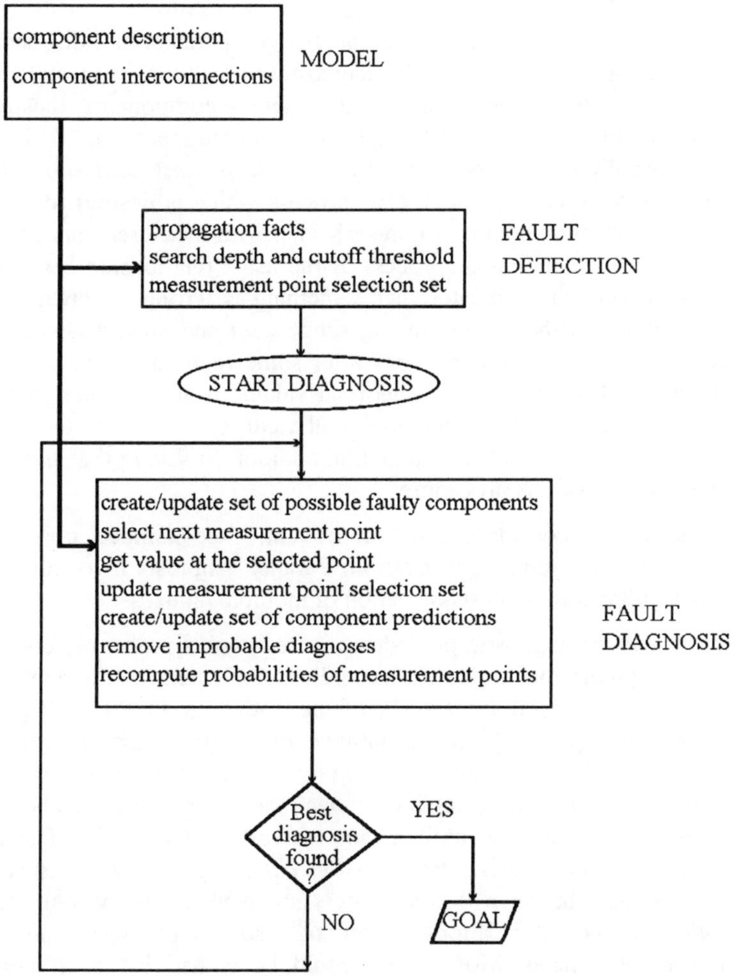

Figure 4.10 The expert system diagnostic process for NPP safety.

A symptom of malfunction is detected when a variation on the variables' qualitative values is observed. The variations of affected variables caused by EFWS malfunctions are:

tank malfunction	dec	std
pipe break	inc	inc
valve malfunction	dec	dec
where dec = decreases, inc = increases, std = steady		

Each possible malfunction affects in a known way (inc or dec), but in an unknown measure, some variables. Thus, the model and its analysis are intrinsically qualitative. The use of simple dynamic models for the system and the components is indispensable for the real-time implementation of the proposed diagnostic procedure. Even quite complicated technical systems, as NPP, are made up of rather simple components. It is usually quite clear what is the input and what is the output of each component and their interactions. Therefore, it is usually easy to conceptually split the process into subprocesses, with simple interactions between them. For each process which is desired to survey, a submodel is written. Each submodel is fed, in real-time, with measurements of the variables that influence the corresponding subprocess. If the usual relationship between the process variables is broken, this indicates that something is wrong. A given submodel receives the same input as the corresponding subprocess and should also give the same output. A fault in a given subprocess may after some time spread its influence over a large part of the total plant and give abnormal values to all variables, but the normal relation between these abnormal variables is still valid, except in the faulty subprocess. The ability to say where the fault is situated, in addition to saying that there is a fault is an important advantage of this procedure.

Each diagnosis can be considered an independent, time-stamped object completing within seconds from the point of invocation. Parallel diagnosis invocations may exist, facilitating the simultaneous analysis-detection of multiple faults.

In implementing expert diagnostic procedures in real-world problems, considering PC-platforms is nothing more than a waste of time. For an efficient real-time implementation, the model-based fault detection part should be coded in FORTRAN on a general purpose number-cruncher, while the qualitative diagnostic reasoning should be performed on a dedicated AI workstation, utilizing for example the convenient and powerful LISP development environment. Coupling between the two modules can be facilitated by means of Ethernet hardware and multivendor network software glue like the TCP/IP Arpanet protocols. Use of *TCP* (***transmission control protocol***) ensures a reliable transfer of critical data between the computers involved in the system. As TCP is a connection-oriented communications protocol, some overhead in connection management can be noticed. More speed could be gained by using connectionless protocols, eg. *UDP* (***user datagram protocol***) or the lower level *IP* (***Internet protocol***) directly, but this would come at the expense of reduced safety and reliability, unless carefully programmed by the application developer.

The methodology described above is applicable in a straightforward manner to other complex plants (chemical plants, conventional power plants, marine equipment, etc.).

4.3.1.2 Fuzzy reasoning diagnosis for NPP safety

Fuzzy relation of symptoms with leaking locations in NPP

The fuzzy diagnosis to be presented, was applied in identifying the leaking location in the cooling system of a Boiling Water Reactor (BWR) plant, in order to ascertain whether the concept of exception was effective or not. Fig. 4.11 shows a schematic representation of the cooling system of this power plant. Typical examples are assumed for the leaking locations and the induced symptoms in conformity with the text. It is also assumed that the leakages inside the dry-well were found generally by eqs. (4.15), (4.16) derived from the implications of eqs. (4.4), (4.5), while those inside the building were identified especially by eqs. (4.26), (4.27) resulting from the exception in eq. (4.14) . Table 4.1 defines the failure and symptom vectors in the cooling system.

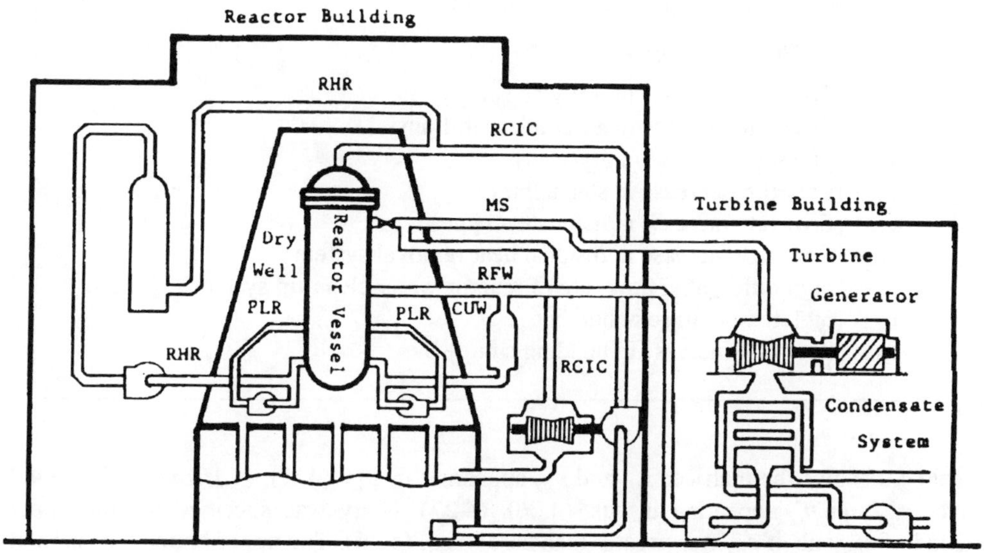

Figure 4.11 Diagram of boiling water reactor cooling system.

MS: Main steam system
RFW: reactor feed water system
PLR: Primary loop recirculation system
RHR: Residual heat removal system
RCIC: Reactor core isolation cooling system
CUW: Reactor water clean-up system

Table 4.1 Definition of failure and symptom vector.

(a) *Elements of failure vector.*

x_1 = {main steam line in dry-well}
x_2 = {residual heat removal system in dry-well}
x_3 = {steam line of reactor core isolation cooling system in dry-well}
x_4 = {reactor water clean-up system in dry-well}
x_5 = {feed water system in dry-well}
$x_6, ..., x_{10}$ = {main steam line in building}
$x_{11}, ..., x_{14}$ = {residual heat removal system in building}
x_{15}, x_{16} = {steam line of reactor core isolation cooling system in building}
x_{17} = {water line of reactor core isolation cooling system in building}
x_{18}, x_{19} = {reactor water clean-up system in building}
x_{20}, x_{21} = {feed water system in building}

(b). *Elements of symptom vector.*

y_1 = {flow rate increase in dry-well sump}
y_2 = {flow rate increase in air condenser drain}
y_3 = {pressure increase in dry-well}
y_4 = {pressure decrease in steam line}
y_5 = {flow rate increase in main steam line}
y_6 = {flow rate increase in residual heat removal system}
y_7 = {high differential flow rate of reactor water clean-up system}
y_8 = {high temperature in building}
y_9 = {flow rate increase in building sump}

Table 4.2 shows the matrices $r_{ij\ell}$ and $p_{ij\ell}$ appearing in eqs. (415), (4.16) while Table 4.3 gives $e_{ij\ell}$ and $h_{j\ell}$ appearing in eqs. (4.26), (4.27). Unity was specified for the upper bound of e_{iju} only if $e_{ij\ell}$ was greater than zero in eq. (4.26). The upper bound was zero at the zero lower bound. The domain ($1 \leq i \leq 5$, $1 \leq j \leq 7$) corresponds to R_{ij} and ($6 \leq i \leq 21$, $4 \leq j \leq 9$) to E_{ij}.

Examples of diagnosis through exceptions.

To ascertain whether the arithmetic given by eqs.(4.B-26), (4.B-27) is capable of detecting a failure or not, presents a problem. Several examples were solved by utilizing only the exceptions, where an input to this diagnostic method was provided by the linguistic truth value of the proposition "the *j*-th symptom is recognized", which determined both the lower and upper bound.

Table 4.2 Matrices of fuzzy relation of failures with symptoms.

(a) Lower bound of $\boldsymbol{R}_{ij\alpha}$ appearing in eq. (4.15):

$$
r_{ij\ell}=\begin{bmatrix}
1.0 & 1.0 & 1.0 & 1.0 & 1.0 & 0.0 & 0.0 \\
1.0 & 1.0 & 0.0 & 0.0 & 0.0 & 1.0 & 0.0 \\
1.0 & 1.0 & 1.0 & 1.0 & 0.0 & 0.0 & 0.0 \\
1.0 & 0.0 & 0.0 & 0.0 & 0.0 & 0.0 & 1.0 \\
1.0 & 1.0 & 0.0 & 0.0 & 0.0 & 0.0 & 0.0
\end{bmatrix}
$$

(b) Lower bound of $\boldsymbol{P}_{ij\alpha}$ appearing in Eq. (4.16):

$$
P_{ij\ell}=\begin{bmatrix}
0.9 & 0.9 & 0.6 & 0.6 & 0.3 & 0.0 & 0.0 \\
0.9 & 0.6 & 0.0 & 0.0 & 0.0 & 0.3 & 0.0 \\
0.9 & 0.6 & 0.3 & 0.3 & 0.0 & 0.0 & 0.0 \\
0.9 & 0.0 & 0.0 & 0.0 & 0.0 & 0.0 & 0.6 \\
0.9 & 0.6 & 0.0 & 0.0 & 0.0 & 0.0 & 0.0
\end{bmatrix}
$$

Table 4.3 Matrix of alternative fuzzy relation $\boldsymbol{E}_{ij}$ and vector of exceptional proposition $\boldsymbol{H}_j$

(a) Lower bound of $\boldsymbol{E}_{ij\alpha}$ appearing in eq. (4.27).

$$
\begin{array}{ccc}
 & \begin{array}{ccccccc} & \mathrm{j}=4 & & & & & \mathrm{j}=9 \end{array} & \\
e_{ij\ell} = & \left[\begin{array}{c:cccccc}
0 & & & 0 & & & \\
\hdashline
 & 0.6 & 0.0 & 0.0 & 0.0 & 0.6 & 0.6 \\
 & 0.6 & 0.0 & 0.0 & 0.0 & 0.6 & 0.0 \\
 & 0.9 & 0.0 & 0.0 & 0.0 & 0.0 & 0.9 \\
 & 0.0 & 0.9 & 0.0 & 0.0 & 0.9 & 0.0 \\
 & 0.0 & 0.9 & 0.0 & 0.0 & 0.0 & 0.9 \\
 & 0.9 & 0.0 & 0.0 & 0.0 & 0.6 & 0.0 \\
 & 0.6 & 0.0 & 0.9 & 0.0 & 0.0 & 0.6 \\
0 & 0.0 & 0.0 & 0.9 & 0.0 & 0.6 & 0.0 \\
 & 0.0 & 0.0 & 0.9 & 0.0 & 0.0 & 0.6 \\
 & 0.9 & 0.0 & 0.0 & 0.0 & 0.6 & 0.0 \\
 & 0.9 & 0.0 & 0.0 & 0.0 & 0.0 & 0.6 \\
 & 0.0 & 0.0 & 0.0 & 0.0 & 0.0 & 0.9 \\
 & 0.0 & 0.0 & 0.0 & 0.9 & 0.9 & 0.0 \\
 & 0.0 & 0.0 & 0.0 & 0.6 & 0.0 & 0.6 \\
 & 0.0 & 0.0 & 0.0 & 0.0 & 0.9 & 0.0 \\
 & 0.0 & 0.0 & 0.0 & 0.0 & 0.0 & 0.9
\end{array}\right] & \begin{array}{l}
i=1 \\ i=6 \\ \\ \\ \\ \\ \\ \\ \\ \\ \\ \\ \\ \\ \\ \\ i=21
\end{array}
\end{array}
$$

(b) Lower bound of $H_{i\alpha}$ appearing in eq. (4.27).

$$
h_{j\ell} = \left[0 \;\vdots\; \overset{j=4}{0.6} \quad 0.3 \quad 0.3 \quad 0.6 \quad 0.6 \quad \overset{j=9}{0.6}\right]
$$

Example 4.1. "No symptom has been recognized". This example should take the linguistic value "completely false" for all j and consequently the range of the α-cut is [0, 0] for all α so that,

$$b_{j\ell}= 0.0, \;\; b_{ju} = 0.0; j = 1, \ldots, 9$$

The calculated failures were,

$$A_{i\alpha} = \varnothing; i = 1, \ldots, 21$$

This resulted in producing the announcement that no failure exists.

Example 4.2. This next example will diagnose a hypothetical state. Suppose a large leakage occurs at the main steam line not in the dry-well but in the reactor building, and recall that such an event was assumed to be recognized as the exception in the present exercise. In this example, the isolation valve for the main steam shall be closed due to a decrease in pressure on the steam line, and the reactor may be stopped if the pressure in the reactor vessel decreases. Consequently the temperature of the atmosphere rises and the flow rate of the sump increases in the building. However, the flow rate of the main steam line decreases rapidly because of the main steam isolation valve being closed.

The symptoms "pressure decrease in the main steam line", "high temperature in the building" and "an increase in the flow rate of the sump in the building" are sharply observed. The values $b_{4\ell} = 1.0$, $b_{8\ell} = 1.0$ and $b_{9\ell} = 1.0$ were substituted for the lower bounds, as these symptoms. The solution A_i (i = 1, ..., 21) is obtained as:

$$A_{6\alpha} = [1, 1], \quad A_{7\alpha} = [1, 1], \quad A_{8\alpha} = (0.6, 1],$$
$$A_{11\alpha} = (0.6, 1] \quad A_{15\alpha} = (0.6, 1], \quad A_{16\alpha} = (0.6, 1],$$
$$A_{17\alpha} = (0.6, 1], \quad A_{20\alpha} = (0.6, 1], \quad A_{21\alpha} = (0.6, 1],$$
$$A_{i\alpha} = \varnothing \text{ (for other } i\text{'s).}$$

The solution indicates accurately a leakage in the main steam line inside the building by reading the α-cut of A_6, A_7 and A_8. At the same time, it suggests that a warning should be given about leaking in the building of the residual heat removal system (A_{11}), the steam line of reactor core isolation cooling systems (A_{15}, A_{16}), the water line of reactor core isolation cooling system (A_{17}) and the feed water system (A_{20}, A_{21}).

Participation of exception in diagnosis. It is not easy to decide how exception as presented, should participate practically in identifying a failure. It is important for the diagnostic system to use a large amount of fragmentary information concerned with particular events. A small number of sharply aggregated implications become much more efficient when mixed with consideration of the fragmentary information, which serves as the exception in the present sense.

Fig. 4.12 indicates a procedure of reinforcement of the diagnostic system by mixing exceptions with implications. Substitution of the recognized symptoms into eqs. (4.15), (4.16) generally yields a failure in terms of linguistic truth values. When an engineer fails to identify the failure firmly and/or an alternative failure weighs on her(his) mind, (s)he should consider the exceptional proposition. In other words, the truth value of P_j of eq. (4.9) must be close to true, and consequently (s)he may fall back on the exceptions. The next example elaborates this procedure.

Example 4.3. "The flow rate of the building sump has slightly increased and the differential flow rate of the clean-up system for the reactor water becomes large".

Lower bounds $b_{7\ell} = 0.6$ and $b_{9\ell} = 0.6$ were adopted for these symptoms. The present method tries a failure inside the dry-well with the conventional fuzzified implication written by eqs. (4.15), (4.16). The solution is of the form,

$$A_{i\alpha} = [0, 0.1]; \ i = 1, \ldots, 5$$

which indicates "no leakage exists inside the dry-well". This is a case where no failure is obtained although several symptoms are recognized. It is found from calculation that the antecedent of eq. (4.10) is deduced to be very true for the proposition $(\neg\exists\, k(A_k \wedge R_{k1})_\alpha = (0.9, 1];\ k = 1,...,5$ This enables the engineer to decide that there might exist a failure elsewhere and steps may be taken to examine by exception; then the solution is of the form,

$A_{17\alpha} = (0.6, 1], A_{19\alpha} = [1, 1], A_{21\alpha} = (0.6, 1], A_{i\alpha} = \varnothing; i = 1, ..., 21$ and $i \neq 17, 19, 21$.

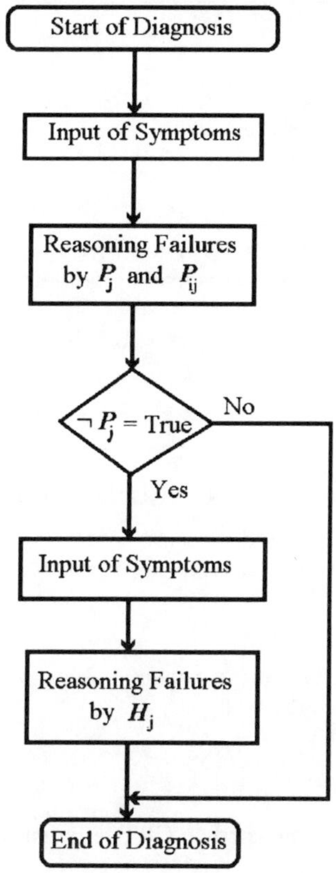

Figure 4.12 Flow of failure diagnosis with implication and exception.

The above equation reveals that a leakage in the clean-up system for the reactor water inside the building (A_{19}) exists, and that of the water line of the reactor isolation cooling system (A_{17}) and the feed water system (A_{21}) in the building are found by a possible grade.

Figure 4.13 shows the interactive procedure of the present diagnosis on a CRT terminal of a personal computer. The truth value of $b_{7\ell} = 0.6$ serves as a symptom for "the differential flow rate increased in the clean-up system a little". This method builds up a hypothesis from the symptom that the clean-up system is leaking in the dry-well, and asks the engineer to ascertain whether the hypothesis generates the various symptoms or not. But, no symptom is recognized being based on the hypothesis (fig. 4.13(a)), and then no failure is identified by the implications (fig. 4.13(b)). In the case that he is able to recognize the symptom "the flow rate increased in the building sump", the input $b_{9\ell} = 0.6$ and all symptoms yield the solution on the CRT (fig. 4.13c)

```
Start
    Flow rate increase in dry-well sump                                   **0.0
    Flow rate increase in air condenser drain                             **0.0
    Pressure increase in dry-well                                         **0.0
    Pressure decrease in steam line                                       **0.0
    Flow rate increase in main steam line                                 **0.0
    Flow rate increase in residual heat removal system                    **0.0
    High differential flow rate of reactor water clean-up system          **0.6

            Possibly leaking in
            reactor water clean-up system in dry-well

            "Check follows"

    Flow rate increase in dry-well sump                                   **0.0
```

(a) First segment of a session with diagnostic system. User responses follow the double asterisks.

```
                    **** Kind of failure ****
                    ****     Possibly    ****

    main steam line in dry-well                                           [0.0 0.1)
    residual heat removal system in dry-well                              [0.0 0.1)
    steam line or reactor core isolation colling system in dry-well       [0.0 0.1)
    reactor water clean-up system in dry-well                             [0.0 0.1)
    feed water system in dry-well                                         [0.0 0.1)
```

(b) Inferred failures by implication listed on CRT terminal with their truth values.

```
                    ****     Possibly    ****
    water line of reactor core isolation cooling system in building       [0.6 0.1)
    reactor water clean-up system in building                             [1.0 1.0)
    feed water system in building                                         [0.6 1.0)
```

(c) Final segment of a diagnosis. Inferred failures by exception are displayed with their truth values.

Figure 4.13 Example of fuzzy fault diagnosis by CRT terminal

It is the subject of further research to integrate fragmentary information about failures in a refined implication by means of repeated experiences of the diagnosis and/or the learning process.

4.3.2 Automatic expert fault diagnosis incorporated in a process SCADA system

During a complex industrial process (e.g. power system, chemical plant, etc.) disturbance, many alarms are presented to the operator making it difficult to determine the cause of the disturbance and delaying the corrective action needed to restore the power system to its normal operating state.

In order to provide continuous real-time analysis of the alarms generated by a *SCADA* (*Supervisory, Control And Data Acquisition*) system, a knowledge-based system, being immune to emotional factors, can be used to assist the operators in analyzing a crisis situation so that an optimal solution may be found as rapidly as possible.

A knowledge-based alarm processor can replace a large number of alarms with a few diagnostic messages that describe the event(s) that generated the alarms. It may also present advice to the operators when certain situations occur. The knowledge-based system performs much of the analysis that a power system operator would have to perform. Since it can quickly analyze the alarms and present a small number of concise messages, the operator is given a clearer picture of the condition or conditions that caused the alarms making it easier for the operator to take corrective action in a timely manner.

Because the system operator (in a power system for example) is very busy during a disturbance, a basic requirement of a knowledge-based alarm processor is that it may not query the operator for any type of information. Since the SCADA system is also busy processing alarms, collecting disturbance data and performing its normal functions such as Automatic Generation Control, the knowledge-based system should not strain the computer resources of the SCADA system. The knowledge-based system must be able to handle multiple, independent power system disturbances, presenting diagnostic messages to the operators within a short period of time. Also, a diagnostic message must be retracted if the conditions that caused the message to be generated are no longer valid.

Two basic approaches are possible in incorporating a ***knowledge-based alarm processor*** (*KBAP*) into an ***Energy Management System*** (*EMS*) or other complex industrial process environment, an embedded approach and an appended approach. In an *embedded approach*, the knowledge-based system is incorporated in the SCADA system. In an *appended approach*, a separate computer is used with a data link connecting the KBAP with the SCADA computer.

The appended approach is selected here, mainly because a knowledge-based system is processor and memory intensive. By implementing the KBAP on a separate computer,

SCADA resource contention is minimized. Also, implementation, maintenance and testing of the KBAP would not disrupt normal SCADA operations. The two main disadvantages of this approach are that a data link must be established between the two computers and there is no direct access to either the SCADA database nor the SCADA man/machine interface.

Since the system operators are comfortable with the SCADA alarm displays, the KBAP uses the SCADA alarm format to present diagnostic messages to the operators. A new message appears as an alarm. When the conditions that caused the message to be generated are no longer valid, the message is presented in a manner similar to an alarm returning to its normal state.

Fig. 4.14 shows the components of a general KBAP. The SCADA alarm processor sends each change of state packet across the data link to the KBAP computer. The KBAP uses the data link to present diagnostic messages to the operator and, on operator request, an explanation on how a KBAP conclusion was reached. The data link is also used by the KBAP to retrieve database information from the SCADA system. When sending alarm packets and database information, the SCADA system sends the symbolic value of a point so that the KBAP does not have to be concerned with the various operating ranges of analog values and the normal positions of digital values. Examples of analog symbolic values are low, normal and high.

In the case of a power system, since information related to a station is maintained on a station-by-station basis, a separate node must be created in memory for each station. Each node contains information about the station including current facts, working hypotheses and validated conclusions. The station node is also the head node of the configuration database for the station. Information is maintained in the station node because the order of incoming alarms cannot be predicted. Each station node is used to keep track of the facts related to the station during the inference process. General rules along with the configuration database simplify the management of the rules and also simplifies the KBAP analysis process.

The system is intended to run on a standalone computer under a multiprocessing operating system. The KBAP consists of the following components: A Rule Preprocessor, Configuration Preprocessor, Alarm Preprocessor, Inference Engine, Conclusion Processor and an Explanation Facility. The Rule Preprocessor and Configuration Preprocessor execute only during KBAP initialization. Once initialized, the Alarm Preprocessor, Inference Engine, and Conclusion Processor each runs as a separate process. This allows the Alarm Preprocessor to queue incoming SCADA alarm packets because alarms could arrive at a rate faster than the Inference Engine is able to process them.

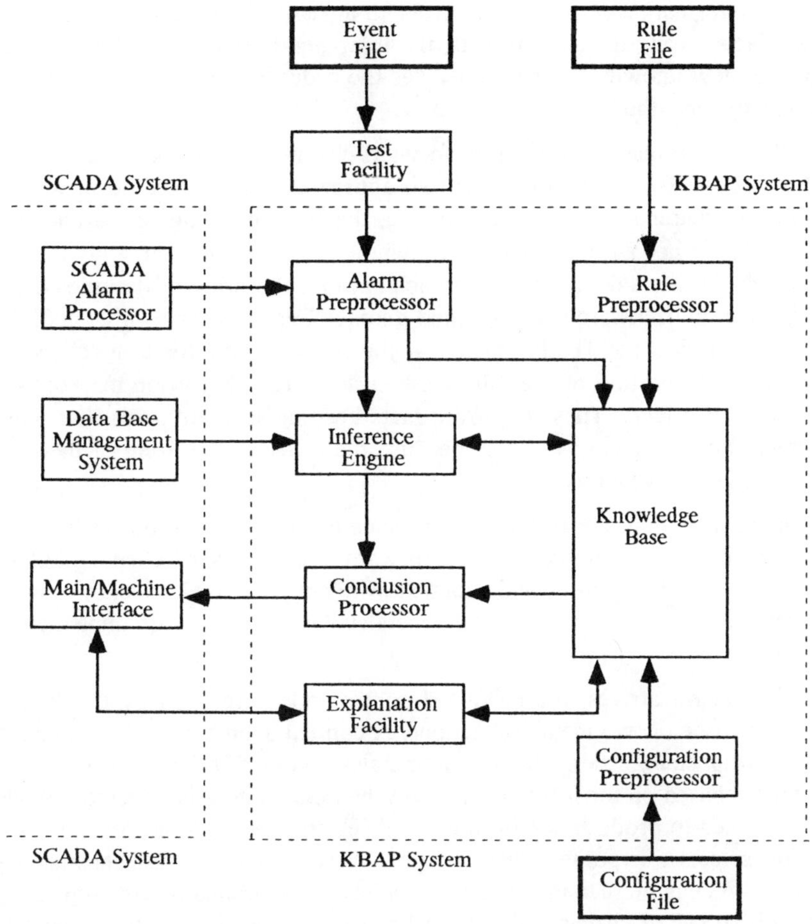

Figure 4.14 The general appended KBAP configuration.

The processing speed of the KBAP depends on both the hardware that the KBAP is implemented on as well as on the rate that the SCADA system supplies alarms to the KBAP. In other words, the limitation of the KBAP in presenting diagnostic messages to the operators is mainly due to the limitation of the SCADA system in detecting and generating alarms. The SCADA limitation is a result of varying *RTU* (***Remote Terminal Units***) scan rates as well as power system relay actions.

The collection of knowledge in the KBAP is referred to as the knowledge base. One way of organizing the knowledge is to form rules and facts. The rules contain accumulated knowledge in the form of IF-THEN constructs. The facts in the knowledge base are collected pieces of information related to the problem at hand. Rules express the relationships between facts. Using the current facts, the Inference Engine decides how to apply the rules to infer new knowledge. It also decides the order in which the rules should be applied in order to solve the problem.

The rules in the KBAP may be fired using forward chaining or backward chaining (see Section 4.2.1). The difference between the two approaches is the method in which the facts and rules are searched. In forward chaining, the Inference Engine searches the IF portion of the rules. When a rule is found in which the entire IF portion is true, the rule is fired. Forward chaining is a data driven approach because the firing of the rules depends on the current data. In backward chaining, the Inference Engine begins with a goal that is to be proved. It searches the THEN portion of the rules looking for a match. When a match is found, the IF portion of the rule is established. The IF portion may consist of one or more unproven facts. These unproven facts become separate goals that must be proved. Backward chaining is goal driven because the order of the firing of the rules is done in an attempt to prove a goal.

Metalevel control rules improve system performance by selecting the object-level rules. The object-level rules in the KBAP are the forward and backward chaining rules described above. Metalevel actions provide context switching and goal selection. Fig. 4.15 shows a conceptual representation of how the metalevel control rules are implemented in the KBAP.

When a SCADA alarm arrives, the metalevel control rules are used by the Inference Engine to generate one of two metalevel actions. A context switching action selects data driven, or forward chaining rules. That is, the metalevel control rules are used to select the proper context based on the incoming alarm. When the context is selected, metalevel control rules are used to produce one or more goal selection actions (if possible). A goal selection action selects a hypothesis that the goal driven, or backward chaining rules, attempt to prove. As can be seen in figure 4.15, object-level actions result from applying the forward and backward chaining object-level rules. For the KBAP, two object-level actions are possible, diagnostic messages and advice for the operators.

The metalevel control rules contain heuristics that guide the Inference Engine in forming hypotheses. Heuristics are rules of thumb that an expert uses when solving a problem. Heuristics are used to narrow the search space for a solution. Backward chaining object-level rules are generic in nature and are not related to any particular station for the case of a power system, or process subsystem, in the general case.

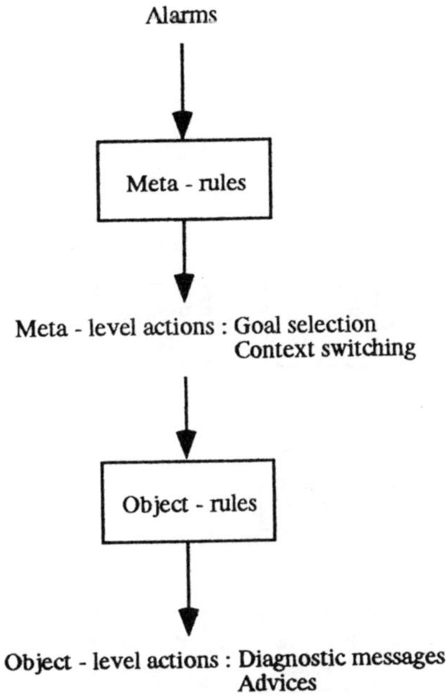

Figure 4.15 Metalevel control in a KBAP.

Fig. 4.16 shows how the metalevel control rules are internally organized. The order of the metalevel control rule nodes is the same order of the rules as found in the rule file. Each metalevel node is the head of a linked list of one or more premise nodes. A premise node contains a premise clause. All of the premise clauses of a rule must be true in order for the rule to fire. The Inference Engine scans the metalevel control list beginning with "metahead", the head node. If all of the premise clauses of a rule are true, the Inference Engine fires the metalevel control rule triggering one of the two metalevel actions, context switching or goal selection. In the case of goal selection, the metalevel node contains the goal, or hypothesis, that the object-level rules attempt to prove. The station node, for the case of a power system, shown in fig. 4.16, has two working hypotheses. Each hypothesis is represented by a hypothesis node linked to the station node. A hypothesis node points to the metalevel node that it is associated with. When a working hypothesis is proved to be valid by the Inference Engine, the hypothesis node is linked to the station node's validated conclusion chain.

Each change of state detected by the SCADA system is reported by the SCADA alarm processor to the KBAP Alarm Preprocessor. A state change includes points going into alarm, points returning to normal, and supervisory control actions. The SCADA change of state packets are converted by the Alarm Preprocessor into a form suitable for the Inference Engine. Incoming packets are queued until the Inference Engine is able to process them. The configuration point nodes contain the current symbolic value of the point. When a change of state occurs on a point, the value of the point in the node is pdated.

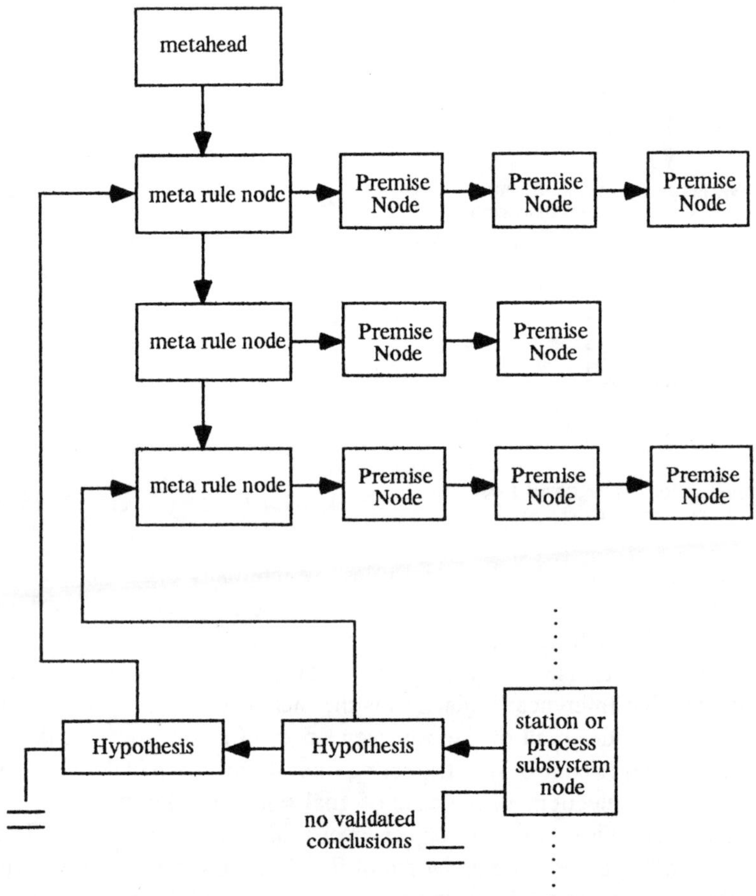

Figure 4.16 The internal organization or the metalevel control rule node.

Because knowledge is often added incrementally to a knowledge-based system, a Test Facility exists so that the firing of the KBAP rules may be observed and validated under controlled conditions. The Test Facility reads each change of state record from an event file and passes the change of state events to the Alarm Preprocessor. The event file is built off-line using a text editor.

If a metalevel hypothesis has been proved true, the Conclusion Processor is invoked to pass the validated conclusion to the SCADA system for presentation to the operator. A validated conclusion may result in one or more new hypotheses being formed. A change of state may also result in a previously validated conclusion becoming no longer true. When this occurs, the Conclusion Processor is invoked to inform the SCADA system that a previous message is no longer valid. This is similar to an alarm returning to normal.

When an operator requests information on how a particular conclusion was reached, the SCADA system man/machine interface sends the request over the data link to the to the KBAP. The Explanation Facility is invoked to process the request (see fig. 4.14). The Explanation Facility passes information back to the SCADA system showing the rules that fired and the facts that caused the rules to fire for presentation to the operators.

System operation.

This part of the section describes the operation of the KBAP for processing different system components. Power system and general industrial plant components, as motor pumps and rotating machinery are discussed.

Low voltage bus and electricity supply networks. For this example, the low voltage bus has three bus sections. Fig. 4.17 shows some of the object-level rules which can be used for fault diagnosis.

Object-level rule names are enclosed in colons and metalevel control rules are surrounded with double colons. Comments are enclosed in /*and*/. Backward chaining is used on the object-level rules in fig. 4.17, in an attempt to prove the goal "G6". If all of the bus sections on a bus have low voltage and the bus breakers are closed, a single low bus voltage message is presented to the operators. On the SCADA system, the operators would receive a separate low voltage alarm for each bus section and possibly other alarms such as under voltage relay alarms. This simple example illustrates how a single message can be presented in place of numerous SCADA alarms. As well as reducing the number of messages that the operators receive, the KBAP also contains rules that diagnose the situation(s) that triggered the alarms.

Tesch *et al.*, (1990), present a case study of a KBAP implementation for Wisconsin's Electric Energy Management System. The KBAP, written in the C Programming Language, uses a configuration database that contains the structure of the power system as well as symbolic data values for each point monitored by the SCADA system. The knowledge-based system continuously analyzes and processes SCADA alarms in real-time, presenting diagnostic messages to the power system operators without requiring any information from the operators.

Brailsford *et al.*, (1990), present a prototype KBAP system, named *FAUST*, for use in 132 kV and 33kV electricity supply networks. All items of the electricity distribution network, especially those at the higher voltages, are telemetered. Each telemetered item is polled regularly (5÷20 secs) and any changes of state and alarms are reported to the operator. FAUST is equipped with a network database, a plant database, advanced graphics facilities and can perform continuously the following tasks: manual input of current network status, input from telemetry stream, telemetry message filtering, fault and outage hypothesis generation and user interface (continuous mode, off-line mode, hard copy mode, fault simulator mode). Causal reasoning based on a model of the operation of the underlying network, heuristic knowledge, complex graph-searching algorithms, the use of blackboards for communication amongst system modules and message passing between objects are employed to resolve the complex real-time diagnosis problem for the distribution network.

```
/* If the first circuit breaker on the device is closed, establish conclusion "C1" as a fact.
The rule name is "Rule 1". The underscore character denotes a blank. The word "first"
is a position indicator.*/

    : Rule-1: If (first circuit-breaker=closed) then C1;

/* Establish conclusion "C2" if the second circuit breaker on the device is closed.*/

    :Rule-2: If (second circuit-breaker=closed) then C2;

/* This rule is fired if the first bus section voltage is low and conclusion "C1" is an
established fact*/

    :Rule-3: If (first bus-voltage=low) and (C1) then C3;

/* The rule order is not important for general rules. Rules are entered free format.
Comments may appear anywhere.*/

    :Rule-6: If (C4) and (C3) then C6;

/* Fire this rule if the third bus section voltage is low and conclusions "C2" and "C5"
are established facts. */

    :Rule-4: If (third bus-voltage=low) and (C5) and (C2) then (C4);

/* If the second bus section voltage is low, establish conclusion "C5".*/

    :Rule-5: If (second bus-voltage=low) then C5;

/* Fire this rule if the bus voltage on the bus at the opposite end of a transmission line
is low and the bus voltage in the current station is normal*/

    :Rule-15: If (opposite bus-voltage=low) and (adjacent bus-voltage=normal) then
    R15;
```

Figure 4.17 General object level rule examples for a low voltage bus.

Power transmission substations. Several additional benefits can be obtained when the substation *integrated control and protection system* (*ICPS*) is used as part of the overall *Energy Management System* (*EMS*). A number of substation ICPSs are being developed around the world today, where the protective relaying, control, and monitoring functions of substation are implemented using microprocessors. In this design, conventional relays and control devices are replaced by clusters of microprocessors, interconnected by multiplexed digital communication channels using fiber optic, twisted wire pairs or coaxial cables. The ICPS incorporates enhanced functions of value to the utility and leads to further advancement of the automation of transmission substations. More powerful processing capabilities can be established if an ICPS is used instead of the conventional SCADA Remote Terminal Units at the substation. In addition, an extensive data base can be available at the substation level. This data can be used to assist dispatcher, protection engineer and maintenance personnel during an emergency.

Fault diagnosis is carried out by operators using information of active relays and tripped circuit breakers. The faulty components are inferred by imagining a protective relay sequence related to the incident and simulating backwards the relay sequence from the observed data. An expert system will be very useful for these type of tasks, since the problem involves a mass of data and uncertainties and cannot be described by a well defined analytical model. For example, a rule to identify a failed breaker is:

```
Rule ?
{
&1 (Relay operated = yes; considered = yes;);
&2 (Breaker name = &1.br1; open = no; failed = no; status = on);
®
modify &2 (failed = yes);
};
```

This rule implies that, if a relay has operated and one of its corresponding breakers connected in the circuit has not opened, identify this breaker as a failed breaker.

Stavrakakis and Dialynas, (1991), describe models and interactive computational techniques that were developed to model and detect automatically the available restoration operations following a fault on substation equipment or a substation abnormality. These abnormalities can be detected before the component enters into the failure stage using the computer-based diagnostic techniques described in Chapter 3. The developed computer-based scheme can be installed easily through a rule-based expert system in power substation ICPSs in order to determine the optimal switching operations which must be executed after a component fault has been detected. The development of a data base containing all the necessary information concerning the component failure characteristics and the average repair or replacement and switching times is possible. A supply reliability index of substation load-points is also evaluated to quantify the reliability performance of the substation.

Underground distribution cables. Kuan and Warwick, (1992), developed a prototype real-time expert system aid for the fault location on high voltage underground distribution cables. Diagnostic tests are performed by measuring the continuity and insulation resistance of the cable.

Different types of faults (e.g. series, low resistance, high resistance, flashing, etc.) produce different transient waveforms which an experienced engineer would be able to distinguish by looking at the certain characteristics of the waveform. The knowledge used is mostly heuristic in nature and in the form of IF-THEN guidelines; e.g. if the second crest of the waveform is smaller than the first crest, then it is most likely to be a short circuit or low resistance fault. Coupled with information from the diagnosis stage, the type of fault can be confirmed; e.g. if the insulation resistance of the cable is low, then it is likely to be a short circuit or low resistance fault. After confirmation of the type of fault, the time interval between the start of the waveform and the fault point is measured and calculations are done to determine the distance to the fault.

Power Systems considered as a whole. Nebiacolomobo *et al.*, (1989), describe *HOLMES*, an ES for automatic fault diagnosis and restoration planning in industrial power systems. The whole power system is divided into modules according to topological and functional criteria. A power substation can be considered as a module. Each module contains a list of all the possible faults that may occur in the power equipment it contains. Each module is connected only to its adjacent ones according to system topology. The only information flowing across the modules are the over-currents and/or over-voltages caused by a fault. With such a model, the propagation of a fault throughout the power system can be easily detected together with the protective relays involved by the fault itself. Petri nets (see Section 4.2.4) are used to solve the representation problem of the protection dynamics under faulty conditions. For every defined module, a Petri Net is built to describe its internal dynamics and its input-output behavior. The causal connections among the modules are represented by input and output conditions. For every output from a module, a corresponding input to another module should exist. By aggregating the Petri net of every module according to the system topology, the global system model is obtained.

After its activation, HOLMES performs pre-fault network analysis, local analysis of received events and possible faults generation for every module, global propagation analysis with correlation of faults generated by different modules, fault validation, results transmission to interface and recording to archives. HOLMES implements a KBAP mechanism to consider the fault diagnosis and propagation in the controlled plant.

In the area of power system security monitoring and control, transient stability is considered to be one of the most important and at the same time most problematic issues, strongly related with the on-line fault diagnosis procedure. Indeed, in a broad sense, transient stability is concerned with the system's capability to withstand severe disturbances and/or faults causing important electromechanical transients. This is a strongly non-linear problem. It also high dimensional, since power systems are by essence large scale.

In general, on-line ***transient stability assessment*** (*TSA*) aims at appraising the power system robustness in the inception of a sudden, severe disturbance or fault, and whenever necessary at suggesting remedial actions. A measure of robustness is the ***critical clearing time*** (*CCT*); this is the maximum time a disturbance may act without causing the irrevocable loss of synchronism of the system machines. In particular, on-line TSA is used in real-time operations and aims at performing on-line analysis and preventive control. Indeed, because the transient stability phenomena evolve very quickly (in the range of very few seconds), a good way to face them efficiently is to prevent them.

Wehenkel *et al*., (1989), proposed a new approach to on-line transient stability assessment of power systems, suitable for implementation in the SCADA system. The main concern of this approach has been the application of an inductive inference method in conjunction with analytic dynamic models and numerical simulations to the automatic building of ***decision trees*** (*DT*s).

A DT is a tree structured upside down. It is composed of *test* and *terminal nodes*, starting at the *top node* (or *root*) and progressing down to the terminal ones. Each test node is associated with a test on the attribute values of the objects, to each possible outcome of which corresponds a successor node. The terminal nodes carry the information required to classify the objects. The methodology developed there is based on inductive inference and more specifically on ID3 which is a member of the TDIDT (top-down induction of decision trees) family (see Section 4.2.1). Most of the inductive inference methods infer decision rules from large bodies of reclassified data samples. The TDIDT aims at producing them in the form of decision trees, able to uncover the relationship between a phenomenon and the observable variables driving it. Adapted and applied to on-line transient stability assessment, the method intends to uncover in real-time the intrinsically very intricate relationships between static, pre-fault and/or pre-disturbance conditions of a power system and their impact on its transient behavior in order to discover the appropriate control actions needed.

Industrial motor pumps.

To investigate the applicability of expert systems in the area of equipment diagnostics of industrial plants, the principles of a knowledge-based system for real-time diagnosing of motor pump malfunctions is presented.

The diagnostic approach used in the development of the knowledge base is based on the method of decision tree (DT) analysis (see previously, Wehenkel (1989)). Using information derived from the equipment maintenance manual and mechanical-electrical drawings, and through observation of the procedures used by efficient engineering mechanics, a decision tree can be developed to mimic the way a human expert makes decisions and arrives at conclusions. The decision tree is then translated directly into the IF-THEN rules which make up the expert system's knowledge base. The manner in which a decision tree can be translated into ***production rule language*** (*PRL*) rules is illustrated in fig. 4.18 which shows a small portion of a decision tree developed to diagnose a faulty pump starting switch and the corresponding rules written in PRL.

Using on-line DT analysis, the system leads the user through the appropriate procedures required to quickly identify the faulty pump circuit component. Graphical displays can be incorporated within the system to assist the user in locating the various components and test points. Once the faulty component has been isolated, the system is capable of accessing a database which can provide personnel with information concerning specific component part numbers, the availability and location of spare parts and the proper repair action to be taken.

Rotating machinery. During the past few years, condition monitoring of gas turbine and other industrial engines has blossomed into an economically viable activity for some of the larger constructors and commercials. These gains have been spurred to a significant degree by the development of sophisticated software algorithms for the interpretation of the limited available sensors.

Software packages are being employed to realize many of these gains using artificial intelligence techniques (Doel, 1990). Classical tools that are currently being employed for commercial engine condition monitoring as *EGT* (***exhaust gas temperature***) margin trending, vibration monitoring, oil monitoring, under cowl leak detection etc. can be expressed analytically and performed automatically using the statistical aids and signal analysis techniques described in Chapter 1.

These basic fault occurrence knowledge sources can be used for the creation of the knowledge base and the inference engine of an expert fault diagnosis tool for rotating machinery.

Before designing a failure diagnostic system, damage statistics of the specific system should be regarded, indicating these sections, in which failures are occurring with a certain frequency. Taking the rotating system of a turbomachine as an example, the defects on moving blades occupy the highest percentage, concerning the number (37%) and the repair costs (26%), followed by casing failures (24%), where the financial effect is not as severe (13%).

In contrary, less frequent failures of bearings (5%) are classified at a high cost level (27%). During the last years therefore, research in monitoring and diagnosing of systems focuses on the improvement of detecting damages of the rotating system, including blades, rotor unbalance, cracks, bearings and others.

Pattern recognition methods have also proven to be very effective in failure detection (see Section 1.3.2.C). For diagnostic purposes, time signals are utilized, which are obtained at the machine with special sensors, like pressure transducers mounted directly above moving or stator blades, accelerometers placed at the casing of the machine, and shaft displacement measuring systems, which detect the shaft oscillations.

RULE #1
IF specific pump problem IS pump will not start from machine
AND control power is available
AND switches are set properly
AND NOT red LED on remote box lights when trying to start pump
AND pump will start from remote box
THEN probable detective pump starting switch
AND end of pump diagnosis

RULE #2
IF operating node IS machine
AND main control switch is ON
AND cutter switch is OFF
THEN switches are set properly

RULE #3
IF 12 VDC LED is ON
OR 110 VDC LED is ON
THEN control power is available

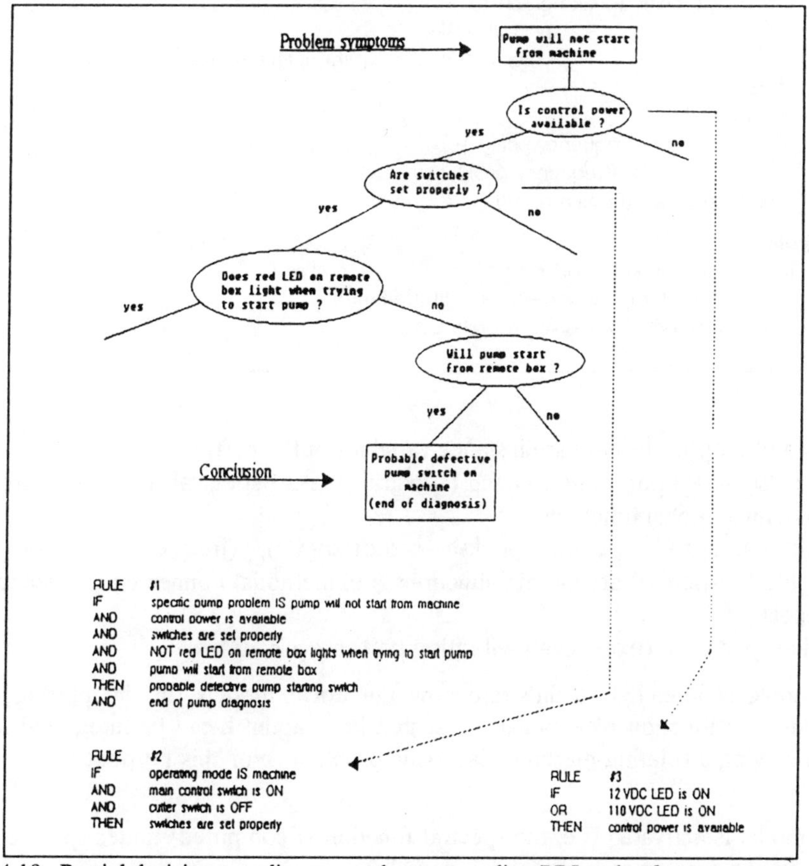

Figure 4.18 Partial decision tree diagram and corresponding PRL rules for a motor pump fault diagnosis.

Consider a vibrating part of a rotating machine. The behavior of the part can be modeled in real-time by the auto spectral density function of the signal of an accelerometer attached to it. The goal is to make diagnosis about the state of the vibrating part. Two diagnoses can be made locally for this part using the peak on the spectra representing its behavior approximately at 10.5 Hz. For example if this peak is,

a) missing, then the part is probably broken;
b) too large, then the part is excessively vibrating.

The following rules can be defined for the part:

```
(rule-1
  if (null ? peak)
  then (set !  peak  (peak-nearest-to
                          (spectral-peaks
                             (apsd <signal> <parameters>)) 10.5)))
(rule-2
  if (and peak
           (or (< (frequency peak) 10)
                (> (frequency peak) 11))
  then (conclude ' (broken part)))
(rule-3
  if (and (not (broken ' part))
            (> (amplitude peak)  amplitude-limit))
  then 'conclude " (excessively-vibrating part)))
```

where,

- peak and amplitude-limit are instance variables of the part;
- (apsd <signal> <parameters>) and (spectral-peaks <spectral function>) are externally defined global functions;
- (peak-nearest-to <spectral peaks> <frequency>), (frequency <peak>), and (amplitude <peak>) are internal functions (i.e. methods) connected to instance variable peak,
- (broken part) and (excessively-vibrating part) are assertions.

This example is intended to illustrate how numerical models (in the present case a spectral density function of a signal) presented in Chapter 1 can be integrated in rules aimed at solving a rotating machine diagnostic problem. For this purpose two steps are taken:

1. The model is activated (i.e. the spectral function is computed) under specific conditions (in the present case if peak has no value).

2. The results (having numerical nature) are compared to limit values (frequency and amplitude limits in the present example) resulting in Boolean values, which can be combined with the assertions.

Given these features, it is not difficult to envision very powerful, linked diagnostic and maintenance systems to diagnose engine problems using all available input in assisting the maintenance process.

Detecting and masking transient failures of the SCADA system itself.

The effect of lightning poses significant problems to computer systems and electronic equipment used in process supervision due to the possibility of a direct strike. Protection against a direct strike is very expensive and seldom used (Noore *et al.*, 1992).

When lightning strikes the plant, it produces both electric and magnetic fields that vary with distance, frequency, and time. These fields can couple into power transmission and communication lines and adversely affect the performance of computers and other electronic equipment. Protection from equipment damage and power supply interruptions through shielding, insulation, lightning arrestors, and filtering circuits help in reducing the effects due to lightning but are not completely effective. The field-induced voltages and currents cause transients that can create an upset in a digital or computer system. Problems due to transients have not been a significant area of research until recently. The reason for this is that electronic equipment that use vacuum tubes or power semiconductor devices is relatively immune to some transients. In addition, with these devices, the short duration of its occurrence usually is not of serious consequence, but the rapid advancement of technology in recent years has led to the emergence of several complex VLSI devices used in the design of computer-based data acquisition systems, process monitoring systems, surveillance systems, and control systems to support both critical and non critical industrial plant activities. With the attendant problems that are exacerbated as a direct consequence of the submicron geometry of advanced microelectronics semiconductor technology, transients of smaller magnitudes and shorter time duration can have a dramatic effect and cause a system upset in computers.

To detect transient faults efficiently within a short time period, the proposed detection mechanism focuses on a set of hardware modules and control paths that are utilized by the microcomputer system. For example, if a transient fault alters the contents of a register that is not used by the microcomputer for performing process monitoring or control operations, it is not necessary to detect this fault. On the other hand, if a transient fault causes a change in the contents of a register used by a specific application, then it is imperative to detect this critical fault. Thus, by monitoring only the control paths of the application program and selected modules used by the microcomputer, the proposed approach detects all critical transient faults and maintains a high fault coverage.

Let the global memory space $M(0, n\text{-}1)$ be partitioned into several disjoint contiguous physical blocks of memory $M(0, i)$, $M(i+1, j)$, $M(j+1, \ldots)$, ... $M(\ldots, k)$, $M(k+1, n\text{-}1)$. Each memory partition $M(x, y)$ is bounded by the lower memory address x and the upper memory address y, and the size of each memory partition is y (= x+1). Each physical memory block $M(x, y)$ is logically mapped to a set of instruction sequences $I(x, y)$ that are executed within the boundary (x, y). Depending on the type of application program, the logical execution of the instruction sequences can, in general, be globally distributed in a noncontiguous manner. Each instruction sequence $I(x, y)$ is modular and represents a subtask of the application program or a subroutine. Although instructions in the instruction sequence $I(x, y)$ are physically contained in $M(x, y)$ sequentially, the program can logically execute several loops within the boundary (x, y). Instructions contained in $I(x, y)$ can either be of single word length or multiple word lengths. The time $T(x, y)$ for executing $I(x, y)$ is exactly computed from the type of processor used in the design, the operating frequency of the clock, the response time of the memory or peripheral device selected, the number of cycles required to execute a single or multiword length conditional or unconditional instruction, the number of loops in $I(x, y)$, the manner in which the loops are structured, and the number of passes per loop.

Given an application-specific program that is structured as described, the starting address x or the lower boundary of each logical set of instruction sequence $I(x, y)$ can be extracted. Associated with each $I(x, y)$, the execution time $T(x, y)$ can be obtained from the program structure. This value is stored in the EPROM, whose address is given by x. Similarly, all execution times are stored under the corresponding addresses. Addresses that do not correspond to the starting address of the instruction sequences are used to store all 0s in the EPROM. The transient fault-detection circuitry is shown in fig. 4.19, and the detection algorithm is outlined as follows:

```
<μc address> = <EPROM address>;
        if <C> = 0 then begin
                enable EPROM ;
        if <E> ≠ 0 then begin
                <timer> = <timer> - 1;
        end;
        else transient-fault-detected;
        else begin
                disable EPROM;
                <timer> = <timer> -1;
        end;
```

If the system clock-out frequency of the microcomputer system is denoted by f_{system} and the countdown timer clock frequency is denoted by f_{timer}, then the length of the counter is given by:

$$\log_2\left\{\frac{f_{timer}}{f_{system}}[T(x,y)]\max\right\} bits.$$

For real time monitoring and control operations, rollback of either instructions, segments of programs, or entire programs is necessary after a transient fault is detected. A rollback operation makes use of the concept of checkpoints. A checkpoint is a preselected point in the execution sequence when the state of the system is saved. Program rollback forces the execution to restart at the last checkpoint. The selection of the number of checkpoints is important. If too many checkpoints are used, the program execution time is increased since the status of the processor has to be saved often. On the other hand, if the checkpoints are fewer in number, the recovery time is increased since the program has to be executed, starting from the previous checkpoint. Strategies for introducing optimal checkpoints in a program can be found in Noore *et al.* (1992). Furthermore it is shown that transient fault detection and recovery can be performed in real time, which is an important factor when monitoring, processing, and controlling critical industrial processes operations. The approaches presented have a wider range of applications and can be extended to real-time, high-risk, and life-critical systems used in the aerospace, banking, military, chemical and nuclear industries.

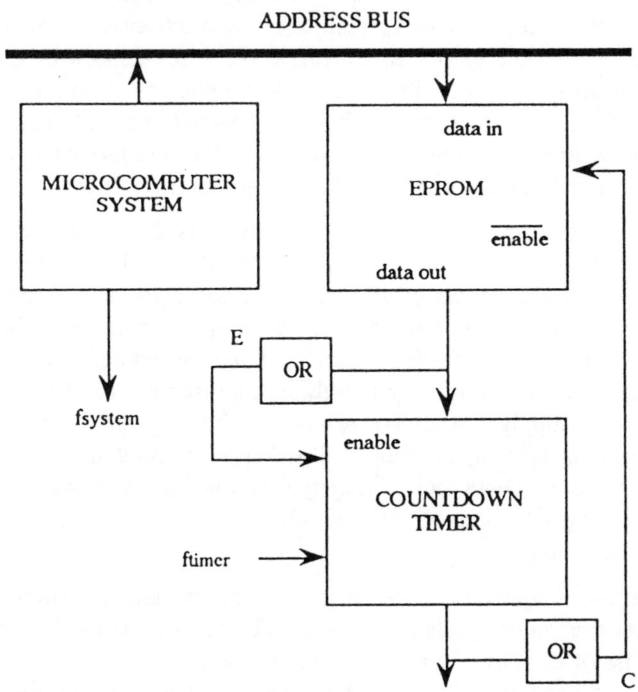

Figure 4.19 Circuit to detect transient faults in a microcomputer system.

4.3.3. Expert systems for quick fault diagnosis in the mechanical and electrical systems domains

A particular mechanical or electrical device works based on certain underlying principles. Due to the components that make it up and on how these components are connected, the device manifests a certain behavior. The connectivity and functionality of the device constitute some of the fundamental knowledge that a human learns first on the road to becoming an expert diagnostician for that device. This knowledge is often the basis on which the human acquires all other knowledge about diagnosing and repairing the device.

Regardless of the ultimate functionality of a mechanical or electrical device, most are built based on a finite set of fundamental building blocks. These building blocks represent some general functionality that can be found within many different devices. For example, no matter what device a copper wire is found in, it is usually used to transport electricity, and regardless of where a battery is found, it usually functions as a reservoir or source of electricity. These building blocks also have a fairly well-defined set of potential fault modes. For example, a copper wire can fail to allow electricity to flow by not making a complete circuit. This could be due to a break in the wire, a short, or a missing ground. A battery can fail to store electricity. This could be due to the age of the battery, to a lack of recharge from the system to which it is connected, etc. A human expert uses this fundamental knowledge about the purpose of the various components within a device to diagnose problems and devices that are unfamiliar to him/her.

Devices are not only composed of primitive components, but they themselves can serve as a functional primitive in a higher level view of the device. For example, the generator belt within the generating system of an automobile can serve as a conduit, transporting mechanical energy from one component to another. At a higher level the entire generating system could be considered a transformer, taking mechanical energy and producing electrical energy. Such a hierarchy can easily be represented using functional primitives (for example, power supply is reservoir, safety switch is regulator, and all links are conduits in a gas heating system) by simply allowing a component at one level to be described as a set of components represented by functional primitives at a lower level. This allows for a hierarchical breakdown of the problem, and a diagnosis can thus be made to any desired level of detail.

This generalization of information can also be used by human experts when they are faced with diagnosing an unfamiliar, but similar, device. Any particular device is usually a member of a class of devices of a particular type. For example, a specific car battery is a member of a set of different kinds of car batteries which are, in turn, members of a more general set that also includes batteries for flash lights and calculators. Each type of battery may look quite different, but they all work on the same basic principles and they all perform a similar job: providing a power source. The different types of batteries create a hierarchy of types of power supplies. An expert can use this knowledge about similarity of function to diagnose problems on devices that may not be familiar but that are similar to one that he/she has expertise in.

Once the device has been represented to the computer using these functional primitives, the resulting model can be used in two ways. First, it can be used as a basis for qualitatively simulating the device. Given specific values for the states of certain components in the system (i.e., switch settings), it can determine the effects on the device by simulation. This can be done both for correct and malfunctioning behavior. Malfunctioning behavior is handled by providing symptoms of the failure at the beginning of the simulation (i.e., the switch is on, but the light is not). This simulation is based on the substances and their qualitative levels as input and output to the components of the device.

The second way that the model can be used is to guide diagnosis. To do this, some very general diagnostic reasoning rules to examine the functional model and isolate the problem to a specific functional unit or units are used by the functional expert system. These rules are based upon the values of the inputs and outputs of the functional units. They are:

1. If an output from a functional unit is unknown, ask the user;
2. If an output from a functional unit appears incorrect, check its input;
3. If an input to a functional unit appears incorrect, check the source of the input;
4. If the input of a functional unit appears to be correct but the output is not, assume that something is wrong with the functional unit being examined.

Once the problem has been isolated, more specific analysis can proceed in the same manner at a lower level, since the knowledge representation can be hierarchical. When a fault hypothesis is confirmed with enough confidence, an attempt is made to repair the believed problem. If the repair fails, then the expert restarts the diagnostic process. This process is followed no matter what type of knowledge is guiding the human at any given point in time, whether it be experiential, fundamental, common sense, or some other type.

In Yoon and Hammer, (1988), an aiding approach has been described and evaluated for novel fault diagnosis in complex electromechanical systems. This approach is an alternative to the one presented in Sectiom 4.3.1.2. The emphasis is on novel rather than routine faults. It contains a qualitative model that may correspond to the human's internal model of the system. This model represents knowledge only of how the system behaves. Therefore, this aiding approach does not rely on proceduralized knowledge. The qualitative model is the basis for much of the aiding that takes place.

Although the definition of novel fault diagnosis was rather narrow in this research, its results may find generalization in any diagnostic task that involves causal reasoning based on deep knowledge. The experimental results (fault diagnosis of an orbital refueling system to refuel orbital satellites with hydrazine) confirmed that a deep-reasoning diagnosis can be aided, without disturbing the human diagnostic procedure, by providing relevant information.

However, the results also suggested that the aiding information should be compatible with the human information processing. Thus, to design a successful information aid, human diagnostic activity should be first understood from the viewpoint of human information processing.

In diagnosis tasks, system behavior in different modes is investifated: normal behavior, actual behavior, and hypothetical behavior. The results suggest that normal system behavior, while intuitively important, should not be presented prominently. This is perhaps because the human is interested in actual behavior. Abnormal system behavior was found to be very important in diagnosis. Simply presenting integrated, actual (observed) system behavior topologically has the comparable effect of improving diagnostic performance.

There are several applications of troubleshooting expert systems which are currently used in the industrial electrical-mechanical sector to perform quick and accurate diagnosis. By examining some of the practical existing systems, one can acquire information which can aid the development of new fault diagnosis systems (see Chen and Ishiko, 1990).

Conceivably, General Electric's (GE's) ***Computer-Aided TroubleShooting*** (*CATS*-1), mainly used at locomotive minor maintenance repair shops, is the earliest and most well known expert system in troubleshooting. In its current status, CATS-1 limits its troubleshooters to the ideas and disciplines of the expert whose knowledge was programmed in the system. What GE intends to add to the system in future development, is a rule editor that facilitates the creation of new rules for the technician without having to concern himself with knowledge engineers. AT&T Bell's ***Automated Cable Expertise*** (*ACE*), a knowledge-based, cable-trouble analysis system to diagnose cable troubles is used in the Texas facilities at Dallas and Houston. The ACE was designed to analyze, automatically, the large amounts of daily complaints and maintenance reports and help the managers to investigate the data contained in the complaint system. The ***Testing Operations Provisioning Administration System-Expert System*** (*TOPAS-ES*) is a real time, distributed, multi-task expert system for switched circuit network maintenance. TOPAS-ES is one of the most eminent troubleshooting expert systems because of its ability to handle the volume of trouble reports, perform its analysis, and continuously support proper trouble resolution. AT&T Bell's ***Interactive Repair Assistant*** (*IRA*) is a knowledge-based (KB) system that provides expert troubleshooting advice to the telephone company's mobile field technicians. Unlike most KB systems, IRA is capable of delivering real time expert advice to many remote users.

For the development of the following three expert systems, the vendors chose commercial expert system shells. ***Termi-point Expert Maintenance System*** (*TEMS*), is an expert system for supporting wire-bonding machine maintenance. TEMS can diagnose certain machine problems in few minutes while the human experts solve the same problems in an hour or more. ***Coordinate Measuring Machine Professional*** (*CMM-PRO*), is used by Rockwell International in Colorado, to assist in the precise calibration and servicing of state-of-the-art computer-controlled coordinate measuring machines. When CMM-PRO was put in use, it became a powerful training and learning tool for novice engineers. The *COOKER, Cooker sterilizer expert system*, used by Campbell Soup Co., can check large sterilizer cookers for malfunctions and can guide the start-up and shutdown procedures of the sterilizer operations. Since the system has been installed in eight plants, COOKER has been used day and night by the plant personnel to diagnose cooker problems without the need of consulting human experts.

EXACT (***Ex**pert system for **A**utomobile air-conditioner **C**ompressor **T**roubleshooting*) is an expert system developed by Toyoda Automatic Loom Work to support field engineers to conduct quickly the effective diagnosis of compressor troubles. EXACT runs under three software environments in the PC: EXSYS, FOXBASE+, and AutoCAD. The expert system module is handled by the EXSYS environment, EXSYS shows the results of the diagnosis using an executable external program created in QuickBASIC. The FOXBASE environment handles the user interface and database management features. On the results screen of the expert system, a pictorial information module is provided under the AutoCAD environment for locating the trouble area in the compressor. Fifty real-world cases have been used to evaluate the system performance (Chen and Ishiko, 1990).

The *ALLY™ Plant Monitoring and Diagnostics* (M&D) system. The integration of various monitoring and diagnostic systems with plant process parameters is realized by the Westinghouse ALLY™ Plant Monitoring and Diagnostic System. It is the environment where diagnostic evaluations are made on plant-wide equipment and systems and the mechanism where information is managed and displayed.

Monitoring and diagnostics (M&D) systems are well recognized as a vital tool to help maintain *structure*, *system* and *component* (*SSC*) integrity and reliability. These systems are generally designed to measure key SSC parameters such that an assessment of their performance can be made. Some of the more advanced systems have the added capability of performing a diagnosis and determining equipment malfunctions and associated contributing factors. For example, not only is it important to detect equipment performance deviations in a timely manner but it is equally necessary to have the information explaining the occurrence. Then, and only then, can proper maintenance actions be prescribed and future failures/degradations be eliminated.

In order for such systems to be effective, they need to have access to a relatively large and distributed set of information. Information directly acquired from the M&D system's local sensor set is necessary but usually insufficient by itself. Global plant process parameters measuring the dynamic behavior of other associated equipment and systems are equally necessary for a complete diagnosis and root cause analysis. Although this information has always been available through plant computers, accessing this information has been difficult because these computers do not typically have sufficient I/O for diverse M&D networks. Moreover, most M&D systems are designed without a common architecture thereby making the connection to the plant computer information systems custom engineered and uneconomic. This difficulty often results in a compromise between standardization and cost. The M&D systems, networked together and linked to the plant information systems, represent a tremendous wealth of information. So much, in fact, that the distributed architecture of the M&D network becomes a necessity, simply because of the quantity of data it must process.

The basic design of ALLY^{TM} implements a Blackboard Architecture, in which all M&D systems and knowledge sources revolve around a common data base of information, the Blackboard. Like the name suggests, it is a mechanism where different knowledge representations of the plant exist and are available for any application. When a knowledge source executes, it reads the Blackboard for its necessary inputs, processes them, and puts the results back on the Blackboard. Other knowledge sources, in turn, use the newly acquired information to further enhance the current model of the plant. The Blackboard architecture consists of the Blackboard itself, a Control Mechanism, Knowledge Sources, and Graphical User Interface.

Along with time based functionality, the system has been designed with high performance and fast execution in mind. This is the means by which the end users are provided with time relevant information. While most other commercial knowledge sources had their beginnings in Lisp versions and then converted to C, ALLY^{TM} plant monitoring and diagnostic system has been exclusively written in C and C++ (an object oriented programming language) with performance as a primary objective. Only standard UNIX system calls, sockets for inter-processor communication, shared memory for intra-processor communications, and signals for interrupts are used. ALLY^{TM} is currently implemented on both Sun Sparc and Hewlett Packard 9000 UNIX platforms. Requirements are: 32 Mb of RAM, 100 Mb swap space, and 600 Mb hard disk. As a benchmark, the HP version of the rule based knowledge source has been measured to fire over 8000 simple rules per second.

JET-X is a PC-based expert system, developed by GE, for support of the TF34-GE-100 engine on the A-10 aircraft. JET-X was designed for use with the existing TF34 Ground Station GEMS IV. GEMS provides for the display of trend and event data acquired by the on board ***Turbine Engine Monitoring System*** (*TEMS*). Both GEMS and TEMS are capable of generating alarms that may indicate engine problems. Maintenance personnel are expected to respond to these alarms by diagnosing potential problems and performing appropriate maintenance actions. JET-X's principal role is to capture the diagnostic procedures in an interactive environment that does not limit the size or complexity of the decision trees used (Doel, 1990).

HELIX (***Helicopter Integrated Expert***) is a helicopter real-time diagnostic expert system using a causal model of the helicopter's engines, transmission, flight control and rotors. At the heart of the HELIX program is a ***Qualitative Reasoning System*** (*QRS*). The QRS is a general mechanism to support the creation of hierarchical device models and reasoning about device behavior using qualitative physics. The HELIX qualitative model is represented as a set of constraints that define the normal behavior of the engines, transmission, flight control, and rotors of the helicopter. Aircraft health is assessed by determining whether observations (sensor readings and pilot control inputs) are consistent with the constraints of the model. If an inconsistency is detected, a process of systematic constraint suspension is used to test various failure hypotheses.

Critical to the efficient operation of the HELIX program is the hierarchical model representation, which enables reasoning at various levels of abstraction. Using a top-down approach, the diagnostic process exploits the hierarchy by beginning fault isolation with the most reduced form of the model. To refine the diagnosis, a branch of the hierarchy may be expanded until a component-level diagnosis is made. The hierarchy also greatly reduces the complexity of multiple failure diagnosis. Rather than considering combinations of failures in all leaf components, the diagnosis can be restricted to combinations of branches in the hierarchy.

HELIX has been successfully tested on a variety of simulated failures. By representing only the normal behavior of the helicopter and testing hypotheses by constraint suspension, HELIX has been able to diagnose single or multiple failures without prior knowledge of failure modes. The approach represents a promising technique for automating the qualitative reasoning required to diagnose novel failures and may form the basis for extensive automation both in airborne and ground-based diagnostic systems (Hamilton, 1988).

ENGEXP is an integrated environment in PC- TurboPascal for the development and application of expert systems to the quick fault diagnosis and repair assistance of equipment and engine. The main components of ENGEXP are the user interface, the expert system core, the feedback and knowledge acquisition tool, and a help facility for the user. As usual, the expert system core contains the inference engine, the knowledge base(s) and the explanation subsystem. A particular feature of the inference engine is that it can resolve situations with multiple causes. The knowledge is organized into three layers as is done by the human experts. The qualitative analysis of the results obtained in a large sample of applications showed a remarkable performance, while the quantitative analysis showed a success rate over 95%. The response time is very good (always ≤ 2 sec., in many cases ≤ 0.5 sec.) despite the overhead caused by the user-friendly user interface (Tzafestas and Konstantinidis, 1992).

FAKS is an expert system for diagnosis and recommendations reports produced on board the ship for the Wartsila Diesel VASA series of ship engines. At the first step in the diagnosing process FAKS creates two mathematical models, one of the "Ideal Engine" based on laboratory data and one of the "Real Engine" based on data from the Diesel engine monitoring system. Every 15 minutes a comparison between these two models is made and the result is "flashed out" through the entire knowledge base. Diagnosis is the result of the automatic continuous evaluation of the engine and contributes to safe, continuous and economic operation. Good possibilities are also given to improve the safety at sea (Ahlqvist, 1990).

SCAR and *SCAR-2* are rule-based systems for assisting electricians in diagnosing faults in a shuttle car quickly. A shuttle car is a vehicle used to transport coal from the working face, where mining takes place, to the primary haulage system, such as a conveyor belt. The shuttle car is a key element in the mining cycle, since no coal can be mined if a car is not available. *Insight 2+*, a microcomputer-based expert system development tool by Level Five Research Inc. is used created the system. The program requires the user to specify the initial symptoms of the failed machine, and the most probable cause of failure is traced through the knowledge base, with the software requesting additional information, such as voltage or resistance measurements, as needed. A causal-reasoning approach was used to develop the production rules. Generalized systematic procedures for creating and organizing the knowledge base, the incorporation of on-screen presentations of the shuttle car circuit schematic, the development of reference and tutorial programs and a microcomputer implementation that resists moisture, oil and dust are some features of the more advanced SCAR-2 system (Novak *et al.*, 1989).

HEPHAESTOS is an interactive expert system for quick fault diagnosis of electric machines. Electric machines play a very important part in furnishing power for all types of domestic and industrial applications. Any fault on a electric machine installed in a line production process, results off-line repairing which disturbs the line production process. HEPHAESTOS is a rule-based expert system with efficient and quick reasoning, containing as much knowledge as possible concerning the fault occurrence, repairing and maintenance of electric machines. The knowledge required to build this expert system has been acquired from expert engineers working for years at workshops of the Public Power Corporation of Greece, construction companies, many instruction manuals, maintenance handbooks and trouble charts. One year real life tests of the system have been performed. The knowledge base can be expanded according to the comments of the users (Protopapas *et al.*, 1990).

IDM (***Integrated Diagnostic Model***) is a hybrid expert system for real-time fault diagnosis and repair assistance of mechanical and electrical devices which integrates shallow and deep knowledge. The shallow knowledge is stored in the experiential knowledge base and the deep knowledge is stored in the physical knowledge base. The physical knowledge base contains functional models of subsystems. Each of these two types of knowledge is structured and represented in a way that is natural for that type of knowledge. Each of the two knowledge bases has its own inference engine. Thus, each is an expert system in its own right. These two expert systems are then integrated into a single expert system via an executor that can draw on either expert system in diagnosing a particular problem. The executor contains its own knowledge base that maintains a global representation of what is known and unknown so far about the device under diagnosis. This knowledge base also provides a means of "translating" the knowledge between the experiential and fundamental knowledge bases. The result is an expert system that can solve problems even when no experiential knowledge exists to handle the situation, resulting in a more graceful degradation of capabilities at the periphery of its knowledge. It can also provide different levels of explanation. Two prototype systems have been implemented, the first for fault diagnosis in a simple thermostat-controlled gas heating system, the second for fault diagnosis in the electrical system of an automobile. Recent work on the IDM involves implementation to handle truly analog devices (Fink and Lusth, 1987).

Finally, *DESPLATE* is an expert system designed by Ng *et al.*, (1990), to diagnose abnormal plan view shapes of steel plates in a plate mill. A brief overview of the production process in the plate mill and the problems that motivated the development of DESPLATE are provided. The goal of DESPLATE is to locate the possible causes such as electrical failures, mechanical breakdowns or wear, operational errors, and pre-rolling conditions for a particular abnormal shape, and to suggest appropriate remedies. Some of the issues arisen in developing DESPLATE are addressed, including the use of graphics for user interface, forward and backward chaining techniques, and knowledge acquisition methodologies from multiple experts from different disciplinary backgrounds.

4.3.4. Automatic expert fault diagnosis for machine tools, robots and CIM systems

Computer Integrated Manufacturing (*CIM*) is an up-to-date research and engineering application field involving many different fields of technology: manufacturing management, industrial engineering, computer aided design-engineering-manufacturing, material handling, control and automation.

Although control methodologies are being practiced for some time and experience has been gained, this is not the case for the monitoring requirements of CIM projects. In controlling a production system workstation, it is required to integrate the monitoring functions of all components and of the global controller (the components being the machine tools, the robots, local transport system and tool management, etc.). At the process level, part scheduling and dispatching are also to be monitored.

Monitoring was, in the past, limited to reading sensors (monitoring sensors) for detection of abnormal conditions often followed by invoking an emergency action such as machine shutdown. Advanced techniques for monitoring are required to cope with actual system complexity (Computers in Industry, 1986). Some requirements in monitoring such as diagnosis and decision can be built using AI techniques. The integration of such techniques is not a new concept in complex automation projects, but it gives rise to a deep discussion on the use of such tools for specific problems, also relating some equivalent available tools in a complementary way. One of the available tools is Petri nets, a commonly used tool for FMS control system specification and implementation. An approach has been developed by Sahraoui *et al.*, (1987), to design a production system workstation monitoring system. The approach is best illustrated by fig. 4.20.

Fuzzy classification is also applicable in on-line machine tool and wear fault diagnosis (Computers in Industry, 1986). Classification is a subdivision of a set of elements or objects specified by features into a set of different subsets each with special importance. In the case of complex systems, one should use all possible information about the system. This is primary information from measurements and expert consultation and secondary information that is a result of signal processing (FFT, regression analysis, statistical aids, parameter estimation and so on). All information together is called *features*.

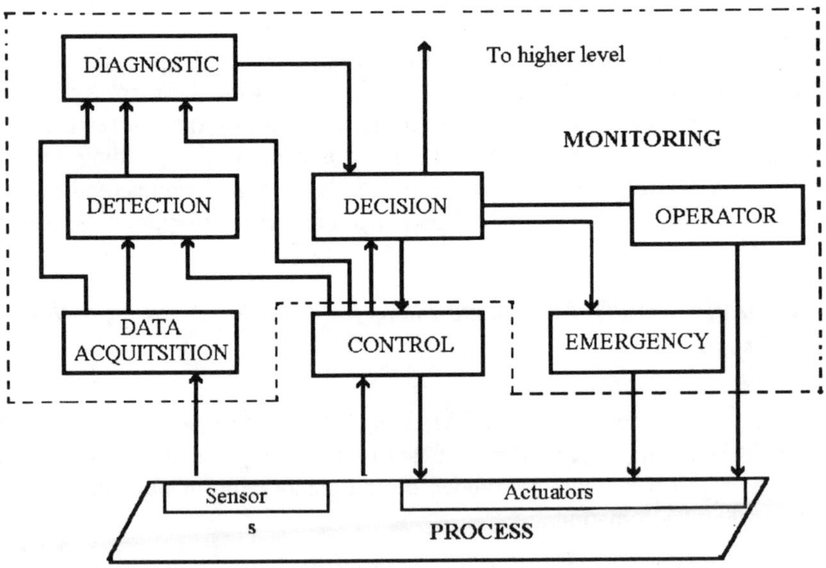

Figure 4.20 A production system workstation monitoring system.

The capabilities of classification are *fundamental ordering* and *structure building*. In the structure building step one gets a cluster configuration. Clusters found by cluster methods are of a natural type. Classical non-fuzzy clustering methods have the disadvantage that the classifier yields only crisp membership values (0 or 1) and also a well defined result for an object with a great distance from every class. This is very risky because there could be an unknown class. The fuzzy concept is then suitable for the description of overlapping classes and is able to refuse objects. This is possible because the membership of every class in such a case is very small. The essence of the fuzzy view is that the membership value of a set element cannot only be 0 or 1 but lies between 0 and 1 (in diagnosis this is the grade of membership to the normal of a special failure class). In Computers in Industry, (1986), a wear fault diagnosis system of conveyor belt rolls based on fuzzy classification is described.

Automatic monitoring is not limited to cell and plant monitoring as explained in the preceding section. In fact there are already some embedded monitoring functions in actual machine tools such as:

- Drilling machines with detection of tool breakage at the end of cycle.
- Turning machines with torque measurement.
- Lathe machines using consumed electric power measurement in order to estimate tool wear.

Concerning the tool wear and prediction of breakage with pattern recognition techniques, Neumann, (1990), has already published promising results.

In order to maintain continued operation, it is essential to detect a tool breakage because such an event can occur at any time and can cause damage to the tool and the part. In certain operations such as milling, small breakages cannot be recognized by monitoring an increase of the torque. A common event is the breakage of the cutting edge and this also can hardly be detected since the average torque in one turn does not change.

An on-line breakage detection algorithm which can be exploited from the fault diagnosis ES can be the following (Computers in Industry, 1986):

The cutting torque discrete signal $x(k)$ where k=0, 1, ... is the integer sample interval, is considered. It is assumed that $x(k)$ can be expressed by a linear combination of the preceding sampled cutting torque signals as,

$$x(k) = a_0 + a_1 x(k\text{-}1) + \ldots + a_n x(k\text{-}n) + w(k)$$

where $w(k)$ is gaussian white noise whose mean value and variance are zero and σ^2 respectively. This expression is an AR (autoregresive) model of order n. The coefficient vector,

$$\boldsymbol{\varphi}(k) = [a_0 \;\; a_1 \;\; \ldots \;\; a_n]^{\mathrm{T}}$$

of the above AR model can be estimated in real time from the preceding observations by the recursive least squares estimation algorithms extensively described in Chapter 3.

The predicted value at the next observation $x(k)$ is obtained by the previous observations $\boldsymbol{f}^{\mathrm{T}}(k) = [\, x(k\text{-}1) \quad x(k\text{-}2) \quad \ldots \quad x(k\text{-}n) \,]$ and the estimated coefficient vector $\hat{\boldsymbol{\varphi}}(k)$ at the (k-1)*th* instance of time as: $\hat{x}(k) = \boldsymbol{f}^{\mathrm{T}}(k)\hat{\boldsymbol{\varphi}}(k)$.

The observed value $x(k)$ is compared with the predicted, and the residual $e(k) = x(k) - \hat{x}(k)$ is calculated. When tool breakage occurs, the cutting state changes suddenly. This leads to an unexpected change of the cutting torque and thus in a sudden increase of the residual. In order to judge the breakage of the tool, the "normalized residual" is used, which is given by the following equation:

$$d(k) = |e(k)| \{\{e^2(k-m) + e^2(k-m+1) + \ldots + e^2(k-1)] / m\}^{-1/2}\}^{-1}$$

When plotting the normalized residual against the cutting time, tool breakage is predicted by a peak. The *Generalized likelihood ratio* (*GLR*) techniques, described in Chapter 2 and applied also in Chapter 3 for automatic detection of abrupt signal or parameter changes, can be used here.

A second type of detection which can be designed, is the signal processing of the acoustic signal (see Chapters 1 and 6). It has been shown that it is relatively easy to analyze the cross-correlation existing between the frequency of the signal due to the parts turning and the cutting force measurement. The conclusions of such analysis can be a basis for implementing an adaptive algorithm to improve the quality of machining, since the forming of parts turnings is an important parameter (Sahraoui, 1987, Neumann, 1990).

Tool wear can also be detected through power consumption using the empirical formula,

$$kWb = T \cdot P$$

where kWb is the power required, T is the material removal rate (in $10^{-6}\ m^3$) and P is the amount of removed material per unit time.

The monitoring of sequential steps can be ensured by using a Petri net-based specification. Three modules $M1$, $M2$ and $M3$ can be envisaged: $M1$ to monitor the initialization of the machine, $M2$ for monitoring during operations and $M3$ for exception handling.

Faults in the machine tool motor (load, electric components) can be detected via pattern recognition and spectrum of current (see also Section 1.3.2). The quotient of amplitudes at the odd harmonics of 50 Hz in the current spectrum shows the form of the magnetic field. Changes in load or geometric changes in the machine influence the waveform of the magnetic field. Furthermore, the current amplitudes give some information about the momentary load and position in the working cycle. Faults in other mechanical components (gear box, belt drive, excenter) can be detected by evaluation of rotation speed and parameter estimation (see Chapter 3).

A rule-based system can be designed to detect on-line tool wear and machine malfunction from such information. The action on the machine tool is to correct the tool in advance and to invoke an emergency shutdown. These issues and implementation aspects are fully discussed by Arreguy *et al.*, (1990), and Monostori *et al.* (1990).

Freyermuth, (1991), introduced a computer assisted incipient fault diagnosis system for industrial robots, with the objective to detect and diagnose faults in the mechanical part of the devices at a relatively early stage to prevent subsequent damages. For this purpose a suitable combination of analytical and heuristic tools was developed. The analytical symptom generation procedure comprises of the detailed mathematical modeling of the robot's different axes. The parameters of the models directly represent characteristic physical quantities (process coefficients) being identified by specific continuous-time parameter estimation algorithms. The estimated quantities then undergo a statistical evaluation. Deviation of coefficients from nominal values are considered as symptoms (see also Chapter 3). The subsequent heuristic evaluation processes these symptoms based on specific fault-symptom-trees and knowledge of fault statistics and process history using a specific inference mechanism (see also Appendix 4.A). The developed diagnosis system can be realized on a PC with a process interface.

Pouliezos and Stavrakakis, (1989), proposed a fault detection mechanism for a robot/controller combination to achieve optimum robot performance at all times. A detection mechanism based on logarithmic likelihood ratios combined with parameter estimation through RLS with forgetting factor, making it essentially a moving window estimator, has been shown to work quite satisfactorily (see Chapters 2 and 3). Fault symptom trees can then be elaborated to built an ES fault diagnosis tool as previously (see Appendix 4.A).

An application example concerning the finding of cause(s) for an observed symptom in a CIM setting, using hybrid expert diagnosis procedures (see Appendix 4.A) is described by Lee, (1990). As shown in fig. 4.21, this CIM system consists of 3 workstations (i.e. machines), 1 main conveyor, 3 subconveyors, 2 material handling robots, 1 programmable controller, and 1 computer terminal. The three machines are a *CNC* (*Computer Numerical Control*) machine, a GCA robot and a General Electric P50 robot. The two material handling robots are a Unimation PUMA 761 and a General Electric MH33.

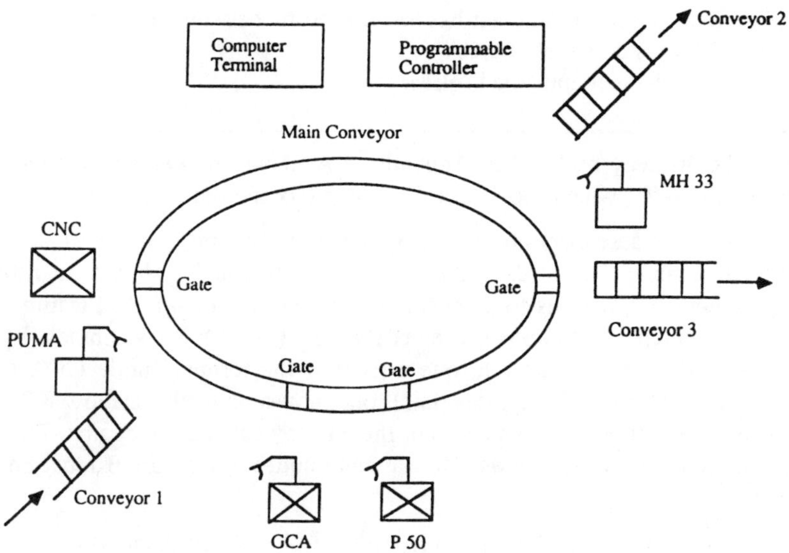

Figure 4.21 CIM system layout.

As the raw material comes in on Conveyor 1 (in which there is no pallet), the PUMA picks it up and places it on the CNC. The CNC processes the raw material into the specific pattern by performing the turning operation. When the CNC finishes its operation, the PUMA picks it up and places it on a pallet on the main conveyor. The pallet arrives at the GCA on the main conveyor and is first identified as type x or y depending on its bar code to be read by the laser bar code reader located at the GCA. If the pallet is of type *x*, then the GCA does operation *A* and releases the pallet *x* onto the main conveyor. Similarly, if the pallet type is *y*, then operation *B* is performed. At the P50, operation *C* is performed if the pallet is of type *y*, and no operation if it is of type *x*. When the processed material (or part) arrives at the MH33, it is unloaded from the main conveyor to Conveyor 2 if it is of type *x*, and to Conveyor 3 if it is of type *y*. Finally empty pallets travel to the CNC (at which point the cycle begins) via the main conveyor.

For the purpose of the diagnosis model, the CIM system is modeled in the way shown in fig. 4.21. Since there are 9 terminal nodes in the functional hierarchy, there are 9 shallow KBs. One of these shallow KBs is shown in the lower half of fig. 4.22 for the GCA. Every node in this shallow KB is represented by a pseudo-rule, defined as a rule of which the consequent part is not explicitly specified. In this sense, this shallow KB is a pseudo-shallow KB. To be complete, one has to specify an antecedent and a consequent. For example,

> R_{111}: If GCA will not respond
> then select the rule that has lower entropy between R_{211} and R_{212}.
> R_{311}: If there is no pallet at GCA
> then the gate must be broken.

Referring to the strategy outlined in Appendix 4.A, it will be explained how it works through use of fig. 4.21, which shows both the deep KB and the shallow KB.

Suppose that one (or a sensor) observes a symptom indicating that a part is not being processed on a machine. This leads to the deep KB starting at F_{00} (however, if there is a good mapping heuristic, it leads to a specific node other than F_{00}). At Stratum 0, using the breadth-first search, q_{11} (=0.65) is greater than q_{12} (=0.35). So Machines (F_{11}) is the next destination. By the same search, one arrives at the terminal node GCA (F_{32}) via Special-purpose Machines (F_{22}). At this point, one moves on to the shallow KB attached to F_{32}. That is, one is at R_{111}, at which point the entropy calculation begins. To calculate the entropy at R_{211} (i.e. H_{211}), the data for all child nodes at Echelon 3 must be known. H_{211} is given by,

$$\begin{aligned} H_{211} &= \{w_{311}p_{311}\ln p_{311} + w_{312}p_{312}\ln p_{312} + w_{313}p_{313}\ln p_{313}\} \\ &= -\{(0.2)(0.2)\ln(0.2) + (0.2)(0.4)\ln(0.4) + (0.2)(0.4)\ln(0.4)\} \\ &= 0.2110 \end{aligned}$$

where,

$$\begin{aligned} w_{311} &= \text{test cost of } R_{311}/\text{maximum test cost of shallow KB} \\ &= \$10/\$50 \text{ (from the test cost of } R_{323}) \\ &= 0.2 \end{aligned}$$

and w_{312} and w_{313} are obtained similarly. In the same manner, H_{212} can be obtained (i.e. $H_{212} = 0.9130$).

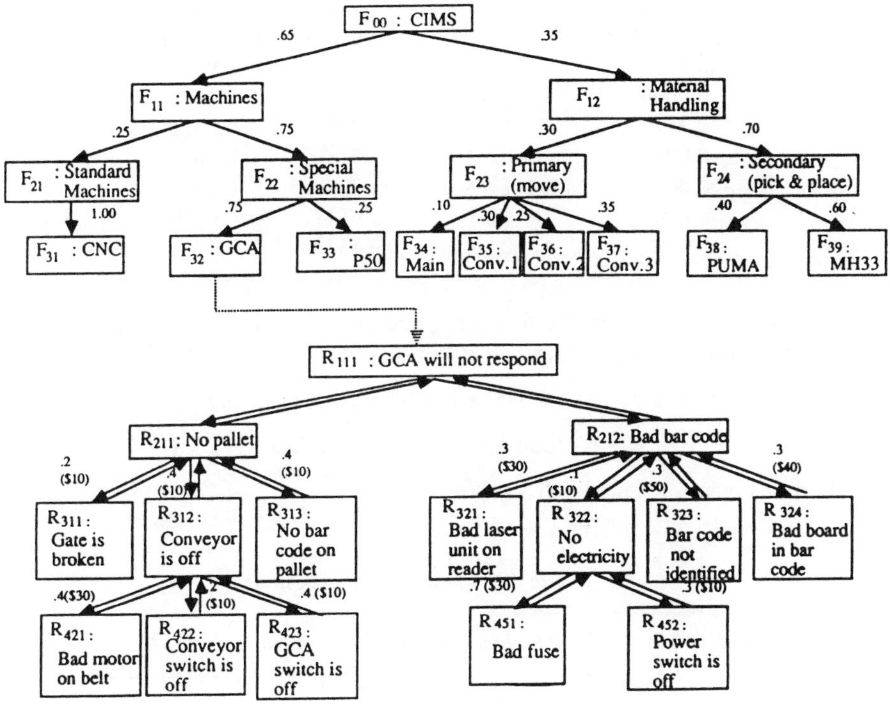

Figure 4.22 CIM system example diagnosis.

Thus the decision can be made at R_{111}. Since H_{211} is less than H_{212}, the search moves down to R_{211}. This procedure is repeated until the terminal node is reached. The results of entropy calculations (and cost/belief ratios if there is a tie in entropy values) are shown in Table 4.4 where (a) is for the R_{211} group and (b) is for the R_{212} group. Note that the entropy for a terminal node is zero, because the terminal node has no children.

From Table 4.4, the first terminal node to be tested is R_{313}. If it turns out that there is no bar code on the pallet after the test, then this pallet is faulty. The diagnostic session stops at this point. However, if the problem still exists although this pallet is already taken care of, the observed symptom may have multiple faults. In this case, the diagnostic session resumes from R_{311} which has the next lowest entropy (actual cost/belief ratio) at the same echelon. If the test result for R_{313} indicates no faults, the diagnostic session also resumes from R_{311}. Should the entire shallow KB contain no faults, the diagnostic procedure has to go back to the terminal node of the deep KB (i.e. F_{32}) and update q_{32}.

Table 4.4 Calculation results for all nodes.

(a) R_{211} group			(b) R_{212} group		
Node (*R*)	*Entropy* (*H*)	*Ratio* (*r*)	*Node* (*R*)	*Entropy* (*H*)	*Ratio* (*r*)
211	0.2110		212	0.9130	
311	0.0000	1.0	321	0.0000	2.0
312	0.3576		322	0.2220	
313	0.0000	0.5	323	0.0000	3.3
421	0.0000	1.5	324	0.0000	2.7
422	0.0000	1.0	451	0.0000	0.9
423	0.0000	0.5	452	0.0000	0.7

By the updating heuristic in the foregoing section, one will find the parent node of F_{32} (where $b = 3$ and $c = 2$) as shown in Table 4.5 (see Appendix 4.A).

The second subscript $GN(2)$ is 2. Therefore the parent node is F_{22}. Also,

$$\text{EN}(3) = c - (\text{cumulative } N \text{ before } GN(2)) = 2 - 1 = 1$$

Based on this result, it follows that $q'_{32} = 0$.

Table 4.5 Finding the parent node.

Node (*F*)	*No. of elements* (*N*)	*Cumulative N*
21	1	$1 < c = 2$
22	2	$3 > c = 2$
23	4	7
24	2	9

(since $N_{b\text{-}1,\, GN(b\text{-}1)} = N_{22} = 2$ *and* $t \neq \text{EN}(b) = \text{EN}(3) = 1$, $t = 2$). Also considering

$$k \neq c = 2,\ k = [c - \text{EN}(b)] + t = (2 - 1) + 2 = 3$$

$$q'_{33} = q_{33}/(1 - q_{32}) = 0.25/(1 - 0.75) = 1.0$$

At Stratum 2, the updated responsibility probability for the parent node F_{22} becomes,

$$q'_{22} = q_{22} - q_{22} \times (q_{32} - q'_{32})$$
$$= 0.75 - 0.75 \times (0.75 - 0) = 0.1875.$$

By applying the same methods, the parent node of F_{22} is F_{11}. The updating for F_{21} is,

$$q'_{21} = q_{21} + [q_{21} / (1\text{-}q_{22})\} \times q_{22} \times (q_{32}\text{-}q'_{32})$$
$$= 0.25 + \{0.25 / (1\text{-}0.75)\} \times (0.75\text{-}0) = 0.8125$$

At Stratum 1, similarly one has,

$$q'_{11} = q_{11} - q_{11} \times (q_{22} - q'_{22})$$
$$= 0.65 - 0.65 \times (0.75 - 0.1875)$$
$$= 0.2844$$

$$q'_{12} = q_{12} + \{q_{12}/(1 - q_{11})\} \times q_{11} \times (q_{22} - q'_{22})$$
$$= 0.35 + \{0.35/(1 - 0.65)\} \times 0.65 \times (0.75 - 0.1875)$$
$$= 0.7156$$

The updating procedure ends here because the current stratum is Stratum 1. This update is summarized in fig. 4.23 where the number in the parenthesis shows the previous responsibility probability.

Hence the diagnostic session continues from F_{12} on since, after updating, q'_{12} far greater than q'_{11} (note that before updating, $q_{12}<q_{11}$).

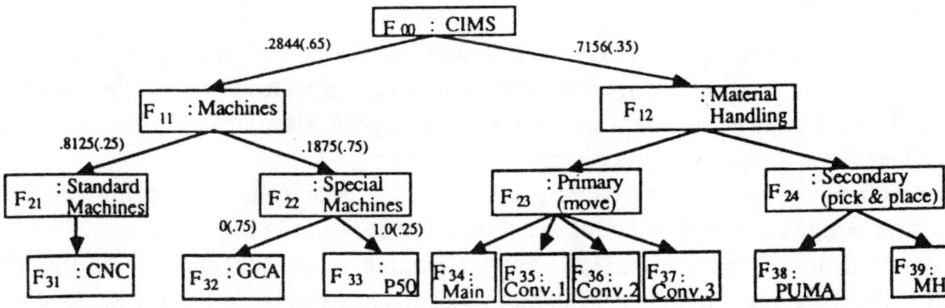

Figure 4.23 Updated probabilities in deep KB for the CIM system diagnosis.

The model is implemented by Lee *et al.*, (1990), on IntelliCorp's *Knowledge Engineering Environment* (*KEE*). The implementation is realized by specifying each of KBs including one deep KB and its corresponding shallow KBs.

4.4 Conclusions

Methodologies for modeling failure knowledge about complex industrial systems and for utilizing these models for real-time reasoning about, and diagnosing, failures have been presented in this chapter. Their key component is the modular framework used for expert knowledge-based failure analysis. This modularity permits easy implementations of fault diagnosis programs, based on this framework, for different industrial systems, since the only domain-dependent components that must be replaced are the specific failure models.

The necessary theoretical background regarding the more recent methods, is given in the Appendices of the Chapter for the interested reader. Many application examples from a wide industrial practice have been cited, with sufficient details giving the reader the ability to develop (his)her own applications.

In parallel with the development of automatic expert diagnosis and knowledge-based *decision support systems* (*DSS*), concern has been growing regarding the potential for catastrophic errors created by these systems and, what is worse, the potential for catastrophes whose causes cannot be established. Concern for the risks associated with expert systems is now so strong, that it has spilled over into public discussion, such as a British television program in which American and British practitioners and critics argued the dangers of using expert systems in medical, industrial, military and other applications. Also, there have been calls for restrictions on the deployment of unsupervised or autonomous system in safety critical situations (The Boden Report, 1989).

Additional issues for the designers of ES, whether these are statistical, knowledge-based or hybrid systems, fall into two main categories: performance and responsibility.

Performance issues.

1. The decision procedure used by the ES must perform well (make or recommend good decisions), even in the face of degraded data. Robustness entails being able to assess the reliability of information sources and to seek alternatives where necessary, as well as merely cope with uncertainty.
2. Few practical situations involve just one class of decision (e.g. diagnosis); ES theory must surely address the problem of deciding what decision is required.
3. Many practical automata must face rapidly changing situations, not only in the information available but also in the problem that needs to be solved. Expert systems must incorporate capabilities for altering their decision goals as circumstances develop.

The central requirements for meeting these demands include the ability of an ES to be rationally flexible, i.e.:

1. Recognize that a decision is needed.
2. Identify the kind of decision it is.
3. Establish a strategy for making it.
4. Formulate the decision options.
5. Revise any or all of the above in the light of new information.

An expert decision system should be capable of autonomously invoking and scheduling these processes as circumstances demand. Classical ES theory offers little guidance for developing the necessary techniques.

Responsibility issues.

One must also achieve a high level of communication between human supervisors and/or auditors wishing to examine and potentially to intervene in any aspect of the automatic expert decision process.

1. If the automatic expert decisions lead to errors, it must be possible to establish the reasons for those errors.
2. Where it is practical and appropriate, provision should be made for a skilled supervisor to exercise overriding control.

In general an automatic expert decision maker needs to be able to reflect on the decision procedure, to be able to examine the:

3. Decision options (what choices exist).
4. Data (the information available that is potentially relevant to a choice).
5. Assumptions (about viability of options, reliability of data etc.).
6. Conclusions (in light of data and knowledge of the setting).

Reflective capabilities should extend to the decision process itself, including:

7. The goals of the decision (what is the decision supposed to achieve).
8. The methods being pursued (what justifies the current strategy).
9. Characteristics of specific procedures (applicability conditions, reliability, completeness etc.).

A theory of expert rational decision making must acknowledge these requirements. Classical expert decision procedures may be optimal in the sense that they promise to maximize the expected benefits to the decision maker, but they must be viewed as unsatisfactory in other ways. It seems that connectionist models are drawing increasing interest as useful tools for expert decision making which can accomplish the above requirements. These methods allow an automatic clever combination of different type knowledge bases for a specific application, thereby reducing the effort required for an equivalent expert system development. The connectionist models also resemble the type of fine-grained parallelism of conventional symbol processing. The theory and practice for their application to robust real-time fault diagnosis are given in Chapter 5.

Expert systems can be written in conventional languages like C, FORTRAN or Pascal, which have the advantage of being widely known and of being the language in which other programs are written, so that no integration problems occur. However, they are poorly adapted to the expression and the handling of knowledge expressed by words. Languages specific to artificial intelligence, like Lisp or Prolog, do not have this disadvantage, but they are not yet widely used, and their standards are not well established. All these languages have the common drawback that they lack programming aids. In contrast, expert system building tools, especially shells, have the advantage that the knowledge representation, the inference engine, and many other facilities, such as reports, windows, menus and forms are largely preprogrammed, facilitating the accomplishment of the above requirements. In the early days, these tools were written in special languages, but nowadays conventional languages are used more and more often.

References

Adelman L. (1989). Measurement issues in knowledge Engineering. *IEEE Transactions in Systems Man and Cybernetics*, **19**, 3, p. 483.

Ahlqvist I. (1990). An expert system based diagnosis system for ship machinery. *Proceedings, 5th. International Conference on Marine Technology - Athens 90*, May 1990, Athens, Greece.

Al-Jaar R. V. and A.A. Desrochers (1990)./ Petri nets in automation and manufacturing. In G. Saridis (Ed.), *Advances in Automation and Robotics, 2*, JAI Press. Inc., CN, p. 153.

Arreguy D. *et al.* (1990). Monitoring machine tools. IEEE Ind. Electr. Conf., November 1990.

Bandekar V. R. (1989). Causal models for diagnostic reasoning. *Artificial Intelligence in Engineering*, **4**, 2, p. 79.

Boden M. (1989). Benefits and risks of knowledge-based systems. *Report of Council for Science and Society*, Oxford University Press,Oxford, U.K

Boose J. H. (1986). Expertise transfer for Expert Systems Design. Elsevier, New York.

Boose, J. H., Bradshaw, J. M., (1988) "Expertise transfer and complex problems: Using AQUINAS as a knowledge acquisition workbench for knowledge-based systems", Int. J. Man-Machine Studies, 2, p. 39.

Boose J. H. and B.R. Gaines (1988). Knowledge acquisition tools for expert systems. In B. R. Gaines and J. H. Boose (Eds), *Knowledge Acquisition Tools for Expert Systems, Vol. 2*, Academic Press, London.

Brailsford J. *et al.* (1990). FAUST: a 132 kV fault diagnosis system. *The Electricity Council Research Center, ECRC/M2460*, Capenhurst, Chester, CH1 6ES, U.K.

Brule J. F. and A. Blount (1989). Knowledge acquisition. Mc Graw Hill, N.Y.

Carriero N. and D. Gelernter (1989). How to write parallel programs: a guide to the perplexed. *ACM Computing Surveys*, **21**, 3, p. 323.

Chen J. G. and K. Ishiko (1990). Automobile air-conditioner compressor troubleshooting - an expert system approach. *Computers in Industry*, **13**, p. 337.

Computers in Industry (1986). Special Issue on Computer Aided Testing and Diagnosis, 7, p. 3-82.

Contini S. (1987). SALP-PC. A fault-tree analysis package on PC. *Technical Note I. 87.165 - PER 1427/87*, JRC-Ispra Establishment, Commission of European Communities.

Doel D. L. (1990). The role for Expert Systems in Commercial gas turbine engine monitoring. *Proceedings, Gas turbine and Aeroengine Congress*, June 11-14, 1990. Brussels, Belgium.

Dolins S. B. and J.D. Reese (1992). A curve interpretation and diagnostic technique for industrial Processes. *IEEE Transactions on Industry Applications*, **28**, 1, p. 261.

Dounias G., Vassilakis P. and V. Moustakis (1993). A model of fault diagnosis using inductive learning techniques. *Proceedings, IEEE "Athens Power Tech '93"*, Sept. 5-8, Athens, Greece.

Fink K. P. and J. Lusth (1987). Expert Systems and diagnostic expertise in the mechanical and electrical domains. *IEEE Transactions on Systems, Man and Cybernetics*, **17**, 3, p. 340.

Forbus K. D. (1987). Interpreting observations of physical systems.*IEEE Transactions on Systems, Man and Cybernetics*, **17**, 3, p. 350.

Forsythe D. E. and B.G. Buchanan (1989). Knowledge Acquisition for Expert Systems: some pitfalls and suggestions. *IEEE Transactions on Systems, Man and Cybernetics*, **19**, 3, p. 435.

Frank P. M. (1990). Fault diagnosis in dynamic systems using analytical and knowledge-based redundancy - a survey and some new results. *Automatica*, **26**, 3, p. 459.

Freiling M. J. *et al.* (1986). The ontological structure of a troubleshooting system for electronic instruments. In D. Sriram and R. Adey (Eds.), *Proceedings, 1st International Conference on Applications of AI in Engineering Problems*, Springer-Verlag, Berlin, p. 609.

Freyermuth B. (1991). Knowledge based incipient fault diagnosis of industrial robots. *Proceedings, IFAC/IMACS Symposium "SAFEPROCESS '91"*, September 10-13, Baden-Baden, Germany.

Gertler J. J. and K.C. Anderson (1992). An Evidential Reasoning Extension to Quantitative model-based failure diagnosis.*IEEE Transactions on Systems, Man and Cybernetics*, **22**, 2, p. 275.

Gray N. A. B. (1990). Capturing Knowledge through top-down induction of decision trees. *IEEE Expert*, June 1990, p. 41.

Grogono P. *et al.* (1991). Expert System evaluation techniques: a selected bibliography. *Expert Systems*, **8**, 4, p. 227.

Gruber T. R. and P.R. Cohen (1987). Design for acquisition: Principles of knowledge system design to facilitate knowledge acquisition. *International Journal of Man-Machine Studies*, **26**, p. 143.

Gupta A., Forgy C. and A. Newell (1989). High-speed implementations of rule-based systems. *ACM Transactions on Computer Systems*, **7**, 2, p. 119.

Hamilton T. P. (1988). HELIX: a helicopter diagnostic system based on qualitative physics. *Artificial Intelligence in Engineering*, **3**, 3, p. 141.

Hickman F. R. *et al.* (1989). Analysis for knowledge-based systems. A practical guide to the KADS methodology. Ellis Horwood Ltd., West Sussex, England.

Hudlicka E. and V. Lesser (1987). Modelling and Diagnosing problem-solving system behavior. *IEEE Transactions on Systems, Man and Cybernetics*, **17**, 3, p. 407.

Isermann R. and B. Freyermuth (1991). Process fault diagnosis based on process model knowledge - Part I and Part II. *ASME Journal of Dynamic Systems, Measurement and Control*, **113**, p. 621.

Johannsen G. and L. Alty (1991). Knowledge engineering for industrial expert systems. *Automatica*, **27**, 1, p. 97.

Johnson L. E. and N.E. Johnson (1987). Knowledge elicitation involving teachback interviewing. In A. Kidd (Ed.), *Knowledge Acquisition for Expert Systems: a practical Handbook*, Plenum Press, New York, p. 91.

Kaiser G. E. *et al.* (1988). Database support for knowledge-based Engineering Environments. *IEEE Expert*, Summer 1988, p. 18.

de Kleer J. and B.C. Williams (1987). Diagnosing multiple faults. *Artificial Intelligence*, **32**, p. 97.

de Kleer J. (1990). Using crude probability estimates to guide diagnosis. *Artificial Intelligence*, **45**, p. 381.

Klein G. A., Calderwood R. and D. Mac Gregor (1989). Critical decision method for eliciting knowledge. *IEEE Transactions on Systems, Man and Cybernetics*, **19**, p. 462.

Kuan K. K. and K. Warwick (1992). Real-time expert system for fault location on high-voltage underground distribution cables. *IEE Proceedings-C*, **139**, 3, p. 235.

Lee W. Y., Alexander S. M. and J.H. Graham (1990). A hybrid approach to a generic diagnosis model. *Proceedings, Fourth International Conference on Expert Systems in Production and Operations Management*, May 14-16, Hilton Head Island, SC, p. 264.

Luger G. F. and W.A. Stubblefield (1989). Artificial Intelligence and the design of Expert Systems. Benjamin/Cummings, N. Y.

MacDonald B. A. and I.H. Witten (1989). A framework for knowledge acquisition through techniques of concept learning. *IEEE Transactions on Systems, Man and Cybernetics*, **19**, p. 499.

Majstorovic V. D. (1990). Expert Systems for Diagnosis and Maintenance: the state of the art. *Computers in Industry*, **15**, p. 43.

Marsh C. A. (1988). The ISA expert system: a prototype system for failure diagnosis on the space station. *Proceedings, 1st International Conference on Industrial and Engineering Applications of AI and ES '88*, **Vol. I**, June 1-3, p. 60.

Maruyama Y. and R. Takahashi (1985). Application of fuzzy reasoning to failure diagnosis. *Journal of Atomic Energy Society of Japan*, **27**, p. 851.

Maâberg W. and H.J. Seifert (1991). Petri net based system for monitoring, diagnosis and therapy of failures in complex manufacturing systems. *Proceedings, IFAC Fault Detection, Supervision and Safety for Technical Processes SAFEPROCESS '91*, September 10-13, Baden-Baden, Germany 1991.

Milne R. (1987). Strategies for Diagnosis. *IEEE Transactions on Systems, Man and Cybernetics*, **17**, 3, p. 499.

Monostori L. *et al.* (1990). Concept of a knowledge based diagnostic system for machine tools and manufacturing cells. *Computers in Industry*, **15**, p. 95.

Mussi S. and R. Morpurgo (1990). Acquiring and representing strategic knowledge in the diagnosis domain. *Expert Systems*, **7**, 3, p. 157.

Narayanan N. H. and N. Viswanadham (1987). A methodology for knowledge acquisition and reasoning in failure analysis of systems. *IEEE Transactions on Systems, Man and Cybernetics*, **17**, 2, p. 274.

Nawab H., Lesser V. and E. Milios (1987). Diagnosis using the formal theory of a signal-processing system. *IEEE Transactions on Systems, Man and Cybernetics*, **17**, 3, p. 369.

Nebiacolombo G. *et al.* (1989). Methodi di formalizzazione della conoscenza basati sulle Reti di Petri per la realizzazione di Sistemi Esperti di diagnosi. *Proceedings, ANIPLA '89, 33rd National Annual Conference on Automation*, Nov. 21-23, 1989, Rome, Italy.

Neumann D., (1990). A knowledge based fault diagnosis system for the supervision of periodically and intermittent working machine tools. *Proceedings, IFAC 11th World Congres*, Tallinn, Estonia, 1990.

Ng T. S., Cung L. D. and J.F. Chicharo (1990). DESPLATE: an ES for abnormal shape diagnosis in the plate mill. *IEEE Transactions on Industry Applications*, **26**, 6, p. 1057.

Noore A., Cooley W. L. and R.S. Nutter (1992). Detecting and masking transient failures in computers used for coal mining operations. *IEEE Transactions on Industry Applications*, **28**, 1, p. 186.

Novak T. *et al.* (1989). Development of an Expert System for Diagnosing component-level failures in a shuttle car. *IEEE Transactions of Industry Applications*, **25**, 4, p. 691.

Obreja I. (1990). Diagnosis of Power Plant Faults using Qualitative models and heuristic rules. *1990 ACM 089791-372-8/90/007/0047*.

O 'Leary T. *et al.* (1990). Validating Expert Systems. *IEEE Expert*, June 1990, p. 51.

Pandelidis I. (1990). Generalized entropy and minimum system complexity. *IEEE Transactions on Systems, Man and Cybernetics*, **20**, 5, p. 1234.

Pappis C. P. and G.I. Adamopoulos (1992). A software routine to solve the generalized inverse problem of fuzzy systems. *Fuzzy Sets and Systems*, **47**, p. 319.

Passino K. M. and P.J. Antsaklis, (1988). Fault detection and identification in an intelligent restracturable controller. *Journal of Intelligent and Robotic Systems*, **1**, p. 145.

Peng Y.and J.A. Reggia (1987). A probabilistic causal model for diagnostic problem solving. Part II: Diagnostic Strategy. *IEEE Transactions on Systems, Man and Cybernetics*, **17**, 3, p. 395.

Pomeroy B. D., Spang H. A. and M.E. Dausch (1990). Event-based architecture for diagnosis in control advisory systems. *Artificial Intelligence in Engineering*, **5**, 4, p. 174.

Pouliezos A. and G.S. Stavrakakis (1987). Linear state estimation in the presence of sudden system changes - An Expert System. *Applied Modeling and Simulation of Technological Systems*, North-Holland, Amsterdam, p. 41.

Pouliezos A. and G.S. Stavrakakis (1989). Fast fault diagnosis for industrial processes applied to the reliable operation of robotic systems. *International Journal of Systems Science*, **20**, 7, p. 1233.

Prasad P. R. and J.F. Davis (1993). A framework for Knowledge-Based Diagnosis in Process Operations. In P. I. Antsaklis and K. M. Passino, "An introduction to intelligent and autonomus control", Kluwer Academic Press, Massachusetts.

Prock J. (1991). A new technique for fault detection using petri nets. *Automatica*, *27*, 2, p. 239.

Protopapas C. A. *et al.* (1990). An Expert System for fault repairing and maintenance of Electric Machines. *IEEE Transactions on Energy Conversion*, **5**, 1, p. 79.

Rasmussen J. (1984). Strategies for state identification and diagnosis in supervisory control tasks, and design of computer-based support systems. In W. B. Rouse (Ed.), *Advances in Man-Machine Systems Research*, *Vol. 1*, JAI Press, Greenwich, CN, pp. 139-193.

Rhodes P. C. and G.J. Karakoulas (1991). A probabilistic model-based method for diagnosis. *Artificial Intelligence in Engineering*, **6**, 2, p. 86.

Roth E. M. and D.D. Woods (1989). Cognitive task analysis: an approach to knowledge acquisition for intelligent system design". In G. Guida and C. Tasso (Eds), *Topics in Expert System Design*, North Holland, Amsterdam.

Rouse W. B., Hammer J. M. and C.M. Lewis (1989). On capturing human skills and knowledge: algorithmic approaches to model identification.*IEEE Transactions on Systems, Man and Cybernetics*, **19**, p. 558.

Sahraoui A. E. K. *et al.* (1987). Combining Petri nets and AI techniques for monitoring. *Proceedings, IEEE Conference on Robotics and Automation*, Raleigh, U.S.A., April 1987.

SIGART-Newsletter (April 1989). Knowledge Acquisition Special Issue, 108.

Soumelidis A. and A. Edelmayer (1991). Modeling of complex systems for control and fault diagnostics: a knowledge based approach. In. S. Tzafestas (Ed.), *Engineering Systems with Intelligence*, Kluwer Academic Publ., The Netherlands, p. 147.

Stavrakakis G. S. and E.N. Dialynas (1991). Efficient Computer-based scheme to improving the reliability performance of power substations. *International Journal of Systems Science*, **22**, 9, p. 1527.

Takahashi R. and Y. Maruyama (1987). Practical expression of exception to diagnosis. *Bull Research Laboratory for Nuclear Reactors, Tokyo Institute of Technology*, **12**, p. 50.

Tesch D. B. *et al.* (1990). A knowledge-based alarm processor for an energy management system. *IEEE Transactions on Power Systems*, **5**, 1, p. 268.

Torasso P. and L. Console (1981). Diagnostic problems solving: Combining Heuristic, approximate and causal reasoning. Van Nostrand Reinhold, N. Y.

Trave-Massuyes L., Missier A. and N. Piera (1990). Qualitative models for automatic control process supervision. *IFAC 11th World Congress*, Tallinn, Estonia, p. 211.

Tsoukalas L. H., Upadhyaya B. R., and N.T. Clapp, N. T., (1991). Hypertext-based integration for nuclear plant maintenance and operation. *Proceedings, AI91 Confernce*, September 15-18, 1991, Jackson Lake, Wyoming, U.S.A.

Tzafestas S. G. (1987). A look at the knowledge-based approach to system fault diagnosis and supervisory control. In S. Tzafestas *et al.* (Eds), *System fault diagnosis, reliability and related knowledge-based approaches*, D. Reidel Publishing Company.

Tzafestas S. G. (1989). Knowledge Engineering approach to system modeling, diagnosis, supervision and control. *Syst. Anal. Model. Simul*, **6**, 1, p. 3.

Tzafestas S. G. (1991). Second Generation Diagnostic Expert Systems: Requirements, Architectures and Prospects. In R. Isermann and B. Freyermuth (Eds.), *Fault detection supervision and safety for technical Processes*, IFAC Symposia Series, No 6, !992.

Tzafestas S. G. and N.I. Konstantinidis (1992). ENGEXP - an integrated environment for the development and application of expert systems in equipment and engine fault diagnosis and repair. *Advances in Engineering Software*, **14**, p. 3.

van Soest D. C., Bakker R. R. and N.J.L. Mars (1990). Model-based diagnosis of Large Systems. *The Second AAAI workshop on model based Reasoning*, Boston, Massachusetts, July 30, 1990.

Wehenkel L. *et al.* (1989). Inductive inference applied to on-line transient stability assessment of Electric power systems. *Automatica*, **25**, 3, p. 445.

Wiele H., Liermann G. and H. Raschke (1988). Monitoring, diagnosis and operator leading systems for flexible automated manufacturing. *Computers in Industry*, **10**, p. 61.

Yoon, W. C. and J.M. Hammer (1988). Deep-Reasoning fault diagnosis: an aid and a model. *IEEE Transactions on Systems, Man and Cybernetics*, **18**, 4, p. 659.

Appendix 4.A A generic hybrid reasoning expert diagnosis model

The diagnosis model developed here consists of a single ***Deep Knowledge Base*** (*DKB*) and a number of ***Shallow Knowledge Bases*** (*SKB*s) attached to each of the terminal nodes in the DKB. The DKB is constructed by viewing the whole system under consideration as a hierarchical system. The SKB is also organized hierarchically, based on the level of decision complexity. The reasoning process employed is a D-S type of hybrid reasoning as shown in fig. 4.24.

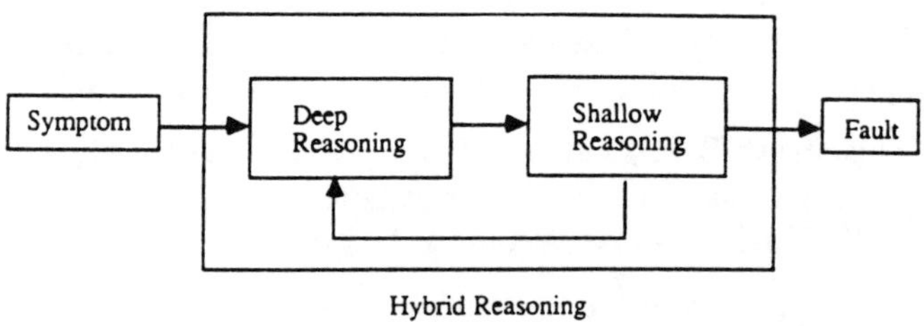

Figure 4.24 D-S (deep-shallow) type of expert hybrid reasoning.

Deep knowledge base. A system is represented by functional blocks within a deep KB in the form of tree (i.e. hierarchy) as shown in fig. 4.25. Even though the entire figure resembles a multiechelon system, each level is called a *stratum* and not an echelon because the figure represents levels of (functional) description or abstraction of the system. In the deep KB, a "functional block" is referred to as a "node". No test is associated with every node in the deep KB and, in turn, "no test" implies "no decision-making". For this reason, every node is connected downwards by a one-way arrow. Except for F_{00} which represents the given system as a whole, it is assumed that functional blocks are independent of each other within each stratum. Every terminal node has its own shallow KB.

With the exception of F_{00}, a functional block F_{ij} is defined as follows:

- i indicates the stratum number, i=1, 2, .., v
- j indicates the "overall" element number, j=1, 2, ..., n

where v and n are arbitrary integers. This definition of a functional block is shown in fig. 4.25.

Let f_{ij} denote a failure probability of a node F_{ij}. An f_{ij} value is obtained from a historical failure data set from a system under consideration. However, one can convert F_{ij} to q_{ij}, termed a responsibility probability, that is, a probability that F_{ij} will be responsible for causing the symptom. Since the f_{ij} value represents an absolute probability from the historical data, the summation of failure probabilities of all the children nodes of any node may not be unity. Thus, for convenience in probability updating (explained later in this chapter), this sum is forced to unity by normalization of f_{ij} into q_{ij}:

$$\sum_{j \in K} q_{ij} = 1 \quad \text{for} \quad i = 1,2,\ldots,v \quad \text{and} \quad j = 1,2,\ldots,n$$

where K is the set of children nodes which have the same parent node.

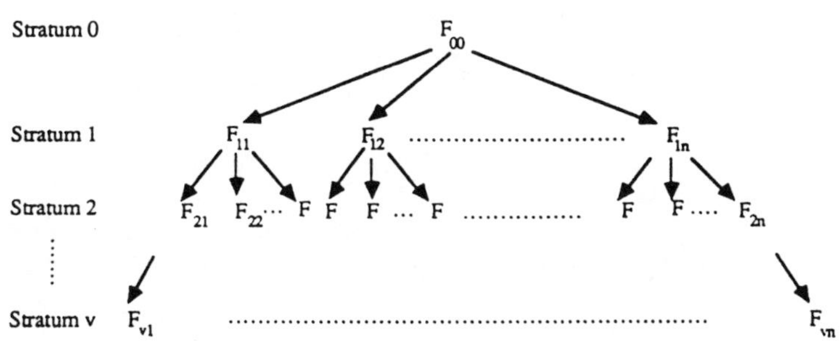

Figure 4.25 Functional hierarchy as deep knowledge base.

Shallow knowledge base. A shallow KB, attached to each of the terminal nodes in the deep KB, is organized in multiechelon form because every rule involves its own decision-making. This decides whether or not the rule is responsible for the observed symptom after testing the consequent of the rule. For this reason, every rule is connected by a two-way arrow. Each level is called an echelon, as shown in fig. 4.26. Unlike F_{ij}, a rule R_{ijk} is defined as follows:

- i indicates echelon i; $i = 1, 2, ..., e$.
- j indicates group j; $j = 1, 2, ..., s$.

where s is the total number of elements in echelon i-1 and k indicates element k; k=1, 2, ..., t; e, s, and t being arbitrary integers.

In the shallow KB, two attributes are associated with every rule R_{ijk} at and below echelon 3 (i.e. R_{ijk} where $i \geq 3$, and j and k are arbitrary). One is p_{ijk}, defined as a degree of belief, e.g. obtained from the options of experts, which acts as a probability that R_{ijk} is believed to be responsible for the observed symptom. Furthermore, similar to that of q_{ij} in the deep KB, p_{ijk} is designed to have the following property:

$$\sum_{k=1}^{t} p_{ijk} = 1$$

where $i \geq 3$ and j is arbitrary.

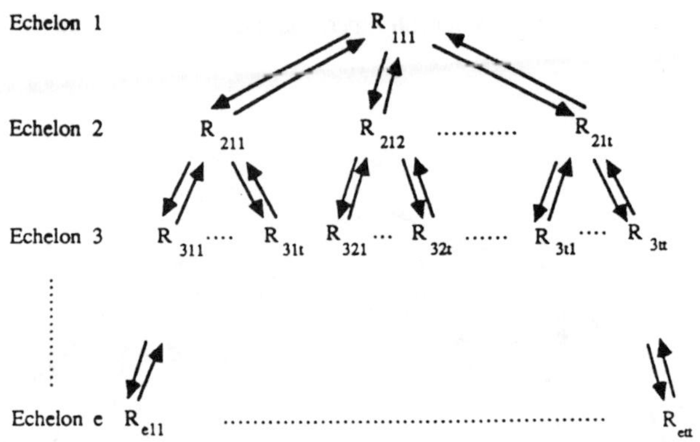

Figure 4.26 Rule hierarchy as shallow knowledge base.

In other words, at and below echelon 3, the sum of the degrees of belief of all the elements within any arbitrary jth group is unity. The other is c_{ijk}, defined as a test cost (in dollars), which accounts for the cost of testing the consequent part of R_{ijk}.

Diagnostic strategy. The diagnostic strategy is an extension of the strategy described by Lee *et al.* (1990). The strategy requires the failure probabilities of functional blocks for the deep KB, the degrees of belief, and the associated test costs of rules for the shallow KB. Given these requirements, this model identifies and isolates a fault by searching the deep KB first and then the shallow KB. Backtracking is allowed if needed. The outline of the diagnostic strategy is illustrated in fig. 4.27. Given a single symptom, the basic cycle of the strategy is to find out a single fault for this symptom. The basic cycle consists of 8 steps which are described below.

A single symptom means that a symptom is observed only once and which component(s) is (are) faulty is detrmines. Multiple faults for the symptom observed can be handled by testing the possible faulty components one by one, sequentially. That is, multiple faults are found by continuing the current diagnostic session even if one fault has already been found. Multiple symptoms can also be treated similarly. However, it will be much better if we can classify these multiple symptoms, based on functional levels. The reason is that the further we go down in the functional hierarchy, the more we reduce the search time required to find the fault.

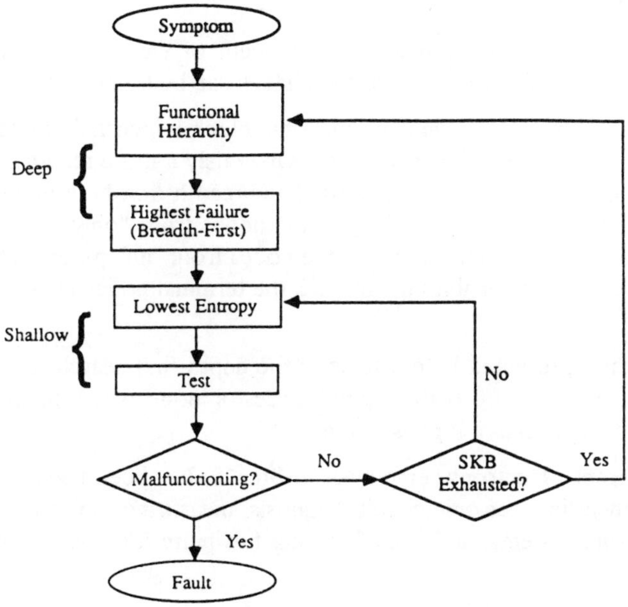

Figure 4.27 Schematic diagram for diagnostic strategy.

The basic cycle of 8 steps embedded in fig. 4.27 is as follows:

Step 1. Construct a functional hierarchy of the given system.
Step 2. When a symptom is observed, go to the deep KB and start at the root F_{00}.
Step 3. Conduct a breadth-first search using the responsibility probability based on the failure probability of each of the functional blocks at the immediate lower stratum. Select the most unreliable branch.
Step 4. Repeat Step 3 with the functional block selected until the terminal node is reached. Then go to the shallow KB attached to the functional block finally found.
Step 5. Use entropy to determine the rule yielding the lowest entropy. Repeat the entropy calculations until the terminal node is reached.
Step 6. Test the "consequent" part of the rule (i.e. the terminal node) found in the previous step. If the consequent is malfunctioning, then stop and assume it is faulty. If it is functioning, then continue the search.
Step 7. Back up to the immediate higher echelon and select the rule having the next lowest entropy. Repeat Step 6.
Step 8. If the right rule is not found from the shallow KB, then go back to the terminal node of the deep KB and update the responsibility probabilities. Starting at the root F_{00}, conduct the same breadth-first search until the terminal node is reached. Then go to Step 5.

In the following, the breadth-first search, the concept of entropy, and the probability updating method are described. Details for these steps can be found in Lee *et al.* (1990).

Breadth-first search. In Step 3, since a functional block which fails most frequently is to be found, a breadth-first search is performed. Obviously a more reliable branch and its own sub-branches do not have to be explored. The breadth-first here however, is somewhat restricted. The reason is that, looking at the nodes from "only one" parent at any level while in a usual breadth-first search, all the nodes from "all" parents at any level are evaluated. Hence, this search enables us to reach the terminal nodes faster than the usual breadth-first search.

This search procedure presented here resembles the depth-first search, but it is different in that it does not go back up to the nearest ancestor node with unexplored children during the backtracking process; it goes to the root.

Entropy. The concept of entropy, employed in Step 5, has been used in selecting the best next measurement in electronic circuit diagnosis, measuring the routing flexibility in flexible manufacturing systems, and manufacturing flexibility (de Kleer, 1987 and 1990, Pandelidis, 1990).

Shannon's entropy, by definition, does not incorporate any other attributes like test cost but estimated probabilities. In order to accommodated other attributes, Shannon's entropy must be modified. This modified form of entropy is called "useful" information on which the entropic measure applied here is based. Given the degrees of belief and the test costs, the entropy of a rule R_{ijk} is defined as follows:

$$H_{ijk} = H_{ijk}(w,p) = -\sum_{x=1}^{t} w_{i+1,k,x} p_{i+1,k,x} \ell n p_{i+1,k,x}, \quad i \geq 2$$

where,

$$w_{i+1} \geq 0, \quad \sum_{x=1}^{t} p_{i+1,k,x} = 1.$$

Here j just indicates the group number from the immediate higher echelon and x denotes the element k from echelon $i+1$. Moreover, note that the two units are non commensurate, i.e. c_{ijk} is in dollars and p_{ijk} is non-dimensional. For this reason, c_{ijk} is normalized as follows:

$$w_{ijk} = \frac{c_{ijk}}{\max\limits_{i,j,k} \{c_{ijk}\}}$$

where w_{ijk} represents the weight of a rule R_{ijk}.

When w_{ijk}=1 for all i, j, and k (i.e. all test costs are equal), the above entropy equation reduces to Shannon's entropy. Notice that all the distributed or assigned beliefs within a group sum to unity.

In short, the rule which has the lowest entropy must be determined. Referring to the shallow KB in fig. 4.26, start by making a decision at R_{111} to select the rule to be tested next. What is needed at this point are entropies of the rules at echelon 2. This, in turn, means that the probabilities (degree of belief) and the test costs at echelon 3 must be known. This procedure is repeated until the terminal rule is reached. A tie occurs at the terminal rules since the entropies for these rules are zero. The ratio, weight/probability, is computed in order to break the tie. This ratio is defined as,

$$r_{ijk} = \frac{w_{ijk}}{p_{ijk}}.$$

So, the tie is broken by selecting the lowest ratio which gives the lowest test cost. A tie at other rules can be broken "arbitrarily" since these rules do not directly involve the testing (i.e. the entropy of each of these rules is not zero). The overall feature of the search in the shallow KB is similar to the best first search because the search is performed in increasing order of entropy.

In Step 8, if after going through the shallow KB, a fault cannot be found, one goes back to the deep KB and update q_{ij}. The next subsection describes an updating heuristic (Lee *et al.*, 1990).

Heuristic for updating probabilities. Since the search path in the shallow KB did not yield any faults, the responsibility probability of the terminal node (called F_{bc}) which owns this shallow KB changes from the current value to zero. This enables one to update the probabilities in the deep KB.

In the light of the test results, the updated responsibility probability of F_{bc}, q'_{bc}, becomes 0. The ripple effect of this updating spreads out to the rest of elements within K, that is, the set whose elements have the same parent node. This K can be interpreted as a group. Then, in this fashion, the updating is propagated upwards through the parent nodes until the root is reached. In the following this updating heuristic is described.

To apply the updating heuristic, the number of elements of each of the nodes (i.e., how many children each node has) in the deep KB must be known. So let,

N_{00} = Number of elements that F_{00} owns

N_{ij} = Number of elements that F_{ij} owns, where i=1,2,..., and j=1,2,...,n.

This data can be obtained when the hierarchy of the deep KB is constructed.

Finding the parent node: First, the last element (called $F_{b\text{-}1,g}$) at Stratum b-1 is searched, by summing $N_{b\text{-}2,j}$ over all j. That is,

$$g = \begin{cases} \sum_{j=1}^{n} N_{b-2j} & if\ 2 < b \le v \\ N_{00} & if\ \ b = 2 \end{cases}$$

This g signifies the total number of elements at Stratum b-1 and is used in determining the group number for F_{bc} at Stratum b (i.e the overall element number for the parent node at Stratum b-1).

Next, compare c (from F_{bc}) with the group-wise cumulative number of elements to determine the second subscript (i.e. the element number) of the parent node at Stratum b-1. The second subscript will be the element number of the node whose cumulative number of elements is equal to or greater than c. Note here that the first subscript (i.e. the stratum number) is automatically known because the parent node is in the immediate higher stratum (i.e. just one stratum higher; b-1) than the current stratum (i.e., b). The second subscript is denoted by GN(b-1). Hence the parent node sought is $F_{b\text{-}1,\,GN(b\text{-}1)}$.

Finding the element number at the current stratum: Let EN(b) be the element number of F_{bc} within the group GN(b-1). EN(b) is calculated by subtracting the cumulative number of elements in the groups just before GN(b-1) [i.e. GN(b-1)-1] from c.

Updating the probabilities at Stratum b: In the following, the symbol ' is used to represent the updated value.

$$q'_{bc} = 0,\ q'_{bk} = \frac{q_{bk}}{1 - q_{bc}}, \qquad k \neq c, \text{ where } k = [c - EN(b)] + t$$

for $t = 1, 2, \ldots, N_{b-1,GN(b-1)}$ and $t \neq EN(b)$. Here, q'_{bk} is proportionally increased, depending on its original weight (i.e. q_{bk}), and thus normalized.

Updating at Stratum b-1: The parent node of F_{bc} will be less responsible for the observed symptom since one of its children turned out to be innocent. Thus the heuristic equation tht represents this amount of reduction in responsibility (probability) is given below:

$$q'_{b-1,GN(b-1)} = q_{b-1,GN(b-1)} - q_{b-1,GN(b-1)} \times (q_{bc} - q'_{bc}).$$

Apply the methods described above to determine the parent node and the element number at Stratum b-1. Let this parent node be $F_{b-2,GN(b-2)}$ and this element number be EN(b-1).

Since one of elements is innocent, the rest of the elements at Stratum b-1 will be more suspect. That is, the amount of reduction used in updating the parent node of F_{bc} will be proportionally distributed to the rest of the elements at Stratum b-1 with the positive sign. Thus the following equation is given:

$$q'_{b-1,k} = q_{b-1,k} + \frac{q_{b-1,k}}{1 - q_{b-1,GN(b-1)} \times q_{b-1,GN(b-1)} \times (q_{bc} - q'_{bc})}, \quad k \neq GN(b-1)$$

where $k = [GN(b\text{-}1) - EN(b\text{-}1)] + t$ for t=1, 2, ... $N_{b-2,GN(b-2)}$ and $t \neq EN(b\text{-}1)$.

Updating at Stratum *u* (*b*-2≤*u*≤1): Similarly to the methods above, one has,

$$q'_{u,GN(u)} = q_{u,GN(u)} - q_{u,GN(u)} \times (q_{u+1,GN(u+1)} - q'_{u+1,GN(u+1)})$$

$$q'_{u,k} = q_{u,k} + \frac{q_{u,k}}{1 - q_{u,GN(u)}} \times q_{u,GN(u)} \times (q_{u+1,GN(u+1)} - q'_{u+1,GN(u+1)}), k \neq GN(u)$$

where, k=[GN(u)-EN(u)]+t for t=1, 2, ..., $N_{u-1}GN(u\text{-}1)$ and $t \neq EN(u)$.

The updating ends at Stratum 1 because the responsibility probability is not involved in the root itself (one did not define q_{00}). The root at Stratum 0 has the number of elements (i.e. N_{00}) only.

By this way, a generic diagnosis model, which takes advantage of a hybrid approach, using the theory of hierarchical systems for both deep and shallow knowledge is developed. In case that a deep KB needs to be updated, the updating heuristic is devised.

The requirements of the model are failure probabilities for deep knowledge and degrees of belief and test costs for shallow knowledge. The technique employed are the breadth-first search and the updating heuristic for deep knowledge, and the concept of entropy for shallow knowledge. Also backtracking is allowed if necessary, and the updating is carried out if one shallow KB is not sufficient to locate a fault.

The developed diagnostic model is applicable to many domains. Once the hierarchies for both a deep KB and its dependent shallow KBs are constructed, the diagnostic strategy of fig. 4.27 can work.

The probability updating is simple and straightforward relative to other techniques, such as Bayes' Rule and Dempster's Rule.

Multiple faults can be handled by multiple diagnosis sessions. In addition, it is desirable to have a scheme to classify symptoms when multiple symptoms are observed.

The main difficulty is that it is not obvious how to delimit the level of abstraction in a deep KB (i.e. how many strata are needed). The number of strata in the deep KB can affect the size (in terms of the number of rules) of each shallow KB because, as the deep KB is specified in more detail, the degree of specificity of the shallow KBs might decrease. This, in turn, implies the smaller size of the shallow KBs. There may be a trade-off between the larger deep KB (the smaller shallow KBs) and the smaller deep KB (the larger shallow KBs).

One of the critical factors with respect to the speed of diagnosis here is a mapping scheme which relates a symptom to a certain node of the deep KB. This node will serve as the start node in the entire diagnostic process. The lower the stratum of the start node, the faster the speed of diagnosis. Unless a mapping scheme is provided, the diagnostic process begins with the root node of the deep KB (see also Section 4.2.1.4.).

Appendix 4.B Basic definitions of place/transition Petri nets and their use for on-line process failure diagnosis

A Petrit net (pt net)is a 6-tuple $N=(S, T, F, K, W, M(0))$ where,

(i) S is a finite set of places $S=\{s_1, s_2, ..., s_{|s|}\}$;

(ii) T is a finite set of transitions $T=\{t_1, t_2, ..., t_{|T|}\}$;

(iii) F is a binary relation $F\subseteq(S\times T)\cup(T\times S)$ which is represented by directed arcs between the places and transitions;

(iv) K is a mapping, $K:S\rightarrow\mathbb{N}\cup\{\infty\}$ which describes the capacity of each place K(s);

(v) W is a mapping, $W:F\rightarrow\mathbb{N}\backslash\{0\}$ which attaches a weight to each arc; and

(vi) $M(0)$ is a mapping, $M(0):S\rightarrow\mathbb{N}$ which gives the initial marking of the places, taking the capacity K(s) of each place into account.

A pt net is a bipartite graph consisting of 2 types of elements, a finite number /S/ of places [definition (i)] and a number /T/ of transitions (ii). Definition (iii) states that these places and transitions are coupled alternately by arcs, which are directed from a place to a transition or vice versa. Pt nets are introduced to describe the transport of a quantity, such as mass, information or a sort of goods. To remain in a system independent formulation, this transport quantity will be called a token. Places, the passive elements of the net, store a certain number (finite or infinite) of tokens according to their capacity K [definition (iv)]. Under special conditions a transition transports a number of tokens from the previous to the next place(s), the transition fires. The number of tokens transferred depends on the transport capacity [the weight W, definition (v)] of the connecting arcs. The transport process starts at time 0 with an initial distribution $M(0)$ of the tokens per place (vi).

The static structure of the pt net, defined in (i), (ii), (iii) and (v) can also be described in an algebraic manner by the so-called $/S/$-row $/T/$-column incidence matrix N. The element N_{ij} of this matrix indicates if place s_i is reached by a weighted arc coming from transition t_j (+sign) or is left by an arc which is directed to transition t_j (-). Assuming that no arc simultaneously starts and ends at the same place (the no-loop condition $(s, t)\in F \Rightarrow (t, s)\notin F$), N_{ij} is defined:

$$N_{ij} = \begin{cases} +W(t_j, s_i), & \text{if} (t_j, s_i \in F) \\ -W(s_i, t_j), & \text{if} (s_i, t_j \in F) \\ 0 & \text{otherwise} \end{cases} \tag{4.1}$$

The dynamic behavior of the pt net is represented by the firing rule. A transition t_j will be able to fire, if the following relation holds, where $\boldsymbol{M}(k)$ is a marking vector, k is a discrete time point ($\boldsymbol{M}(k):S\rightarrow\mathbb{N}$)

$$0 \le \boldsymbol{M}(k\text{-}1) + t_j \le \boldsymbol{K} \tag{4.2}$$

This means that firing will be only possible if the token content per place after firing does not surpass its capacity. $\boldsymbol{K}$ is the capacity vector of dimension $/S/$ and the transition t_j is the jth column vector of the incidence matrix $\boldsymbol{N}$. After firing the subsequent marking $\boldsymbol{M}(k)$ arises,

$$\boldsymbol{M}(k) = \boldsymbol{M}(k\text{-}1) + \boldsymbol{t}_j \tag{4.3}$$

With the help of equation (4.3), and beginning at time k=0, follows:

$$\boldsymbol{M}(k) = \boldsymbol{M}(0) + \boldsymbol{N}\boldsymbol{v} \tag{4.4}$$

where the vector $\boldsymbol{v}$ describes the firing frequencies of each transition which lead from the initial marking to the actual state. Of particular importance are special sets of places of the pt net called S-invariant. They are integer solutions of the linear equation,

$$\boldsymbol{i}_S^T \boldsymbol{N} = 0 \tag{4.5}$$

With the help of these S-invariants a new principle for fault detection in complex systems can be formulated.

Supposing it is possible to map the structure of the total process as a pt net, the transport of the physical conservation quantity is represented by the firing of tokens. If the conservation quantity takes only a few discrete values and the signals measuring the number of tokens are not noisy, the process monitoring is easy: Using eq. (4.4) it can be tested at each scanning time point if the actual marking vector $\boldsymbol{M}(k)$ beginning from the initial marking $\boldsymbol{M}(0)$ is reachable. If $\boldsymbol{M}(k)$ is not reachable it can be concluded that an error has occurred. The algorithmical evaluation of this failure detection criterion is simple. Keeping in mind that the marking vector is integer valued, eq. (4.4) is therefore a linear diophantic equation system. It is sufficient to test the existence condition of (4.4) at each time step. Examples for systems governed by such well defined unnoisy physical quantities are industrial production systems or automatic shunting yards.

In the case of plant fault monitoring, the measurement signals are noisy and their domain of definition is much larger than in the former case. Because of this the simple evaluation of eq. (4.4) must fail.

Multiplying (4.4) with the transpose of the S-invariant and taking (4.5) into account, yields:

$$i_S^T M(k) = i_S^T M(0) \tag{4.6}$$

For applications in power and other industrial plants, it is correct to assume that the net token flow across the envelope surface of the total process under consideration vanishes or is zero in the mean. Otherwise continuous plant operation is not possible. Moreover it is assumed that the transitions fire without changing the number of tokens, in other words, the sum of arc-weights in front and after a transition should be equal:

$$\sum_{s \in {}^*t} W(s,t) = \sum_{s \in t^*} W(t,s) \tag{4.7}$$

with ${}^*t{:=}\{s{\in}S{:}(s, t){\in}F\}$, $t^*{:=}\{s{\in}S{:}(t, s){\in}F\}$. Eq. (4.7) is a conservation law for the firing of tokens. Under both these conditions it is clear that an S-invariant exists which does not contain any other elements than 1 because each column sum vanishes. Such an S-invariant is called hereafter a covering S-invariant. Therefore equation (6) can be rearranged as

$$\sum_{i=1}^{i=|S|} M_i(k) - \sum_{i=1}^{i=|S|} M_i(0) = 0 \tag{4.8}$$

The second sum in eq. (4.8) must be calculated only once at the initial time k=0.

Taking the noisy nature of the measurement values into consideration, a new fault criterion for continuous total processes can be formulated:

$$\left| \sum_{i=1}^{i=|S|} [M_i(k) - M_i(0)] \right| < \varepsilon \tag{4.9}$$

Eq. (4.9) is well suited for on-line process monitoring. The actual number of tokens per place $M_i(k)$ is compared with the initial token content of the total process. This is possible because the continuous total process is naturally an initial boundary problem in contrast to the partial processes of the analytical redundancy methods. Therefore each slow varying fault can be detected as soon as it surpasses the threshold ε. The height of ε depends on the sensor noise and can easily be determined in an initial learning period.

If eq. (4.9) does not hold one of the following reasons must be true:

(i) A sensor fault has occurred, and one of the measured token numbers is erroneous;
(ii) Inside the total system a source or sink of tokens has arisen, which means the structure of the pt net has changed; or
(iii) The net token flow across the envelope surface of the total process is no longer zero mean, and the operation of the plant has become discontinuous.

Which exactly of these different faults has occurred cannot be recognized by using eq. (4.9), that is, a fault location is not possible. For this to be done more knowledge about the total process under consideration in the form of quantitative or qualitative physical models is needed. Details for the rule-based techniques and the way that can be combined with the present technique can be found in Section 4.2.1. It should be noted here that the process description in terms of pt nets is analogous to the state space formulation

$$\boldsymbol{M}(k)=\boldsymbol{A}\boldsymbol{M}(k\text{-}1)+\boldsymbol{B}\boldsymbol{u}(k\text{-}1) \qquad (4.10)$$

With the assumption that the state $\boldsymbol{M}$ is totally measurable and that it only represents physical quantities of the same kind (i.e. only masses or only temperatures) the state transition matrix $\boldsymbol{A}$ is equal to the identity matrix. Eliminating all previous time points k-1, k-2, ..., 1 in eq. (4.10) one gets,

$$\boldsymbol{M}(k)=\boldsymbol{M}(0)+\boldsymbol{B}\sum_{j=0}^{k-1}\boldsymbol{u}(j) \qquad (4.11)$$

This condition of observability (4.11) will be equivalent to the condition of reachability (4.4) if the sum of the control vector in (4.11) is set to a vector $\boldsymbol{v}$ and the input matrix $\boldsymbol{B}$ is named by incidence matrix $\boldsymbol{N}$. The analogy demonstrates that it is suitable to use the Petri net description to problems of process fault monitoring. Moreover, the possibilities of the methods of Chapter 3 could be exploited through on-line monitoring of the evolution of the vector $\boldsymbol{M}(k)$.

Appendix 4.C Analytical expression for exception using fuzzy logic and its utilization for on-line exceptional events diagnosis

Definition of exception. Human thinking is characterized by the argument that the symbol ($\rightarrow$) always appears in the universal proposition while the symbol ($\wedge$) occurs in the existential proposition. This can be represented formally by

$$\forall x \ [P(x) \rightarrow Q(x)] \tag{4.12}$$

$$\exists x \ [P(x) \wedge S(x)] \quad (S(x) \neq Q(x)) \tag{4.13}$$

where $P(x)$, $Q(x)$ and $S(x)$ are appropriate dictative functions. These equations give the interpretation that human thinking depends dominantly on the principle and simultaneously permits inconsistent remarks.

Conventional fuzzy diagnosis consists of two implications which take the form of eq. (4.12), viz., "For all j there exists a failure x_i that correlates with a symptom y_j, if y_j is recognized" and "For all i there is a symptom y_j such that should be observed if x_i appears". The sets of failures and symptoms X, Y are given in the form,

$$X = \{x_i \mid 1 \leq i \leq m\}; \quad Y = \{y_j \mid 1 \leq j \leq n\} \tag{4.14}$$

where x_i, y_i are elements of X, Y respectively. Here the diagnosis system is given by the propositions,

$$P_j\!: \ B(y_j) \rightarrow \exists x_i \, (A(x_i) \text{ and } R(x_i, y_j)); \ j = 1, \ldots, n \tag{4.15}$$

$$P_{ij}\!: \ A(x_i) \rightarrow \exists y_j \, (B(y_j)); \ i = 1, \ldots, m; \ j = 1, \ldots, n \tag{4.16}$$

where the function $A(x_i)$ means that the failure x_i appears, and $B(y_i)$ gives that the symptom y_j is being recognized. $R(x_i, y_j)$ indicates the correlation between x_i and y_j.

The exception is defined by the logical form of eq. (4.13). The aphorism "*Exceptio probat regulam*" means actually that the exception can serve as an examiner of the rule. This emphasizes that the exception has a large capability of testing rules in diagnosis. This aphorism motivated Maruyama and Takahashi, (1985, 1987), to introduce the exception into simply-structured diagnosis while neglecting a hierarchy on rules and/or a classification of the failures. How should the exception be represented and utilized to reinforce the diagnosis?

Suppose a fully-experienced engineer who makes the comment "there can be special logic for finding a failure while it is identified generally by such implications as eq. (4.12) or eqs. (4.15), (4.16)". When he was averse to use eq. (4.12), his logic might be transformed in the following manner:

$$\neg \, (\forall x \, [P(x) \rightarrow Q(x)]) = \exists x \neg \, (\neg P(x) \vee Q(x)) = \exists x \, [P(x) \wedge \neg \, Q(x)] \tag{4.17}$$

where "$\neg$" denotes negation.

Since he possessed a subconscious proposition "If $\exists x[P(x) \wedge \neg Q(x)]$ then there would exist an alternative proposition $S(x)$ which satisfied $\exists x[P(x) \wedge S(x)]$ ", a negation of eq. (4.12) can produce the relation,

$$\exists x[P(x) \wedge \neg Q(x)] \rightarrow \exists x[P(x) \wedge S(x)] \tag{4.18}$$

where the proposition,

$$\exists x[\neg Q(x) \rightarrow S(x)] \tag{4.19}$$

should be assumed. This argument indicates that a negation of eq. (4.12) is transformed to such a form capable of serving as an exception as eq. (4.13).

This formal definition for the exception enables its practical expression for diagnosis purposes by the same procedure of eqs. (4.17)-(4.19).

The negation of eq. (4.15) is given by,

$$\neg P_j\text{:}\ \exists y_j[B(y_j) \wedge \neg \exists x_i(A(x_i) \wedge R(x_i, y_j))];\ i=1, \ldots, m;\ j=1, \ldots, n \tag{4.20}$$

In order for eq. (4.20) to hold, the condition requires an alternative relation $E(x_i, y_j)$ consistent with the implication,

$$\neg \exists x_k(A(x_k) \wedge R(x_k, y_\ell)) \rightarrow \exists x_i(A(x_i) \wedge E(x_i, y_j)) \tag{4.21}$$

for $k=1, \ldots, m$; $\ell=1, \ldots, n$; $i=1, \ldots, M$; $j=1, \ldots, N$, as shown in eq. (4.19). The logical expression of eq. (4.21) could then be read as "in the special case where an engineer cannot identify the failure x_k with $A(x_k)$ and $R(x_k, y_\ell)$; $1 \leq k \leq m$, $1 \leq \ell \leq n$, there may exist another failure x_i that is correlated to $E(x_i, y_j)$; $1 \leq i \leq M$, $1 \leq j \leq N$, $M > m$, $N > n$, with the symptom y_j". A range was assumed to be larger with $E(x_i, y_j)$; $1 \leq i \leq M$, $1 \leq j \leq N$, $M > m$, $N > n$) than with $R(x_i, y_j)$; $1 \leq i \leq m$, $1 \leq j \leq n$). The perfect sets of failure and symptom might be in the range of ($1 \leq i \leq M_0$, $1 \leq j \leq N_0$) as shown in fig. 4.28. Connecting eq. (4.20) with eq. (4.21), the practical form of the exception H_j to P_j is given by:

$$H_j\text{:}\ \exists y_j[B(y_j) \wedge \exists x_i(A(x_i) \wedge E(x_i, y_j))];\ i=1, \ldots, M,\ j=1, \ldots, N \tag{4.22}$$

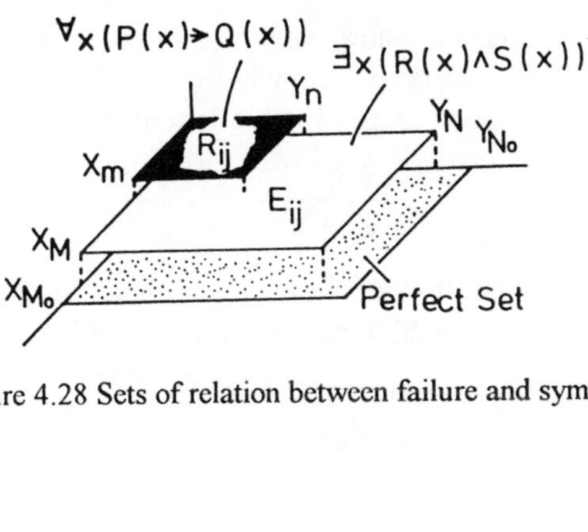

Figure 4.28 Sets of relation between failure and symptom

For simplicity, derivation of eq. (4.22) has been discussed on the binary logical point of view. Here, fuzzification of these equations shall be performed in accordance with fuzzy set theory, since recognition depends on the subjective tasks of a human. Defining the fuzzy sets on the spaces X, Y and $X \times Y$:

$$\begin{aligned} f &= (\text{set of appearing failures}) \\ s &= (\text{set of recognized symptoms}) \\ e &= (\text{set of relations } (x_i, y_j)) \end{aligned} \qquad (4.23)$$

One can obtain fuzzy propositions A_i , B_j and E_{ij} from $A(x_i)$, $B(y_j)$ and $E(x_i, y_j)$ in the form,

$$A_i = (x_i \text{ is } f), \quad B_j = (y_j \text{ is } s), \quad E_{ij} = ((x_i, y_j) \text{ is } e) \qquad (4.24)$$

where the truth values of these propositions are represented by linguistic truth values of $\boldsymbol{A}_i$, $\boldsymbol{B}_j$ and $\boldsymbol{E}_{ij}$. The linguistic truth values are fuzzy sets defined on the truth values space. Substituting eq. (4.24) into eq. (4.22), the fuzzification of eq. (4.22) is written in the logical form,

$$H_j\!: \ \exists_j [B_j \wedge \exists_i (A_i \wedge E_{ij})]; \ i=1, \ldots, M, \ j=1, \ldots, N \qquad (4.25)$$

Utilization of exception to diagnosis.

Cancellation law. Usually diagnosis terminates when the failures are identified using the fuzzified form of eqs. (2.15), (2.16), which is capable of providing the two equations

$$\boldsymbol{B}_{j\alpha} = \left(\bigvee_{i} (\alpha_{i\ell} \wedge r_{ij\ell}), \ 1 \right] ; \ j = 1, \ldots, n \qquad (4.26)$$

$$\boldsymbol{A}_{i\alpha} = [0, (1 + b_{ju} - p_{ij\ell}) \wedge 1)] \ ; \ i = 1, \ldots, m \qquad (4.27)$$

where $a_{i\ell}$, $r_{ij\ell}$ and $p_{ij\ell}$ stand for the lower bound of the linguistic truth values $\boldsymbol{A}_{i\alpha}$, $\boldsymbol{R}_{ij\alpha}$ and $\boldsymbol{P}_{ij\alpha}$ respectively and b_{ju} is the upper bound of $\boldsymbol{B}_{j\alpha}$. However, only when an engineer fails to find out the failure by eqs. (4.26), (4.27), the exceptional proposition of eq. (4.25) should be utilized.

An effective technique for inferring failure should be introduced to deal with exceptions in the actual condition. The cancellation law will be applied as a tool, since eq. (4.25) takes the form $P \wedge Q$ basically. The cancellation law is written

$$\frac{P, \ Q}{Q} \qquad (4.28)$$

which is then read practically as "If $P \wedge Q$ is true then Q must be true". Defining the truth values of P, Q and $P \wedge Q$ by $\boldsymbol{P}$, $\boldsymbol{Q}$ and $\boldsymbol{T}$ respectively, eq. (4.29) is transformed into the fuzzified expression,

$$\boldsymbol{T}_{\alpha} = \boldsymbol{P}_{\alpha} \wedge \boldsymbol{Q}_{\alpha} \qquad (4.29)$$

where the symbol α stands for an α-cut defined by the ranges,

$$\boldsymbol{T}_\alpha = (t_\ell, t_u), \quad \boldsymbol{P}_\alpha = (p_\ell, p_u), \quad \boldsymbol{Q}_\alpha = (q_\ell, q_u) \tag{4.30}$$

and the suffixes ℓ and u stand for the lower and upper bound respectively. With the use of eq. (4.30), eq. (4.29) is transformed into,

$$(t_\ell, t_u) = (p_\ell \wedge q_\ell, p_u \wedge q_u) \tag{4.31}$$

The cancellation law in fuzzy logic serves as the axiom to determine both the lower and upper bounds of $\boldsymbol{Q}_\alpha$, which satisfy eq. (4.31) at given $\boldsymbol{T}_\alpha$ and $\boldsymbol{P}_\alpha$. Thus a solution of $\boldsymbol{Q}_\alpha$ is described by,

$$q_k = \begin{cases} t_k, & p_k > t_k \\ (t_k, 1], & p_k = t_k, \quad (k = \ell, u) \\ \varnothing, & p_k < t_k \end{cases} \tag{4.32}$$

where $\varnothing$ represents the empty set.

It may be meaningful for reinforcement of the present diagnosis to consider only the case where the linguistic truth value of $P \wedge Q$ is close to "true". Then, $\boldsymbol{T}_\alpha$ is given by,

$$\boldsymbol{T}_\alpha = (t_\ell, 1] \tag{4.33}$$

Connecting eq. (4.33) with eq. (4.32), the inferred upper and lower bound of $\boldsymbol{Q}_\alpha$ are obtained as,

$$q_u = \begin{cases} 1, & p_u = 1 \\ \varnothing, & p_u < 1 \end{cases} \tag{4.34}$$

$$q_\ell = \begin{cases} t_\ell, & p_\ell > t, \quad p_u = 1 \\ (t_\ell, 1], & p_\ell = t, \quad p_u = 1 \\ \varnothing, & \text{otherwise} \end{cases} \tag{4.35}$$

Diagnosis by utilization of exception. Assuming that the exception of eq. (4.35) holds, the truth value of "very true" is given by,

$$\boldsymbol{H}_{j\alpha} = (h_{j\ell}, 1] \tag{4.36}$$

and the failure, symptom and fuzzy relation are written,

$$\boldsymbol{A}_{i\alpha} = (a_{i\ell}, a_{iu}), \quad \boldsymbol{B}_{j\alpha} = (b_{j\ell}, b_{ju}), \quad \boldsymbol{E}_{ij\alpha} = (e_{ij\ell}, e_{iju}) \tag{4.37}$$

then, substitution of eqs. (4.33)-(4.35) into eq. (4.25) generates the solution,

$$\bigvee_{i=1}^{m} (a_i \wedge e_{ij})_u = \begin{cases} 1, & b_{ju} = 1 \\ \varnothing, & b_{ju} < 1 \; ; \; j = 1, \ldots, N \end{cases} \tag{4.38a}$$

$$\bigvee_{i=1}^{m}(\alpha_i \wedge e_{ij})_\ell = \begin{cases} h_{j\ell}, & b_{j\ell} > h_{j\ell}, \quad b_{ju} = 1 \\ (h_{j\ell}, 1], & b_{j\ell} = h_{j\ell}, \quad b_{ju} = 1; \; j = 1, \ldots, N \\ \varnothing, & \text{otherwise} \end{cases} \quad (4.38b)$$

Equations (4.37) and (4.38) show that the truth value of the failure becomes meaningful only when $\boldsymbol{B}_j$ is closer to "completely true" than $\boldsymbol{H}_j$. Satisfaction of this condition allows one to obtain A_i by solving the inverse problem of the form (Pappis and Adamopoulos, 1992),

$$\bigvee_{i=1}^{m}(\boldsymbol{A}_i \wedge \boldsymbol{E}_{ij}) = \boldsymbol{H}_j \; ; \; j = 1, \; \ldots, N \quad (4.39)$$

This can be illustrated conceptually in fig. 4.29. The failure is calculated from $\bigvee_{i=1}^{M}(\boldsymbol{A}_i \wedge \boldsymbol{E}_{ij})$ at the given smooth line of "true" $\boldsymbol{H}_j$ and the broken line of "very true" $\boldsymbol{B}_j$, but there is no solution for the failure, since $\bigvee_{i=1}^{M}(\boldsymbol{A}_i \wedge \boldsymbol{E}_{ij})$=∅ at the given broken line of "more or less true" $\boldsymbol{B}'_j$.

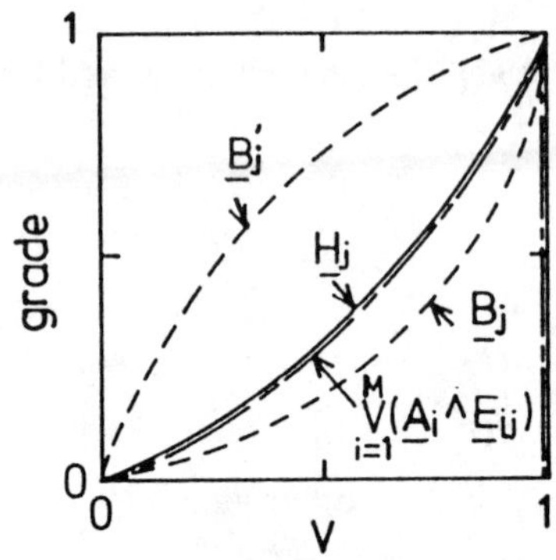

Figure 4.29 Linguistic truth value of failure derived from exception.

CHAPTER 5

FAULT DIAGNOSIS USING ARTIFICIAL NEURAL NETWORKS (ANNs)

5.1 Introduction

The theory and practical applications of *Artificial Neural Networks* (*ANNs*) are expanding with very high rates, and the fields of application are increasing. It is not surprising, therefore, that fault diagnosis is one of the main areas that ANNs have been used with promising results, along with similar progress in control and identification of non-linear dynamical systems.

Fault diagnosis using neural networks has the same structure as model-based methods: a set of signals, carrying fault information is fed into a neural machine which outputs a vector fault signal indicating normal or faulty system operation. Thus, it can be seen that the main difference between the two approaches is in the diagnosis engine. The selection of the input set (called training set), neural machine and output signal/classification method will be the central theme of this Chapter and will be examined in detail in subsequent sections.

ANN-based fault diagnosis is aimed at overcoming the shortcomings of model-based techniques. These techniques require mathematical process models that represent the real process satisfactorily. The model should not, however, be too complicated, because calculations easily become very time-consuming. In methods relying on state-variable estimation, state variables are seldom measurable and so nonmeasurable state variables have to be estimated. For estimation, a non-linear dynamic process model must be linearized around an operating point. This approach requires a relatively exact knowledge of the parameters of the linearized or linear model. In addition the process must operate near the point where linearization was done because the model is valid only in the neighborhood of the operating point.

In fault detection based on parameter estimation the process model parameters have a complicated relationship to physical process coefficients. Malfunctions usually effect the physical coefficients and the effect can also be seen in the process parameters. Because not all physical process parameters are directly measurable, their changes are calculated via estimated process model parameters. The relationship between the model parameters and the physical coefficients should be unique and preferebly exactly known. This is seldom the case.

The performance of model-based methods depends strongly on the usefulness of the model. The model must include every situation under study. It must be able to handle changes in the operation point. If the model fails, the whole diagnostic system fails. The sensitivity to modeling errors has become the key problem in the application of model-based methods. On the other hand, ANN methods do not require an analytical model of the process but need representative training data. The idea is that the operation of the process is classified according to measurement data. Formally this is a mapping from measurement space into decision space. Thus, data play a very important role in this method.

Model-based methods are usually computationally demanding, especially when nonlinear models are used. On-line modeling of the process requires even more computational expenditure. ANN methods are usually computationally easier but the calculation task depends very much on the data and the actual problem.

Model-based methods are mostly difficult to change afterward. The build-up of a model-based diagnostic system requires a lot of effort and changing one equation leads easily to changes in many other equations or parameters. ANN methods are more flexible in the sense that changing the data means the same as changing the properties of the diagnostic system. On the other hand changing the data may lead to repeating the diagnostic task all over again.

ANN-based fault diagnosis can be seen as a ***P**attern **R**ecognition* (*PR*) problem. Traditional pattern recognition and classification can be divided into three stages: measurements, feature extraction, and classification. First the appropriate data is measured. Then a feature vector is computed. The extraction should remove redundant information from measurement data and create simple decision surfaces. Finally the feature vector is classified into one or more classes. When fault detection and diagnosis are combined, the classes are the following: normal operation, fault number 1, fault number 2, etc. Traditionally, pattern recognition concentrates on finding the classification of features. The problem is, how to calculate the features. There is no *a priori* basis for the choice of the calculation. It's difficult to know which of the features are essential and which are irrelevant. Inappropriate choices lead to the need for complex decision rules, whereas good choices result in simple rules (Himmelblau, (1978), Pao, (1989)).

A human being has an amazing skill at recognizing patterns. A human often uses very complex logic in recognizing patterns and in classifying them. A human can pick up some examples of the classes but cannot determine any formal law for the classification. Mathematically this can be seen as an opaque mapping from pattern space into class-membership space. In computer pattern recognition, the opaque mapping has to be replaced by an explicitly described procedure - a transparent mapping (Pao, (1989)).

Like human beings, neural networks are also trained with a group of examples. When a classification is realized with neural networks, the whole mapping from measurement space into decision space is done at the same time and the mapping is learned by training examples.

It should be evident from the introducing comments that ANN-based fault diagnosis methods were first applied to complex, non-linear processes where previous attempts using conventional methods had failed. Chemical processes are such an example: Hoskins and Himmelblau, (1988), illustrated an artificial neural network approach to the fault diagnosis of a simple example process composed of three continuous-stirred-tank reactors in series. Watanabe *et al.* (1989) presented a network architecture to estimate the degree of faulures. They used an example system of three measurements and five faults. Venkatasubramanian and Chan (1989) used a binary-input network to diagnose faults of a fluidized catalytic cracking process. They also compared the neural network approach with a knowledge-based approach. Sorsa *et al.* (1993) applied radial basis networks to detecting deactivation of catalysts in jacketed reactors.

In the Computer Integrated Manufacturing (CIM) field, relevant work has been reported by Barschdorff *et al.* (1991) who applied back propagation and condensed nearest neighbour techniques to wear estimation and state classification of cutting tools. Miguel *et al.* (1993) investigated the applicability of an Adaptive Resonance Theory (ART)-3 based neural network to the detection and diagnosis of rotative machine failure. Syed *et al.* (1993) used Kohonen-maps for the real-time monitoring and diagnosing of robotic assemblies. Chow *et al.* (1991) implemented a three-layer feed-forward neural network for the real-time condition monitoring of induction motors. Yamashina *et al.* (1990) considered neural networks as a failure diagnosis tool for servovalves. Suna and Berns (1993) considered a back-propagation structure for pipeline fault diagnosis.

In the aerospace industry Rauch *et al.* (1993) performed fault detection, isolation and reconfiguration of an F-22 aircraft using neural networks, while Feng *et al.* (1993) applied an ART-2 neural network for automatic diagnosis of a variable thrust liquid rocket engine. Passino *et al.* (1989) used a multilayer perceptron, as a numeric-to-symbolic converter in a failure diagnosis application on an aircraft example.

Finally, Naidu *et al.* (1990) and Konstantopoulos and Antsaklis (1993) implemented a neural-model of a four-parameter controller suitable for sensor and actuator failures.

The outline of this chapter is as follows: in the first sections the theory of neural networks is outlined together with the basic topologies of ANN-based fault diagnosis. This is followed by a detailed examination of the principles of ANN-based fault diagnosis. Finally specific applications from representative fields are presented.

5.2 Introduction to neural networks

Biological systems implement pattern recognition computations via interconnections of physical cells called neurons. This provides motivation to consider emulation of this computational mechanism for automated PR applications. Researchers from such diverse areas as neuroscience, mathematics, psychology, engineering, and computer science are attempting to relate underlying models for pattern recognition, the computation that is desired, the potential parallelism that emerges, and the operation of biological neural systems. In fact, a whole field of study, centered around the creation and study of intelligent systems by recreating the computational structures of the human (or animal) brain, has fully emerged in only the last decade. This movement is known by several names, including *Connectionist Modeling*, *Neuromorphic Modeling* and ***Parallel Distributed Processing*** (*PDP*).

The idea that computations underlying the emulation of intelligent behavior may be accomplished via interactions of large numbers of simple processing units is hardly new. For one, the neuron models of the behavior of brain cells provides biological inspiration and plausibility for ANNs. The adaptable, context-sensitive, error-tolerant, large memory capacity, and real-time capability of the human information processing system (mostly the brain) suggests an alternative architecture to emulate. The mere fact that the basic computing element of the human information processing system is relatively slow (in the millisecond range), but the overall processing operation is achieved in a few hundred milliseconds, suggests that the basis of the biological computation is a small number of serial steps, each massively parallel. Furthermore, in this inherently parallel architecture, each of the processing elements is locally connected and relatively simple. Thus, connectionist or *neural* computing is not new or revolutionary. It is rather, evolutionary, with roots in a number of well-understood concepts, including biological pattern recognition, perceptrons and linear machines, adaptive networks, and fine-grained parallel computing paradigms.

Basically, three entities characterize an ANN:

1. The characteristics of individual units or artificial neurons;
2. The network topology, or interconnection of neural units; and
3. The strategy for pattern learning or training.

To some extent, the ANN approach is a *nonalgorithmic, black box strategy*, which is trainable. One hopes to train the neural black-box to learn the correct response or output (e.g. classification) for each of the training samples. This strategy is attractive to the system designer, since the required amount of a priori knowledge and detailed knowledge of the internal system operation is minimal. Furthermore, after training the internal (neural) structure of the artificial implementation it is hoped that it will *self-organize* to enable extrapolation when faced with new, yet similar, patterns, on the basis of experience with the training set.

The connectivity of a neural network determines its *structure*. Groups of neurons could be locally interconnected to form *clusters* that are only loosely, weakly, or indirectly connected to other clusters. Alternately, neurons could be organized in groups or *layers* that are directionally connected to other layers.

Several different generic neural network structures, are useful for ANN-based fault detection. Examples are:

- The ***Pattern Associator*** (*PA*). This neural implementation is exemplified by feedforward networks. A sample feedforward network is shown in Fig. 5.1a. In section 5.4.2.1 this type of network structure is explored in detail. Its learning (or training) mechanism (the backpropagation approach and the generalized delta rule) are considered and the properties of the approach are explored.
- The *Content-Addressable Memory* or *Associative Memory* model *(CAM or AM).* This neural network structure, is best exemplified by the *Hopfield* model. A sample structure is shown in Fig. 5.1b.
- *Self-Organizing Networks.* These networks exemplify neural implementations of unsupervised learning in the sense that they typically cluster, or self-organize input patterns into classes or clusters based on some form of similarity.

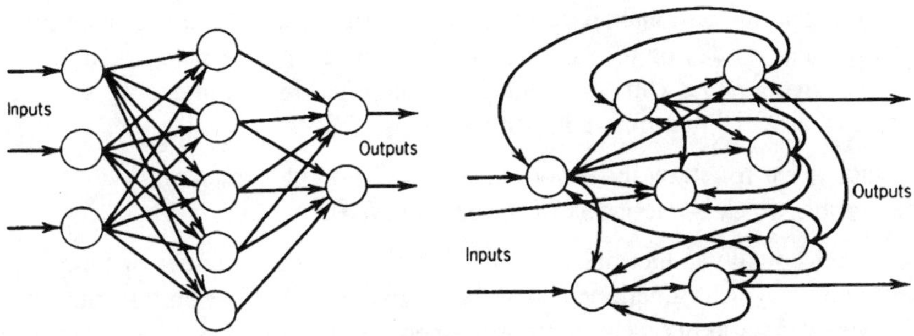

Figure 5.1

(a) Feedforward NN structure (b) CAM/AM neural network structure

Although these network structures are only examples, they seem to be receiving the vast amount of attention.

The feedback structure of a recurrent network shown in Fig. 5.1b suggests that network *temporal dynamics*, that is change over time, should be considered. In many instances the resulting system, due to the nonlinear nature of unit activation-output characteristics and the weight adjustment strategies, is a highly nonlinear dynamic system. This raises concerns with overall network stability, including the possibility of network oscillation, instability, or lack of convergence to stable state. The stability of nonlinear systems is often difficult to ascertain.

Learning in ANN-based fault detection may be either *supervised* or *unsupervised*. An example of supervised learning in a feedforward network is the generalized delta rule. An example of supervised learning in a recurrent structure is the Hopfield (CAM) approach. Unsupervised learning in a feedforward (nonrecurrent) network is exemplified by the Kohonen self-organizing network, whereas the ART approach exemplifies unsupervised learning with a recurrent network structure.

5.3 Characteristics of Artificial Neural Networks

An ANN is a distributed information processing system composed of many simple computational elements (*nodes*) locally interacting across very-low-bandwidth channels (connections). Most ANNs can be described as a *directed graph* (i.e., digraph) where the vertices are called *processing nodes* and the edges are called *connections*. Inputs and outputs of the network are simply additional connections originating or terminating in the external environment. The structure of these connections to the external environment is determined by the problem to which the network is applied.

Nodes in artificial neural networks are simple processors inspired by their biological counterparts. They can be described by a simple function that provides a mapping from n-dimensional space (inputs received via connections to other nodes or from the external environment) to one-dimensional space (a single output value, which becomes the input connections to other nodes or to the external environment). The network is inherently parallel, so that many nodes can carry out their computations simultaneously. Two features of a typical artificial neuron are illustrated in Fig. 5.2:

1. The activity α (or state of the node) resulting from the inputs, and
2. the output, which is a function of the activity, $f(\alpha)$.

Processing nodes contain an internal structure that consists of a number of functions and related data items. Typical functions include an activation function and an output function, while typical data items include input connection weights and local memory variables (e.g. activity value, output value, learning parameters, etc). All operations performed by a processing element are determined by its *transfer function*. The transfer function is usually composed of at least two functions: the *activation function*, which is a rule for combining the input connection signals as modulated by the connection weights to provide a state of activation, and the *output function*, which maps (using the local memory values) the current state of activation to an output signal.

Activation functions. Examples are shown in Figs. 5.3a-5.3b. This activation characteristic may, for example, be simply a threshold test, thus emulating a relay characteristic. Conversely, the possibility of external (i.e. not the output of other neurons) inputs to the neural unit, inhibitory inputs (as in the perceptron, Minsky and Papert, 1969), and weighted and nonlinear combinations of inputs are also possible. The perceptron and other types of threshold logic were early attempts to employ single-layer, linear neural-like networks to classification problems in pattern recognition. In a more general sense, neural units may be thought of as programming objects. A number of characterizations of this concept are described in Feldman and Ballard, (1982). Although there exist significant performance differences, analog and discrete firing characteristics are not distinguished.

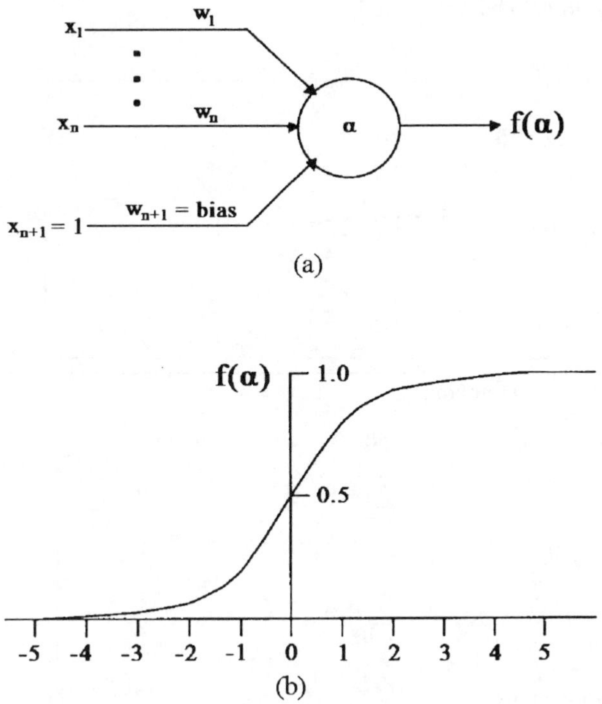

Figure 5.2 Features of artificial neurons

(a) processing node with adjustable weighted inputs.

(b) node output, as function of its activity

Output functions. As shown in Fig. 5.4, a variety of functions that map neuron input activation into an output signal are possible. The simplest example is that of a linear unit, where,

$$o = f(a) = a \tag{5.1}$$

One particular functional structure that is often used is the *sigmoid characteristic,* where,

$$o = f(a) = \frac{1}{1+e^{-\lambda a}} \tag{5.2}$$

Equation (5.2) yields $o \in [0,1]$. Then λ is an adjustable gain parameter that controls the "steepness" of the output transition as shown in Fig. 5.4e. Typically, λ=1 and (5.2) is often referred to in the literature as a *logistic function.* A computational advantage of the activation function of (5.2), which is useful in training, is that when λ=1,

$$\frac{\partial o}{\partial a} = o(1-o) \tag{5.3}$$

An interesting observation is that the average firing frequency of biological neurons as a function of excitation follows a sigmoidal characteristic.

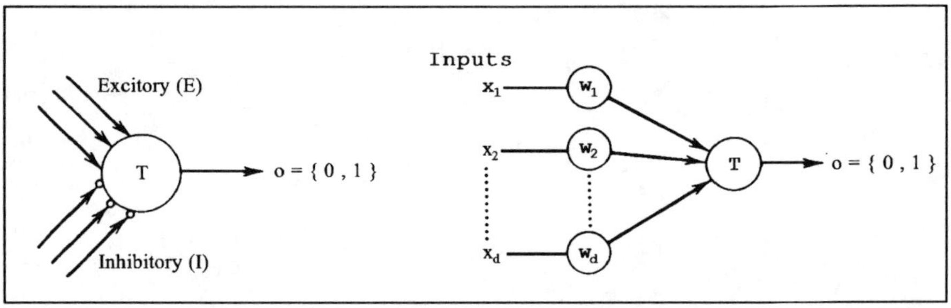

T: Threshold

Inputs	Output	Inputs	Output
$E \geq T$; $I=0$	1	$\sum_{i=1}^{d} x_i w_i < T$	0
$E \geq T$; $I>0$	0		
$E<T$; $I=0$	0	$\sum_{i=1}^{d} x_i w_i \geq T$	1
$E<T$; $I>0$	0		

(*E*: sum of activated excitory inputs
I: sum of activated inhibitory inputs)

(a) (b)

Figure 5.3 Neuron activation characteristics.
(a) McCulloch-Pitts model (b) Linear weighted threshold model

Another particularly interesting class of activation functions are *bilevel* mappings or thresholding units. For example,

$$o = f(a) = \begin{cases} 1 & a \geq 0 \\ 0 & a < 0 \end{cases} \tag{5.4}$$

Fig. 5.4b shows this unit characteristic using a general threshold, T. Alternatively,

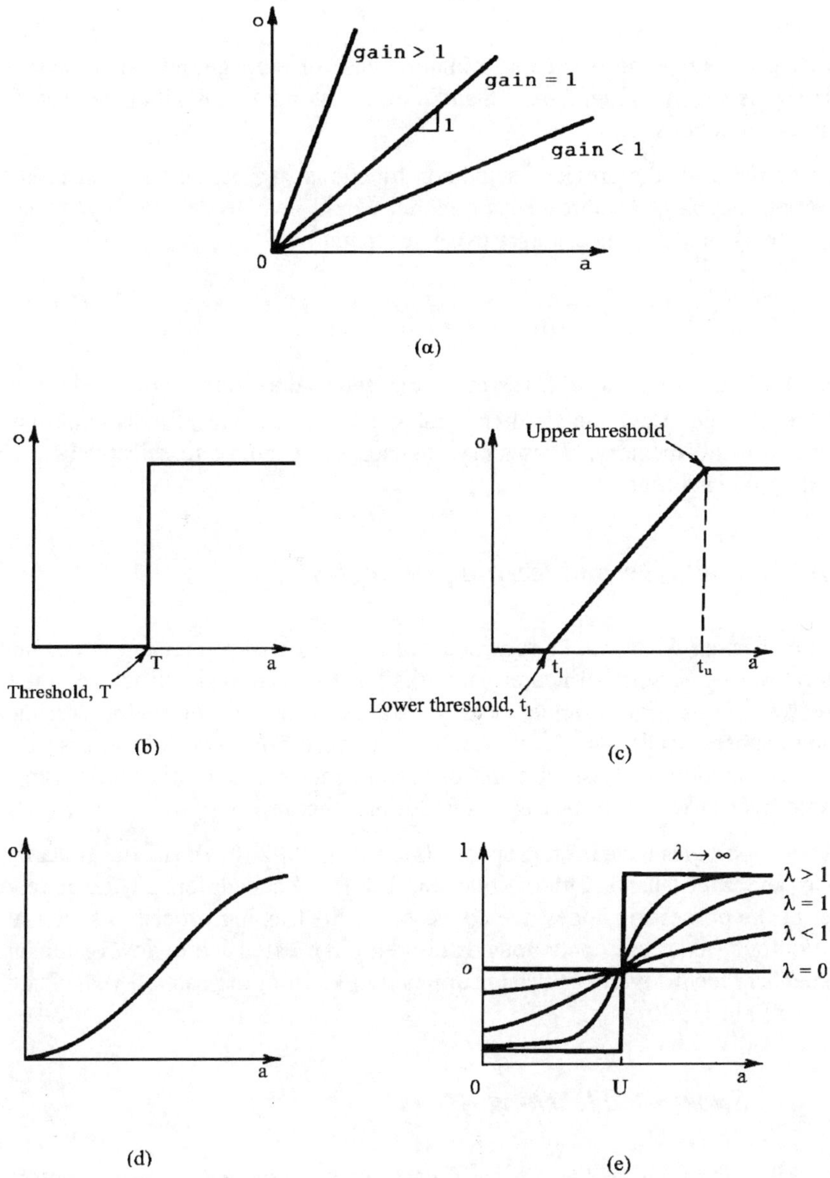

Figure 5.4 Neuron output function characteristics.
(a) Linear, (b) threshold (relay), (c) threshold (linear), (d), (e) sigmoidal.

$$o = f(a) = \begin{cases} +1 & a \geq 0 \\ -1 & a < 0 \end{cases} \tag{5.5}$$

Thresholding units may be viewed as a limiting case of the sigmoidal unit characteristic of (5.2). This is shown in Fig. 5.4e. In addition, thresholding units may be used to compute Boolean functions.

The characteristics of the previous activation functions suggest an instantaneous activation-to-outpu mapping. A more realistic model would involve delay or dynamics in the unit response. A model to incorporate dynamics might be,

$$\frac{\mathrm{d}a_i(t)}{\mathrm{d}t} = -\frac{1}{\tau_i}a_i(t) + \frac{1}{\tau_i}a_i^{\tau}(t) \tag{5.6}$$

where $\alpha_i(t)$ is the activation, $a_i^{\tau}(t)$ is the actual input activation and $\hat{o}_i$ is the time constant of the *i*th unit. Equation (5.6) constrains the time change of individual unit states and enables a local "memory." Biases may be used, for example to selectively inhibit the activity of certain neurons.

5.4 ANN topologies and learning strategies

As noted earlier, an ANN can be regarded as a collection of processing nodes and connections. However, almost all neural networks have considerable structure beyond this simple representation. For example, most ANN architectures group the processing nodes into disjoint subsets, called *layers,* in which all the processing nodes have essentially the same transfer function. Processing nodes can send connections to other processing nodes in the same layer as well as to processing nodes on other layers.

Many ANN topologies have been proposed (Hopfield, 1982; Feldman and Ballard, 1982; Rumelhart and McClelland, 1986; Kohonen, 1984). Each differs in the number and character of the processing nodes, the connections, the training procedures, and whether the input/output values are continuous or discrete. An extensive review of fundamental developments in feedforward artificial neural networks from the past 30 years is given by Widrow and Lehr (1990).

5.4.1 Supervised learning ANNs

Supervised learning requires the pairing of each input vector with a target vector representing the desired output; together these are called a *training pair*. Usually a network is trained over a number of such training pairs. An input vector is applied, the output of the network is calculated and compared to the corresponding target vector, and the difference (error) is fed back through the network and weights are changed according to an algorithm that tends to minimize the error. The vectors of the training set are applied sequentially, and errors are calculated and weights adjusted for each vector, until the error for the entire training set is at an acceptably low level.

5.4.1.1 Multilayer, feedforward networks

In this section a neural network with a layered, feedforward structure and error gradient-based training algorithm is presented. Although a single-layer network of this type, known as the *perceptron*, has existed since the late '50s (Minsky and Papert, 1969), it did not see widespread application owing to its limited classification ability and the lack of a training algorithm of the multilayer case. Furthermore, the training procedure evolved from the early work of Widrow (Widrow and Hoff, 1960) in single-element, nonlinear adaptive systems such as *ADALINE* (Widrow and Lehr, 1990).

The feedforward network is composed of a hierarchy of processing units, organized in a series of two or more mutually exclusive sets of neurons or layers. The first, or input, layer serves as a holding site for the values applied to the network. The last, or output, layer is the point at which the final state of the network is read. Between these two extremes lie zero or more layers of hidden units. Links, or weights, connect each unit in one layer to only those in the next-higher layer. Fig. 5.5 illustrates the typical feedforward network. The network as shown consists of a layer of d input units (L_i), a layer of c output units (L_o), and a variable number of internal or *hidden* layers (L_{hi}) of units.

The topology of the multilayer, feedforward network is not its essential point. What is more important is the role by which the topology acquires intelligence, i.e. its learning rule. In fact, the delay in developing a proper rule can be blamed for the early stalling of neural network research. Even though an explicit learning formula was not developed until around 1986 by Rummelhart and McClelland, an existence proof by Kolmogorov laid the theoretical foundations for such a rule.

Kolmogorov's Mapping Neural Network Existence Theorem: Given any continuous function $\varphi: \boldsymbol{I}^d \rightarrow \boldsymbol{R}^c$, $\varphi(\boldsymbol{i})=\boldsymbol{o}$ where $\boldsymbol{I}$ is the closed unit interval [0,1] (and therefore $\boldsymbol{I}^d$ is the d-dimensional unit cube), φ can be implemented exactly by a three-layer neural network having d processing elements in the input layer, $(2d+1)$ processing elements in the (single) hidden layer, and c processing elements in the output layer. The processing elements in the hidden layer implement the mapping function,

$$z_k = \sum_{j=1}^{d} \lambda^k \psi(i_j + \varepsilon k) + k \qquad (5.7)$$

where i_j are the network inputs and the real constant λ as well as the continuous real monotonic increasing function ψ are independent of φ (although they do depend on d). The constant ε is a rational number, $0<\varepsilon\leq\delta$, where δ is an arbitrarily chosen positive constant. Further, it can be shown that ψ can be chosen to satisfy a Lipschitz condition

$$|\psi(i)-\psi(o)| \le c|i-o|^{\beta} \text{ for any } 0<\beta\le 1.$$

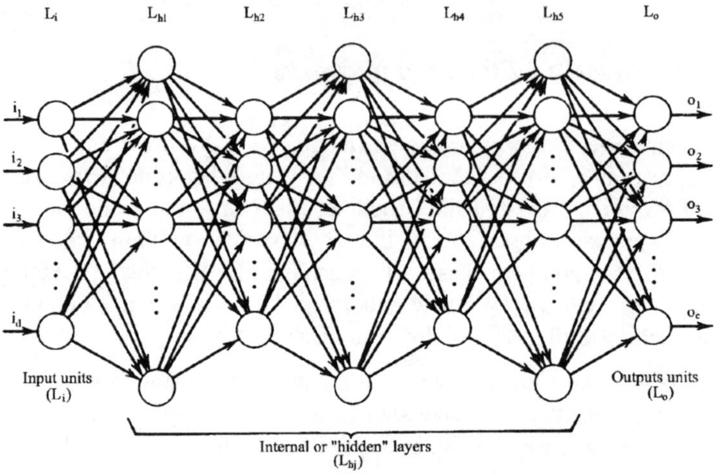

Figure 5.5 Structure of multiple-layer feedforward neural network.

The output layer elements implement the following mapping:

$$o_i = \sum_{k=1}^{2d+1} g_i(z_k)$$

where the functions g_i, $i = 1, 2..., c$ are real and continous (and depend on φ and ε).

The utility of this result is somewhat limited however, since no indication of how to construct the ψ and g_i functions is given. For example, it is not known whether the commonly used sigmoidal characteristics even approximate these functions.

The ***Genaralised Delta Rule*** (*GDR*) or *back-propagation* learning rule is the most widely used procedure for "tuning" multilayer, feedforward networks. This was proposed, as stated earlier, by Rummelhart and Mc Clelland in 1986. At about the same time, Mecht-Nielsen reintroduced Kolmogorov's Theorem, thus opening a new era in neural network research. Since then, a number of variants of the GDR have been produced, aiming at overcoming certain shortcomings of the original approach or trying to fine-tune the algorithm to specific tasks such as system identification, control or pattern recognition. The back propagation algorithm belongs to the class of *back-coupled error correction* procedures, which use a term that has to be computed from the networks' output mode in question, and is then transmitted back to adjust the weights.

Before giving the details of the rule, let us describe the basic operations in step format:

1. Apply input (stimulus) vector to network.

2. "Feed forward" or propagate input pattern to determine outputs.
3. Compare outputs in output layer with desired pattern response.
4. Compute and propagate error measure backward (starting at output layer) through network.
5. Minimize error at each stage through weight adjustments.

The error measure or objective function to be minimized in GDR learning is the mean-squared error between the actual outputs from the output layer and the desired or target outputs for all the input patterns. This function is given by,

$$J(\boldsymbol{w}) = \sum_{p} E_p \tag{5.8}$$

where,

$$E_p = \frac{1}{2}\left\| \boldsymbol{y}_p - \boldsymbol{d}_p \right\|^2 \tag{5.9}$$

and $\boldsymbol{w}$ is the vector containing the network weights. Given a training input vector $\boldsymbol{x}_p$, $\boldsymbol{y}_p$ is the output vector generated by the forward propagation of the activation through the network, and $\boldsymbol{d}_p$ represents the desired output vector associated with $\boldsymbol{x}_p$.

Whether this objective function is appropriate in specific situations, e.g. in classification problems, is debatable, and modifications may be made if this is not the case.

The objective function is minimised by a gradient descent technique. Applying minimisation criteria to the output node yields the following weight adjustment rule:

$$\Delta_p w_{ii} = \eta \delta_{pj} y_{pj} + \alpha \Delta_{p-1} w_{ji} \tag{5.10}$$

where $\Delta_p w_{ji}$ is the change in the weight from node i to node j after training sample p, η is a gain term usually called the *learning rate* and α is a *momentum* term that smoothes the effect of dramatic weight changes by adding a fraction of the weight change, $\Delta_{p-1} w_{ji}$, from the previous training sample p-1. The error signal δ_{pi} is a measure of the distance from the activation level of node j to its desired level after training sample p.

The GDR provides two rules for calculating the error signal of a node. For an output node,

$$\delta_{pj} = (d_{pj} - y_{pj}) f'(\alpha_{pj}) \tag{5.11}$$

where d_{pj} is the desired activation level for node j with respect to activity generated by the input pattern p.

The major contribution of the GDR lies in its formula for computing the delta values for the hidden nodes, since before the GDR, there was no learning procedure for multilayer, feed-forward networks. For an arbitrary node j in a hidden layer, the rule for calculating the error signal is,

$$\delta_{pj} = f_j'(\alpha_{pj})\sum_k \delta_{pk} w_{kj} \tag{5.12}$$

where the summation is over all the k nodes to which the node j sends output.

As the name back propagation suggests, the basic idea behind this computation of error signals for the hidden nodes, it to propagate errors back through the system that are based on observed discrepancies between the values of output nodes and the expected output for a training pattern. The error signal focuses on adjusting connection strengths that are most responsible for the output error. Then, if,

$$f_j(a_j) = \frac{1}{1+e^{-a_j}}$$

performing the differentiation, yields for the error signal,

$$\delta_{pj} = \begin{cases} (d_{pj} - y_{pj})y_{pj}(1-y_{pj}) & \text{if } j \in L \\ y_{pj}(1-y_{pj})\sum_k \delta_{pk} w_{kj} & \text{otherwise} \end{cases} \tag{5.13}$$

where L is the set of indices for the output nodes.

Learning involves two phases. First, the inputs are propagated in a feed-forward fashion through the network to produce output values that are compared to the desired outputs, resulting in a error signal for each of the output nodes. Second, the error is propagated backward through the network using the following procedure. The deltas (error signals) for the output layer are calculated first, and these deltas are used recursively to calculate the deltas for the adjacent hidden layer using (5.13). These deltas can then be used to update all the weights in the network.

Cybenko (1989), amongst others, has proved that a two-layer perceptron network can approximate any continuous mapping with an arbitrary accuracy, provided sufficiently many hidden units are available. He has also shown that arbitrary decision regions can be arbitrarily well approximated by a two-layer perceptron network. Unfortunately the papers do not tell how these accurate approximations can be achieved.

The back-propagation algorithm does not always find the global minimum but may stop at a local minimum. Baba (1989), proposed a random optimisation method for finding the global minimum of error function in neural networks. In practice the type of minimum has little importance as long as the desired mapping or classification is reached with a desired accuracy. The optimization criterion of the back-propagation algorithm is not very good from the pattern recognition point of view. The algorithm minimizes the square error between the actual and the desired output - not the number of faulty classifications, which is the main goal in pattern recognition. The algorithm is too slow for practical applications especially if many hidden layers are used (Pao, 1989, Wasserman, 1989). To improve on speed, Cho and Kim (1993) present three categories for rapid backpropagation learning. Their performance is evaluated and their merits and shortcomings are briefly discussed. Apart from this, a back-propagation network has poor memory. When the network learns something new it forgets the old one. Despite of its shortcomings the back-propagation is very widely used.

5.4.1.2 Recurrent high-order neural networks (RHONNs)

Recently, there has been a concentrated effort towards the design and analysis on learning algorithms that are based on the Lyapunov stability theory. Polycarpou and Ioannou (1992) present a general formulation for modeling, identifying and controlling nonlinear dynamical systems using various neural network architectures is presented, and analytical results concerning the stability of these schemes are obtained. Gaussian radial-basis-function networks have also been used for adaptively controlling dynamical systems with unknown nonlinearities.

A special architecture of RNN is that instead of having so-called hidden layers, one can enhance the input pattern with additional high-order terms and then usually find that a flat net with no hidden layers suffices for the purpose. This kind of architecture is called ***Recurrent High-Order Neural Networks*** (*RHONNs*).

High-order networks are expansions of the first-order Hopfield and Cohen-Grossberg models (described later) that allow higher-order interactions between neurons. RHONNs have a superior storage capacity, while stability properties of these models for fixed weight values have been proved. Furthermore, several authors have demonstrated the feasibility of using these architectures in applications such as grammatical inference and target detection.

Kosmatopoulos *et al.* (1992) present efficient learning algorithms for recurrent high-order neural models and analyze their stability properties.

The performance of RHONNs during the learning phase is superior to the performance of conventional architectures (backpropagation, etc). Convergence is achieved in much less iterations, e.g., for a task of learning 75 input/output pairs (with each input to be a pattern of 30 real number feature values and each associated output a single scalar), a backpropagation architecture needs 50-80 iterations, while a RHONN architecture only 6 iterations in order to converge with a system error in the order of 10^{-3}.

Recurrent neural network (RNN) models are characterized by a two way connectivity between neurons. This distinguishes them from feedforward neural networks, where the output of one neuron is connected only to neurons in the next layer. In the simple case, the state history of each neuron is determined by a differential equation of the form

$$\dot{x}_i = -a_i x_i + b_i \sum_j w_{ij} y_j$$

where x_i is the state of the i-th neuron, a_i , b_i are constants, w_{ij} is the weight connecting the j-th input to the i-th neuron, and y_j is the j-th input to the above neuron. Each y_j is either an external input or the state of a neuron passed through a sigmoidal function, i.e., $y_j=S(x_j)$ where $S(.)$ is a sigmoidal nonlinearity.

The dynamical behavior and stability properties of neural network models of this form have been extensively studied by Hopfield, as well as other researchers. These studies showed encouraging results in application areas such as associative memories, but they also revealed limitations of this simple model.

In a recurrent *second-order* neural network the total input to the neuron is not only a linar combination of the components y_j, but also of their products $y_j y_k$. Moreover, one can pursue along this line to include higher-order interactions represented by triplets $y_j y_k y_l$, quadruplets, etc. This class of neural networks forms a recurrent higher-order neural network (RHONN).

Consider now a RHONN consisting of n neurons and m inputs. The state of each neuron is governed by a differential equation of the form

$$\dot{x}_i = -a_i x_i + b_i \left[\sum_{k=1}^{L} w_{ij} \prod_{j \in I_k} y_j^{d_j(k)}(t) \right]$$

where $\{I_1, I_2, \ldots, I_L\}$ is a collection of L not-ordered subsets of $\{1, 2, \ldots, m+n\}$, a_i, b_i are real coefficients, w_{ij} are the (adjustable) synaptic weights of the neural network, and $d_j(k)$ are non-negative integers. The state of the *i-th* neuron is again represented by x_i, and $\boldsymbol{y} = \begin{bmatrix} y_1 & y_2 & \ldots & y_{m+n} \end{bmatrix}^{\mathrm{T}}$ is the vector consisting of inputs to each neuron, defined by,

$$\boldsymbol{y} = \begin{bmatrix} y_1 \\ \vdots \\ y_n \\ y_{n+1} \\ \vdots \\ y_{n+m} \end{bmatrix} = \begin{bmatrix} S(x_1) \\ \vdots \\ S(x_n) \\ u_1 \\ \vdots \\ u_m \end{bmatrix}$$

where $\boldsymbol{u} = \begin{bmatrix} u_1 & u_2 & \ldots & u_m \end{bmatrix}^{\mathrm{T}}$ is the external input vector to the network. The function $S(.)$ is a monotone increasing, differentiable sigmoidal function of the form

$$S(x) = \alpha \frac{1}{1+e^{-\beta x}} - \gamma$$

where a, b are positive real numbers and g is a real number. In the special case that a=b=1, g=0, we obtain the logistic function, and by setting a=b=2, g=1, one obtains the hyperbolic tangent function; these are the sigmoidal functions most commonly used in neural network applications.

The weights of the RHONN are adjusted using the learning algorithm proposed by Kosmatopoulos *et al.* (1992),

$$w_k(t+1) = w_k(t) + \eta\left[\prod_{j \in I_k} y_j^{d_j(k)}(t)\right] e(t)$$

where $e(t)=x(t)-c(t)$ denotes the state error.

An important advantage of the RHONN model is that one can obtain efficient learning algorithms that also guarantee the stability of the overall system. In addition, real time performances can be achieved due to the fast learing phase (e.g. exponential weights convergence) as well as due to their dynamical and adaptive nature (dynamic adjustment of the weights) see Kosmatopoulos *et al.*, (1993).

5.4.2 Unsupervised learning

The multilayer, feedforward and the Hopfield neural networks both exemplify supervised learning. In this section, neural networks based on *unsupervised learning* are shown. Specifically, networks that are used to determine natural clusters or feature similarity from unlabeled samples are explored. The "cluster discovery" capability of such networks leads to the descriptor *self-organizing*.

Fundamentally, unsupervised learning algorithms (or "laws") may be characterized by first-order differential equations (Kosko, 1990). These equations describe how the network weights evolve or adjust over time. Often, some measure of pattern associativity or similarity is used to guide the learning process, which usually leads to some form of network correlation, clustering, or competitive behavior.

5.4.2.1 Adaptive Resonance Architectures (ART)

Neural self-organizing architectures based on *Adaptive Resonance Theory* (*ART*) (Carpenter and Grossberg, 1987) consist principally of a pair of interacting neural subsystems, as shown in Fig. 5.6. Neural unit interconnections are both intra- and inter-system. The concepts of competitive learning and interactive activation are fused in this approach, in a manner that leads to a stable learning algorithm. The intersystem feedback structure is apparent from Fig. 5.6. The control signals shown are used to regulate the system operational mode, as described subsequently, and distinguish this system from a simple Hopfield network.

Fig. 5.7 shows an expanded view of the neural subsystems of Fig. 5.6. Note that the FA and FB layers are totally interconnected, that is, the activation of each FA unit is fed to all FB units and vice versa. This interlayer feedback structure is used to facilitate *resonance* when a match between an encoded pattern and an input pattern occurs. This typifies an *ART1* architecture. The FA subsystem may be viewed as the "bottom" layer that both "holds" the input pattern and, through the bottom-up weights b_{ij} (the interconnection strength from FA unit j to FB unit i), forms the FB layer excitation.

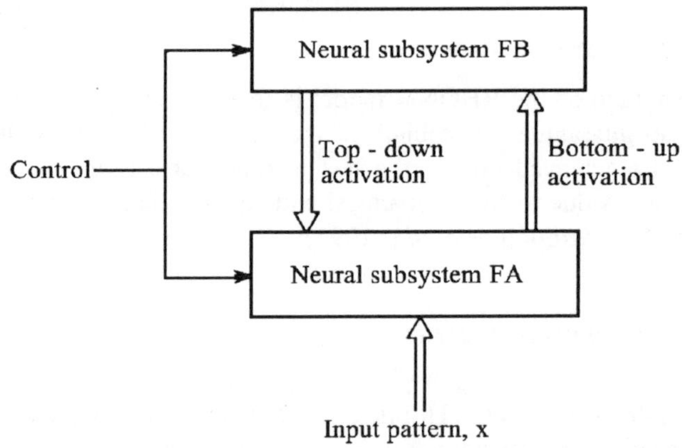

Figure 5.6 Structure of ART network.

The FB layer is composed of "grandmother" cells, each representing a pattern class. FB unit activations are fed back to the FA layer units via the interlayer t_{ij} interconnections. These may be viewed as long-term memory (pattern storage) interconnections. Most importantly, the FB layer units employ a self-exciting, competitive, neighbor-inhibiting interconnection structure, whereby each FB unit reinforces its own output through a positive interconnection between its output and one of its inputs, while maintaining a negative (inhibitory) connection to every other FB layer unit. A sample implementation of this structure and resulting action is as follows:

Interconnection weights w_{ij}, for units totally within layer FB, are determined by,

$$w_{ij} = \begin{cases} 1 & i = j \\ -\varepsilon & i \neq j \text{ and } i,\ j \text{ correspond to units in FB} \end{cases} \tag{5.14}$$

This corresponds to an "on-center, off-surround" interconnection strategy. Inter-layer interconnection weight values t_{ij}, where i corresponds to units in the FA layer, are described later. The competition or inhibition paramater ε is a design parameter, with the constraint,

$$\varepsilon < \frac{1}{N_{FB}} \tag{5.15}$$

where N_{FB} is the number of units in layer FB. Competition dynamics in layer FB are modeled via,

$$o_i(n+1) = f_i\left[o_i(n) - \varepsilon \sum_{i \neq k} o_k(n)\right] \quad i = 1,\ 2,\ \ldots,\ N_{FB} \qquad (5.16)$$

where $o_i = f_i(a_i)$ and $f_i(\ )$ is a unit activation-output mapping function that must be monotonically nondecrreasing for positive a_i and zero for negative a_i, and is a fundamental part of the *MAXNET* structure (Lippman, 1987). Thus, only one pattern class is designed to "win" if the overall network converges for a given input pattern. The reader will note some similarity of this local competition or inhibition-based structure with the Kohonen structure discussed subsequently. The overall ART1 architecture, then, is a cooperative-competitive feedback (recurrent) structure.

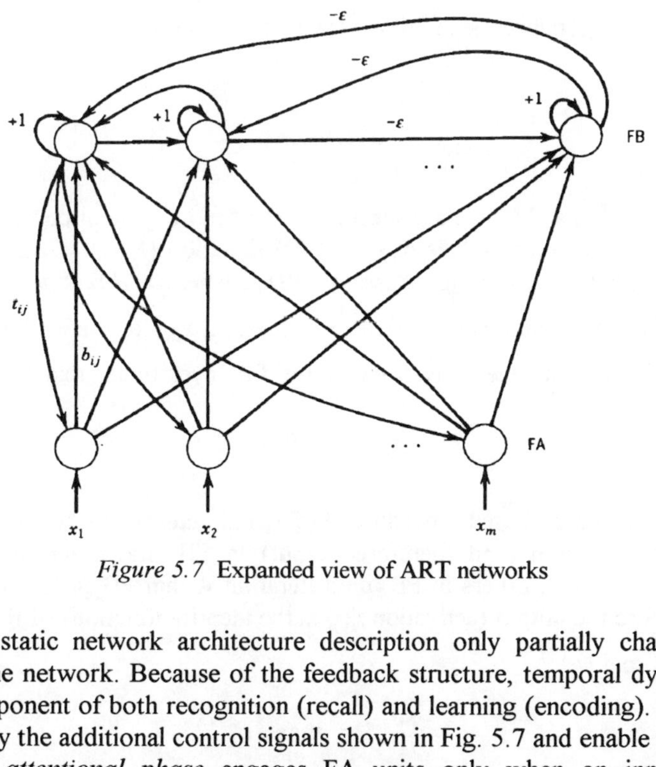

Figure 5.7 Expanded view of ART networks

The previous static network architecture description only partially characterizes the operation of the network. Because of the feedback structure, temporal dynamics are an important component of both recognition (recall) and learning (encoding). These actions are governed by the additional control signals shown in Fig. 5.7 and enable two phases of operation: an *attentional phase* engages FA units only when an input pattern is presented; an *orienting phase* successively "thins" units in FB, until a *winner*, or pattern class is found. If it is not possible to determine a winner, an uncommitted FB unit is used to represent this new pattern class, thus facilitating learning.

When presented with an input pattern, the ART network implements a combined recognition/learning paradigm. If the input pattern is one that is the same as, or close to, one previously memorized, desired network behavior is that of recognition, with possible reinforcement of the FB layer on the basis of this experience. The recognition phase is a cyclic process of bottom-up adaptive filtering (adaptive since the weights b_{ij} are changeable at each iteration) from FA to FB, selection of a stored pattern class in FB (the *competition*), and mapping of this result back to FA, until a consistent result at FA is achieved. The top-down feedback of the competition winner output from FB to form FA activations that may be viewed as encoded or learned expectations. This is then the network state of *resonance* and represents a search through the encoded or memorized patterns in the overall network structure. If the input pattern is not recallable, desired behavior is for the FB layer to adapt or learn this class, by building or assigning a new node henceforth representing this pattern class.

An algorithm that accomodates binary input {-1, 1} features is as follows:

1. Select ε, ρ and initialize the interlayer connections as follows:

$$t_{ij}^{0}=1 \quad \forall i,j \tag{5.17}$$

$$b_{ij}^{0}=\frac{1}{1+n} \tag{5.18}$$

   Equations (5.18) and (5.19) are specific cases of more general constraints that must be placed on the initial values of t_{ij} and b_{ij}. Equation (5.17) satisfies the so-called *template learning inequality*, whereas (5.18) satisfies the *direct access inequality*.

2. Present d-dimensional binary pattern $\boldsymbol{x} = (x_1, x_2, ..., x_d)^T$ to the FA layer.

3. Using b_{ij}, determine the activations of the FB layer units, that is, each unit has activation,

$$a_i^{FB}=\sum_j b_{ij}x_j \tag{5.19}$$

4. Use the competition-based procedure of (5.16) to determine a "winner" or unit with maximum activation (and therefore output) in FB. Each unit in FB therefore "competes" with all others in FB, until iteration within FB yields only one active unit. Denote the output (activation $f_i(\ )$ is the identify function) of the winning unit as $o_j^{FB_{win}}$, that is,

$$o_j^{FB_{win}}=\max_{o_k \in FB}\left\{o_k^{FB}\right\} \tag{5.20}$$

   Related to the winner unit in (5.20) is the function $m(j)$, used for weight updates and shown in (5.25).

5. The top-down verification phase begins. Using the winner unit found in step 4, this result is then fed back to FA via the top-down or t_{ij}, interconnections using,

$$a_i^{FA}=t_{ij}o_j^{FB_{win}} \tag{5.21}$$

for each unit in FA. The fed-back FA unit activations (or outputs) are then compared with the given input pattern. This is an attempted confirmation of the winning unit class found in step 4. Numerous comparisons are possible, with the overall objective of determining whether the top-down and input activations are sufficiently close. For example, since the inputs are binary, the comparison

$$\sum_i a_i^{FA} > \rho \|\boldsymbol{x}\| \tag{5.22}$$

may be used. In ART1, $\|\boldsymbol{x}\| = \sum_i |x_i|$. Here ρ is a design parameter representing "vigilance" of the test, that is, how critically the match should be evaluated.

6. If (5.22) is true, that is, the test succeeds, the b_{ij} and t_{ij} interconnections are updated to accommodate the results of input $\boldsymbol{x}$ using discrete versions of the slow learning dynamic equations:

$$\dot{t}_{ij} = \alpha_1 m(j)\left[-\beta_1 t_{ij} + f_i(x_i)\right] \tag{5.23}$$

where we recall that t_{ij} is the strength of interconnection from unit j in FB to unit i in FA. α_1 is a positive parameter that controls the learning rate, and $\hat{a}_1$ is a positive constant that allows gradual "forgeting" or decay. $f_i(x_i)$is the output of FA unit i using input x_i as activation, and $m(j)$ is described more fully below and in (5.25). Note that the competitive interconnections within FB defined in (5.14) are not adjusted through (5.23); only the top-down interlayer weights t_{ij} are modified. Similarly, the bottom-up weights are adjusted by,

$$\dot{b}_{ji} = \alpha_2 m(j)\left[-\beta_2 b_{ji} + f_i(x_i)\right] \tag{5.24}$$

where b_{ji} is the strength of the interconnection from unit i in layer FA to unit j in layer FB. Similarly, α_2 and β_2 are analogous to those in (5.23). Function $m(j)$ is used to restrict the updating of weights to those involving only the winning class $o_j^{FB_{win}}$ as defined in (5.20), using,

$$m(j) = \begin{cases} 1 & \text{if } o_j = o_j^{FB_{win}} \\ 0 & \text{otherwise} \end{cases} \tag{5.25}$$

If the test of (5.22) fails, this unit is ruled out and step 4 is repeated until a winner can be found or there are no remaining candidates.

Equations (5.23)-(5.25) represent one example of a learning strategy in the ART approach. Carpenter and Grossberg (1987), provide separate "slow" and "fast" learning procedures. Parameters α_1 and α_2 control the rate at which the system learns or adapts and must be chosen carefully. Learning rates that are too slow, yield systems that are rigid (or nonadaptive in the extreme case). Conversely, learning rates that are too fast cause the system to display chaotic (or what is termed "plastic") behavior. In the extreme case, the system tries to learn every input pattern as new class. Thus, a trade-off exists between system insensitivity to novelty (truly new patterns) and an overly plastic behavior. Simplified versions of these updating strategies are shown in Pao, (1989).

5.4.2.2 *Kohonen maps*

Kohonen, (1984), has suggested an alternative unsupervised neural learning structure involving networks that perform dimensionality reduction through conversion of feature space to yield topologically ordered similarity graphs or maps or clustering diagrams (with potential statistical interpretations). In addition, a lateral unit interaction function is used to implement a form of local competitive learning.

Fig. 5.8 shows possible 1-*D* and 2-*D* configurations of units to form feature or pattern dimensionality reducing maps. For example, a 2-*D* topology yields a planar map, indexed by a 2-*D* coordinate system. Of course, 3-*D* and higher dimensional maps are possible. Notice that each unit, regardless of the topology, receives the input pattern $x=[x_1 \ x_2 \ \dots \ x_d]^T$ in parallel. Considering the topological arrangement of the chosen units, the *d-D* feature space is mapped into 1-*D*, 2-*D*, 3-*D*, and so on. The coordinate axes used to index the unit topology, however, have no explicit meaning or relation to feature space. They may, nevertheless, reflect a similarity relationship between units in the reduced dimensional space, where topological distance is proportional to dissimilarity.

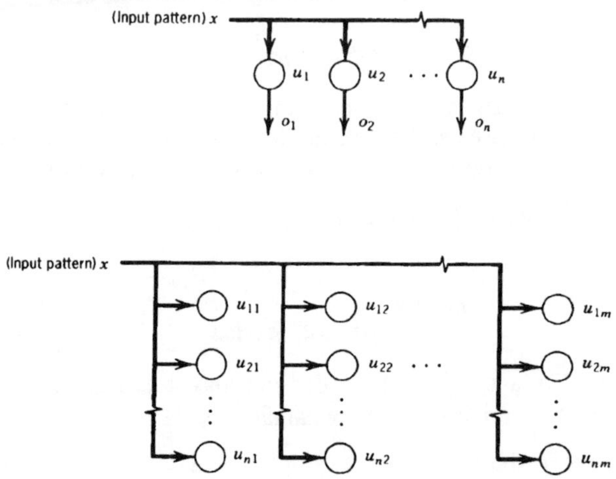

Figure 5.8 Topological map configurations.

Choosing the dimension of the feature map involves engineering judgment. Some PR applications naturally lead to a certain dimension; for example, a 2-*D* map may be developed for speech recognition applications, where 2-D unit clusters represent phonemes. The dimensions of the chosen topological map may also influence the training time of the network. It is noteworthy, however, that powerful results have been obtained by just using 1-and 2-D topologies.

Once a topological dimension is chosen, the concept of an equivalent dimension neighborhood (or cell of bubble) around each neuron may be introduced. An example for a 2-D map is shown in Fig. 5.9. This neighbourhood, denoted N_c, is centered at neuron u_c, and the cell or neighborhood size (characterized by its radius in 2-D, for example) may vary with time (typically in the training phase). For example, initially N_c may start as the entire 2-D network, and the radius of N_c shrinks as iteration proceeds. As a practical matter, the discrete nature of the 2-D net allows the neighborhood of a neuron to be defined in terms of nearest neighbors; for example with a square array, the 4 nearest neighbors of u_c are its N, S, E, and W neighbors; the 8 nearest neighbors would include the "corners." In 1-D, a simple distance measure may be used.

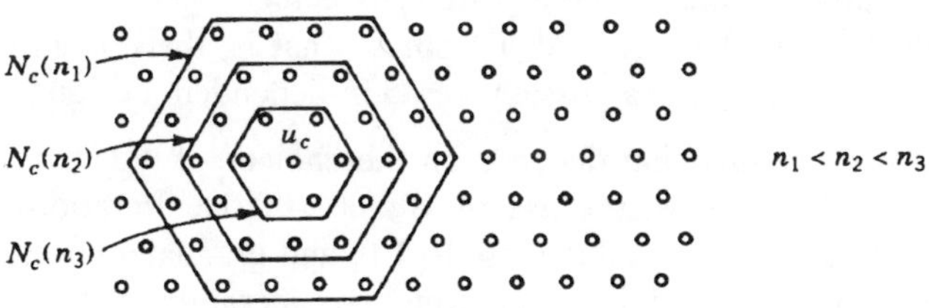

Figure 5.9 A topological neiborhood N_c of unit u_c showing shrinking training iteration n_i.

Each unit u_i in the network has the same number of weights as the dimension of the input vector and receives the input pattern $\mathbf{x}$ in parallel. The goal of the self-organizing network, given a large, unlabeled training set, is to have individual neural clusters self-organize to reflect input pattern similarity. Defining a weight vector for neural unit u_i as $\boldsymbol{w}_i=(w_{i1}, w_{i2}, ..., w_{id})^T$, the overall structure may be viewed as an array of matched filters, which competively adjust unit input weights on the basis of the current weights and goodness of match. A useful viewpoint is that each unit tries to become a matched filter, in competition with other units.

Assume that the network is initialized with the weights of all units chosen randomly. Thereafter, at each training iteration k and for an input pattern $\boldsymbol{x}(k)$, a distance measure $d(\boldsymbol{x}, \boldsymbol{w}_i)$ between $\boldsymbol{x}$ and $\boldsymbol{w}_i$, $\forall i$ in the network is computed. This may be inner product measure (correlation), Euclidean distance, or another suitable measure. For simplicity, the Euclidean distance is adopted. For a pattern $\boldsymbol{x}(k)$, a matching phase is used to define a winner unit u_c, with weight vector $\boldsymbol{w}_c$, using,

$$\|\boldsymbol{x}(k)-\boldsymbol{w}_c(k)\| = \min_i\{\|\boldsymbol{x}(k)-\boldsymbol{w}_i(k)\|\} \tag{5.26}$$

Thus, at iteration k, given $\boldsymbol{x}$, c is the index of the best matching unit. This affects all units in the currently defined cell, bubble or cluster surrounding u_c, $N_c(k)$ through the global network updating phase as follows:

$$\boldsymbol{w}_i(k+1) = \begin{cases} \boldsymbol{w}_i(k) + \alpha(k)[\boldsymbol{x}(k) - \boldsymbol{w}_i(k)] \;; & i \in N_c(k) \\ \boldsymbol{w}_i(k) & ; \; i \notin N_c(k) \end{cases} \tag{5.27}$$

Note that (5.27) corresponds to a discretized version of the differential adapation law:

$$\frac{\mathrm{d}\boldsymbol{w}_i}{\mathrm{d}t} = \alpha(t)\left[\boldsymbol{x}(t) - \boldsymbol{w}_i(t)\right]; \; i \in N_c(t) \tag{5.28}$$

$$\frac{\mathrm{d}\boldsymbol{w}_i}{\mathrm{d}t} = 0 \; ; \; i \notin N_c(t) \tag{5.29}$$

Clearly, (5.28) shows that $d(\boldsymbol{x}, \boldsymbol{w}_i)$ is decreased for units inside N_c, by moving $\boldsymbol{w}_i$ in the direction ($\boldsymbol{x}$-$\boldsymbol{w}_i$). Therefore, after the adjustment, the weight vectors in N_c are left unchanged. The competitive nature of the algorithm is evident since after the training iteration, units outside N_c are relatively further from $\boldsymbol{x}$. That is, there is an opportunity cost of not being adjusted. Again, α is a possibly iteration-dependent design parameter.

The resulting accuracy of the mapping depends on the choices of $N_c(k)$, $\alpha(k)$, and the number of iterations. Kohonen cites the use of 10,000-100,000 iterations as typical. Furthermore, $\alpha(k)$ should start with a value close to 1.0, and gradually descrease with k. Similarly, the neighborhood size $N_c(k)$ deserves careful consideration in algorithm design. Too small a choice of $N_c(0)$ may lead to maps without topological ordering. Therefore, it is reasonable to let $N_c(0)$ be fairly large (Kohonen suggests 1/2 the diameter of the map) shrinking $N_c(k)$, perhaps linearly, with k to the fine-adjustment phase, where $N_c(k)$ consists only of the nearest neighbors of unit u_c. Of course, a limiting case is where $N_c(k)$ becomes one unit.

5.5 *ANN-based fault diagnosis*

As pointed out in the introduction, an ANN-based fault monitoring scheme has, in a way, the same structure as a corresponding model-based scheme: a number of signals, which are deemed typical of the state of the process that is being monitored, is fed into a neural machine. The neural machine outputs a fault vector which is manipulated by a decision logic to decide if a fault has occured and possibly isolate it and estimate its size.

5.5.1 *Choice of neural topology*

As far as the choice of neural topology is concerned, no general guidelines exist. It is true however that feed-forward, multilayered networks employing the GDR (or back propagation) learning rule are used in the majority of published applications. Its elegant structure, is however offset by two factors:

- It may not converge.
- It has a slow convergence of $O(N^3)$ where N is the number of weights.

The first problem can be usually overcome by multiple starts with different random weights and by a low value of the learning rate η (Lippman, 1987). To accelerate the learning procedure dedicated parallel hardware can be used for the computations. The extent of both drawbacks seems to depend on parameters η (learning rate) and α (momentum factor). Unfortunately, their optimum values cannot be determined a priori and furthermore they may change during the training (i.e. they are time-varying). Their adaptive setting is the subject of ongoing research (Cho and Kim, 1993).

Since the appropriate choice of a network topology cannot be made a-priori, it is good practice to compare the performance of various topologies and choose the best performer. This procedure is not straightforward, however. Not only one has to compare different topologies, but also different configurations of the same topology. In a feedforward, multilayered ANN the values have to be found empirically. Moreover, the number of hidden layers and the number of nodes per hidden layer must also be found by experiment. This is a well known drawback in implementing this kind of ANN. Node activator functions must also be chosen amongst the class of possible alternatives (threshold, sigmoid, hyperbolic, Gauss etc.). It follows therefore that a logical procedure for optimum network topology is to search each proposed topology for its best configuration and then choose the best amongst the best. Sorsa et al. (1991) have used this idea in comparing three topologies: a single-layer perceptron, a multilayer perceptron and a counter-propagation network which combines a Kohonen layer for classification with an ART architecture for mapping. Results were obtained on a simulated model of a heat exchanger and a continuous stirred tank reactor. Their results will be detailed in the examples section.

5.5.2 Choice of output fault vector and classification procedure

What is really asked from an ANN-based fault diagnosis system is to recognize fault patterns inherent in signals carrying fault information. Thus, as already pointed out, a fault diagnosis problem can be viewed as a pattern recognition problem.

Neural networks have been used for pattern recognition for sometime and there exist some powerful theorems in this area. In fact, Mirchandani and Cao (1989), have shown that in a d-dimensional space, the maximum number of regions that are linearly separable using h hidden nodes is given by,

$$M(h,d) = \sum_{j=0}^{d} \binom{h}{d} \quad \text{if } h > d$$

$$= 2^h \qquad \text{if } h \leq d$$

This theorem holds for hard-limiting nonlinearities, i.e. binary outputs. However, the conclusions can be extended to other types of nonlinearities (sigmoid etc). The behaviour of such networks is of course more complex, because the decision regions are bounded by smooth curves instead of straight line segments.

In traditional pattern recognition techniques the pattern classification is carried out throught a series of decision functions. A classification of d-dimension pattern space with M clusters, may be viewed as a problem of defining hyperplanes to divide the d-dimensional Euclidean space into M decision regions. More complex decision functions will be needed for linearly unseparable decision regions. Moreover, probability models are often employed under the premise of prior probabilities, because perfect typical reference pattern examples are not easy to obtaine. How to select the suitable decision function forms and how to modify the concerned parameters of the decision functions are not easy to be determined for the traditional pattern recognition methods.

Similarly to traditional decision theory, neural networks perform the classification by creating decision boundaries to separate the different pattern classes. However, unlike traditional classifiers, when a classification is realized with neural networks, the decision functions are not needed to be given beforehand. The whole mapping from sample space into decision space is developed automatically by using the learning algorithm. The knowledge of fault patterns is stored distributively in the highly interconnected nonlinear neuron-like elements. Moreover, it is these nonlinear activations in the network that lead to the strong classification ability of the artificial neural networks for the high dimension pattern space.

The usual pattern vector employed in ANN-based fault diagnosis has dimension equal to the number of faults that must be detected. In theory, an 1 in the ith position indicates an ith type fault, while a zero pattern vector signals normal operation. In practice, however, the network is trained for the values of 0.9 and 0.1 for fault and no-fault cases respectively, since 0 and 1 are limiting cases for sigmoid activators (usually employed), thus stalling the learning procedure if used. After training, a fault of type i is declared if the ith element of the output pattern vector exceeds a threshold. This threshold must be defined considering false alarm rates and it is usually calculated by simulation. A value of 0.5 is a safe guess.

Note that with this formulation, multiple faults can be detected, if the network has been trained with this situation. An alternative procedure, mimicking parameter estimation techniques, would be to produce a system parameter vector as an output pattern vector. In this way, the neural network would act as a parameter estimator. Fault decision would then be accomplished, using any of the methods discussed in Chapter 2. This decision phase could also be implemented by a neural network. Thus the inputs to this second network would be the parameter estimates, while its output would be a pattern vector having the structure discussed previously.

5.5.3 Training sample design

The appropriate selection of input training data is a very important stage in the development of an ANN-based fault diagnosis system. There is little guidance in the literature regarding the choice of representative sets of examples for training with continuous inputs, because most studies involve binary inputs. Most studies also used a closed set of possible inputs. Training dynamical systems, however, requires continous signals.

The first step of the procedure is to decide on the system parameters that are representative of the system's condition. This is of course application-dependent, but it may be safely assumed that the input/output signals of a state-space representation of the plant will be adequate. It may be neccessary to do some pre-processing on the input signals, such as scaling or filtering. The total number of samples needed depends on the network's characteristics, i.e. topology, activator functions, learning rules etc. It is evident that a small training sample is a desired system characteristic.

The training sample must contain signals from every possible fault situation of the plant and in a representative range of values. This may be impractical or even dangerous in certain situations of critical faults (eg. nuclear reactors, aircraft) and simulation data is then needed. This in part offsets the comparative superiority of neural networks regarding the point of model necessity. Even more, it is true that most published research in ANN-based fault diagnosis relies on simulated process models. Is this a sign that this approach is not implementable? This question cannot be answered now, since it is acceptable to use simulated models in early stages of development of new ideas.

5.6 Application examples

In this section, it is hoped to clarify many of the points discussed earlier and illustrate the applicability of the various methods.

The cited examples span a considerable part of industrial fields where ANN-based fault diagnosis is proposed as an alternative to other techniques. The presentation is structured in such a way as to highlight and enlighten the following crucial points:

- process model and fault models
- network topology, configuration and learning rule
- input training signals
- output patter vector
- results

The examples that follow are only a representative sample of available literature and additional references are cited at the end of the chapter.

5.6.1 Applications in Chemical Engineering

The field of Chemical Engineering is especially suited for applying ANN-based fault diagnosis systems. The nature of chemical processes, i.e. nonlinear, nonstationary and uncertain dynamic plants, can be accomodated by neural network structures.

Because modern chemical plants are extremely complex, they are susceptible to equipment malfunction and operator error. The complexity hampers the operator's ability to diagnose and eliminate potential process upsets or equipment failures before they can occur (Himmelblau, 1978). Hence, a continuing question in chemical engineering is how to use the process state vector to make or aid decisions about possible action or control at each time increment. Current techniques rely on expert systems, modeling using classical techniques in the time or frequency domains, and statistical analysis.

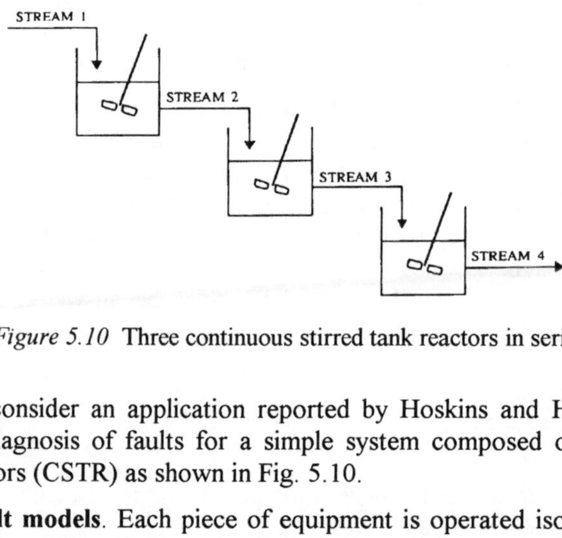

Figure 5.10 Three continuous stirred tank reactors in series.

As an example consider an application reported by Hoskins and Himmelblau (1988), illustrating the diagnosis of faults for a simple system composed of three continuous stirred tank reactors (CSTR) as shown in Fig. 5.10.

Process and fault models. Each piece of equipment is operated isothermally and with constant fluid volume (i.e., no fluid accumulation is permitted). The state variables used as fault indicators are the flow rate, the temperature and the concentration of components A and B in streams 1 and 4. All six state variables are monitored, and their sensors are assumed to function properly. All readings are taken at steady state.

Table 5.1 lists six selected faults, labeled A through F, each affecting the operation of this process. The possible faults involve the system flow rate, the temperature and the inlet concentration of component A. Table 5.2 shows the sensor measurement patterns of the values of the six state variables associated with the six selected faults used for training the network.

Table 5.1 List of selected faults

A	Inlet Concentration of Component *A*	Low
B	Inlet Concentration of Component *A*	High
C	Inlet Flow Rate	Low
D	Inlet Flow Rate	High
E	Temperature	Low
F	Temperature	High

Table 5.2 Sensor measurement patterns of six selected faults
FR=flowrate (ft^3-min), *T*=temperature, (^{O}F), *C*=concentration (lb-moles/ft^3)
superscript denotes stream; subscript denotes component

	Measurements					
Faults	*FR*	*T*	C_A^1	C_B^1	C_A^4	C_A^4
A	18.0	190	0.3	3.18	0.2275	3.252
A	18.0	190	0.6	2.88	0.3755	3.104
B	18.0	190	1.3	2.18	0.5990	2.881
B	18.0	190	1.6	1.88	0.6681	2.812
C	13.0	190	1.0	2.48	0.4475	3.033
C	15.0	190	1.0	2.48	0.4777	3.002
D	22.0	190	1.0	2.48	0.5600	2.920
D	26.0	190	1.0	2.48	0.5958	2.884
E	18.0	150	1.0	2.48	0.8960	2.584
F	18.0	210	1.0	2.48	0.3102	3.170
F	18.0	230	1.0	2.48	0.1703	3.310

Network architecture. Fig. 5.11 shows the network architecture used for fault detection and diagnosis. It consists of six inputs corresponding to the six state variables of the system, three hidden nodes, and six output nodes corresponding to the six respective process faults listed in Table 5.1. The GDR rule with various learning rates was used to train the network ($0.25<\eta<0.9$, $\alpha=0.99$).

Input training signals. Twelve measurement patterns were used to train the network: the 11 measurement patterns listed in Table 5.2, plus one measurement pattern for a normally operating system. These measurement patterns were obtained from a digital computer simulation program designed to model the dynamics of an arbitrary combination of reactors and/or vessels. A second-order reaction ($2A \rightarrow 2B$) was assumed to occur between components *A* and *B* in each tank with a frequency factor of 5.0×10^{14} (lbmole)/(ft^3)(min) and an activation energy of 4.47×10^4 (Btu/lbmole). Because the values of the sensor readings in the input patterns to the network are factors in the equations that update the values of the weights in the learning procedure, measurements with larger magnitudes exert a greater influence on learning. To remove this bias, the simulated measurement data were scaled (as indicated by the preprocessing boxes depicted in Fig. 5.11) so that the inputs to the network varied continuously over the range of -1.0 to 1.0.

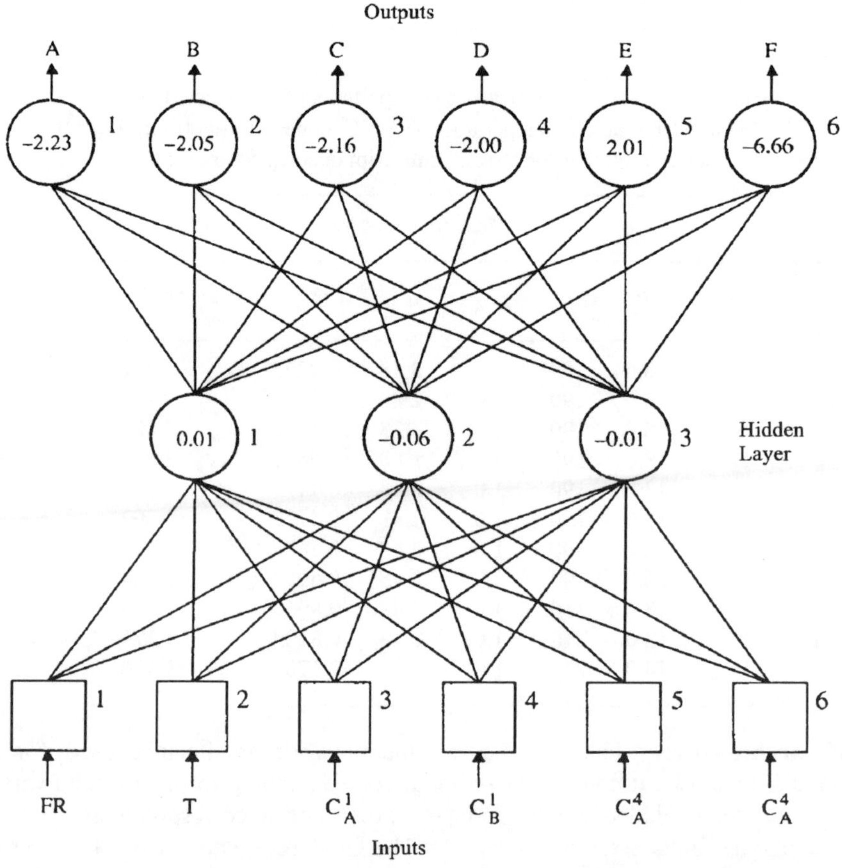

Figure 5.11 Trained network (numbers in circles represent biases of nodes).

Since one of the handicaps that fault detection and diagnosis must overcome is imprecise sensor measurements random noise was added to the inputs, modifying them to,

$$\boldsymbol{x}' = \boldsymbol{x} + \boldsymbol{n}$$

where $\boldsymbol{n}$ was a noise vector in which each element contained a random real value in the range of 0-10% of the input, and $\boldsymbol{x}'$ was the input vector with the added noise.

Output pattern vector. During training, the target values used for the output nodes were set to 0.1 and 0.9 rather than 0 and 1. A learning criterion of 0.01 for each pattern error E_p (of Eq. 5.9) was set to terminate the learning process.

Results. Noisy data require more complex decision regions. The learning results are shown in Fig. 5.12. By using a sigmoid discriminant function, a multilayer feedforward ANN and the GDR learning procedure, the network could properly classify inputs. The plots in Fig. 5.12 exhibited the same general trends as in the linearly separable case. Note, however, that the range of the number of hidden nodes for low convergence rates becomes smaller, especially for the high learning rates.

To demonstrate the generalization capabilities of the ANN, the percent misclassification was examined for this example with the three hidden nodes and the learning rate parameter set to 0.40. To test the ability of the network to recognize new input patterns, 35 measurement patterns not used in the training process were presented to the trained network. These new measurement patterns were chosen to be representative of the measurement space for the six faults and the normal operating system. Fig. 5.13 shows the percent of correct classifications as a function of the number of input patterns (with added noise) used to train the network.

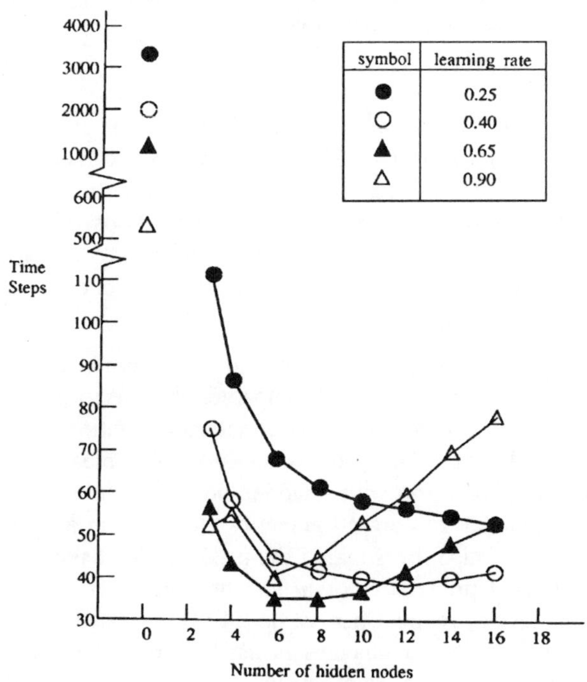

Figure 5.12 Experimental results.

Perfect generalization occured when only two measurement patterns in the training set had been used for each fault group (i.e., 12 input patterns total in the training set, two of which were used for the normal operating system). Increasing the number of measurement patterns maintained the same excellent performance level. However, some failure to generalize efficiently was observed for the training set containing only seven training patterns because the training set was so restricted that it was not representative of the general mapping. This result was not surprising since the seven-input pattern pattern training set included no example patterns that were representative of the crossover between the normal and faulty region.

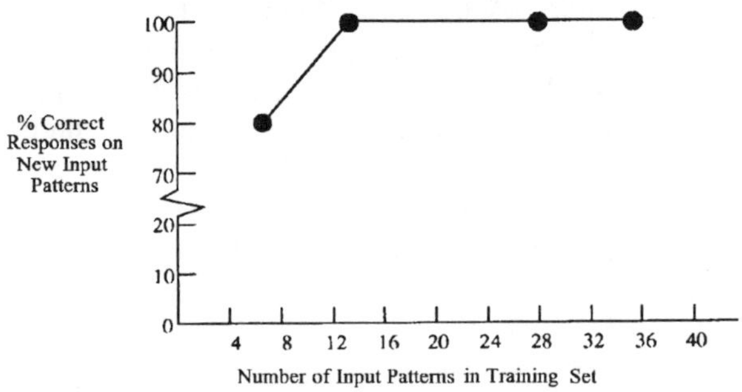

Figure 5.13 Generalization capacity vs. training set size.

Additional references. Venkatasubramanian (1989, 1990) and his research team have reported experiments on a similar CSTR. They have used a feedforward neural network with backpropagation and compared two methods of presenting input patterns: raw time-series data and moving-average data values. Two methods of discretizing the desired output were also compared: a linear and an exponential. Extensive experimentation showed better performance of the "linear" discretization, while raw time-series data input produced slightly better results.

Sorsa *et al.* (1991) conducted a comparison study on a simulated process consisting of a heat exchanger and a CSTR. They compared the performance of a single-layer perceptron, a multilayer perceptron and a counterpropagation network consisting of a Kohonen and an ART layer. The process had 14 noisy measurements and 10 typical faults. The multilayer perceptron with 4 hidden nodes, using a hyperbolic tangent as the nonlinear element was able to correctly identify the faults in all cases. The same group (Sorsa *et al.* 1993) have also investigated the use of radial basis function ANNs to fault diagnosis for dynamical processes not in steady state operation. An orthogonal least squares algorithm developed by Chen *et al.* (1991) is used to train the network. A simulated CSTR with set-point changes is used to test the validity of the proposed approach and preliminary results are promising.

5.6.2 Applications in CIM

Due to their efficient problem solving capabilities, parallel processing model, and ability to spontaneously react to environment changes, neural networks have prompted interest in their application to various real-time dynamic manufacturing systems (Moon, (1990), Lo and Bavarian, (1991), Gien *et al.* (1993)).

Diagnostic problem solving methods, based on either deep reasoning (from the first principles) or shallow reasoning (Davis, 1984), are considered to be unsuitable for domains with changing and short-lived processes (Reed *et al.*, 1988). In a typical application of a robotic assembly, a sequence of short-lived processes (typical of robot operations) brings about continuous changes to the state of the assembly components. Such characteristics necessitate flexible and adaptable solutions with efficient real-time response capabilities for the detection of and the recovery from unexpected problems during execution. The commonly used expert system solution for monitoring and diagnosing can be inefficient and inflexible (Schutte *et al.*, (1987), Zeilingold and Hoey, (1990)), particularly when it involves a large number of rules (leading to a large and computationally expensive search space) which require frequent updates due to the environment changes.

Specifically, the monitoring and diagnosing of assembly execution errors, although recognized as an important problem in robotic assembly (de Mello and Sanderson, (1990), Kusiak and Finke, (1988), Chang and Wee, (1988)), has not been addressed adequately. Solution to this problem is generally difficult due to the real time assembly constraints and the complex dynamic interactions between various components, such as robot, conveyor system, tools, and parts. The real-time constraints are particularly critical to the solution when computationally intensive sensory information (tactile, vision, etc.) is to be processed and accessed for the purpose of monitoring and diagnosing during each assembly step.

Syed *et al.* (1993) have proposed a neural network approach to the solution of this problem, by implementing an unsupervised map, namely the Kohonen map, in a robotic subassembly involving a fastener in a dishwasher power module.

Process and fault models. As just mentioned, a subassembly involving a fastener in a dishwasher power module is considered. A tactile sensor is attached to the end-effector of the robot arm. A generic robot operation, such as (PICKUP fastener FROM table), may have several execution instances for the same part. These instances can differ from one another in terms of the robot-part surface contact point and/or the approach angle of the robot end-effector. Therefore, each operation instance has its own particular part handling error characteristics.

To construct a neural map, one needs to establish a correlation between the error characteristics and the observed assembly interaction data. For example, a mean part surface contact area of between 0.0 and 0.5, measured using a tactile sensor, is correlated with the possible error in the execution of the (LIFT part) operation. The size of the surface contact area indicates how properly the part is grasped.

The input to each dimension in the neural map must be represented numerically. The output data from sensory systems are normally in numerical form, and, thus, can be directly used as input to neural maps. However, robot operation identifiers have to be converted into corresponding numerical values. This can be accomplished by assigning each operation an equal numerical interval. Table 5.3 summarizes the numerical representation for a 2-*D* type-II map, when the fastener is used in the subassembly. For the input vector (ξ_1, ξ_2) the value of ξ_1 denotes the operation type and ξ_2 represents the mean measurement value.

In Fig. 5.14, the shaded areas in the 2-*D* space correspond to the abnormal regions. The abnormal regions are defined by all possible values of the input vector (ξ_1, ξ_2).

Table 5.3 An example 2-D input space

	PICKUP fastener FROM table	ξ_1	ξ_2
1	TOP {loc=(0.5, 10.0, 15.4), angle=90.0}	0.0-0.2	0.75-1.00
2	TOP {loc=(0.5, 19.0, 15.4), angle=90.0}	0.0-0.4	0.75-1.00
3	SIDE {loc=(0.5, 10.5, 15.4), angle=0.0}	0.4-0.6	0.00-0.25
4	SIDE {loc=(0.5, 20.5, 15.4), angle=0.0}	0.6-0.8	0.00-0.25

Network architecture. A self-organizing Kohonen map is used for diagnosis. Details for its learning are discussed in section 5.4.2.2. The learning factor $\alpha(k)$ is given by:

$$\alpha(k) = \varepsilon \cdot e^{\frac{dist^2}{2\Delta^2}}$$

where,

- *dist* is the distance between the winning neuron and the actual one in the neural grid. As the nodes are plotted in the input space by using their input weights, the distance associated with nodes must be relative to their positions in the grid and not to their positions in the input space.
- ε and Δ are two parameters that have to be adapted to optimize the training. Their values decrease with the number of iterations to improve and optimize the training process. The following equations give the relation between the initial parameter values and their values at iteration ($10 \times p \times q$):

$$\varepsilon(10 \times p \times q) = (\mu_\varepsilon)^q \times \varepsilon(0)$$

$$\Delta(10 \times p \times q) = (\mu_\Delta)^q \times \Delta(0)$$

where $0<\mu_\varepsilon<1$ and $0<\mu_\Delta<1$.

Training. Fig. 5.15 shows the topology of output nodes after 100,000 training iterations. The incoming weights, w_{ij}, are indicated by the intersections of lines on the array of output nodes. Fig. 5.16 displays the final organization of the array of output nodes after 200,000 iterations. The final organization preserves the topology of the abnormal regions of the input space.

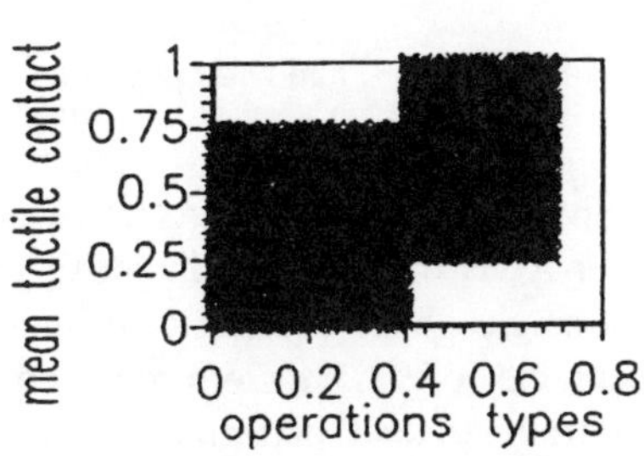

Figure 5.14 Network training inputs

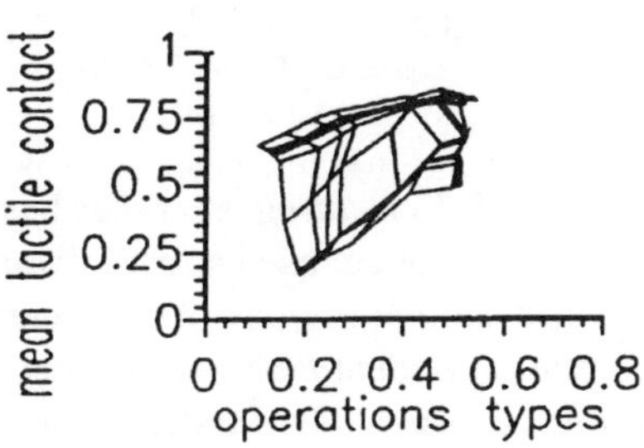

Figure 5.15 Intermediate node positions

Results. The final map obtained through the training is utilized during the assembly process to monitor the robot operation. A winning node is determined, whenever the sample of sensory input vectors (ξ_1, ξ_2) are applied to the network. The correctness of the robot operation is deduced from the winning node j, the density of nodes (d_j) in its neighborhood region, and the density threshold t_j.

The threshold value and the neighborhood region are defined heuristically by an operator. An example size for the neighborhood region is a rectangle of 0.08×0.15 relative to the ξ_1 and ξ_2 axis, respectively. For this example, the value of density threshold, t_j, is set to 10. With these values for threshold and neighborhood region, let us consider two different operation instances from Table 5.3. For the operation instance in column 2, an input vector (0.38, 0.5) will indicate an execution error since the d_j will be 15. On the other hand, an input vector (0.25, 0.9) for the operation instance in column 4 will not show an execution error as the value of d_j will be zero.

Additional references. Yamashina *et al.* (1990), used a feedforward, multilayer ANN to diagnose failures in a pneumatic servovalve used in automated production systems. A time-series vibration signal is monitored by an accelerometer and resulting data are summarized by six characteristic parameters. Four types of failures were considered, and a separate neural network was designed for each case. A conjugate gradient method coupled with a variable metric method was seen to produce more reasonable search directions and avoid oscillations. The diagnosis performance was very promising, reaching false alarm probabilities of 0.01 and lower.

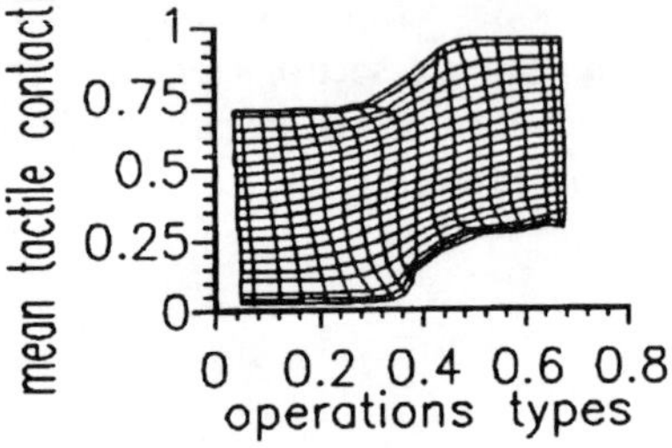

Figure 5.16 Final node positions.

Fault diagnosis in *rotating machines* has also been an application area of ANNs. Chow *et al.* (1991), have again used a feedforward, multilayer ANN to study the diagnosis of two of the most common types of incipient faults in a single phase squirrel-cage induction motor: stator winding fault and bearing wear. Accuracies of 97,3% were reported using a network of 16 hidden nodes, trained from 35 training data patterns by the back-propagation algorithm.

Tinghu *et al.* (1993), have used similar techniques to analyze and diagnose five types of typical faults in rotating machinery (unbalance, seal rub, misalignment, rotor crack, oil whirl) based on the standard frequency spectrum waveform features which are represented by power ratios in nine different frequency intervals.

Barschdorff *et al.* (1993), have investigated the ability of neural networks to diagnose tool wear in cutting processes like turning or grinding. They have used cutting force components and vibrations of the workpiece holder as suitable indicators of tool wear. A typical feedforward network is compared to a developed *Condensed Nearest Neighbor* (*CNN*) network (Barschdorff and Bothe, 1991). Results showed some benefits of the CNN against back-propagation networks. They also indicated that process parameters can be used as inputs to increase the variety of cutting conditions under which the system operates efficiently.

5.6.3 Power systems diagnosis

Fault location in power systems is defined as the identification of a fault or double fault from the system components such as trasmission lines, buses, transformers and circuit breakers in substations through analyzing the on/off status of several relaying systems or tripped order of circuit breakers. The difficulties in the estimation are derived from the malfunction of relaying systems or circuit breakers themselves, that is they sometimes do not operate when they should do, or they do the switching when they should not do.

Conventional methodologies which have been applied so far to the problem in power systems, include:

- Logical expression (Wake and Sakaguchi, 1984).
- Expert systems (Matsumoto and Sakaguchi, 1983).
- Parameter estimation techniques (Stavrakakis and Dialynas, 1991).

Ogi *et al.* (1991), have used a modular neural network approach for power system and equipment diagnosis. Despite its shortcomings the GDR was used as the learning rule in this application as well.

Plant and fault models. An example system with six buses, two transformers and two transmission lines with their protective relaying systems is used as shown in Fig. 5.17. A fault location related to buses, transmission lines and transformers must be estimated from the on/off status of relaying systems or circuit breakers in addition to the hypothesis of malfunctions in the relaying systems or circuit breakers themselves. The following is a list of the components of the sample power system:

• Relays (A_{*m}, B_{*m} C_{*m}, T_{1*}, T_{2*}, L_{1*}, L_{1*})		26
• Circuit breakers (CB_{*})		11
• Fault components given		10
Bus (A_1, A_2, B_1, B_2, C_1, C_2)	6	
Transmission line (L_1, L_2)	2	
Transformer (T_1, T_2)	2	

The names of relays include a suffix which has the following meanings:

m: Main protective
p: Primary backup protective
s: Secondary backup protective
t: Third backup protective (which covers opposite direction of *s*)

A 3-layer feedforward network was used for fault location. Its input layer received the on/off status of relays and circuit breakers and its output layer indicates which components have failed. This indication is shown by the largest element of the output vector.

Training. To train the network, a back-propagation learning algorithm with epoch training was used, that is weight updates were performed after presentation of the entire data set. The convergence criterion used was the usual absolute maximum error between desired and actual output. Training patterns which satisfied the criterion were progressively excluded from the weight update sequence. The training patterns consisted of 41 input/output pairs. The first 10 of them were concerned with normal operation, 22 with a circuit breaker malfunction and 9 with a relaying system malfunction (Table 5.4)

Results. To test the efficiency of the network, its response to non-trained patterns was examined. Two experiments were conducted: in the first, two fault components with all relays and circuit breakers operating normally was simulated while in the second a single fault with two circuit breakers malfunctioning was applied. The results showed that ANNs with more than 50 hidden units were able to classify the non-trained patterns correctly.

Table 5.4 Input patterns

A_{1m}	CB_1	CB_2								$/A_1/$
A_{2m}	CB_1	CB_3								$/A_2/$
B_{1m}	CB_6	CB_4	CB_7							$/B_1/$
B_{2m}	CB_6	CB_5	CB_8							$/B_2/$
C_{1m}	CB_9	CB_{11}								$/C_1/$
C_{2m}	CB_{10}	CB_{11}								$/C_2/$
T_{1m}	CB_2	CB_4								$/T_1/$
T_{2m}	CB_3	CB_5								$/T_2/$
L_1B_m	CB_7	L_1C_m	CB_9							$/L_1/$
L_2B_m	CB_8	L_2C_m	CB_{10}							$/L_2/$
A_{1m}	CB_1	T_{1t}	CB_2	CB_2						$/A_1/$
A_{1m}	CB_2	T_{2t}	CB_3	CB_5						$/A_1/$
A_{1m}	CB_1	T_{2t}	CB_3	CB_5						$/A_2/$
A_{2m}	CB_3	T_{1t}	CB_2	CB_4						$/A_2/$
B_{1m}	CB_4	CB_6	L_1C_s	CB_9						$/B_1/$
B_{1m}	CB_4	CB_7	L_2C_s	CB_{10}	T_{2s}	CB_3	CB_5			$/B_1/$
B_{1m}	CB_6	CB_7	T_{1s}	CB_2	CB_4					$/B_1/$
B_{2m}	CB_6	CB_8	L_2C_s	CB_{10}						$/B_2/$
B_{2m}	CB_6	CB_8	L_1C_s	CB_9	T_{1s}	CB_2	CB_4			$/B_2/$
B_{1m}	CB_6	CB_8	T_{2s}	CB_3	CB_5					$/B_2/$
C_{1m}	CB_6	L_2B_s	CB_8							$/C_1/$
C_{1m}	CB_{11}	L_1B_s	CB_7							$/C_1/$
C_{2m}	CB_{10}	L_1B_s	CB_7							$C_2/$
C_{2m}	CB_{11}	L_2B_s	CB_8							$/C_2/$
T_{1m}	CB_2	T_{1p}	CB_4							$/T_1/$
T_{1m}	CB_4	T_{1p}	CB_2							$/T_1/$
T_{2m}	CB_3	T_{2p}	CB_5							$/T_2/$
T_{1m}	CB_5	T_{2p}	CB_3							$/T_2/$
T_{1m}	CB_2	T_{1p}	CB_4							$/T_1/$
L_1B_m	CB_7	L_1C_m	L_1C_p	CB_9						$/L_1/$
L_1C_m	CB_9	L_1B_m	L_1B_p	CB_7						$/L_1/$
L_2B_m	CB_9	L_2C_m	L_2C_p	CB_{10}						$/L_2/$
L_2C_m	CB_{10}	L_2B_m	L_2B_p	CB_8						$/L_2/$
T_{1t}	CB_2	CB_4	T_{2t}	CB_3	CB_6					$/A_1\ A_2/$
T_{1s}	CB_2	CB_4	T_{2s}	CB_3	CB_5	L_2C_s	CB_{10}	L_1C_s	CB_{10}	$/B_1\ B_2/$
L_1B_s	CB_7	L_2B_s	CB_8							$/C_1\ C_2/$
T_{1p}	CB_2	CB_4								$/T_1/$
T_{2p}	CB_3	CB_5								$/T_2/$
L_1B_p	CB_7	L_1C_m	CB_9							$/L_1/$
L_1B_m	CB_7	L_1C_p	CB_8							$/L_1/$
L_2B_p	CB_8	L_2C_m	CB_{10}							$/L_2/$
L_2B_m	CB_8	L_2C_p	CB_{10}							$/L_2/$

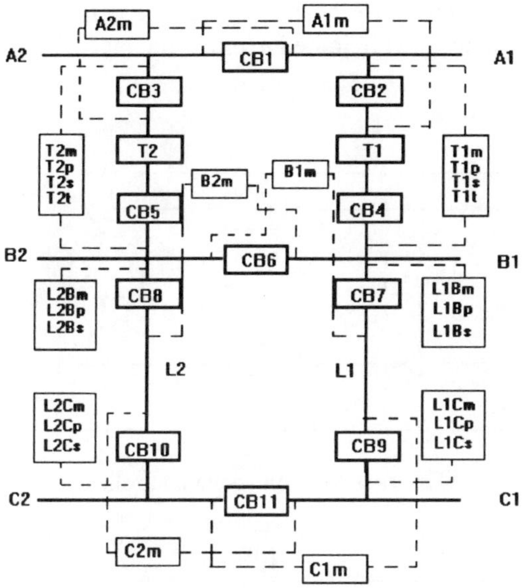

Figure 5.17 Example power system.

5.6.4 Neural four-parameter controller

The four parameter controller, (Nett, 1988), illustrated in Figs. 5.18 and 5.19, is a generalization of the familiar two-parameter controller (Antsaklis, 1992). This controller has two vector inputs and two vector outputs, resulting in a controller with four matrix parameters. Its various elements are: r is the reference input, α the diagnostic controller output designed to reproduce the failures, y_c the controller output that is manipulated by the plant and should be considered the ideal actuator input, u_c the manipulated controller input, n_α is an exogenous input accounting for unmodeled sensor signals, u is the actual actuator output, y is the plant output fed into the sensor, z is the plant variables not used by the controller and w is the unmanipulated plant input.

The linear controller can be described by the following relation,

$$\begin{bmatrix} \alpha \\ y_c \end{bmatrix} = \begin{bmatrix} K_{11} & K_{12} \\ K_{21} & K_{22} \end{bmatrix} \begin{bmatrix} r \\ u_c \end{bmatrix}$$

The objective of the additional controller output α is to identify and reproduce sensor and actuator failures f_s and f_a. Thus, the overall control objectives are:

- Achieve set point tracking

- Reject unmeasured disturbance w at the plant output z.
- Reject both actuator and sensor noises (η_a, η_s) and failures (f_a, f_s) at z.
- Achieve above objectives with limited u and in the face of plant modeling errors.

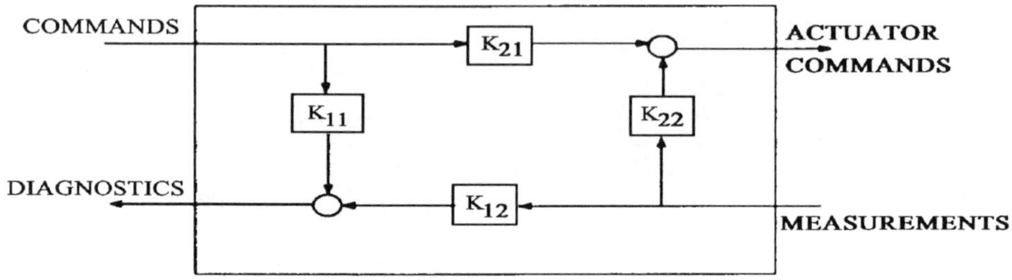

Figure 5.18 Four-point controller.

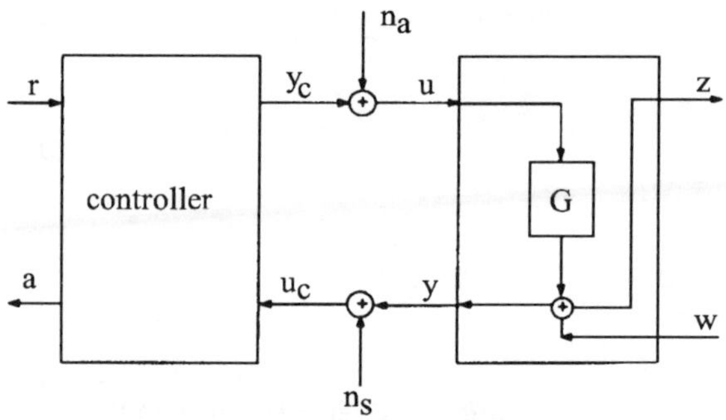

Figure 5.19 General controller and plant.

These requirements lead to certain conflicts (Nett *et al.*, 1988): reproducing sensor or actuator failures at the diagnostic output contradicts the requirement for noise and disturbance rejection. Also, sensor diagnostic performance has to be traded against actuator diagnostic performance.

Relying on the fact that a nonlinear controller should outperform its linear version, Konstantopoulos and Antsaklis (1993) implemented a four-parameter controller in a neural network. The general structure of a system designed for actuator failures is shown in Fig. 5.20.

The neural controller has two main inputs: a reference signal r and the output of the plant y_p. Experience has shown that delayed system signals enhances training performance. For this reason, delayed reference inputs, plant outputs and controller outputs were input to the neural network. It also has two outputs: a diagnostic output α_{act}, and y_c, which can be considered the ideal actuator input. The controller was trained with the following objectives.

- To achieve set point tracking and isolate and reproduce actuator faults.

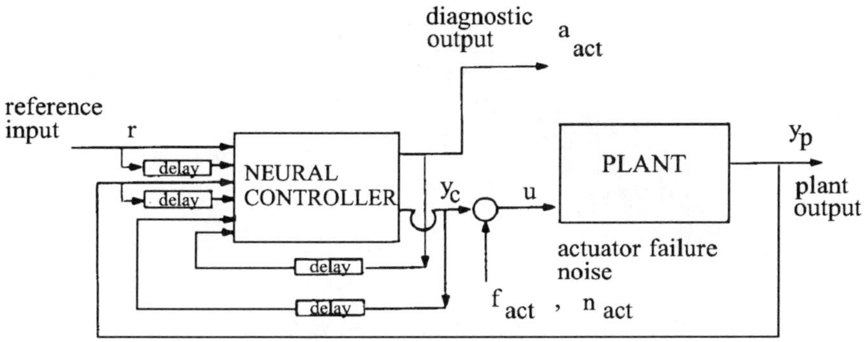

Figure 5.20 Neural network CDC controller for actuator failures.

The nonlinear plant model used is described by,

$$y_p(k+1) = \frac{y_p(k)y_p(k-1)(y_p(k)+2.5)}{1+y_p^2(k)+y_p^2(k-1)} + u(k)$$

A neural model of the above process was used to train the neural controller as described in the sequel.

Network architecture. The usual multilayer, feedforward neural network was employed. Experimentation revealed that a good performer was a network with two hidden layers consisting of 20 and 5 nodes respectively. The network was trained with the back-propagation GDR rule having momentum of 0.9 and learning rate of 0.01.

Training. The reference input used was a random signal in [-2, 2]. Failures were introduced as a sinusoidal function at specific time instants. The number of delayed signals was experimentally found to be 3. Training was done in two stages. First the network was trained as a conventional controller. The resulting weights were used as initial weights for the second stage where the network was trained to meet the diagnostic requirements as well.

Results. The overall structure met the design requirements satisfactorily. In Fig. 5. 21 some sample results are given. In this example, the real plant was used and a ramp failure occured at the actuator.

Design trade-off was accomplished by assigning different weights to the diagnostic and control outputs. In this way, by assigning a larger weight to the control objective, better reference tracking was achieved, whereas better reproduction of actuator failures was obtained by assigning larger weight to the diagnostic output.

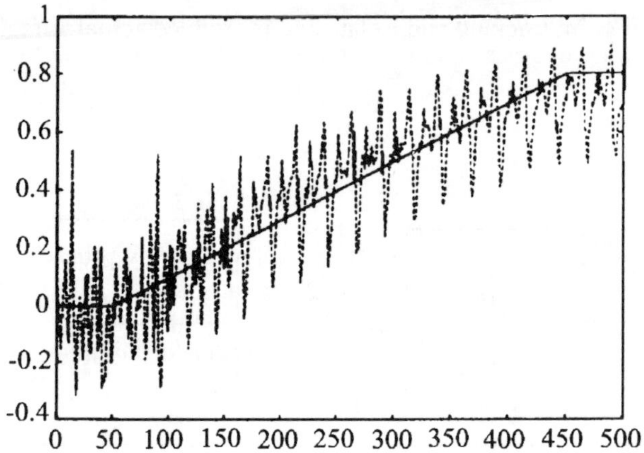

Figure 5.21 Real plant during test.
Actuator failure (solid) and diagnostic output (dotted line).
3 delays, 200,000 iterations, actmean=0.1059, ramp failure.

Additional references. Naidu *et al.* (1990) have developed a neural network sensor failure detection system along the lines suggested by Nett's work on the four-parameter controller just discussed. The back-propagation topology was used and compared to ***Finite Integral Squared Error*** (*FISE*) diagnostics as well as to the nearest neighbor classifier for an ***Internal Model Control*** (*IMC*)-controlled system involving an uncertain linear, time-invariant, first-order plant and linear or nonlinear plants that lie within the model uncetainty bounds. Detailed studies produced promising results.

5.6.5 Application of neural networks in nuclear power plants monitoring

Historically, utilities and other operators of nuclear plants have relied on human operators to monitor the plants and to diagnose any problems which occur. With the notable exceptions of Three Mile Island in the United States and Chernobyl in the former Soviet Union, this approach appears to have worked reasonably well. However, there is clear evidence linking these accidents and a number of troublesome operational incidents ("near accidents") over the years to "operator error". One possible solution is to automate the plants and "take the operator out of the loop". For a variety of reasons (legal, regulatory, and other), such a solution is not practical at the present time. The alternative approach of "backstopping" the operators by providing them with the results of automated surveillance (including diagnostics) of the overall plant, is considered here.

The large number of process parameters and system interactions pose difficulties for the operators, particularly during abnormal operation or emergencies. During such situations, individuals can be affected by stress or emotion which may influence their performance in varying degrees. Taking some of the uncertainty out of their decisions, by providing real-time diagnostics and assistance, has the potential of increasing plant availability, reliability and safety by avoiding errors that lead to trips or that endanger the safety of the plant. The emerging technology of ANNs offers a method of implementing real-time monitoring and diagnostics in a nuclear power plant. The various advanced technologies, generally regarded as being within the scope of artificial intelligence, especially neural networks, are believed to be appropriate for these tasks. The overall objective is to provide the operator with necessary information about the power plant in a way that would be useful, timely and non-intrusive. Special emphasis is given to the early detection of abnormalities and deviations from normal operation, with the intent that the operator could take corrective or compensating action, if appropriate.

Generally, the developed technology involved three specific tasks that were undertaken using a variety of methods.. These where,

1. Diagnostics based on pattern recognition in time-records and related representations of variables, (e.g. spectral densities).
2. Feature detection based on recognition of patterns in data, and
3. Modeling of phenomena and systems with interpretation of input-output relationships.

Many projects involved both pattern recognition and modeling. In most cases, comparison of predicted results (based on models developed from data taken when the system was working properly) or patterns (learned by neural network models from data presented to it) with actual results or patterns is involved. Often, data had to be preprocessed to put it into an acceptable form (e.g., a fast Fourier transformation of the time-series to produce a spectral plot of the data) before it can be introduced into a neural network. Once a neural network has been trained to recognize the various conditions or states of a complex system, it takes only one recall cycle of the neural network, typically a few milliseconds, to detect or identify a specific condition or state. If the neural network is implemented in hardware, the detection or identification is almost instantaneous. Typically, the measured variables from the nuclear power plant systems are analog signals that must be sampled, digitized, preprocessed, and normalized to expected peak values before they are introduced into neural networks. The neural networks are usually simulated on modern high-speed digital computers that carry out the calculations serially. However, it is possible to implement neural networks using specially designed microchips where the network calculations are truly carried out in parallel, thereby providing virtually instantaneous outputs (microsecond response times) for each set of inputs.

Monitoring of nuclear power plant sensors and systems.

Sensor validation. A neural network-based process for automated sensor validation during both steady-state and transient operations has been developed by Upadhyaya and Eryurek (1992). A neural network model of the process (or a portion of the process) can be developed using experimental data. The input layer of artificial processing elements (PEs) has several input signals, and the output layer usually has just one signal which is to be predicted. The neural network is trained over the range of anticipated operating conditions while the system is known to be operating properly. In operation, the neural network model repeatedly predicts the output value, based on the actual inputs. As a sensor deteriorates or fails, the measured value (which is erroneous) deviates from the predicted value (based on the input values and the neural network model). A comparison of the predicted and actual values will indicate the validity of the sensor reading. Tests with data from the Experimental Breeder Reactor - II (EBR-II) and a commercial four-loop PWR nuclear power plant, have shown that this technique can be used to validate data from sensors placed in nuclear power plants. Networks for the validation of readings of hot leg temperature, reactor power and control rod position for a four-loop PWR have demonstrated the usefulness of this approach.

Plant-wide monitoring using neural networks. The technique for sensor validation has been expanded into a plant-wide monitoring system by Upadhyaya and Eryurek, (1992), using an autoassociative neural network where the inputs and the outputs are the same variables. The number of artificial neurons in the intermediate layer (usually about two or three times the number of input nodes) was selected to minimize the training time and maximize the ability of the neural network to generalize. The neural network was trained over the range of operation using the same data for the input vector and the desired output vector. Backpropagation using a sigmoidal function with an adjustable coefficient was used to train the network when the system was operating properly. Under these conditions, the neural network outputs represent estimates of the instantaneous values of the output variables, and all of these estimates are virtually identical to the actual outputs. When a sensor begins to drift or a failure is introduced into a data channel, the actual value (neural network input) changes, but the corresponding predicted value (neural network output) remains virtually unchanged. Hence, monitoring the differences between the estimates predicted by the neural network (outputs) and the actual values from the system (inputs) provides a method of identifying drift or instrumentation system (or sensor) failure. An alternative interpretation of these differences might be that the input-output relationship of the system from which the signals come, may have changed due to system failure or changes of some sort in the system. This technique was applied to data from eighteen signals from the Experimental Breeder Reactor - II (EBR - II) during increase in power from 45% to 100%. Errors between estimates predicted by the trained neural network and the actual values during normal operation were usually less than 0.5%. When one of the sensors failed or an error was introduced into one of the sensor outputs (the inputs to the neural network), the corresponding output of the neural network changed only slightly. Hence, a difference between the network output (the predicted value of the variable) and the actual variable identified that sensor or instrumentation channel as the one with a problem.

Monitoring of check valves for operability. Although there are many possible failure mechanisms for check valves, the most common problems associated with check valve failures are due to system flow oscillations or system piping vibrations which induce check valve component wear and thus component failure. A technique involving the use of a neural network for the analysis of acoustical data from check valves to evaluate their status has been reported by Ikonomopoulos, Tsoukalas and Uhrig (1992). The ***power spectral density*** (*PSD*) of the sampled time-series at a point on the check valve body near the hinge pin is used as the input to the neural network, and the PSD of the sampled time-series at another point on the check valve body near the backstop is the desired output of the neural network. The network is trained while the flow varies over the normal range of operation when the valve is known to be operating properly. The neural network is then used in a monitoring mode to predict the output sensor PSD from the input PSD and a comparison is made between the predicted and actual output PSDs. Deviations indicate that the interrelationship between the input and output signals has changed due to a change (failure) of the valve. Analysis of time-records from two piezoelectric accelerometers attached to the body of a check valve on a large Boiling Water Reactor Nuclear Power Plant has been used to demonstrate this process. Comparison of spectra between identical 30-inch check valves (one broken and one normal), operating under identical conditions clearly demonstrated that this technique can identify the failed valve. The index of normal system behavior is the mean square difference, obtained by summing the square of the difference in individual spectral values between the predicted and actual spectra of the failed valve, divided by the number of spectral values. Records for three 6-inch check valves (one normal and two that failed for different reasons) operating under identical conditions indicated that failures with different degrees of severity give different values of mean square differences.

Loose parts monitoring. The detection of loose parts in the primary or secondary system of a nuclear power plant is based on the identification of sounds produced by tumbling parts hitting a pipe wall, the tube sheet of a steam generator, or other surfaces that are part of the coolant system boundary. The spectrum of the sound is dependent upon the size and shape of the loose part and the materials of construction of both the part and the system boundary. Once the sound spectrum for parts of different sizes and shapes (those most likely to break loose, e.g., the hinge pin of an upstream check valve) has been measured, a pattern matching technique can then be used to identify the part or parts. To date, neural networks have not been used for this purpose, but they have been used in Germany to negate false alarms of a loose parts monitor. In one of the German plants, there is a metal to metal contact (not caused by a loose part) that produces a sound that is detected by the loose parts monitor. To overcome this problem, a neural network was trained to identify the unique sound of the metal contact. Then when this sound occurs, the neural network identifies it and enables the loose parts monitor alarm.

Monitoring of performance and efficiency. Of the hundreds of values of quantities used to calculate indices of performance (efficiency, heat rate, specific fuel/energy consumption etc.) of power plants, many remain constant because they are related to the design of the system. With neural networks, it is possible to model the dynamic portion of a process using only those variables that change during operation. Once the neural network model has been developed and trained, it is then possible to determine the sensitivity of each output (e.g., efficiency or an appropriate performance index) with respect to changes in each input. The larger sensitivity coefficients indicate the input variables that mostly influence efficiency and performance. Hence, efforts to improve performance should concentrate on these quantities. This technique was applied to TVA's Sequoyah Unit #1 Nuclear Power Plant by Guo and Uhrig (1992). Of approximately 130 quantities used to calculate the heat rate (inversely related to efficiency), only 26 actually changed with time. Records of these variables for 45 weekly measurements were used as inputs to a neural network to develop a model of the thermodynamic processes of the plant. The training of the network involved a technique developed by Pao, (1989), in which a Kohonen neural network was used to cluster the data into 22 clusters with an Euclidean distance of about 1.0. Then the centroids of these clusters were used as inputs to a neural network. Backpropragation was used for training until the overall system error was about 0.01. When the original data (not the centroids of the clusters used in training) were introduced into this model for recall, the results were usually within 0.1% of the measured values.

Optimization of efficiency using sensitivity analysis. The neural network model for the thermodynamic behavior of the Sequoyah Nuclear Power Plant described above was then used for a sensitivity analysis in which the partial derivative of each output variable with respect to each individual input variable was evaluated. The larger the partial derivative, the more influence the particular input variable has on the output (efficiency or heat rate) and the sign (+ or -) indicates the direction of the influence. The ratio of the largest sensitivity coefficient to the smallest was greater than 10 for the Sequoyah Nuclear Power Plant heat rate output. The sensitivity of the heat rate with respect to different variables can suggest which efforts that improve efficiency, are likely to be most effective (i.e., the utility should devote its efforts to improving the quantities that have the largest sensitivity coefficient).

On-line thermal margin (DNBR) estimation. On-line thermal margin monitoring in nuclear power plants using neural networks has recently been described by Kim, Lee and Chang, (1992), where a neural network model was introduced to predict the *DNBR* (***departure from nuclear boiling ratio***) for a given set of operating conditions in a Korean nuclear power plant. Since DNBR is used to estimate the core thermal margin (i.e., the difference between the predicted DNBR and the limiting DNBR), it is a parameter that is critically important for safety in nuclear power plants. The approach used was to train a neural network to map the plant variables being monitored to the DNBR as calculated by the computer code COBRA. The neural network was trained over the range of input variables that was expected to occur during the fuel cycle. A fully connected three-layer feedforward neural network was used for estimating DNBR performance of the core. The output layer had a single processing element (PE) representing the DNBR for the given plant operating parameters, which were the input variables. A statistical sensitivity analysis relating the DNBR to the various parameters indicated that the major parameters affecting the DNBR of a PWR during plant operation were the core inlet temperature, the core power (of heat flux), the enthalpy rise peaking factor, the core inlet flow rate and the system pressure; hence, the input layer had five PEs. The DNBR obtained from the neural network using data not used in the training process showed that under steady state conditions, the results agreed with those obtained from COBRA calculations within $\pm$ 2.5% in virtually all cases.

Diagnosis of nuclear power plant transients. When a nuclear power plant is operating normally, the readings of the instruments in a typical control room form a pattern (or unique set) of readings that represents a normal state of the plant or system. When a disturbance occurs, the instrument readings undergo a transition to a different pattern, representing a different state that may be normal or abnormal depending upon the nature of the disturbance. The fact that the pattern of instrument readings undergoes a transition to a new state that is different, is sufficient to provide a basis for identifying the transient or the change of the state of the system. In implementing such a transient diagnosis system in a nuclear power plant a large number (perhaps 20 to 200) of output variables from the plant are sampled simultaneously, normalized to expected maximum values, preprocessed if necessary, and transmitted to the input layer of a neural network. The unique pattern among these 20 to 200 variables represents the condition of the plant at that particular instant. When the system is operating at a steady state or changing slowly, the pattern of variables at each sampling instant remains the same or changes slightly, and the output of the neural network remains the same. However, at a time Δt after a transient begins, the sampled values form a different pattern (i.e., the relationship between the variables has changed and continues to change as the transient progresses). When the sampled values are fed to a trained neural network, it gives an indication of the state of the system. Successive sets of sampled inputs will indicate the same transient is under way if the pattern is adequately developed. Indeed, there is a whole group of patterns associated with each unique transient that must be included in the training set.

Neural networks trained on simulator transients. Work by Bartlett *et al.*, (1992), has demonstrated the validity of the concepts discussed above. Data from the training simulator at TVA's Watts Bar Nuclear Power Plant provided data for some 22 to 27 variables for seven different accident transients (loss of coolant in the hot leg of the reactor coolant system (RCS), loss of coolant in the cold leg of the RCS, main steam-line break in containment, main feedwater line break in containment, total loss of off-site power, control rod ejection, and steam generator tube leak). Simultaneously sampled values of these time records at equally spaced time intervals, constituted the input vectors to the neural network for training. An "auto-adaptive" stochastic learning technique was developed by Bartlett and Uhrig, (1992), and applied to a special neural network with a dynamic node structure (e.g., the number of nodes in each of the three hidden layers used was optimized). A new method for the stochastic optimization of these inter connections using a Monte Carlo training procedure was developed to train this network to identify these seven different nuclear power plant transients. This general approach has been continued by Kim, Aljundi and Bartlett, (1992), using data from the San Onofre Nuclear Power Training Simulator.

Guo and Uhrig, (1992), simplified the diagnostic neural network by using a modified backpropagation technique to train a neural network with extensive lateral inhibition in the middle layer. Twenty two inputs and seven outputs (one for each transient used in the training) were used. The middle layer had sixteen PEs arranged in a four by four array with negatively weighted connections in both directions between adjacent (including diagonal) PEs and self-feedback on each node with positive weights. In all cases, this neural network was able to detect the transient before the plant tripped, even in the presence of 2% noise. For fast transients, the diagnosis was almost instantaneous. This work was extended through the use of a sensitivity analysis to determine the most important input variables for each transient. Then individual modular neural networks with the five or six most important input variables and a single output were used to detect each transient. These modular networks were much simpler, did not require lateral feedback in the middle layer, and gave equally good, if not better, results. The problem was that it was necessary to develop the complex neural network with lateral feedback in order to utilize sensitivity analysis to identify the most important variables for each transient. To overcome this problem, a genetic algorithm optimization was performed to select the most important variables for each transient.

Using neural networks to identify abnormal events. Neural network techniques have been applied by Ohga and Seki, (1991), to identify an abnormal event that caused a trip in a BWR in Japan. A primary feature of the system was that the result of the neural network analysis was confirmed using the knowledge base on plant status when each event occurs. The neural network recognized the change patterns of the state variables and output the event code corresponding to the abnormal event. The neural network had three layers with 40, 4, and 3 nodes in the three layers. Five kinds of state variables were used in the neural network. For each state variable, eight data values were acquired before, at, and after a plant trip. Sampling times were different before and after the trip.

Data were normalized and sent to the neural network. The event identification method was tested using a workstation. The test data were prepared, based on the simulated results of a transient analysis program. Data were prepared for different plant configurations (changing fuel burn-up, beginning or end of fuel cycle, and abnormal progression speed of any variable, etc.). Test results showed that the neural network could identify a trained event even when the plant conditions were different from those used during training and when the data acquisition system contained noise.

Prediction of plant parameters using artificial neural network model. Artificial neural networks were applied to the prediction of nuclear reactor parameters in load-following operations by Roh, Cheon, Kim and Chang (1992). The system used consisted of four parameters processing banks of neural networks. Each bank contained two types of neural networks, a general multi-layer network and a hybrid functional link network, that attempted to learn or to infer signal behaviors. The overall prediction results agreed well with the actual plant data, thereby providing the plant operators with accurate indications of system state variables. Such information is necessary in control problems and for validating instrumentation outputs and monitoring system performance.

Data from the Korean Kori-4 Nuclear Power Plant were used for training the neural networks. The five input parameters were axial offset, control rod position, critical boron rates, burn - up rates and a code-calculated trend search parameter. Each parameter corresponded to the desired output for the only two load change modes (100% to 80% and 70% to 100%). A total of 48 load change patterns were used for learning by the neural networks. The results obtained from the neural networks agreed with calculations within 3.4% and 4.4% for the hybrid functional-link network and the general multilayer network respectively.

Connectionist expert system. An expert system that has a neural network in its knowledge bases is called a connectionist expert system. A backpropagation neural network model was applied to a connectionist expert system for the identification of transients in a nuclear power plant by Cheon *et al.* (1992). Connectionist expert systems that incorporated neural networks into the diagnostic process yield great benefits in terms of speed, robustness, and knowledge acquisition and demonstrate the feasibility of connectionist expert system's applications to the identification of transients in nuclear power plants. When a transient disturbance occurs, the sensor outputs or instrument readings undergo a transient from the existing pattern to a different pattern that represents a different state of the plant. The transient identification is approached from a pattern-matching perspective in that an input pattern is constructed from symptoms, and that symptom pattern is matched to an appropriate output pattern that corresponds to the transient that occurred.

The connectionist expert system has significant advantages over the traditional rule-based expert system. Results showed that once the network had been fully trained with various patterns, it could identify the transient easily, even with incomplete or distorted patterns. Furthermore, multiple transients were identified.

The connectionist expert system approach is most appropriate for classification problems in environments where data are abundant and noisy, and where humans tend to generate brittle and perhaps contradictory IF-THEN rules. Since connectionist expert systems are very fast, they are well suited for real-time applications.

Severe accident management. *Severe accident management system on-line network* (*SAMSON*) is a computational tool used in the event of a nuclear power plant accident. Doremus, (1992), claims that SAMSON examines over 150 status points monitored by plant process computers after a severe accident has begun and makes predictions about when core damage, support plate failure, reactor vessel failure, and containment failure will occur. These predictions are based on the current state of the plant assuming that all safety equipment not already operating will fail. The status points analyzed include radiation levels, flow rates, pressure levels, temperatures and water levels.

SAMSON uses neural networks trained using backpropagation learning to make predictions. Previous training on data from an accident analysis code has allowed SAMSON to associate different states in the plant with different times to critical failures. The accidents currently recognized by SAMSON include steam generator tube ruptures and loss of coolant accidents, with breaks ranging from less than 0.001 square feet in size to breaks greater than 3 square feet. SAMSON contains several neural networks for each accident type and size, and chooses the correct network after the accident classification by an expert system. SAMSON also provides information concerning recovery strategies and the status of plant sensors.

Hybrid (neural-fuzzy) system for transient identification. A unique hybrid system has been developed by Tsoukalas, Ikonomopoulos, and Uhrig, (1991), for the identification of transients in complex systems. It couples a rule-based expert system using fuzzy logic to a pretrained artificial neural network and uses a model-reference approach to help in the identification of noisy data. The expert system performs the basic interpretation and processing of the model data. A set of pretrained artificial neural networks provides the models of the transients. Membership functions (from fuzzy logic) that condense information about a transient into a form convenient for a rule-based identification system and characterize the transients are the outputs of the neural networks. After training the system is capable of performing faster than real-time. To demonstrate the use of this system, two classical transients, (a) a rupture of the main steam line and (b) a rupture of a main feedwater line, were simulated on a computer. Three parameters, pressurizer pressure, hot leg temperature and steam level indication were chosen for differentiating these transients. Time series of these three variables during the transient were the inputs to the neural network. The output is a membership function of a fuzzy system. Tests showed that this system is capable of differentiating between the two accidents, even when the three inputs are corrupted with up to 20% random noise.

Control rod wear recognition. Wear of cladding on power control rods of nuclear power plants cause rod clusters to be replaced prematurely. To mitigate wear by repositioning the control rods, it is necessary to identify the location of each wear scar and to measure its depth during inspections. Boshers *et al.*, (1992), have described a method that combines algorithmic, rule-based and neural network methods to perform the inspection (currently performed by human operators) to identify the wear scars and find certain quantitative properties of the scar. The functions of the neural network based software are:

1. To identify and extract wear sections.
2. To determine wear information including wear position, peak wear depth, and cross-sectional are loss, and
3. To organize this information into a data base.

Initial feature extraction and neural network processing for wear recognition had been implemented. The neural network grouped all types of wear (single peak, double peak, etc.), into one class for the initial recognition purpose. The prototype neural network used eight features, initially extracted from the time records, as inputs, and the proper values for these eight quantities as the desired outputs. A threshold was established, which if exceeded, indicated that a specific fault feature has been detected. In preliminary tests, output values have exceeded 50% of the threshold value, but there were no misclassifications.

5.7 The integration of neural networks in real-time expert systems

Expert systems are especially useful in situations where the knowledge of an expert is explicitly accessible. In other words, one must be capable of translating the knowledge into a model which consists of data and rules and which describe sufficiently the behavior of the real world. However, if an expert cannot explain how (s)he solves a certain problem (when intuition is used), then an expert system to model this knowledge is not of much help. This is not the case for neural networks: they are able to find by self-learning the solving path of the problem. However typical problems that exist when using neural networks are:

(a) Finding the right architecture and (b) the right pre-processing of the inputs.

Therefore it is recommended before solving a complex problem using neural networks, to look for appropriate literature on similar topics. In general though, neural nets can be quite successful in certain areas, and are especially, more flexible than other methods.

From this perspective one could say that in certain cases it would be appropriate to partition the knowledge in "formalisable" (precise) knowledge and in "unformalisable" (vague) knowledge. The formalisable knowledge could be modeled by an expert system, the unformalisable knowledge by a neural network. A global project cycle is shown in fig. 5.22.

Problems which are likely to occur during the knowledge acquisition are:

1. Knowledge seems to be intuitive, but in reality is not. The reason is that the expert has difficulties in formalising knowledge;
2. There is no example data available for the intuitive part of the expert's knowledge.

The first problem is very difficult to recognise. A good preparation of the interviews is usually very important (time is very often a bottleneck).

The second problem can be very difficult. If there are no (or not sufficient) examples available, then there should be at least a possibility to "judge" the neural net outputs on their correctness. If this is not the case, the use of a neural net is certainly not the correct solution.

Topology of integration. There are several possibilities for integrating neural networks in expert systems:

a. Neural net functions called by the inference engine of the expert system.
b. Neural net functions to pre-process (and post-process) the data in the inference engine.

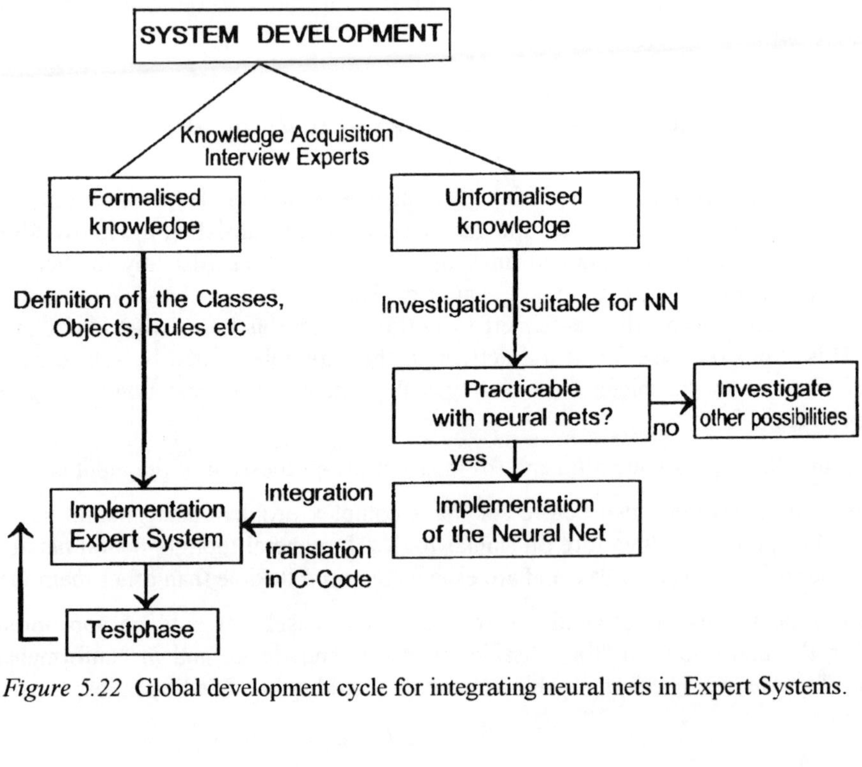

Figure 5.22 Global development cycle for integrating neural nets in Expert Systems.

5.7.1 The AI components

A number of reasons for justifying the need for real-time expert systems are the following:

- There exist too many inputs to be monitored effectively by humans.
- There exist too many complex relations between the inputs, which are essential to make a decision.
- Decisions should be taken faster than a human could do.
- The system should run 24 hours a day, 7 days a week without loss of quality.
- Hard to find qualified personnel.
- Too many personnel staff needed to run system effectively.

Some real-world examples are:

- Monitoring of the Hubble Space Telescope (NASA) (6000 sensors).
- Network monitoring (Bank of Canada).
- Monitoring car manufacturing.
- Nuclear Power Plant Risk simulator.
- Satellite monitoring and control (ESA).
- Process monitoring and control of chemical processes.
- Traffic Control.

One of the tools that can be used for these kind of applications is *RTworks*. RTworks is a software development tool for real-time monitoring and control applications. RTworks makes use of the client/server concept: it breaks its major tasks into three types of processes: inference engine processes, data acquisition processes and human interface processes (graphical interface). With a traditional expert system shell, the inference engine, data acquisition, and user interface would be all grouped together into one large process, potentially tying up resources, such as memory and CPU, and making it difficult to react quickly to critical events. In this case, one could distribute processes over several computers (e.g. LAN).

The inference engine of a real-time expert system should be adapted to copy with real-time applications. In addition to the inference strategies' forward chaining and backward chaining, the inference engine should also offer time-driven rules. For instance a rule could run periodically (every 10 seconds). Another relevant needed feature is the use of temporal reasoning in rules (e.g. if during the last 5 minutes the power has been decreased for 15 seconds, then...). The speed of the inference engine should be fast enough to be used by the real-time application. Typically, the inference engine of RTworks processes about 12.000 rules per second.

What does real-time mean? There exist many definitions of real-time. It is commonly assumed to mean "fast", in the sense that a system is considered real-time if it processes data quickly. A better definition states that "the system responds to incoming data at a rate faster than it arrives". An overview of typical response times is shown in Table 5.5.

It shows that real-time has not to be real fast to mean real-time; many traditional business applications have "real-time" elements to them.

Table 5.5 Response times of typical real-time problems.

Turning valves at a paint factory	200 mS
Satellite Ground Station monitoring	1-10 Seconds
Network Monitoring Alarm System	1 Minute
Ship tracking	1 Hour
Movie Theater Advertising	Daily
Tracking movement of continents	Yearly

Integration of user-defined processes into real-time expert systems. The RTWorks concept is one of an open system, whereby the integration of other languages (like C) should be possible. User-defined programs in C can be directly linked to the inference engine. In this case, a compiled neural net in C-code could be integrated as a new function inside the inference engine.

Neural network software suitable for integration. The situations in which neural networks are appropriate, is already discussed above. The technical integration that a neural network tool should offer, will be discussed next.

A neural network software should be capable of translating the (trained) neural net in C-code (or any other common language) for further use in other tools. One of the neural network development tools that are available and offer this, is NeuralWorks Professional II Plus from NeuralWare Corp. NeuralWorks has proven to be a good neural network tool by his great flexibility and capabilities. NeuralWorks supports about 35 learn network topologies and it is possible to implement its own learning functions.

An important feature of NeuralWorks as already mentioned, is its ability to translate networks in C Code. This makes it possible to integrate neural nets in a wide variety of tools. In this case, it can be easily linked to the inference engine of RTworks.

Technical integration of NeuralWorks in RTworks. A typical example would be the following situation: a system which consists of a monitoring subsystem and a diagnosing subsystem (fault diagnosis). The monitoring subsystem could be typically handled by an expert system while the diagnosing subsystem by the use of a neural network (assuming this cannot be comprehended in formal rules). Namely, one can often detect errors easily with formalised rules. A fault diagnosis, especially where there exist many influencing factors to the cause of the problem, is very difficult to describe in a couple of simple rules. The use of neural networks could be quite useful as a fault diagnosis system.

A "real world" example could be the monitoring of manufacturing in a paper factory using a real-time expert system. The supporting diagnosis in case of failure can be done with a neural network. In this case the presence of an expert is less needed. Example data on diagnosis cases could be gathered and could be further used to train the neural net.

The integration of RTworks (or any other relevant software) and NeuralWorks can be accomplished as follows: first the expert system performing the monitoring task is built. Rules must be defined that will detect an error during monitoring.

Next a neural network is trained for the fault diagnosis task, which should be able to give an analysis of the error (cause, impact etc.). To accomplish this, which inputs are needed must be determined first and second, relevant example data must be gathered. After successful training one can integrate the net in RTworks (or any other relevant software) by translating it in C-code. These next steps are as follows:

1. Translate the neural net in C-code.
2. Link the compiled C-code to the inference engine as a new user-defined function (rtlinkie nn.c).
3. Build rules that activate the diagnosing subsystem (function).

Some typical example rules are:

```
IF "deviated behavior"                          && Error found?
        THEN error = TRUE;                      && yes
IF error                                        && when error is detected
        THEN cause = NEURALNETDIAG(input1, input 2...);
Info("_hci", Cause of the problem is:", cause); && send a message to the operator
```

In RTworks it would be furthermore possible to divide the incoming data in a monitoring and a diagnosis datagroup. On startup only monitoring data will be received. In case there is an error, only data relevant to analyse the problem will be received. In this way one can reduce the data transfer.

References

Antsaklis P.J. (1992). Neural networks for the intelligent control of high autonomy systems. *Intelligent Systems Technical Report 92-9-1*, Department of Electrical Engineering, University of Notre Dame.

Barschdorff D., Monostori L. and T. Kottenstede (1993). Wear estimation and state classification of cutting tools in turning via artificial neural networks. *Proceedings, International Conference on Fault Diagnosis TOOLDIAG '93*, Toulouse, France, April 5-7, 669-677.

Baba N. (1989). A new approach for finding the global minimum of error function of neural networks. *Neural networks,* **2**.

Barschdorff D., Monostori L., A.F. Ndenge and G.W. Wöstenkühler (1991). Multiprocessor systems for connectionist diagnosis of technical processes. *Computers in Industry*, **17,** 131-145.

Bartlett E. and R.E. Uhrig (1992). Nuclear Power Plant Status Diagnostics Using An Artificial Neural Network. *Nuclear Technology*, **97**.

Boshers J. A., Saylor C., Kamadolli S., Wood R. and C. Isik. (1992). Control Rod Wear Recognition Using Neural Nets. In D.J. Sobajic, (Editor), *Proceedings of the 1992 Summer Workshop on "Neural Networks Computing for the Electric Power Industry*, Stanford, CA., August 17-19.

Carpenter G.A. and S. Grossberg (1987). A massively parallel architecture for a self-organizing neural pattern recognition machine. *Computer Vision, Graphics and Image Processing*, **37**, 54-115.

Carpenter G.A. and S. Grossberg (1987). ART2: Self-organization of stable category recognition codes for analog input patterns. *Applied Optics*, **26**, 3, 4919-4930.

Chang K. and W.G. Wee (1988). A Knowledge-based planning system for mechanical assembly using robots. *IEEE Expert*, 18-30.

Cheon S.W., Kang G.S. and S.H. Chang (1992). Application of Neural Networks to Connectionist Expert System for Identification of Transients in Nuclear Power Plants. *Proceedings, 2nd International Forum. Expert Systems and Computer Simulation in Energy Engineering*, Erlangen, Germany, March 17-20, pp 22-1-1 to 22-1-5.

Chen S., Cowan C.F.N. and P.M. Grant (1991). Orthogonal least squares learning algorithm for radial basis function networks. *IEEE Transactions on Neural Networks*, **2**, 2, 302-309.

Cho S.B. and J.H. Kim (1993). Rapid back-propagation learning algorithms. *Circuits, Systems and Signal Processes*, **12**, 2.

Chow M.-y., Mangum P.M. and S.O. Yee (1991). A neural network approach to real-time condition monitoring of induction motors. *IEEE Transactions on Industrial Electronics*, **38**, 6, 448-453.

Cybenko G. (1989). Approximation by superpositions of a sigmoidal function. *Mathematics of Control, Signals and Systems*, **2**, 303-314.

Davis R. (1984). Diagnostic reasoning based on structure and behavior. *Artifical Intelligence*, **24**, 347-410.

Doremus R. (1992). SAMSOM: Severe Accident Management System On-Line Network. In D.J. Sobajic, (Editor), *Proceedings of the 1992 Summer Workshop on "Neural Network Computing for the Electric Power Industry*, Stanford, CA, August 17-19.

Feldman J.A. and D.H. Ballard (1982). Connectionist models and their properties. *Cognitive Science*, **6**, 205-254.

Feng X., Zhang Y. and Q. Chen (1993). Fault simulation of a variable thrust liquid rocket engine based on neural networks. *Proceedings, International Conference on Fault Diagnosis TOOLDIAG '93*, Toulouse, France, April 5-7, 787-791.

Gien D. *et al.* (1993). A neuro-fuzzy approach for real-time diagnosis on flexible manufacturing cells. Proceedings, Sixth International Conference on Neural networks and other industrial and cognitive applications, Nimes, France, October 25-29.

Guo Y. and K.J. Doodley (1992). Identification of change structure in statistical process control. *International Journal of Production Research*, **30**, 7, 1655-1669.

Guo Z. and R.E. Uhrig R.E. (1992). Using Modular Neural Networks to Monitor Accident Conditions in Nuclear Power Plants. *Proceedings of the SPIE Technical Symposium on Intelligent Information Systems, Application of Artificial Neural Networks III*, Orlando, FL, April 20-24.

Guo Z. and R.E. Uhrig (1992). Use of Artificial Neural Networks to Analyze Nuclear Power Plant Performance. *Nuclear Technology*, **99**.

Himmelblau D.M. (1978). Fault detection in chemical and petrochemical processes. Elsevier Publishers, Amsterdam.

Hopfield J.J. (1982). Neural networks and physical systems with emergent computational abilities. *Proceedings of the National Academy of Sciences (Biophysics)*, **79**, 2554-2558.

Hoskins J.C. and D.M. Himmelblau (1988). Artificial neural network models of knowledge representation in chemical engineering. *Comput. Chem. Eng.*, **12**, 881-890.

Ikonomopoulos A., Tsoukalas L.H., Mullens J.A. and R.E. Uhrig (1992). Monitoring nuclear reactor systems using neural network and fuzzy logic. *Proceedings, 1992 Topical Meeting in Advances in Reactor Physics*, March 8-11, Charleston, U.S.A.

Ikonomopoulos A., Tsoukalas L. H. and R.E. Uhrig (1992). Use of Neural Networks to Monitor Power Plant Components. *Proceedings of the American Power Conference*, Chicago, IL., April 13-15, 1992

Kim K., Aljundi T.L. and E. Bartlett (1992). Confirmation of Artificial Neural Networks: Nuclear Power Plant Fault Diagnostics. *Transactions of the American Nuclear Society*, **66**, Chicago, IL, November 15-20.

Kim H.K., Lee S.H. and S.H. Chang (1992). Neural Network Model for On-Line Thermal Margin Estimation of a Nuclear Power Plant. *Proceedings of the Second International Forum, Expert Systems and Computer Simulation in Energy Engineering*, Erlangen, Germany, March 17-20, pp 7-2-1 to 7-2-6.

Kohonen T. (1984). Self-organisation and associative memory. Springer-Verlag, Berlin.

Konstantopoulos I.K. and P.J. Antsaklis (1993). The four parameter controller: A neural network implementation. *Proceedings, IEEE Meditteranean Symposium on New Directions in Control Theory and Applications*, Chania, Greece, June 21-23.

Kosko B. (1990). Unsupervised learning in noise. *IEEE Transactions on Neural Networks*, **1**, 1, 44-57.

Kosmatopoulos E. B., Ioannou P. A. and M.A. Christodoulou (1992). Identification of Nonlinear Systems Using New Dynamic Neural Network Structures. *Proceedings, 31st IEEE Conference on Decision and Control*, Tucson, Arizona, USA, December 16-18.

Kosmatopoulos E. B., Christodoulou M. A. and P.A. Ioannou (1993). Learning laws that ensure exponential error convergence. Proceedings, 32nd IEEE Conference on Decision and Control, San Antonio, Texas, USA, December 15-17.

Kusiak A. and G. Finke (1988). Selection of process plans in automated manufacturing systems. *IEEE Transactions of Robotics and Automation*, **4**, 4.

Lippmann R.P. (1987). An introduction to computing with neural nets. *IEEE ASP Magazine*, **4**, 4-22.

Lo A. and B. Bavarian (1991). Scheduling with neural networks for flexible manufacturing systems. *Proceedings, IEEE International Conference on Robotics and Automation*, 818-823.

Mirchandani G. and W. Cao (1989). On hidden nodes for neural nets. *IEEE Transactions on Circuits and Systems*, **36**, 661-664.

Moon Y.B. (1990). Forming part-machine families for cellular manufacturing: A neural-network approach. *Internatioanl Journal of Advanced Manufacturing Technology*, **5**, 278-291.

Matsumoto K. and T. Sakaguchi (1983). Methods to determine the restoration plan of power system by a knowledge based system. *Transactions, IEE of Japan*, **103 B**, 3.

de Mello L. S. H. and A.C. Sanderson (1990). AND/OR Graph representation of assembly plans. *IEEE Transactions on Robotics and Automation*, **6**, 2, 188-199.

Miguel L.J., Baeyens E. and J.L. Coronado (1993). Application of an ART-3 based neural network to fault diagnosis in dynamic systems. *Proceedings, International Conference on Fault Diagnosis TOOLDIAG '93*, Toulouse, France, April 5-7, 713-717.

Minsky M. and Papert S. (1969). Perceptrons - An introduction to computational geometry. MIT Press, Cambridge, Mass.

Naidu R.S., Zafiriou E. and T.J. McAvoy (1990). Use of neural networks for sensor failure detection in a control system. *IEEE Control Systems Magazine*, **10**, 49-55.

Nett C.N., Jacobson C.A. and A.T. Miller (1988). An integrated approach to controls and diagnostics: the 4-parameter controller. *Proceedings, 1988 American Control Conference*, 824-835.

Ogi H., Tanaka H. and Y. Akimoto (1991). Module neural network application for power system/equipment diagnosis. *Proceedings of ESAP '91*, April, 1991.

Ohga Y. and H. Seki (1991). Using a Neural Network for Abnormal Event Identification in BWRs. *Transactions of the American Nuclear Society*, **63**, 110-111.

Pao Y. (1989). Adaptive pattern recognition and neural networks. Addison Wesley, N.Y.

Passino K.M., Sartori M.A. and P.J. Antsaklis (1989). Neural computing for numeric-to-symbolic conversion in control systems. *IEEE Control Systems Magazine*, **9**, 44-52.

Peng Y. and J.A. Reggia (1989). A connectionist model for diagnostic problem solving. *IEEE Transactions on Systems, Man and Cybernetics*, **19**, 2, 285-298.

Polycarpou M. M. and P.A. Ioannou (1992). Neural Networks as On-Line Approximators of Nonlinear Systems. *Proceedings, 31st IEEE Conference on Decision and Control*, Tucson, Arizona, USA, December 16-18.

Rauch H.E., Kline-Schoder R.J., Adams J.C. and H.M. Youssef (1993). Fault detection, isolation and reconfiguration for aircraft using neural networks. *Proceedings, AIAA Conference on Guidance, Navigation and Control*, August '93.

Rauch H.E. and D.B. Schaechter (1992). Neural networks for control, identification and diagnosis. *Proceedings, World Space Congress*, Washington, D.C., August 28-September 5.

Ray A.K. (1991). Equipment fault diagnosis - A neural network approach. *Computers in Industry*, **16**, 169-177.

Reed N.E. *et al.* (1988). Specialized Strategies: An alternative to first principles in diagnostic problem solving. *AAAI*, 364-368.

Roh M. S., Cheon S.W., Kim H.G. and S.H. Chang (1992). Prediction of Nuclear Reactor Parameters using Artificial Neural Network Models. *Proceedings of the 2nd International Forum on "Expert Systems and Computer Simulation in Energy Engineering"*, Erlangen, Germany, March 17-20.

Rummelhart D.E. and J.L. McClelland (1986). Parallel Distributed Processing - Explorations in the microstructure of cognition, Volume 1: Foundations. MIT Press, Cambridge, Mass.

Rummelhart D.E. and J.L. McClelland (1986). Parallel Distributed Processing - Explorations in the microstructure of cognition, Volume 2: Psychological and biological models. MIT Press, Cambridge, Mass.

Schutte P. *et al.* (1987). An evaluation of a real-time fault diagnosis expert system for aircraft applications. *Proceedings, 26th IEEE Conference on Decision and Control.*

Sorsa T. and H.N. Koivo (1991). Applications of artificial neural networks in process fault diagnosis. Proceedings, IFAC Fault Detection, Supervision and Safety for Technical Processes, Baden-Baden, Germany, September 10-13.

Sorsa T., Koivo H.N. and H. Koivisto (1991). Neural networks in process fault diagnosis. *IEEE Transactions on Systems, Man and Cybernetics*, **21**, 4, 815-825

Sorsa T., Suontausta J. and H.N. Koivo (1993). Dynamic fault diagnosis using radial basis function networks. *Proceedings, International Conference on Fault Diagnosis TOOLDIAG '93*, Toulouse, France, April 5-7, 160-169.

Stavrakakis G.S. and E.N. Dialynas (1991). Efficient computer-based scheme for improving the reliability performance of power substations. *International Journal of Systems Science*, **22**, 9, 1527-1539.

Suna R. and K. Berns (1993). Pipeline diagnosis using backpropagation networks. *Proceedings, of Sixth International Confernece on Neural networks and other industrial and cognitive applications*, Nimes, France, October 25-29.

Syed A., El-Maraghy H.A. and N. Chagneux. Application of Kohonen maps in real-time monitoring and diagnosing of robotic assembly. *Proceedings, International Conference on Fault Diagnosis TOOLDIAG '93*, Toulouse, France, April 5-7, 780-786.

Tinghu Y., Binglin Z. and H. Ren (1993). A neural network methodology for rotating machinery fault diagnosis. *Proceedings, International Conference on Fault Diagnosis TOOLDIAG '93*, Toulouse, France, April 5-7, 170-178.

Tsoukalas L. H., Ikonomopoulos A. and R.E. Uhrig (1991). Hybrid Expert System-Neural Network Methodology for Transient Identification. *Proceedings of the American Power Conference*, Chicago, IL, April 29-May 1.

Wake T. and T. Sakaguchi (1984). Method to determine the fault components of power system based on description of structure and function of relay system. *Transactions, IEE of Japan*, **101 B**, 10.

Wasserman P.D. (1989). Neural Computing: Theory and Practice. Van Nostrand Reinhold, N.Y.

Widrow B. and M.A. Lehr (1990). 30 years of adaptive neural networks: Perceptron, Madaline and Back-propagation. *Proceedings of the IEEE*, **78**, 9.

Upadhyaya B. R. and E. Eryurek (1992). Application of Neural Networks for Sensor Validation and Plant Monitoring. *Nuclear Technology*, *97*, p 170.

Vaidyanathan R. and V. Venkatasubramanian (1990). Process fault detection and diagnosis using neural networks: Dynamic processes. *Proceedings, AIChE Annual National Meeting*, Chicago, U.S.A.

Venkatasubramanian V. and K. Chan (1989). A neural network methodology for process fault diagnosis. *AIChE Journal*, **35**, 1993-2002.

Watanabe K., Matsuura I., Abe M., Kubota M. and D.M. Himmelblau (1989). Incipient fault diagnosis of chemical processes via artificial neural networks. *AIChE Journal*, **35**, 1803-1812.

Widrow B. and M.E. Hoff (1960). Adaptive switching circuits. *1960 IRE WESCON Conv. Record*, Part 4, August 1960, 96-104.

Widrow B. and R.Winter (1988). Neural nets for adaptive filtering and adaptive pattern recognition. *IEEE Computer Magazine*, March, 25-39.

Yamashina H., Kumamoto H., Okumura S. and T. Ikesaki (1990). Failure diagnosis of a servovalve by neural networks with new learning algorithm and structure analysis. *International Journal of Production Research*, **28**, 6, 1009-1021.

Zeilingold D. and J. Hoey (1990). Model for a space shuttle safing and failure detection expert system. *Proceedings, 5th Conference on Artificial Intelligence for Space Applications.*

CHAPTER 6

IN-TIME FAILURE PROGNOSIS AND FATIGUE LIFE PREDICTION OF STRUCTURES

6.1. Introduction

The in-time failure prognosis and safety assessment of today's high risk industrial structures implies the accurate estimation of the residual lifetime of the structure in the course of its service. Reduction of the operation cost, estimation of the structural aging, structure life extension, prevention of catastrophic accidents, environmental protection, are some of the aspects that have to be considered in the management of complex high risk industrial systems as nuclear power plants, chemical plants, off-shore structures, marine structures, gas (LNG, LPG) installations etc.

To achieve a realistic safety assessment, the capability of modeling correctly the uncertainties, of updating the estimates on the basis of any new data available and of using field expert knowledge and heuristics is required.

Thus, a collection of data information has to be achieved on material properties, defect distribution (position and size), degradation mechanisms affecting the structure, record and forecast about loads and environment, assumptions about the states which are considered as dangerous for the component.

A whole series of inspection instruments and techniques have evolved over the years and new methods are still being developed to assist in the process of assessing the integrity and reliability of parts and assemblies. Non-destructive testing (NDT) evaluation methods are widely used in industry for checking the quality of production, and also as part of routine inspection and maintenance in service.

Because of the obvious importance of the subject, and the fact that most of the inspection methods are based on well-established scientific principles, there is a great number of publications suitable for use in the engineering practice (see periodicals as "NDT International", "Materials evaluation", etc.). In the present chapter the concept of the in-time failure prognosis and realistic safety assessment of, mainly, metallic structures will be first defined and the recent NDT methods with some of their representative applications will be presented. The concept of inspection will also be clarified. Analytical modeling and expert knowledge modeling approaches will be described for damage mechanism analysis and in-time failure prediction in structures with the ability to use fresh data and information continuously or periodically coming from the component or the structure during operation for modifying and improving the prediction.

Application examples from the nuclear, marine, mechanical and manufacturing sectors will be presented to delve into the matter.

6.2 *Recent non-destructive testing (NDT) and evaluation methods with applications*

6.2.1 Introduction

Any safety assessment of metallic structures can be thought of as a global procedure relying on a number of steps or single procedures to be concatenated: non-destructive testing (NDT) and evaluation (NDE), material characterization, analysis of loads, stress analysis, fracture mechanics analysis, fatigue crack growth (FCG) analysis, failure analysis.

Any system for in-time structural safety assessment and early failure prognosis must have the ability to analyze and interpret large quantities of information and potentially allow a time dependent updating of the interpretations in order to achieve the following goals in-time :

- Identification of the actual state of the structure and of the damage process actually taking place.
- Prevision of the future behavior of the structure.
- Decision and planning of appropriate actions.

From the above formal definition of the in-time safety assessment and failure prognosis concept, NDT and evaluation (NDE) is obviously the first important procedure which must be performed successfully in order to achieve the above goals.

Engineers are well used to assessing the properties of a material by means of standardized tests on prepared test pieces. Much valuable information is obtained from these tests including data on the tensile, compressive, shear and impact properties of the material, but such tests are of a destructive nature. In addition, the material properties, as determined in a standard test to destruction, do not necessarily give a clear guide to the performance characteristics of a complex-shaped component which forms part of some larger engineering assembly.

Defects of many types and sizes may be introduced to a material or a component during manufacture and the exact nature and size of any defects will influence the subsequent performance of the component. Other defects, such as fatigue cracks or corrosion cracks, may be generated within a material during service. The origins of defects in materials and components are shown in fig. 6.1 (from Hull and Vernon, 1988). It is therefore necessary to have reliable means for detecting the presence of defects at the manufacturing stage and also for detecting and monitoring the rate of growth of defects during the service life of a component or assembly.

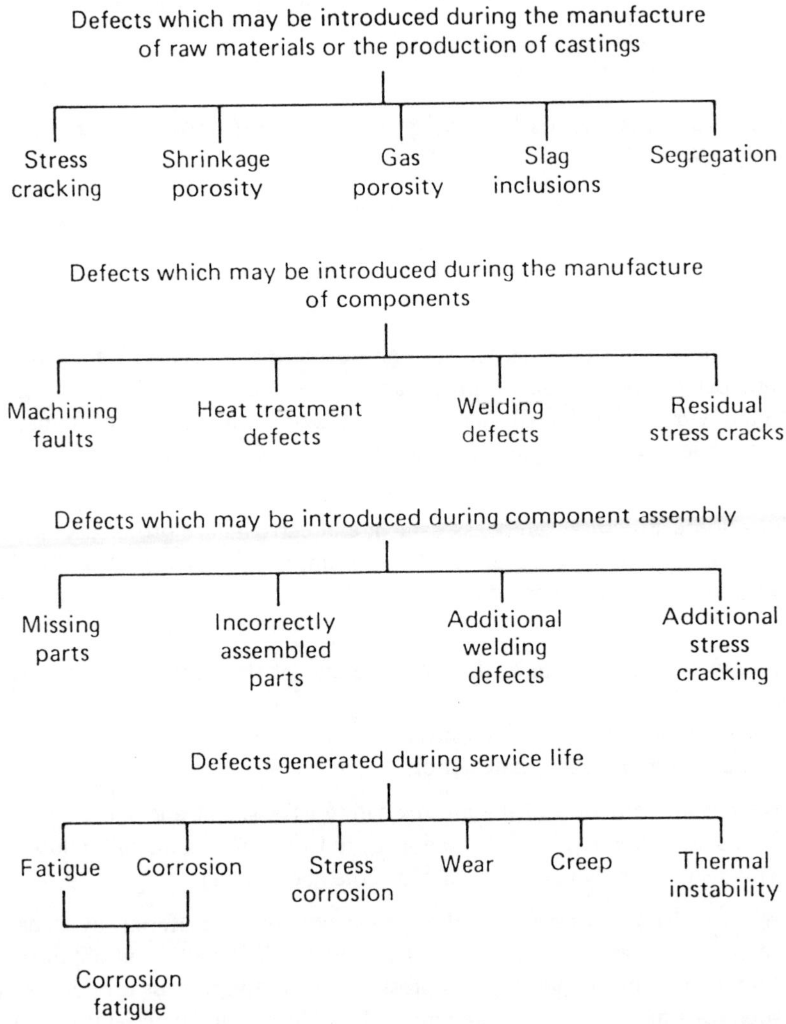

Figure 6.1 Origins of some defects found in materials and components.

Using well-established physical principles, a number of non-visual inspection systems have been developed which will provide information on the quality of a material or component and which do not alter or damage the components or assemblies which are tested. The basic principles and major features of the main non-destructive testing (NDT) systems are given in Table 6.1

Table 6.1

System	*Features*	*Applicability*
Liquid penetrant	Detection of defects which break the surface	Can be used for any metal, many plastics, glass and glazed ceramics
Magnetic particle	Detection of defects which break the surface and sub-surface defects close to the surface	Can only be used for ferromagnetic materials (most steels and irons)
Electrical methods (Eddy currents, acoustic emission)	Detection of surface de-fects and some sub-surface defects. Can also be used to measure the thickness of a non-conductive coating, such as paint, on a metal	Can be used for any metal
Ultrasonic testing	Detection of internal de-fects but can also detect surface flaws	Can be used for most materials without limi-tations on the maximum material thickness
Radiography	Detection of internal de-fects, surface defects and the correctness of part as-semblies	Can be used for many materials but there are limitations on the maxi-mum material thickness

The various non-destructive test methods can be used, in practice, in many different ways and the range of equipment available is extensive. Compact and portable equipment is available which can be used, both inside a test house or out on site, or the basic test principle can be incorporated in some large inspection system dedicated to the examination of large quantities of a single product or a small range of products or the components of a structure.

This applies to all the test methods described in this Chapter. When non-destructive testing systems are used, care must be taken and the processes controlled so that not only qualitative but quantitative information is received and that this information is both accurate and useful. If non-destructive testing is mis-applied it can lead to serious errors of judgment of component quality.

It is necessary that the most dangerous possible failure modes of a component be anticipated, and from this the types and limiting sizes of potentially dangerous defects deduced. In the first instance this is the responsibility of the product or structure designer, and thus it is he who should specify initially what defects are unacceptable and give guidance on the appropriate method of inspection. Van Dijk and Boogaard, (1992),describe an approach and format which allows effective selection of inspection systems by defining general objectives regarding detection performance and reliability.

For the successful application of non-destructive testing the test system and procedures must be suited to the inspection objectives and the types of flaws to be detected, the operator must have sufficient training and experience, and the acceptance standards must be appropriate in defining any undesirable characteristics of a non-conforming part.

In conventional design, a "design stress" is established by dividing a specific value of yield or proof stress by a suitable safety factor, and this "strength" is assumed to be representative of the material used to make a component. The fracture mechanics approach to design, however, recognises that flaws can exist in a component, before and during life in service, and attempts to describe quantitatively the effects of such flaws on component integrity. Fracture mechanics describes the capacity of critical structural components to resist the onset of rapid crack growth. Components are characterized by a material property called the *critical stress intensity factor* or *fracture toughness* and the largest flaw that can be tolerated in any specific section of a component. In addition, service environment is taken into account in this quantitative assessment.

The role of non-destructive inspection is to guarantee with a level of confidence that cracks corresponding to a critical size for fracture, at the design load, are absent from a component when the component is used in service. It might be necessary to guarantee, with confidence, that cracks smaller than the critical size are absent also. It is important to allow for sub-critical crack growth, especially in components subjected to fatigue loading or to corrosive environments, so that such components can achieve a minimum specified service life before catastrophic failure occurs. In some situations, periodic service inspection or constant monitoring might be necessary to ensure that cracks do not reach a critical size. The use of fracture mechanics concept in design places a premium on the ability of the various non-destructive methods to detect small cracks. The difference between the critical size and the smallest detectable size becomes the level of safety (Lucia 1985, Dufresne *et al.* 1988).

The introduction of any inspection system incurs cost but very often the effective use of suitable inspection techniques will give rise to very considerable financial savings.

While effective quality control inspection can result in financial savings and help to prevent catastrophic failures in service, it is also true to say that the imposition of too many or too sensitive inspection systems can be very wasteful in terms of both time and money. Excessive inspection may not result in an increase in product performance or reliability. Absolute perfection in a product is impossible to achieve and attempting to get very close to the ideal can prove to be very expensive.

The main non-destructive testing systems are briefly described in the next section.

6.2.2 The main non-destructive testing methods

6.2.2.1 Liquid penetrant inspection

Liquid penetrant inspection is a technique which can be used to detect defects in a wide range of components, provided that the defect breaks the surface of the material. The principle of the technique is that a liquid is drawn by capillary attraction into the defect and, after subsequent development, any surface-breaking defects may be rendered visible to the human eye. In order to achieve good defect visibility, the penetrating liquid will either be coloured with a bright and persistent dye or else contain a fluorescent compound. In the former type the dye is generally red and the developed surface can be viewed in natural or artificial light, but in the latter case the component must be viewed under ultra-violet light if indications of defects are to be seen. There are five essential steps in the penetrant inspection method. These are:

Surface preparation. All surfaces of a component must be thoroughly cleaned and completely dried before it is subjected to inspection. It is important that any surfaces to be examined for defects must be free from oil, water, grease or other contaminants if successful indication of defects is to be achieved.

Application of penetrant. After surface preparation, liquid penetrant is applied in a suitable manner, so as to form a film of penetrant over the component surface. The liquid film should remain on the surface for a period sufficient to allow for full penetration into surface defects.

Removal of excess penetrant. It is usually necessary to remove excess penetrant from the surface of the component. Some penetrants can be washed off the surface with water, while others require the use of specific solvents. Uniform removal of excess penetrant is necessary for effective inspection.

Development. The development stage is necessary to reveal clearly the presence of any defect. The development is usually a very fine chalk powder. This may be applied dry, but more commonly is applied by spraying the surface with chalk dust suspended in a volatile carrier fluid. A thin uniform layer of chalk is deposited on the surface of the component. Penetrant liquid present within defects will be slowly drawn by capillary action into the pores of the chalk. There will be some spread of penetrant within the developer and this will magnify the apparent width of a defect. When a dye penetrant is used the dye colour must be in sharp contrast to the uniform white of the chalk-covered surface. The development stage may sometimes be omitted when a fluorescent penetrant is used.

Observation and inspection. After an optimum developing time has been allowed, the component surface is inspected for indications of penetrant "bleed back" into the developer. Dye-penetrant inspection is carried out in strong lighting conditions, while fluorescent-penetrant inspection is performed in a suitable screened area using ultra-violet light. The latter technique causes the penetrant to emit visible light, and defects are brilliantly outlined.

The liquid penetrant process is comparatively simple as no electronic systems are involved, and the equipment necessary is cheaper than that required for other non-destructive testing systems. The establishment of procedures, and inspection standards for specific product parts, is usually less difficult than for more sophisticated methods. The technique can be employed for any material except porous materials, and, in certain cases, its sensitivity is greater than that of magnetic particle inspection. Penetrant inspection is suitable for components of virtually any size or shape and is used for both the quality control inspection of semi-finished and finished production items and for routine in-service inspection of components.

The system is used in the aerospace industries by both producers for the quality control of production and by users during regular maintenance and safety checks. Typical components which are checked by this system are turbine rotor discs and blades, aircraft wheels, castings, forged components and welded assemblies. Many automotive parts, particularly aluminum castings and forgings, including pistons and cylinder heads, are subjected to this form of quality control inspection before assembly. Penetrant testing is also used for the regular in-service examination of the bogie frames of railway locomotives and rolling stock in the search for fatigue cracking.

6.2.2.2 Magnetic particle inspection

Magnetic particle inspection is a sensitive method of locating surface and some sub-surface defects in ferro-magnetic components. The basic processing parameters depend on relatively simple concepts. In essence, when a ferro-magnetic component is magnetised, magnetic discontinuities that lie in a direction approximately perpendicular to the field direction will result in the formation of a strong "leakage field". This leakage field is present at and above the surface of the magnetised component, and its presence can be visibly detected by the utilization of finely divided magnetic particles. The magnetic particles which are used for inspection may be made from any ferro-magnetic material of low remanence and they are usually finely divided powders of either metal oxides or metals. The particles are classified as *dry* or *wet* according to the manner in which they are carried to a component. Dry particles are carried in air or gas suspension while wet particles are carried in liquid suspension. The application of dry particles or wet particles in a liquid carrier, over the surface of the component, results in a collection of magnetic particles at a discontinuity. The "magnetic bridge" so formed indicates the location, size, and shape of the discontinuity.

Magnetisation may be induced in a component by using permanent magnets, electro-magnets or by passing high currents through or around the component. The latter technique is widely used for production quality control application because high-intensity magnetic field can be generated within components. Hence, good sensitivity in flaw indication and detection is attained.

The effectiveness of defect indication will depend on the orientation of the flaw to the induced magnetic field and will be greatest when the defect is perpendicular to the field, as it is shown in fig. 6.2.

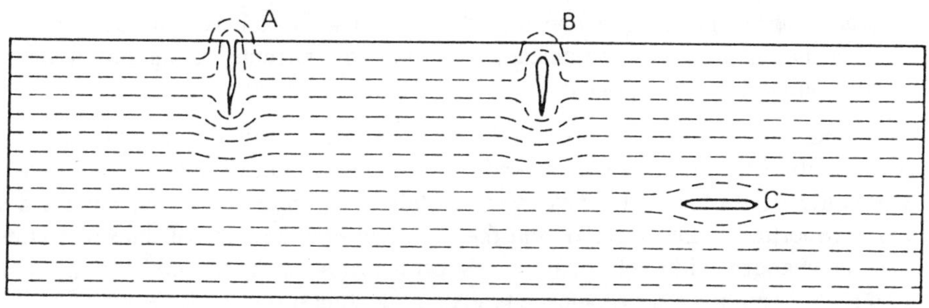

Figure 6.2 Magnetic flaw detection.
Detectable surface leakage fields produced by defects A and B. Defect C likely to remain undetected.

A component with a continuous hole through it can be magnetised by energising a straight conducting cable passing through the hole. This inspection technique is often used in the examination of parts such as pipe connectors, hollow cylinders, gear wheels and large nuts.

The principal industrial uses of magnetic particle inspection are in-process inspection, final inspection, receiving inspection and in maintenance and overhaul.

Although in-process inspection is used to highlight defects, as soon as possible in the processing route, a final inspection gives the customer a better guarantee of defect-free components.

During receiving inspection, both semi-finished purchased parts and raw materials are inspected to detect initial defects. Incoming rod and bar stock, forging blanks and rough castings are inspected in this way.

The transportation industries (road, rail, aircraft and shipping) maintain planned overhaul schedules at which critical parts are inspected for cracks. Crank-shafts, frames, flywheels, crane hooks, shafts, steam turbine blades and fasteners are examples of components vulnerable to failure, in particular fatigue failure. Hence, there is a need for regular inspection.

6.2.2.3 Electrical test methods (eddy current testing (ECT))

If a coil carrying an alternating current is placed in proximity to a conductive material, secondary or eddy currents will be induced within the material. The induced currents will produce a magnetic field which will be in opposition to the primary magnetic field surrounding the coil. This interaction between fields causes a back e.m.f. in the coil and, hence, a change in the coil impedance value. If a material is uniform in composition and dimensions, the impedance value of a search coil placed close to the surface should be the same at all points on the surface, apart from some variation observed close to the edges of the sample. If the material contains a discontinuity, the distribution of eddy currents (and their magnitude) will be altered in its vicinity and there will be a consequent reduction in the magnetic field associated with the eddy currents, so the coil impedance value will be altered.

Eddy currents flow in closed loops within a material and both the magnitude and the timing or phase of the currents will depend on a number of factors. These factors include the magnitude of the magnetic field surrounding the primary coil, the electrical and magnetic properties of the material, and the presence or otherwise of discontinuities or dimensional changes within the material. Several types of search coil are used, two common types being the flat or pancake type coil which is suitable for the examination of flat surfaces, and the solenoid type coil which can be used in conjunction with solid or tubular cylindrical parts. For tubes, a solenoid type coil may placed around the tube or inserted into the bore.

These techniques are highly versatile and, with the appropriate equipment and test method, can be used to detect surface and sub-surface defects within components, determine the thickness of surface coatings, provide information about structural features, such as crystal grain size and heat treatment condition, and also to measure physical properties including electrical conductivity, magnetic permeability and physical hardness.

In the case of ferro-magnetic materials there is a continuous spectrum of effects ranging from predominantly magnetic effects at low frequencies to eddy current effects at the higher frequencies, where the magnetic effect is suppressed. At the higher-frequency end, the techniques used relate to assessment of the distortion and reduction of the eddy current fields induced within the material. Changes in eddy current fields will indicate those defects which affect the flow of eddy currents in the surface layers of the material, such as cracks. At low frequencies it is the magnetic effects which predominate and the effect that the material has on the B-H loop is observed, and this relates to structural properties such as hardness. With non-magnetic materials, only eddy current effects occur, irrespective of the frequency but, in general, inspection techniques for non-magnetic materials use the higher frequencies, that is greater than 1kHz.

For many test and inspection purposes, a coil or coils are mounted in a holder as an inspection probe. As stated earlier, a coil is frequently wound around a ferrite core and, while the search coil is generally protected by a plastic casing, the end of the ferrite core often projects beyond the plastic case. Eddy current test probes do not need any coupling fluid between them and the testpiece, unlike ultrasonic probes, because they are coupled to the material by a magnetic field, and consequently little if any surface preparation is necessary prior to inspection. Many types of inspection probes have been designed but generally they can be divided into surface probes and hole probes.

It is necessary to calibrate eddy current test equipment, and reference testpieces for calibration purposes should be made from material of similar type and quality to that which is to be tested so as to have the same conductivity value. A test-block should contain a series of defects of known size and shape and these are frequently made by making several fine saw-cuts of varying but known depth.

In many situations users will also use defective parts containing, for example, fatigue cracks as reference and calibration pieces.

One method of representing the signals from eddy current inspection probes is by the phasor technique or phase analysis. When it is only necessary to detect changes in one of the parameters which affect impedance and all over factors are constant, then the measurement of a change in impedance value will reflect a change in that parameter. However, there are many instances where it becomes necessary to separate the responses from more than one parameter, and to separate the reactive and resistive components of impedance. This requires the use of more sophisticated instruments but in this way it becomes possible to identify the type of defect present and not merely its position.

There is a phase difference between the reactive and resistive components of the measurement voltage. Consider the voltage as vectors A and B. The frequency is the same for both and, therefore, the radian velocity ω will be the same for both ($=2\pi f$).

The resistive and reactive components of a measurement (probe coil) voltage can be fed to the "X" plates and "Y" plates respectively of a cathode ray oscilloscope and displayed as a two-dimensional representation.

The impedance changes caused by various types of defect or by changes in conductivity will give screen displays as shown in figs. 6.3b and c.

The eddy current system is a highly versatile system and can be used to detect not only cracks but several other conditions, including corrosion. Corrosion of hidden surfaces as, for example, within aircraft structures, can be detected using phase-sensitive equipment. It is a comparative technique in that readings made in a suspect area are compared with instrument readings obtained from sound, non-corroded material (Hagemaier *et al.* 1985).

An eddy current test system can also be used for the routine inspection of aircraft undercarriage wheels. The wheel is placed on a turntable and the probe coil which is mounted at the end of an adjustable arm, is positioned near the bottom of the wheel. As the wheel turns on the turntable so the probe arm moves slowly up the wheel, giving a close helical search pattern. It is necessary to use a second probe to check under the wheel flange. A hand-held probe is used for this part of the inspection.

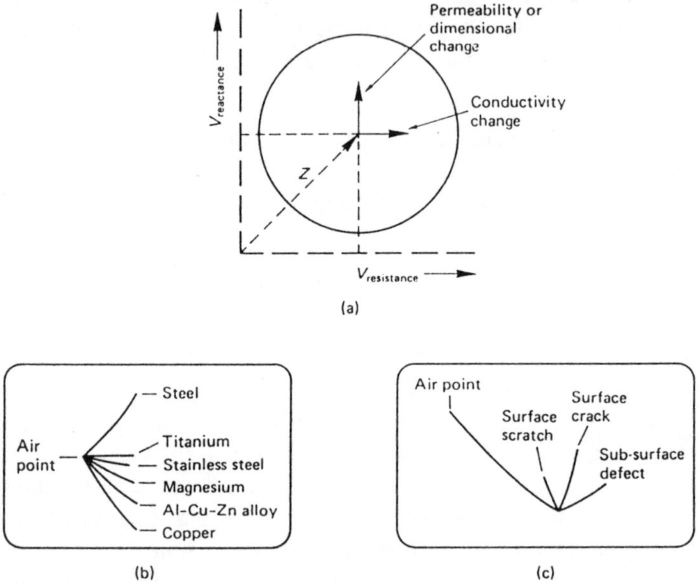

Figure 6.3
(a) Vector point.
(b) Impedance plane display on oscilloscope, showing differing conductivities.
(c) Impedance plane display, showing defect indications.

The ability of eddy current techniques to determine the conductivity of a material has been utilized for the purpose of checking areas of heat-damaged skin on aircraft structures. If the type of aluminum alloy used in aircraft construction becomes overheated it could suffer a serious loss of strength. This is accompanied by an increase in the electrical conductivity of the alloy. The conductivity of sound material is generally within the range of 31 to 35 per cent IACS. Defective or heat-damaged material would show a conductivity in excess of 35 per cent IACS.

6.2.2.4 Ultrasonic testing

Ultrasonic techniques are very widely used for the detection of internal defects in materials, but they can also be used for the detection of small surface cracks. Ultrasonics are used for the quality control inspection of part processed material, such as rolled slabs, as well as for the inspection of finished components. The techniques are also in regular use for the in-service testing of parts and assemblies.

Sound waves are elastic waves which can be transmitted though both fluid and solid media. The audible range of frequency is from about 20 Hz to about 20 kHz but it is possible to produce elastic waves of the same nature as sound at frequencies up to 500 MHz. Elastic waves with frequencies higher than the audio range are described as *ultrasonic*. The waves used for the non-destructive inspection of materials are usually within the frequency range 0.5 MHz to 20 MHz.

Piezo-electric materials form the basis of electro-mechanical transducers for ultrasonic NDT. The original piezo-electric material used was natural quartz. Quartz is still used to some extent but other materials, including barium titanate, lead metaniobate and lead zirconate, are used widely. When an alternating voltage is applied across the thickness of a disc of piezo-electric material, the disc will contract and expand, and in so doing will generate a compression wave normal to the disc in the surrounding medium. When quartz is used the disc is cut in a particular direction from a natural crystal but the transducer discs made from ceramic materials such as barium titanate are composed of many crystals fused together, the crystals being permanently polarised to vibrate in one plane only.

Wave generation is most efficient when the transducer crystal vibrates at its natural frequency, and this is determined by the dimensions and elastic constants of the material used. Hence, a 10 MHz crystal will be thinner than a 5 MHz crystal. A transducer for sound generation will also detect sound. An ultrasonic wave incident on a crystal will cause it to vibrate, producing an alternating current across the crystal faces. In some ultrasonic testing techniques two transducers are used - one to transmit the beam and the other acting as the receiver - but in very many cases only one transducer in necessary. This acts as both transmitter and receiver. Ultrasonic is transmitted as a series of pulses of extremely short duration and during the time interval between transmissions the crystal can detect reflected signals.

The presence of a defect within a material may be found using ultrasonics with either a transmission technique or a reflection technique.

Normal probe transmission method. In this method a transmitter probe is placed in contact with the testpiece surface, using a liquid coupler, and a receiving probe is placed on the opposite side of the material (see fig. 6.4).

If there is no defect within the material, a certain strength of signal will reach the receiver. If a defect is present between the transmitter and receiver, there will be a reduction in the strength of the received signal because of partial reflection of the pulse by the defect. Thus, the presence of a defect can be inferred.

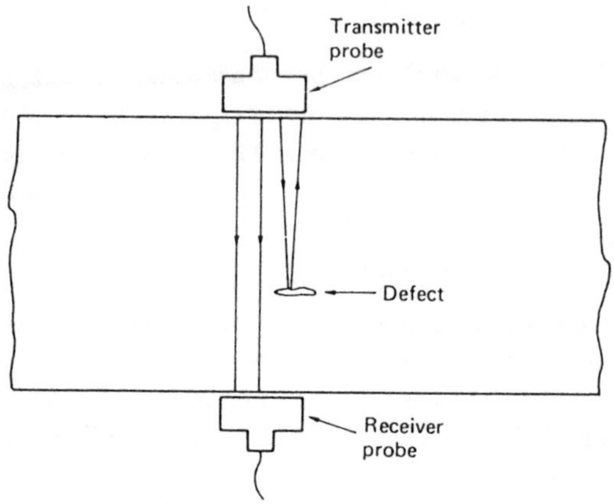

Figure 6.4 Normal probe transmission technique

This method possesses a number of disadvantages. These are:

(a) The specimen must have parallel sides and it must be possible to reach both sides of the piece.
(b) Two probes are required, thus doubling the possibility of having inefficient fluid coupling.
(c) Care must be taken that the two probes are exactly opposite one another.
(d) There is no indication of the depth of a defect.

Angle probe transmission method. There are certain testing situations in which it is not possible to place a normal probe at right angles to a defect and the only reasonable solution is offered by angle probes. A good example of this technique is in the inspection of butt welds in parallel-sided plates. The transmitter and receiver probes are arranged as in fig. 6.5a. If there is any defect in the weld zone, this will cause a reduction in the received signal strength. Distance AB is known as the "skip distance" and for the complete scanning of a weld the probes should be moved over the plate surface as shown in fig. 6.5b. In practice, both probes would be mounted in a jig so that they are always at the correct separation distance.

Angle probe reflection method. Defects can also be detected using one angle probe in the reflection mode, as shown in fig. 6.6. It is important when using an angle probe in this type of test that the flaw detector be accurately calibrated using a reference test-block. The design and use of a calibration block are covered in Hull and Vernon (1988).

A Rayleigh, or surface, wave can be used for the detection of surface cracks (see fig. 6.7). The presence of a surface defect will reflect the surface wave to give an echo signal in the usual way. Surface waves will follow the surface contours and so the method is suitable for shaped components such as turbine blades.

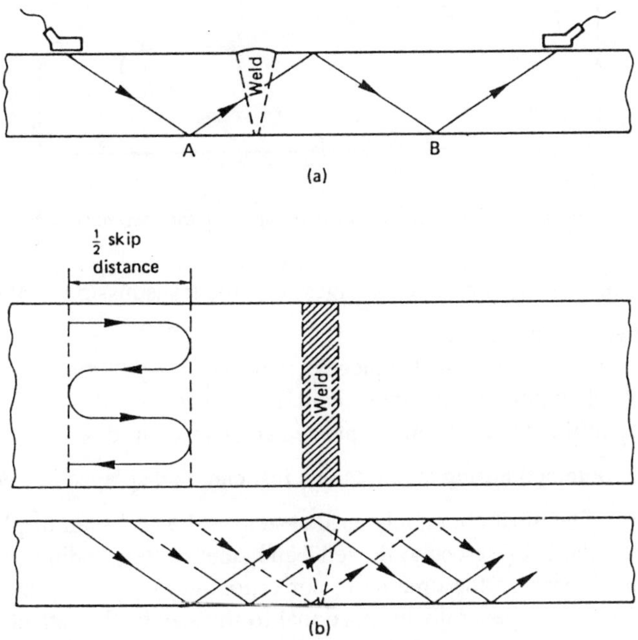

Figure 6.5 Angle probe transmission method
(a) probe positions and skip distance;
(b) scanning method for complete inspection of butt weld.

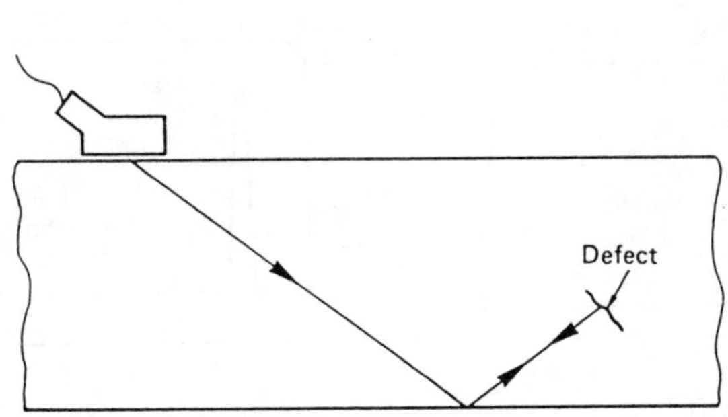

Figure 6.6 Reflective technique with angle probe.

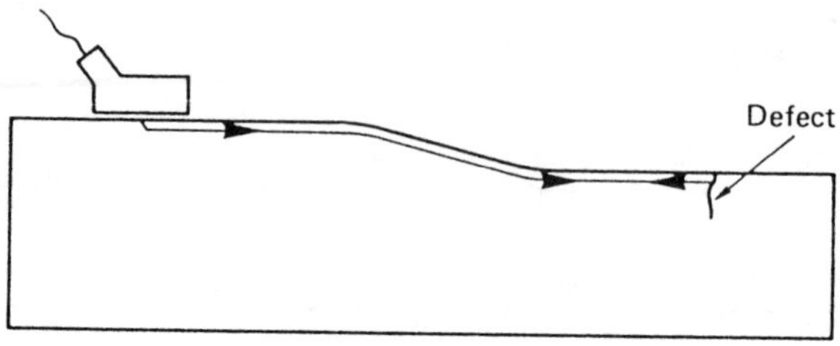

Figure 6.7 Crack detection using a surface wave probe.

The reflection method has certain advantages over the transmission method. These are:

(a) The specimen may be of any shape.
(b) Access to only one side of the testpiece is required.
(c) Only one coupling point exists, thus minimizing error.
(d) The distance of the defects from the probe can be measured.

The information obtained during an ultrasonic test can be displayed in several ways.

"A" scan display. The most commonly used system is the "A" scan display (see fig. 6.8). A blip appears on the CRT screen at the left-hand side, corresponding to the initial pulse, and further blips appear on the time base, corresponding to any signal echoes received. The height of the echo is generally proportional to the size of the reflecting surface but it is affected by the distance travelled by the signal and attenuation effects within the material. The linear position of the echo is proportional to the distance of the reflecting surface from the probe, assuming a linear time base. This is the normal type of display for hand probe inspection techniques.

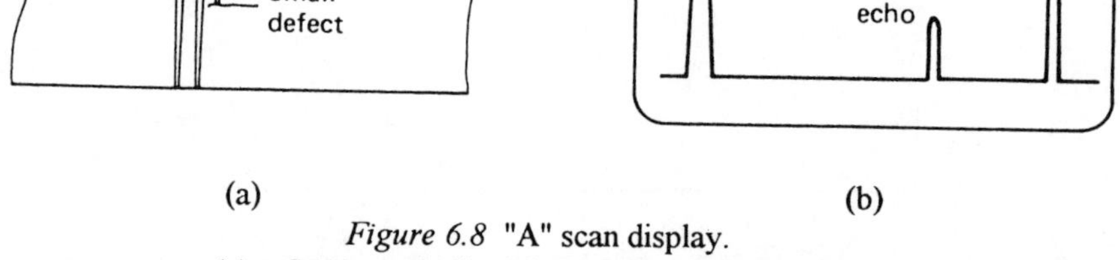

(a) (b)

Figure 6.8 "A" scan display.
(a) reflections obtained from defect and backwall.
(b) representation of "A" scan screen display.

A disadvantage of the "A" scan is that there may be no permanent record, unless a photograph is taken of the screen image, although more sophisticated modern equipment have the facility for digital recording.

"B" scan display. The "B" scan enables a record to be made of the position of defects within a material. The system is illustrated in fig. 6.9. There needs to be co-ordination between the probe position and the trace, and the use of "B" scan is confined to automatic and semi-automatic testing techniques. With the probe in position "1" the indication on the screen is as shown in fig. 6.9, with (i) representing the initial signal and (ii) representing the backwall. When the probe is moved to position "2", line (iii) on the display represents the defect. This representation of the testpiece cross-section may be recorded on a paper chart, photographed, or viewed on a long-persistence screen.

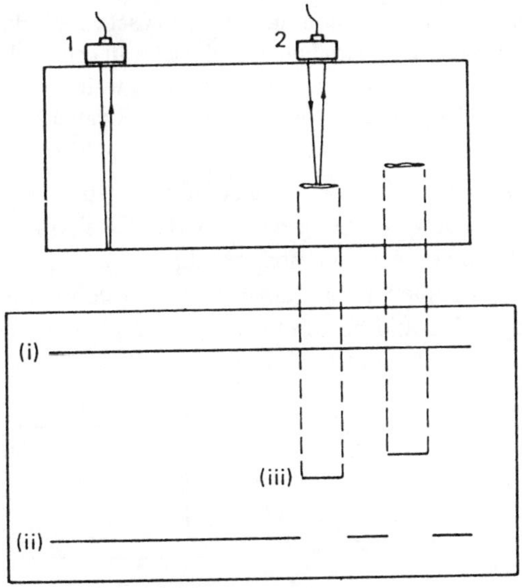

Figure 6.9 "B" scan display.

"C" scan display. While the "B" scan gives a representation of a side elevation of the testpiece, another method, termed "C" scan can be used to produce a plan view. Again, the "C" scan display is confined to automatic testing (Nielsen (1981), Yanagi, (1983)).

Identification of defects.

By means of ultrasonic methods not only can the exact position of internal defects be determined but it is also possible, in many cases, to distinguish the type of defect. In the following, the various types of signal response received from particular types of defect will be considered.

(*a*) *Defect at right angles to the beam direction.* When no defect is present, a large echo signal should be received from the backwall. The presence of a small defect should give a small defect echo and some reduction in the strength of the backwall echo. When the defect size is greater than the probe diameter the defect echo will be large and the backwall echo may be lost (fig. 6.10), depending on the depth of the defect in relation to beam spread in the far zone.

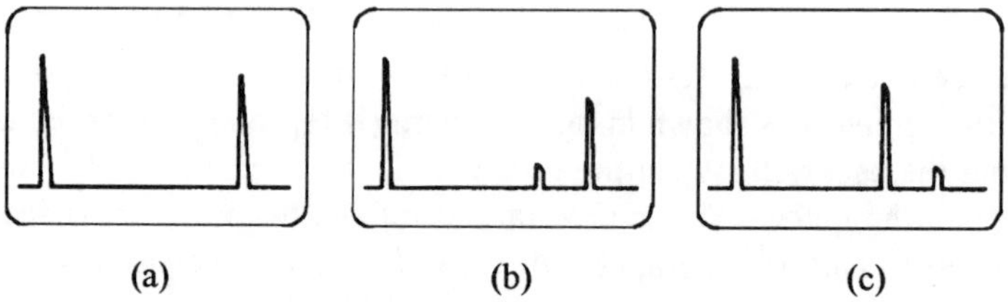

Figure 6.10 Effect of defect size on screen display.
(a) defect free-initial pulse and backwall echo only;
(b) small defect echo but large backwall echo;
(c) large defect echo with small backwall echo.

(*b*) *Defects other than plane defects.* Areas of micro-porosity will cause a general scattering of the beam, giving some "grass" on the CRT trace and with loss of the backwall echo (fig. 6.11a). A large spherical or elliptical inclusion or hole would tend to give a small defect echo coupled with a small backwall echo (fig. 6.11b), while a plain trace showing no echo at all could be an indication of a plane defect at some angle other than normal to the path of the beam (fig. 6.11c).

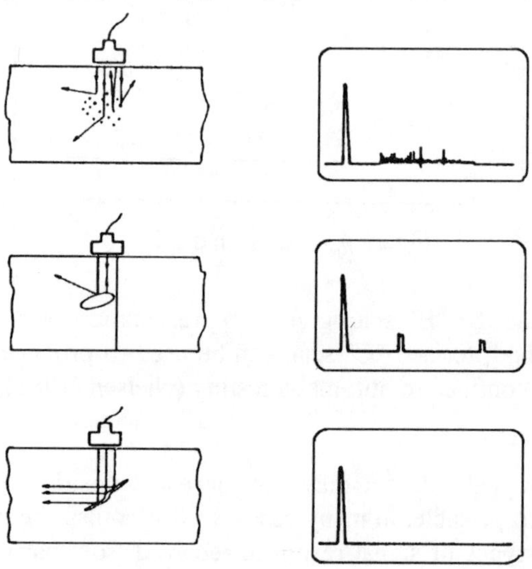

Figure 6.11 (a) Micro-porosity, (b) Elliptical defect, (c) Angled defect.

(*c*) *Laminations in thick plate.* The plate should be completely scanned in a methodical manner, as shown in fig. 6.12. The indications of laminations are a closer spacing of echoes and a more rapid fall-off in the size of the echo signals. Either or both of these indications are signs of lamination (see fig. 6.13).

(*d*) *Lamination in thin plate.*

A thin plate may be considered to be a plate of thickness less than the dead zone of the probe. A sound plate will show a regular series of echoes with exponential fall-off of amplitude. A laminated region will show a close spacing with a much faster rate of amplitude fall-off. The pattern may change from an even to an irregular outline. It is this pattern change which, in may cases, gives the best indication of lamination in thin plate (fig. 6.14).

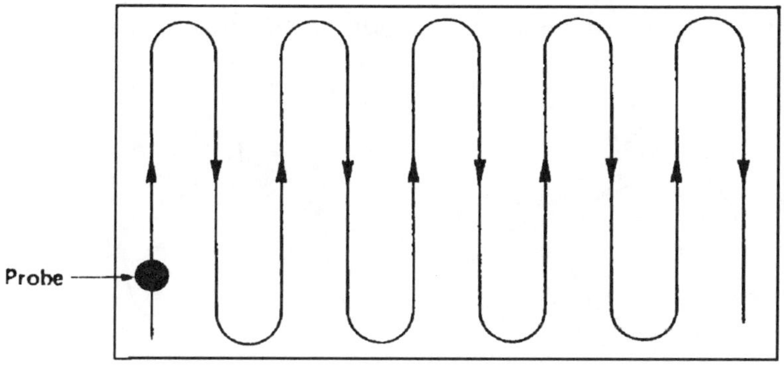

Figure 6.12 Method of scanning a large surface.

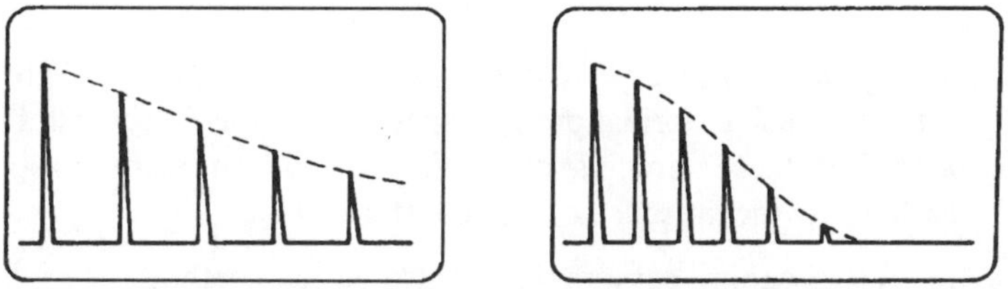

(a) (b)

Figure 6.13 Indication of lamination in thick plate: (a) good plate; (b) laminated plate

(*e*) *Weld defects.* Ultrasonic testing using angle probes in either the reflection or transmission mode is a reliable method for the detection of defects in butt welds and for determining their exact location. It is, however, fairly difficult to determine with certainty the exact nature of the defect, and much depends upon the skill and experience of the operator. If, following ultrasonic inspection, there is any doubt in the mind of the operator about the quality of a weld, then it would be wise to check radiographically the suspect area.

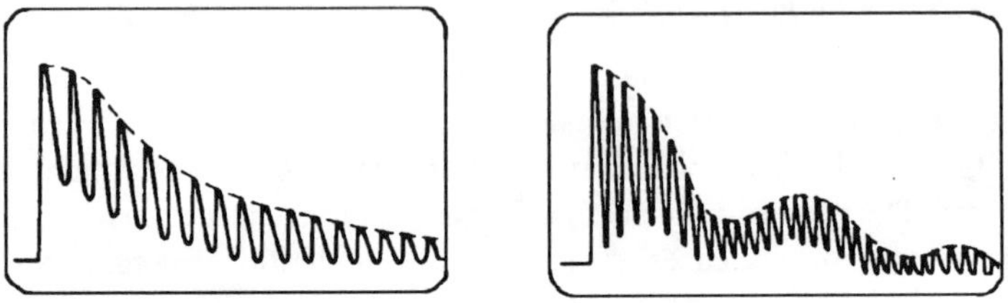

(a) (b)

Figure 6.14 Indication of lamination in thin plate: (a) good plate; (b) laminated plate

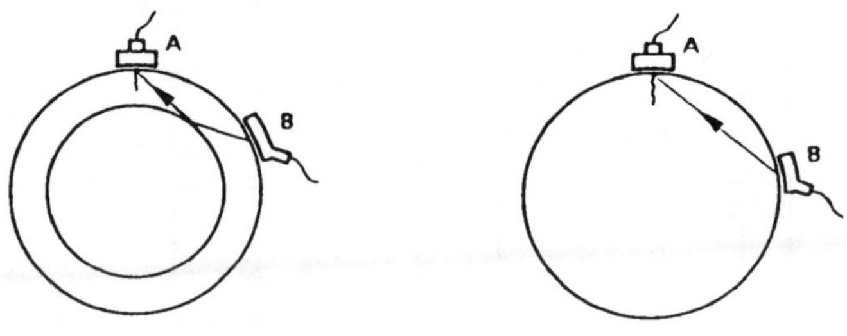

(a) (b)

Figure 6.15 Detection of radial defects in:
(a) tubes,
(b) solid bar; normal probe in position A will not show defect but angle probe at B will.

(f) Radial defects in cylindrical tubes and shafts. A radial defect in a cylindrical member is not generally detectable using normal probe inspection, as the defect will be parallel to the ultrasonic beam. In these circumstances the use of an angle probe reflection technique will clearly show the presence of defects (fig. 6.15).

As has been seen in the foregoing paragraphs, ultrasonic test methods are suitable for the detection, identification and size assessment of a wide variety of both surface and sub-surface defects in metallic materials, provided that there is, for reflection techniques, access to one surface. There are automated systems which are highly suitable for the routine inspection of production items at both an intermediate stage and the final stage of manufacture. Using hand held probes, many types of components can be tested, including in *situ* testing. This latter capability makes the method particularly attractive for the routine inspection of aircraft and road and rail vehicles in the search for incipient fatigue cracks (Yanagi, 1983). In aircraft inspection, specific test methods have been developed for each particular application and the procedures listed in the appropriate manuals must be followed if consistent results are to be achieved. In may cases a probe will be specially designed for one specific type of inspection.

In nuclear plants, chemical plants, pipelines, vessels, off-shore and marine structures material damages due to fatigue loading causes can be detected and dimensioned efficiently using ultrasonic NDT methods (Landez *et al.* 1992). The continuous monitoring by ultrasonic can help to monitor high stress concentration or cracked zones. Ultrasonic probes are permanently stuck on the region to monitor the critical area or (in case of existing crack) the crack tip and the crack root (by diffraction and reflection respectively). Any change in the UT signal amplitude indicates that a modification took place in the inspected zone (formation of a crack or propagation of the existing one).

Under the assumption that no additional non-linear effects affect the measurement, it has been shown that this method can detect damage from nearly 10% of life span.

6.2.2.4. Radiography

Very-short-wavelength electromagnetic radiation, namely X-rays or γ-rays, will penetrate through solid media but will be partially absorbed by the medium. The amount of absorption which will occur will depend upon the density and thickness of the material the radiation is passing through, and also the characteristics of the radiation. The radiation which passes through the material can be detected and recorded on either film or sensitised paper, viewed on a fluorescent screen, or detected and monitored by electronic sensing equipment. Strictly speaking, the term *radiography* implies a process in which an image is produced on film. When a permanent image is produced on radiation-sensitive paper, the process is known as *paper radiography*. The system in which a latent image is created on an electrostatically charged plate and this latent image used to produce a permanent image on paper is known as *xeroradiography*. The process in which a transient image is produced on a fluorescent screen is termed *fluoroscopy*, and when the intensity of the radiation passing through a material is monitored by electronic equipment the process is termed *radiation gauging*.

It is possible to utilize a beam of neutrons rather than X-rays or γ-rays for inspection purposes, this being termed *neutron radiography*. The process of neutron radiography involves the transmission of neutrons through a component or assembly and the production of a radiograph on film.

After an exposed radiographic film has been developed, an image of varying density will be observed with those portions of the film which have received the largest amounts of radiation being the darkest. As mentioned earlier, the amount of radiation absorbed by the material will be a function of its density and thickness. The amount of absorption will also be affected by the presence of certain defects such as voids or porosity within the material. Thus radiography can be used for the inspection of materials and components to detect certain types of defect.

The use of radiography and related processes must be strictly controlled because exposure of humans to radiation could lead to body tissue damage.

Radiography is capable of detecting any feature in a component or structure provided that there are sufficient differences in thickness or density within the testpiece. Large differences are more readily detected than small differences. The main types of defect which can be distinguished are porosity and other voids and inclusions, where the density of the inclusion differs form that of the basis material. Generally speaking, the best results will be obtained when the defect has an appreciable thickness in a direction parallel to the radiation beam. Plane defects such as cracks are not always detectable and the ability to locate a crack will depend upon its orientation to the beam. The sensitivity possible in radiography depends upon many factors but, generally, if a feature causes a change in absorption of 2 per cent or more compared with the surrounding material, then it will be detectable.

Radiography and ultrasonics (see § 6.2.2.3) are the two methods which are generally used for the successful detection of internal flaws that are located well below the surface, but neither method is restricted to the detection of this type of defect. The methods are complementary to one another in that radiography tends to be more effective when flaws are non-planar in type, whereas ultrasonic tends to be more effective when the defects are planar.

Radiographic inspection techniques are frequently used for the checking of welds and castings, and in many instances radiography is specified for the inspection of components. This is the case for weldments and thick-wall castings which form part of high-pressure systems.

Radiography can also be used to inspect assemblies to check the condition and proper placement of components. It is also used to check the level of liquid in sealed liquid-filled systems. One application for which radiography is very well suited is the inspection of electrical and electronic component assemblies to detect cracks, broken wires, missing or misplaced components and unsoldered connections.

Radiography can be used to inspect most types of solid material but there could be problems with very high or very low density materials. Non-metallic and metallic materials, both ferrous and non-ferrous, can be radiographed and there is a fairly wide range of material thicknesses that can be inspected (Grangeat *et al.*, 1992). The sensitivities of the radiography processes are affected by a number of factors, including the type and geometry of the material and the type of flaw.

Although radiography is a very useful non-destructive test system, it possesses some relatively unattractive features. It tends to be an expensive technique, compared with other non-destructive test methods. The capital costs of fixed X-ray equipment are high but coupled with this considerable space is needed for a radiography laboratory, including a dark room for film processing. Capital costs will be much less if portable X-ray sets or γ-ray sources are used for *in situ* inspections, but space will still be required for film processing and interpretation.

The operating costs for radiography are also high. The setting up time for radiography is often lengthy and may account for over half of the total inspection time. Radiographic inspection of components or structures out on sites may be a lengthy process because the portable X-ray equipment is usually limited to a relatively low energy radiation emission. Similarly, portable radio-active sources emitting γ-radiation tend to be of fairly low intensity. This is because high-intensity sources require very heavy shielding and thus cease to be truly portable. In consequence, on-site radiography tends to be restricted to a maximum material thickness of 75 mm of steel, or its equivalent. Even then, exposure times of several hours may be needed for the examination of thick sections. This brings a further disadvantage in that personnel may have to be away form their normal work posts for a long time while radiography is taking place.

The operating costs for X-ray fluoroscopy are generally much lower than those for radiography. Setting-up times are much shorter, exposure times are usually short and there is no need for a film processing laboratory.

Another aspect which adds to radiography costs is the need to protect personnel from the effects of radiation, and stringent safety precautions have to be employed. This safety aspect will apply to all those who work in the vicinity of a radiography test as well as to those persons directly concerned in the testing.

6.2.2.5 Acoustic emission (AE)

High-frequency waves, at frequencies within the range 50 kHz to 10 MHz, are emitted when strain energy is rapidly released as a consequence of structural changes taking place within a material. Plastic deformation, phase transformations, twinning, micro-yielding and crack growth result in the generation of "acoustic" signals which can be detected and analysed. Hence, it is possible to obtain information on the location and structural significance of such phenomena.

Basically there are two types of acoustic emission form materials - a continuous type and an intermittent or burst type. Continuous emission is normally of low amplitude and is associated with plastic deformation and the movement of dislocations within a material, while burst emissions are high-amplitude short-duration pulses resulting from the development and growth of cracks.

Acoustic emission inspection offers several advantages over conventional non-destructive testing techniques. For example, it can assess the dynamic response of a flaw to imposed stresses. When a crack or discontinuity approaches critical size there is a marked increase in emission intensity, and hence, a warning is given of instability and catastrophic failure. Also, it is possible to detect growing cracks of about 2×10^{-4} mm in length. This is a much smaller size than is detectable by conventional techniques. In addition, acoustic emission inspection requires only limited access and may be performed directly on components in service.

The usual type of transducer employed for the sensing of acoustic emissions from components uses a lead zirconium titanate element with a high electromechanical coupling coefficient. The signal from the transducer is amplified, filtered and processed to give an audio and/or visual recording.

The development of new broad-band transducers and the use of powerful tools for signal analysis allow quite important steps forwards in the application of AE techniques. It has been possible to achieve quantitative source characterization and 3-dimensional source location.

This last achievement enabling the through thickness advance of a crack point to be monitored, has been got on laboratory specimens, but its extension to real structures seems feasible (Godfrey *et al*, 1986). Because the information carried by acoustic emission is not explicit, the application of DSP (Digital Signal Processing) and statistical pattern recognition techniques is of great help in order to bring to light the information content of the AE signals. It is in particular possible to discriminate noise and reject it, and also characterize the acoustic emission source (Grangeat *et al.*, (1992), Singh and Udpa (1986), Tukuda and Mitsuoka (1986)).

6.2.2.6 Other non-destructive inspection techniques

Optical inspection probes. Optical inspection probes are a major aid to visual inspection as they permit the operator to see clearly inside pipes, ducts, cavities and other openings to which there is limited access. The basic parts of an inspection probe system are the objective lens head which is inserted into the cavity, the viewing eyepiece, and the illumination system. The development of fiber optical systems has permitted major advances to be made in the design and construction of inspection probes.

Optical inspection probes are of two general types, rigid or flexible, but within both of these categories there are many different sizes and designs available.

A rigid inspection probe comprises an optical system with a viewing eyepiece at one end. Illumination is conveyed to the inspection point through an optical fiber bundle and both the optical and illumination systems are enclosed within a stainless steel tube. Light from an external source, which is usually a variable intensity mains and/or battery-operated quartz-halogen lamp, is conveyed to the probe through an optical fiber light guide.

Rigid probes are produced in many sizes from the smallest, with tube diameters of 2 mm or less, up to large probes with tube diameters of 15 or 20 mm. The maximum usable or working length of a probe is the extent to which it can be inserted into an opening; it is not a constant and it varies with the value of probe diameter. Probes of all diameters are made in a variety of lengths but the maximum working length of a 2 mm diameter instrument is about 150 mm. The maximum working length for an 8 mm diameter probe can be up to 2 m and for larger sized probes usable lengths may extend up to 4 or 5 m. It is not practical, even for large diameter devices, to increase the usable length beyond about 5 m, without incurring considerable loss of quality in the eyepiece image.

Inspection probes may be designed to give either direct viewing or to view at some angle to the line of the probe. Some of the viewing angles catered for by the instrument manufacturers are 15°, 60°, 80°, 90° and 120°. In addition to a range of viewing angles the objective lens system can be designed to give a narrow, intermediate or wide field of view. It is also possible to have instruments which possess adjustable prisms to vary the viewing angle. This is only possible in probes with diameters of about 8 mm or greater, but such probes are often made to give any viewing angle between 60° and 120°.

Rigid inspection probes are extremely useful instruments but, like all delicate instruments, have to be handled carefully. The probes, particularly those of small diameter, can be damaged very easily if mishandled.

The usefulness and versatility of light inspection probes is increased by the use of flexible probes. These incorporate a fiber optic coherent image guide and a separate fiber light guide for illumination, both contained within a flexible plastic or braided metal sheathing. The external diameters of the flexible probes generally range from 4 mm to about 15 mm and the working lengths may be up to about 3 m. Flexible probes are usually designed to provide either direct viewing ahead or viewing at 90° to the probe axis, but the larger diameter probes can be produced with a movable inspection head, the position of which is controlled from the eyepiece end.

Inspection probes, then, are extensions to the human eye, and are used to view areas which would otherwise be impossible to inspect without either dismantling or even cutting open the part or assembly. The images produced at the eyepiece end of a probe may be photographed and a permanent record secured. It is also possible to mount a TV camera as a substitute for the normal eyepiece lens and display the resulting image on a monitor screen. Such an installation is described by Namioka *et al.* (1992) to inspect the inside condition of pipelines.

Laser-induced ultrasonics. A recent development in non-destructive testing is the generation of ultrasonic pulses within a material without the necessity to have a transducer crystal in contact with the testpiece. The ultrasonic pulses are produced by focusing a series of light impulses from a laser on the surface of the testpiece.

The laser, which may be situated up to 10 meters from the testpiece, sends out a series of very short high-energy light impulses and these impulses, each of about 20 nanosecond length, are converted by thermo-mechanical effects into sound impulses at a frequency within the range 1 MHz to 100 MHz. A laser pulse, incident on a solid surface, produces rapid heating of the surface, at the point of incidence, resulting in a localized temperature rise. Thermal expansion of the "hot zone" causes generation of ultrasonic waves, which propagate across the component surface and within the component body. The intensity of the incident laser impulses is such that no damage is caused to the surface of the testpiece. The emission from a second laser illuminates the surface of the testpiece and ultrasonic echoes returning to the testpiece surface cause deflections which cause a modulation of the reflected light from the illuminating laser.

The third component of the system is an interferometer which analyses the modulated reflected light signal and converts it into a signal which can be presented on the screen of a cathode ray tube in a manner similar to the usual type of ultrasonic signal display.

The main advantages of this technique are that no mechanical coupling is necessary and the acquisition of results is rapid. Laser-based ultrasonic interrogation systems are in use, currently, to detect the existence of piping and liquid metal level in cast steel ingots. Although the sensitivity of the system is lower than that of some of the more conventional techniques, for example, ultrasonic pulse-echo testing, the system has attracted some interest for the continuous monitoring of components on process lines in the manufacturing industries.

Time-of-flight diffraction. A new ultrasonic technique has been developed, namely time-of-flight diffraction (TOFD), which relies on the diffraction of ultrasonic waves from crack or defect tips, rather than reflection, as in pulse-echo. The technique is very useful in determining the true size of fatigue cracks, even though a crack may be pressed together by the applied load or residual stress network. With conventional pulse-echo testing, complete or partial transmission of the wave pulses across the "closed" crack can lead to errors in the analysis of crack size, because of the reduction in amplitude of the reflected signals. TOFD is so called because it relies on the wave propagation times to indicate and locate the diffraction source. An example of applying the technique is shown in fig. 6.16.

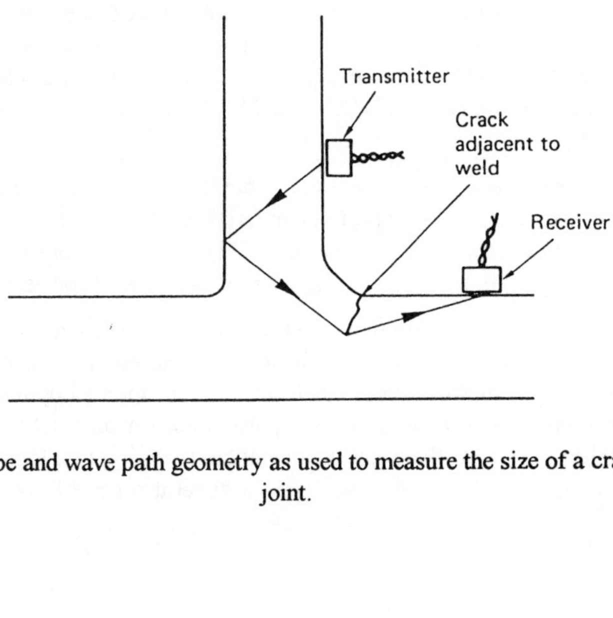

Figure 6.16 Probe and wave path geometry as used to measure the size of a crack in a welded joint.

The low signal-to-noise ratio often necessitates signal averaging, and comparison and subtraction of surface waves also may be necessary.

Crack depth gauges. Cracks which appear at the surface of a material can be readily detected using liquid penetrant of magnetic particle inspection methods, but neither of these methods will give an accurate assessment of the depth of a crack. Crack depth gauges are frequently used in conjunction with these other non-destructive tests to give a measure of the depth of flaws which have been located. One simple but effective device for this consists of two closely spaced electrical contacts. The gauge is placed on the surface of the material and the electrical resistance between the two contact points measured. When the gauge is placed on the testpiece with the contacts on either side of a surface crack, the measured resistance will be greater as current now has to follow an extended path around the crack. The meter scale is generally calibrated to give a direct reading of crack depth.

Thermography. Thermography is concerned with the mapping of isotherms, or contours of equal temperature, over the surface of a component. Heat-sensing materials or devices can be used to detect irregularities in temperature contours and such irregularities can be related to defects. Thermography is particularly suited to the inspection of laminates. The conduction of heat through a laminate will be affected by the presence of flaws in the structure, resulting in an irregular surface temperature profile. Typical flaws which can be detected are unbonded areas, crushed cells, separation of the core from the face plates and the presence of moisture in the cells of honeycomb structures.

Thermographic methods may either be of the direct contact type, in which heat-sensitive material is in contact with the component surface, or indirect contact, in which a heat-sensitive device is used to measure the intensity of infrared energy emitted from the surface.

Pulses of heat energy, from a source, are directed at the component under test. It is usual, but not essential, to direct the incident energy on to one surface of a component and observe the effects at the opposite surface after conduction through the material. Flaws and irregularities in structure will affect the amount of conduction in their vicinity. If it is impossible to have access to both surfaces, the technique can still be used. The heat energy incident on the surface will be conducted away through the material at differing rates, depending on whether or not flaws are present.

Direct contact methods include the use of heat-sensitive paints and thermally quenched phosphors. Indirect contact methods, which offer greater sensitivity, involve the use of infra-red imaging systems with a TV-video output.

Heat-sensitive paints. Heat-sensitive photo-chromic paints are effective over a temperature range from about 40°C to 160°C. Some paints show several color changes within their reaction temperature range and, with careful application, will have a sensitivity of the order of ±5°C. When heat reaches the painted surface by conduction through the material the paint colour changes, usually with a bleaching effect. Where a flaw impedes conduction the colour will be unchanged. On the other hand, if heat energy is directed at the painted surface the reverse effect will show up as heat is conducted away from the surface more rapidly through good regions than through defective areas.

Thermally quenched phosphors. These are organic compounds which emit visible light when excited by ultra-violet radiation. The brightness of the emission decreases as the temperature of the compounds increases. Phosphors are available that are useful at temperatures up to about 400°C and with a resolution of ±1°C.

Thermal pulse video thermography. In this system no physical contact is necessary with the material and very rapid rates of inspection are possible. A high-intensity heating source is used to send pulses of infra-red energy into the material. The surface is scanned by an infra-red thermal imager with a TV-video output. This system, again, can either be used for sensing heat transmitted through the component or for single-sided inspection when only one surface is accessible. Very good sensitivities are possible. Digitized image processing to provide image enhancement is also possible.

6.2.3 Signal processing (SP) for NDT

NDT is gradually adapting to the use of the most recent developments in digital signal and image processing. By signal processing (SP) is meant digital techniques that transform an input signal into an output signal or into parameters. In this very broad sense, not only averaging and Fourier transform but also TOFD, SAFT, 3-D reconstruction, expert system based signal classification, adaptive learning network and neural networks can be classified as SP techniques (more details are given in the following).

Non-destructive testings generally result in an inverse problem: given a set of external measurements compute the location and size of defects inside the material. Although the basic equations differ from one method to the other ("direct problem" formulations are different) their common feature is that they cannot be simply inverted because of i) noise, ii) lack of measurements, iii) incomplete modeling and iv) all together.

NDT noise itself is rarely the widespread additive Gaussian white noise. A first example is the response of coarse grained materials when tested with ultrasonics: each grain behaves as a reflector so that the whole response does not resemble a random white noise and classical averaging does not work. In EC (eddy current) testing of some steam generator tubes, flattening noise can be decomposed into several narrow band components (it is therefore non-random, but it is not "stable" either). Moreover noise and flaw frequency spectra are the same: if this flattening noise is bandpass-filtered, the useful information is also filtered out. Another example originates from gammagraphy testing of thick wall samples. The radiographs are corrupted by a granular noise due to thickness and film, which has to be modeled adequately before processing.

Digital signal processing could be difficult and hence expensive to implement. Therefore one has to implement it only in these cases where all other means have failed. This is why smart experimental setups and acquisition schemes have been developed. Among them ***Synthetic Aperture Focusing Technique*** (*SAFT*), ***Time of Flight Diffraction*** (*TOFD*) and numerous enhancements of these basic techniques have been developed (Ludwing and Roberti, 1989). SAFT is some sort of "beamforming" already known in array processing for underwater acoustics or RADAR. For each pixel of the insonified specimen the A-scan signals received at n-transducers are averaged after time-shifting. The shifts are computed from the different distances between one transducer and the pixel under study. Scanning the sample results in an enhanced image because of constructive addition of waveforms.

The aim is to better understand the content of the measured signals and then to be able to simulate it (together with its accompanying noise) for study and method evaluation purposes. The field of image processing has become a big consumer of sophisticated modelisation based on stochastic processes on one hand (Boolean and Markov models) and Bayesian procedures on the other hand. Boolean models represent the granular noisy part of images by assuming it is a Poisson-type random spatial distribution of some basic pattern (usually the convex part of a Gaussian whose parameters are randomly chosen). This proved appropriate for radiograph modelisation. Markov models account for the reasonable idea that statistical relationships between one pixel and the rest of an image are summarized in a window around this pixel. These models are used in many processing tasks and particularly in reconstruction. Time-frequency domain methods are required for extracting physical parameters of interest when these involve joint variations of time and frequency. Techniques based on the Wavelet Transform appear to be suitable for acoustic signal processing particularly for detection and description of burst, whose arrival time and waveforms are unknown. Lastly the resolution procedure based on the classical Bayes' rule is an elegant way to introduce human "knowledge" on the desired image and on its transformation. To sum up, besides the mathematical aspects that can discourage NDT persons, the important point in this general stochastic approach is its ability to take prior knowledge into account as probability laws and distrurbing noise as random processes. It is amusing (although well known) that introducing such apparently complex tools leads at the end to tractable calculations and interesting results (Singh and Udpa (1986), Ludwing and Roberti (1989), Grangeat *et al.* (1992)).

An important issue in US (ultrasonics) is to restore signals from austenitic welds because they are severely corrupted by noise. Several techniques have become popular. They are called "averaging" although this reference to a "linear" mix could be misleading. *Spatial averaging* consists in selecting for example the minimum values in a number of waveforms produced by different probe locations close to each other, while *frequency averaging* does the same but from different frequency bands of a single signal. Both methods are based on the reasonable assumption that signal (i.e. defects) responses are coherent whereas noise (grain reflections) responses are not. Signal to noise enhancements of up to 10 dB have been reported.

In flattening noise from some steam generator tubes, filtering has to be done by specific digital techniques. They are based on a noise reference either picked from the signal itself or provided by an auxiliary signal (this is the so called correlofilter). When the signal is not stationary, one convenient way to filter it is to let a feedback loop estimate the filter coefficients from the measured samples. The output of the filter can be used as a noise estimation and subtracted from the original signal.

After clean signals have been recovered, an automatic decision about their nature is desired. Besides the statistical techniques (principal component analysis, discriminant analysis and others) a classification scheme got a tremendous favour at the end of the 70's: the Adaptive Learning Network (ALN). ALN is an empirical combination of candidate parameters, in which a non-linear polynomial model is constructed. At each iteration the model "grows", that is the coefficients and the structure of the model are determined simultaneously. The model's output can be either a classification or an estimation of some parameter of interest. ALNs have been tested both for US (Ultrasonics) and EC (Eddy Currents) signals. As revealed, they performed more or less like classical multidimensional statistics and apparently disappeared from reports.

As the question of what are the optimal parameters remains open, another approach has been proposed for EC signals. The idea, first used for hand print character recognition, consists in retaining only the first terms of some sort of Fourier development of the EC complex signal. These terms are then used as features for classification. The experience about these Fourier descriptors applied to support plates discrimination is that they are too global to allow an accurate localization of small flaws.

Since the rediscovery of Rosenblatt's perceptron in the 80's, *NNs* (*neural networks*) have been proposed for numerous tasks. A particular combination has proved to be fruitful:

1. A three-layer architecture (one hidden layer).
2. The back propagation algorithm to estimate the weights.
3. Classification purposes.

Whereas NNs have given an opportunity to revisit once more some classical problems, their new features are:

1. Efficient hardware implementation.
2. Some preprocessing capability (Komatsu *et al.*, (1992), Parpaglione, 1992).

Nevertheless, the amount of relevant examples has a paramount importance in NN approach as well as in other approaches. It is surprising that nobody has compared ALNs and NNs at least from a NDT point of view.

6.2.4 Applications of SP in automated NDT

A few successful applications of SP in automated NDT for defect detection and characterization and for enhancing the defect detection sensitivity, are described here.

Maraging and austenitic stainless steel welds - Defect detection and characterization using pattern and cluster analysis.

High sensitivity defect detection and characterization in weldments of these materials continues to be of interest. This is primarily due to the fact that these weldments are used in large numbers in critical and heavy industry applications. Dendritic (hence anisotropic) microstructures of these weldments, especially in the thickness range of 10 to 40 mm pose problems to ultrasonic testing. Considering these facts, the ASME boiler and pressure vessels code has recommended that in the case of austenitic stainless steel weldments, any defect that is 10% of thickness should be recorded and monitored. DSP procedures, by using very effective cluster and pattern analysis algorithms have been developed (Raj, 1992). These enable detection and characterization of defects down to 1% of weld thickness (14.0 mm weld thickness) in austenitic stainless steel welds. Work on higher thicknesses is in progress. The complexity of this problem is an excellent area for the development of an expert system, for offering advice in carrying out effective NDE on these weldments.

In the case of maraging steel weldments used in the rocket motor casings by the aerospace industry, tight cracks (3 mm × 1 mm) produced by fatigue loading were detected and characterized (Raj, 1992) using similar cluster and pattern analysis principles. Detection of such small defects for this application, enhances the payload capacity of the rocket, resulting significant in economic and technological gains.

In both the above cases, the cluster analysis methods use the crosspower spectrum (between signals from weld noise and those from defects), to obtain cluster elements. The pattern analysis method generates a pattern called the demodulated autocorrelogram (DMAC) pattern from the autocorrelation function of a signal and study its features.

Characterization of foreign inclusions in composite material using pattern analysis. Unwanted foreign inclusions find their way into composite materials during their manufacture. It is known that these inclusions affect the load bearing capacity and performance of these materials. Conventional immersion ultrasonic testing is able to detect these inclusions but fails to characterize them into different categories. C-scan imaging techniques are convenient for characterising such foreign inclusions in composite materials. A simple and effective method utilizing the DMAC pattern analysis, for the characterization of these inclusions has been developed by Kalyanasundaram *et al.* (1991). The procedure functions on template matching and thus can be easily automated and offers an excellent way to use neural networks for pattern matching and classification.

Acoustic emission (AE) signal analysis. Acoustic emission signal analysis has yielded important information in the detection of leaky components under pressure, in pressurized heavy water reactors. In one of the above problems, the ratio of the spectral energies present in different bands of the power spectrum of the AE signal, is used in order to detect the leaking component, since the signal-to-noise (SNR) ratio was very poor. This is an example where problems due to poor SNR were overcome by appropriate use of DSP.

In the NDE of rotating machinery, such as steam turbines and turbine generators, AE is used to detect malfunctions such as rubbing and bearing tilt. In order to detect and transmit AE signals from an operating rotor (to enable on-line processing), a wireless AE monitor has been used, which can detect and transmit AE signals ranging from 50 kHz to 250 kHz. Acoustic emission parameters such as events, energy values, amplitude distribution, frequency components, skewness and kurtosis values have also been correlated with the "health" of cutting tools, used in lathes. Failure prediction in gearboxes by the processing and analysis of its vibration (rotational) signals have been done with success. It has been concluded that the imminent failure could be predicted accurately using cepstrum analysis (see Chapter 1). Vibrations in the gear meshings have been monitored to detect failure in gears, where the tooth meshing vibration components and their harmonics are eliminated from the spectrum of the time domain average. The reconstructed time signal shows the presence of defects (if present) which otherwise cannot be seen in the time domain average. This again underlines the importance and usefulness of SP in the field of acoustic testing.

Time of Flight Diffraction (TOFD) technique for defect sizing. When ultrasonic waves encounter a crack-like defect, not only reflection but also production of scattered and diffracted waves takes place, over a wide angular range from defect tips. The separation of diffracted waves in space and hence in time directly relate to the size of the defect. By knowing the delays for different waves, it is possible to compute the size and location of the defect.

This technique has been used and the results are found to be in conformity with ASME XI with respect to determination of maximum acceptable defect height and depth of upper and lower edges for internal defects which lie deeper than 30% of the specimen thickness in steel exceeding 12 mm thickness. Again, the results are in conformity with modified ASME XI for all defects in steel exceeding 10 mm thickness.

Synthetic Aperture Focusing Technique (SAFT) for increased resolution. In this procedure, a large aperture focused probe is synthesized electronically, thereby increasing the fundamental resolution and defect sizing accuracy of the technique. A wide angle compression probe and a point flaw in the specimen is assumed for the purpose of simplicity. When the transducer scans over the specimen, each reflected echo for various scan positions with respect to the position of closest approach of transducer to the flaw, is delayed in time due to the greater distance travelled by ultrasonic waves. If the individual scans are shifted by an amount equal to their predicted time delays, they will come into coincidence with each other and when they are summed, the resultant will be a large amplitude response. If the same procedure is repeated centered around another position, the above time shift compensation does not produce a set of self-coincidence scans which results in a significantly smaller response. The time shifts can be achieved either electronically or digitally using a computer. This technique is an excellent example of the advantages that accrue form the combination of conventional and advanced techniques. Typical applications of this important technique, apart from radar, is the in-service inspection of pressure retaining boundaries for accurate defect sizing.

Reduction of random noise using split spectrum processing. This technique is implemented by splitting the frequency spectrum of the received signal by using Gaussian, overlapping band pass filters having central frequency at regular intervals. If the inverse Fourier transform is taken for *N* filters, *N* time domain signals are obtained. These *N* number of time domain signals are subjected to algorithms such as minimisation and polarity thresholding for extracting useful information. The *split spectrum processing* technique is widely applied in the analysis of signals form noisy materials like centrifugally cast stainless steels, carbon epoxy composites, welded joints and cladded materials.

The ALOK technique. The *ALOK* technique was conceived and developed by the Fraunhofer Institute for NDE techniques (IzfP), Saarbrucken, Germany. The principle of this technique is to characterize a reflector by its time of flight characteristics rather than on the basis of its reflected amplitudes. A modified version of this technique, developed by Siemens, rapidly acquires a manifold of amplitude and corresponding time of flight values in each A-scan, concentrating on the relevant A-scan information by a specific pattern recognition process. ALOK provides remarkable advantages with respect to general improvement of the inspection, increase in the information density (reduction of documentation) and simplification of data evaluation.

Microstructure and mechanical properties characterization using acousto-ultrasonics. This approach is based on the concept that spontaneously generated stress waves produced during failure interact with material morphology. By introducing ultrasonic waves into the material, simulated acoustic stress waves are produced which are affected by the material condition. The waves are measured in the form of *stress wave factors* (*SWF*), defined as the number of oscillations higher than a chosen threshold in the ringdown oscillations of the output signal. The SWF is correlated to the microstructure and mechanical strength. Damages in the specimen produces corresponding changes in the signal attenuation resulting in lower SWF readings.

6.2.5 Conclusions

The techniques mentioned in the above sections are a selection of some of the more recent developments in NDT and NDE. By the turn of this century, the number of techniques that will be used in the field of NDE, will be significantly large in number and varied in approaches. Development of techniques would be driven by the need to characterize the number of engineering materials and the number of processes used to manufacture these materials. Each of these processes would demand monitoring of a variety of parameters in real-time, in order to ensure high productivity, minimal wastage and acceptable characterization levels.

The knowledge and databases pertaining to these aspects will be very large, but the expertise to assimilate and apply this large amount of knowledge at every stage of the production and NDE of materials and components, will be scarce. Intelligent and timely application of this knowledge in the production and NDE of manufactured materials and structural components in service, would be the key for success in industrial ventures.

The use of neural networks for NDE signal analysis and classification, fuzzy logic for decision making and expert systems for NDE are being increasingly studied and used (see Georgel and Zorgati, 1992, Schlicht and Zhirabok, 1992, Garribba *et al.*, 1988). The motivation behind the use of these new concepts is the need for the acquisition, analysis, storage, retrieval and presentation of very large amounts of possibly imprecise and incomplete data. NDE, being an interdisciplinary area that affects and is affected by a large number of complex variables ("concepts" or "inputs") in any problem, is a fertile ground for these new concepts to grow and contribute.

Stavrakakis and Psomas, (1993), described popular neural network models powerful in performing complex pattern classification for NDT data interpretation. A hybrid system applied to the classification of ultrasonic images of weld defects, was described. Rules and neural networks were used for different aspects of the classification task. Replacing those rule-based modules, which classified echodynamics and potential defect areas, with neural network modules, facilitated successful classification of previously unseen test examples. The case of ECT data evaluation using NN was also examined. A neural network which used optimum input parameters showed good potential as a tool for automatic ECT data evaluation. Finally, experimental results performed at an Acoustic Emission Laboratory were presented, to evaluate the location of an acoustic emission source in a tube which simulates a fault, using a neural network model. Comparative simulation results showed very good performance of the ANN-based acoustic emission data interpretation method.

The ANN-based NDT data interpretation can learn and keep the advantages and avoid the drawbacks of the present methods. Due to the parallel computation mechanism inherent in the ANN structure, hardware IC ANNs are much quicker and more suitable for on-line applications.

Acoustic NDE methods as applied to new materials, on-line acoustic NDE for real-time monitoring of plants and components, life prediction of components and plants, intelligent processing of materials and components are some of the new directions where acoustic NDE methods would be required to contribute in the coming decade. In order to do so, acoustic NDE methods would increasingly rely upon basic and advanced signal analysis methods, pattern and cluster analysis methods and exploit the advances that are being made presently in the field of artificial intelligence (AI).

It can be conclusively stated that acoustic NDE methods at their present state of development would gain significantly by intelligent and balanced use of these advanced concepts.

6.3 Real-time structural damage assessment and fatigue life prediction methods

6.3.1 Introduction

Structural damage assessment is a process that involves decision-making under uncertainty. Under the effect of the operating conditions in industrial plants and structures, potentially dangerous phenomena affecting the structural integrity, are likely to take place as fatigue due to high levels of temperature and pressure, creep-fatigue interaction, stress corrosion, material embrittlement, etc. The analysis of these mechanisms is of basic importance in the evaluation of the structure or the component lifetime. This may be carried out at various levels:

- Damage analysis at the level of the granular and intergranular crystalline structure of the material (physico-metallurgical approach).
- Analysis of the macroscopic effects of damage at the defect propagation level (phenomenological approach).

From an analytical point of view, the reliability of a structure or a component can be thought of as the probability of absence of crossings of a safe level by accumulated damage processes. The best approach to its assessment is constituted by the theory of stochastic processes (Lucia, 1985) and details will be given in the next paragraphs. In a real-time structural damage assessment, starting with the initial defect distribution, which gives the absolute number of defects according to depth after the structure fabrication, real-time, mainly acoustic, NDE results (US and AE) permit to modify the initial defect distribution by the growth of defects during the structure's lifetime. The initial defect distribution and the non-detection probability of the ultrasonic inspection are exponential functions with a negative exponent. This means that they are decreasing with increasing defect depth. The defect distribution will be updated after each in-service inspection and all defects larger than the acceptable size are to be removed with the exception of those which have gone undetected.

NDE data can be processed continuously with current signal analysis computer means (see Section 6.2), while in-service inspection cannot be performed continuously because of obvious economical and engineering limitations. For this purpose, a new strategy for inspection and repair of structural elements and systems is presented by Thoft-

Christensen and Sorensen (1987). The total cost of inspection and repair is minimized with the constraints that the reliability of elements and/or of the structural system are acceptable. The design variables are the time intervals between inspections and the quality of the inspections. Numerical examples are presented to illustrate the performance of the strategy. The strategy can be used for any engineering system where inspection and repair are required.

Although the physico-metallurgical aspect is important for the understanding of rupture mechanisms in structures, in the literature the phenomenological approach is more commonly found, based on laboratory tests and semi-empirical models, using mainly linear fracture mechanics (Lucia, 1985, Kozin and Bogdanoff, 1992). The fracture mechanics relationships allow a link to be established between the defect dimensions, the load level and the stress intensity. The stress intensity can be compared with the material's resistance to rupture. Given the complexity of the phenomena involved however, experimental support is essential for the definition of rupture mechanism models. The problem arises at the moment of transferring the laboratory results, if obtained in over-simplified environment and loading conditions, to component reality and, on this basis, predicting the lifetime.

During the past twenty years a rather extensive effort has been devoted to developing techniques that permit the accurate in-time prediction of structures fatigue life (see in "*Theoretical and Applied Fracture Mechanics*", "*Engineering Fracture Mechanics*", "*International Journal of Fracture*", "*Structural Safety*" and "*International Journal of Fatigue*"). As the knowledge related to fatigue of structures and materials expanded, it became clear that in many cases fatigue could be treated from a propagation point of view. This knowledge has led to the development of phenomenological, fracture mechanics and probabilistic fracture mechanics, stochastic process, time-series analysis and knowledge-based approaches to assess in-time the fatigue crack growth (FCG) and the failure probability in structures. This is the "heart" of the in-time fatigue life prediction leading to increased life of structures subjected to dynamic loads. These approaches will be presented and discussed in the following.

6.3.2 Phenomenological approach for fatigue failure prognosis

The interpretation of damage as the birth and propagation of defects in the elementary structure of the material, caused by the alternating stress field acting on imperfections of the crystalline network, on distortions due to impurities, etc. is generally accepted. Qualitatively one can distinguish:

- An initial nucleation stage (defect generation).
- A transition stage (defects coalescence).
- A propagation stage (unstable growth of the largest defects).

In general, N_F being the number of cycles to rupture and N_O, N_T, N_C the cycles at the end of the nucleation, transition and propagation stages, one has:

$$N_F = N_O + N_T(N_O) + N_C(N_T) \tag{6.1}$$

which expresses the fact that each stage is determined by the level reached in the preceding stage. The damage accumulation mechanisms are presumably different in the three stages and in each one they are significantly dependent on the environmental conditions and the stress field; for this reason the construction of a unified model of interpretation seems unlikely. Furthermore, the relative importance (in terms of number of cycles) of the above three fatigue stages depends on the intensity of the alternating load. This rather complex picture can be partially clarified by the consideration that experimentally evident fatigue failures in operating components are mainly caused by fabrication defects, generally introduced during welding procedures. As a consequence, the nucleation and transition stages are relatively less important than the propagation stage, starting from the fabrication defects. These defects are, in fact, usually larger than the defects of the elementary structure of the material.

To date, efforts have concentrated on development of independent models for cyclic constitutive behavior, cyclic crack initiation and cyclic crack propagation. However the transition between crack initiation and crack propagation has not been thoroughly researched as yet, to be integrated in a unified life prediction method (Bhargava *et al.* (1986), Lankford and Hudak (1987), Halford *et al.* (1989)).

Having said that, the problem is: how to estimate the time to failure of a cyclically loaded structural component. The estimation of the lifetime (expressed, e.g., by the number of load cycles allowed) is more than just a research problem; it is a practical problem of current design, considered by the current standards. For this reason it appears opportune to present first the criteria of fatigue design of the ASME Sect. III standards and the acceptance criteria of the defect propagation rates of the ASME Sect. XI standards.

The fatigue design and residual life time criterion of the ASME standards coincides with the Miner rule. It is still the most commonly used, not only for components subject to ASME standards but also, and especially, for components in the aeronautical industry for which the dimensioning for fatigue is often the main consideration.

The ASME standards for components of conventional (Section VIII, Div. 2) or nuclear (Section III) installations are based on limitations imposed to a "cumulative factor of use" U:

$$U = \sum_i U_i = \sum_i n_i / N_i \leq 1 \tag{6.2}$$

where n_i is the number of cycles envisaged for the cyclic load of type i and N_i the allowed number of cycles as deduced from the fatigue curves (S-N curves). An S-N curve is given as,

$$N = KS^{-m} \; (S > S_0) \tag{6.3}$$

where N is the number of cycles to failure under constant-amplitude stress range S, K and m are the S-N curve parameters, and S_0 is a stress cut-off level below which no damage is accumulated.

The S-N curves used are generally obtained from monoaxial cyclic load tests at constant amplitude, on notched samples: a suitable safety factor, of the order of 3 or 4, is applied to the mean experimental curve to account for the considerable scatter of results for design purposes. With a safety factor of 4 the design reliability is of the order of 99.9% for a standard deviation of 20%.

The fatigue damage accumulation D follows Miner's law as

$$D=\sum_{i=1}^{N_0}\frac{1}{N_i(S_i)}=\frac{1}{K}\sum_{i=1}^{N_0}S_i^m \tag{6.4}$$

in which N_0 is the total number of stress cycles. In a deterministic design, it is assumed that failure occurs when D=1.

To account for model's (6.4) uncertainty by a random variable Δ, the fatigue limit state function is defined as:

$$g(Z)=\Delta-D=\Delta-\frac{1}{K}\sum_{i=1}^{N_0}S_i^m \tag{6.5}$$

If one further defines that any random variable x can be expressed as,

$$x=B_x x_c \tag{6.6}$$

in which x_c is the characteristic value of x and B_x is a normalized random variable associated with x, the limit state function (6.5) is then of the following form:

$$g(Z)=\Delta-\frac{1}{B_K K_c}\sum_{i=1}^{N_0}(B_S S_{i,c})^m \tag{6.7}$$

in which the S-N exponent m is usually treated as a deterministic constant.

Failure is defined by the event $g(\mathbf{Z})\leq 0$, and $g(\mathbf{Z})>0$ identifies a safe state. The failure probability is,

$$P_f=P(g(Z)\leq 0)=\int_{g(Z)\leq 0}f_z(z)dz \tag{6.8}$$

where $f_z(z)$ is the multivariate density function of $\mathbf{Z}$.

For a large class of engineering problems, the basic random variables are generally modeled by continuous probability functions and the failure probabilities are generally small. Hence, it is generally preferable to apply the analytical *FORM/SORM*, as the methods are very efficient and accurate for small failure probability problems. Detailed computational procedures of FORM/SORM are given in Madsen *et al.* (1986), where the reliability index *â* is applied, which is related to the failure probability by

$$\beta = - \Phi^{-1}(P_f) \tag{6.9}$$

in which $\Phi(\ )$ is the standard normal distribution function.

From technical or economical considerations, the required safety level in the design life may sometimes have to be achieved by additional safety measures such as inspection, so that the design safety can be updated to the required safety level. The principal of reliability updating is based on the definition of conditional probability,

$$P(F|I) = \frac{P(F \cap I)}{P(I)} \tag{6.10}$$

where I is the inspection event and F is the structural failure event which is described by $F{=}g(\mathbf{Z}) \leq 0$. Computational methods for updating the failure probability may be found in Madsen *et al* (1986).

The question of the accuracy of the Miner rule has often been raised, although few valid attempts have been made to verify it experimentally because of the high cost of these experiments. During the seventies, many experimental campaigns were undertaken, in particular by NASA and Messerschmitt-Bolkow-Blohm, leading to the conclusion that the Miner rule has some relevant drawbacks, since there are no criteria to estimate a priori whether the prediction will be conservative or not and because the parameters which influence the estimation are not known.

The simplicity of the Miner rule is the result of the main hypothesis on which it is based: the linear accumulation of damage. This hypothesis corresponds to the assumption of a constant failure rate which is in contrast with the experimental evidence of fatigue damage, characterized by an increasing rupture rate.

Recently, Ben-Amoz, (1992), developed a cumulative damage theory to predict the residual fatigue life in two-stage cycling. Based on the concept of bounds it was shown that the mean residual fatigue life, as well as the entire test data scatter, can be bracketed by appropriate bounds. However, the use of such a theory requires prior knowledge of the life fraction spent in initiating fatigue crack. It is shown there that the crack initiation life fraction can be determined from two-stage cumulative damage tests. The bounds are derived from both high-to-low and low-to-high amplitude fatigue tests. Furthermore, it is shown that the two bounds actually coincide to give an exact expression for the crack initiation life fraction.

6.3.3 Probabilistic fracture mechanics approach for FCG life estimation

The prediction of the fatigue damage accumulation relies on the *S-N* curves where the number of cycles to failure is based on a large visible crack size and the remaining fatigue life is conservatively neglected. The prediction of the fatigue crack growth (FCG) using the fracture mechanics, however, describes the crack growth physically and is able to calculate the fatigue life up to fracture accounting for possible inspection effects. This approach is thus more sophisticated and its application is becoming widespread.

Some hundreds of more or less different relationships can be found in the literature for expressing the fatigue growth of cracks (Hoeppner and Krupp (1974), Akyurek and Bilir (1992)). Some of them are purely theoretical or based on microscopic properties of material, but the most widely employed are semi-empirical and have been developed mainly as interpretative models of experi-mental results: they allow the prediction of the behavior of the crack size "α" as a function of "N" the number of stress cycles. This prediction is based on the integration of the growth rate, for fixed initial conditions of the defect. The result arrived at is not, however, in general, representative of the real growth situations. This is due both to the fact that the initial conditions have a considerable scatter of values and to the fact that, for the same initial conditions, there is an intrinsic variability in the process of damage by fatigue which leads to a distribution of values "α", at cycle N (Virkler *et al.*, 1979). It thus appears natural to consider the relationships for defining the growth rate as stochastic (Ghonem and Dore, 1987). In this context, therefore, prediction methods can be seen as being based on the integration of the FCG relationships with the parameters or the initial conditions or the loads represented by random variables. These procedures lead to the determination of a distribution of dimensions for the propagated defect, at cycle N, or of a distribution of the number of cycles N for a given propagation from α_0 to α_f. These distributions are the basis for the prediction of the residual life of structures stressed by fatigue. Three randomization methods are presented by Lucia (1985), as an indication of the vast range of applications of this methodological approach.

Probabilistic models for fatigue crack growth.

As mentioned before, the statistical variability in crack growth depends on many undetermined factors which can be classified as small differences of material intrinsic properties, loading environment, specimen geometry, measuring system, even microstructure and the state of stress, etc. In general, the fatigue crack growth can be expressed by the following nonlinear relation,

$$d\alpha/dN = Q(\Delta K, K_c, R, K_{th}, \alpha, \ldots) \qquad (6.11)$$

where,

- $Q(\ldots)$ a non-negative function,
- α half-crack-length, mm
- N number of fatigue cycles, (cumulative load cycles)
- ΔK stress intensity factor range at the crack tip, given by the relation $\Delta K = S(\pi\alpha)^{1/2}F(\alpha)$, $MPa\ m^{1/2}$
- S applied stress (load) range, MPa
- $F(\alpha)$ crack shape geometrical factor (see Verreman *et al.*, 1987, Dufresne *et al.*, 1988)

$d\alpha/dN$	fatigue crack growth (FCG) rate, m/cycles
R	algebraic ratio of the minimum to maximum load (stress) in a cycle
K_c	material fracture toughness (i.e. the critical value of the stress intensity factor), $MPa\ m^{1/2}$
K_{th}	FCG threshold stress intensity, $MPa\ m^{1/2}$

The problem of determining the fracture toughness parameter for a specific industrial structure material (a double-sided spiral submerged arc welded pipeline) from measurements, is analytically treated by Al-Obaid, (1992). The FCG threshold is the upper bound of a set of amplitudes at a given K_{max}, which, when applied to a fatigue crack, do not produce crack propagation, independent of the way these amplitudes are applied. Cycles with amplitude ΔK greater than threshold produce FCG at that K_{max}.

If the lower limit for fatigue crack growth under zero-tension fatigue is defoned along with $K^*_{\max}$, a FCG threshold has to exist and has to be measured over the fatigue loading range between $K^*_{\max}$ and K_c. It must be noted that K_c illustrates some failure condition. Marci, (1992), clarifies these concepts and discusses the experimental procedures to determine K_{th} for a specific material as well as the requirements to ensure transferability of the experimentally measured threshold to service type fatigue loading.

The empirical laws which are most often used in engineering are the Paris-Erdogan and Forman laws (Hoeppner and Krupp, 1974, Lucia, 1985, Akyurek and Bilir, 1992):

$$\frac{d\alpha}{dN} = C(\Delta K)^m \tag{6.12}$$

$$\frac{d\alpha}{dN} = \frac{C(\Delta K)^m}{(1-R)K_c - \Delta K}, \qquad S > S_0(\Delta K_{th}) \tag{6.13}$$

where the stress cut-off level $S_0(\Delta K_{th})$ is a function of the threshold of the stress intensity factor range (ΔK_{th}), below which there is no crack growth, and C, m=crack growth parameters.

Laws (6.12) and (6.13) are almost universally applied to stage-II FCG, that is, crack growth at alternating stress intensity values somewhat larger than the threshold alternating stress intensity value ΔK_{th}, but below the value of ΔK at which unstable crack propagation begins to occur.

All of the factors and parameters mentioned above are treated as random variables in probabilistic fracture mechanics (PFM). Therefore, the statistical investigation and the accumulation of statistical data of these parameters are very necessary and important for the reasonable and economic design of fatigue structures. In particular, for fatigue reliability analysis of structures, a probabilistic or stochastic model is required for fatigue crack growth. As a result, many probabi-listic or stochastic fracture mechanics models have been developed to deal with the variability of crack growth (Lucia, 1985, Journet and Pelloux, 1987, Ghonem and Dore, 1987, Cortie and Garrett, 1988, Zhu and Lin, 1992, Nisitani *et al.*, 1992). These models have their realistic physical or microstructural basis for some special conditions. A major problem with these models is the difficulty in obtaining sufficient data due to time and money. For this reason, some models are not verified by experimental data, and it is difficult to apply some models in engineering.

The purpose here is to present a simple probabilistic model which is easy for designers to use in predicting crack growth behavior. In the model, crack growth parameters C and m in the Paris-Erdogan and Forman laws are considered as random variables, and their stochastic characterizations are found from a crack growth experiment with small sample size. Furthermore, using the COVASTOL computer program (more details for this program are given later), the statistical distributions of crack growth rate $\mathrm{d}\alpha/\mathrm{d}N$ and cycles to reach a given crack length are obtained. The experimental results are used to verify the theoretical prediction of the statistical properties of fatigue crack growth behavior for aluminum 2024-T3 test specimens.

Material inhomogeneity has long been considered to be an important factor in crack initiation. However, it also has considerable influence on crack growth, which is not commonly perceived in deterministic fracture mechanics. Material inhomogeneity is usually negligible in crack growth under general laboratory conditions, especially under a random spectrum, because the fatigue stress dominates the scatter aspects of crack growth. However, there is considerable variability in a well-controlled test under a constant amplitude spectrum. For a good probabilistic model of crack growth, material inhomogeneity must be involved.

Several different approaches have been followed for the probabilistic modeling of material inhomogeneity. The most common approach is to randomize the crack growth parameters. For the crack growth equations of Paris-Erdogan and Forman there are several randomizations possible: both C and m in the Paris-Erdogan and Forman laws could be random variables; or C could be a random variable and m a constant; or m could be a random variable and C a function of m. However, C is really not a material constant (as was initially assumed by Paris), but it depends on the mean stress or stress ratio, frequency, temperature, etc. In particular, the stress ratio R is recognized to have significant influence on C.

Then, Paris-Erdogan and Forman equations can be transformed respectively to yield:

$$\ln(\mathrm{d}\alpha/\mathrm{d}N) = \ln(C) + m\ln(\Delta K) \rightarrow Y = \ln(C) + m\,X \qquad (6.14)$$

where,

$$Y = \ln(\mathrm{d}\alpha/\mathrm{d}N), \quad X = \ln(\Delta K)$$

and,

$$\ln(\mathrm{d}\alpha/\mathrm{d}N) = \ln(C) + m\ln(\Delta K) - \ln[(1-R)K_c - \Delta K] \rightarrow Y = \ln(C) + m\,X \quad (6.15)$$

where,

$$Y = \ln(\mathrm{d}\alpha/\mathrm{d}N) + \ln[(1 - R)K_c - \Delta K], \quad X = \ln(\Delta K)$$

According to relations (6.14) and (6.15), the C and m values of each specimen can be obtained from the test results of "α" vs. N curves and linear regression analysis, which is identical to the method of least-squares or the method of maximum likelihood.

There is a considerable experimental evidence that the crack length $\alpha(N)$ as a function of the number of accumulated cycles can also be modeled by the following three exponential forms (Hoeppner and Krupp, 1974).

$$\alpha(N) = C_1 N^{m_1}$$
$$\alpha(N) = C_2 (\log_{10} N)^{m_2} \tag{6.16}$$
$$\alpha(N) = C_3 e^{m_3 N}$$

where C_i and m_i; i=1,2,3, are functions of applied load, material characteristics, geometrical configuration of the component and the initial quality of the product being tested. Equations (6.16) can be rewritten as:

$$\ln \alpha(N) = \ln C_1 + m_1 \ln N$$
$$\ln \alpha(N) = \ln C_2 + m_2 \ln [\log_{10} N] \tag{6.17}$$
$$\ln \alpha(N) = \ln C_3 + m_3 N = C^* + m_3 N$$

Thus, regression lines of various types can be obtained for: the crack growth data reported for each test, for all data from a given specimen geometry, and for all data considered as one group.

The raw data from a crack propagation test are the half crack length, α, and the number of cumulative load cycles, N, needed to grow the crack to some crack length, α, from some reference initial crack length.

The current interpretation of these data is to report the FCG rate, $\mathrm{d}\alpha/\mathrm{d}N$, vs. ΔK, ΔK being the stress intensity factor range at the crack tip for each individual test. The graphical representation of these data includes a log-log plot of $\mathrm{d}\alpha/\mathrm{d}N$ vs. ΔK, leading to the best fit straight line on this plot, see fig. 6.17. This data processing method is strictly related to the use of the well-known Paris-Erdogan low or Forman law as a model for the FCG rate.

The overall variability encountered in FCG rate data depends on the variability inherent in both the data collection and data processing techniques. If C and m are taken as random variables, C and m are related. Cortie and Garrett (1988) have shown that the C-m correlation, while present, does not possess any fundamental significance and is purely the result of firstly, the logarithmic method conventionally used to plot the data, and secondly, the nature of the dimensions of the physical quantities used in the Paris-Erdogan equation. In the light of the probabilistic theorem, distribution of crack growth rate $\mathrm{d}\alpha/\mathrm{d}N$ as a function of ΔK can be deduced form stochastic characterizations of C

and m as well as the above logarithmic equations Crack growth rate $d\alpha/dN$ is able to accept a log-normal distribution (i.e. $\ln(d\alpha/dN)$ is able to accept a normal distribution), and its mean and variance are given using the Paris-Erdogan equation, by (two-variable prediction method):

$$E\left[\ln\ \frac{d\alpha}{dN}\right] = E[\ln(C)] + E[m\ \ln \Delta K]$$

$$\sigma\left[ln\left(\frac{d\alpha}{dN}\right)\right] = \{\sigma^2[\ln(C) + \sigma^2[m\ \ln(\Delta K)] + 2\rho_{cm}\sigma[\ln(C)]\ \sigma[\ln(\Delta K)]\}^{1/2} \quad (6.18)$$

where ΔK can be taken as any value, and the means, variances and correlation ρ_{cm} of C and m are taken from the statistical analysis of the raw FCG test data (Virkler *et al.*, 1979, Stavrakakis *et al.*, 1990).

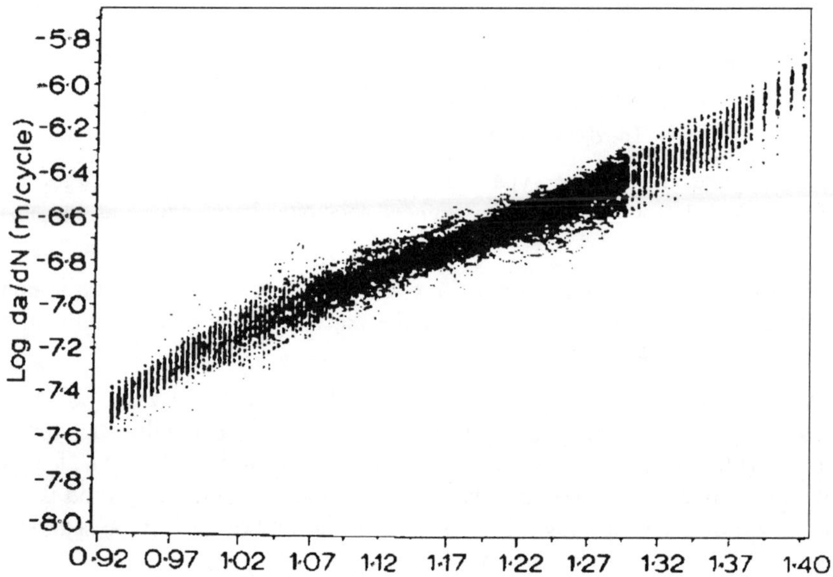

Figure 6.17 Summary of FCG rate data for the Virkler et al. (1979) case calculated by the ASTM E647-83 standard method.

If m is a constant and C a random variable, by the same principle, the crack growth rate $d\alpha/dN$ as a function of ΔK, can be shown to follow a log-normal distribution with mean and variance given by (single-variable prediction method):

$$E\left[\ln\frac{\mathrm{d}\alpha}{\mathrm{d}N}\right]=E[\ln(C)]+m\ \ln\Delta K$$

$$\sigma\left[\ln\left(\frac{\mathrm{d}\alpha}{\mathrm{d}N}\right)\right]=\sigma[\ln(C)+m\ \ln\Delta K]=\sigma[\ln(C)] \tag{6.19}$$

The Forman law can be used similarly.

The Shapiro-Wilk and Kolmogoroff-Smirnov tests can be applied to test if the logarithmic crack growth rate ln(dα/dN) could accept a normal distribution. Virkler *et al.*, (1979), have shown that dα/dN can accept a log-normal distribution, at least at a 10% level of significance, and the mean and variance of the experiment agree with the predictions. The means agree more closely with predictions. Moreover, the two-variable prediction method shows a slightly better correlation with experimental results than the single-variable prediction method, from more comparisons between predicted and experimental results (which are omitted for conciseness).

The distribution of cycles to reach a given crack length. The two-variable prediction method, in which C and m are random variables, is applied to obtain the distribution of cycles to reach a given crack length. The cycles to grow from an initial crack length α_0 to another length α_i, denoted $N_g|_{\alpha_0}^{\alpha_i}$, are inverted to:

$$N_g|_{\alpha_0}^{\alpha_i}=\int_{\alpha_0}^{\alpha_i}\frac{\mathrm{d}\alpha}{C(\Delta K)^m} \tag{6.20}$$

From eq. (6.20), $N_g|_{\alpha_0}^{\alpha_i}$ is a joint random variable of C and m. The Monte-Carlo simulation technique can be applied to get a convenient distribution of $N_g|_{\alpha_0}^{\alpha_i}$ through simulating distributions of C and m.

The distribution of crack lengths after a given service life (number of cycles). This procedure computes the propagation of a given defect or distribution of defects in a given position and the corresponding failure probability during accidental loading. It is thus more meaningful for real-time fatigue life prediction than the previous one.

The competence for facing such problems of cumulative structural damage has been acquired in the Components Diagnostics and Reliability Sector at the Joint Research Center of EEC, Ispra-Italy, from the development of analytical models for the representation of cumulative damage process and for the estimation of lifetime distribution under fatigue loading. Two numerical codes have been developed to this end, namely *COVASTOL* and *RELIEF*. The COVASTOL code has been developed in the framework of a more general study on the in-time estimation of nuclear reactor pressure vessels residual life time and failure probabilities. It is based on the application of probabilistic linear elastic fracture mechanics on statistical distributions of data concerning flaws, material properties and loading conditions (see Dufresne *et al.*, 1988).

The RELIEF code is based on the representation of the process of damage accumulation as a semi-Markovian stochastic process; no assumptions are made about the elementary mechanisms causing the accumulation of damage. The latter approach will be presented in the next section.

The COVASTOL code estimates the FCG rate by Paris' law with statistically distributed coefficients. The probability of onset of unstable crack propagation is estimated through the convolution of the distributions of the stress intensity factor and of the material resistance expressed by the static fracture toughness. The great advantage of this model is its simplicity, while tests are necessary to determine the coefficients m and C. It should, of course, be kept in mind that the Paris relationship does not generally describe correctly the behavior of cracks in the nucleation stage or near fracture; for small ΔK, for example, the propagation rate is overestimated (Nisitani *et al.*, 1992). However, it should be pointed out that no model describes in its entirety the crack propagation phenomenon. Under these conditions, the definition of at least three ranges of ΔK should allow more accurate FCG predictions. In that respect it is also certain that the different methods of treating the original data (α, N) introduce a scatter connected to the more or less pronounced importance of the subjective factor in each method. The method in the COVASTOL code is as follows (see also ASTM E647-83 standard):

- starting with the experimental data in each of the ranges considered, $(\mathrm{d}\alpha/\mathrm{d}N)_{mean}$ is computed for a certain number of ΔK_i levels;
- A linear regression is performed to determine, according to these values, the parameters m and C relative to Paris' law for the ΔK ranges considered. In each of these classes of ΔK a mean value of m is computed and retained and from this value the distribution of C is calculated. This distribution is presented in the form of a histogram of five class intervals.

It is quite important to mention here that in the operations connected with the fatigue crack growth calculation, a special procedure is implemented for the combination of the histograms, as follows:

- a given pair of values (or class intervals) α_0, b_0 (elliptical defects are considered) is combined with every class interval of C (coefficient of Paris law) only for the first stress transient;
- after that, α_0 , b_0 are transformed in one pair of histograms α_1, b_1. In the subsequent transients only combinations among class intervals of the same order are taken into account.

Concerning the width of the defects, because no data are usually available from manufacturers, its distribution is calculated by estimating the probability for two or more defects (assumed with one weld bead width) to overlap, both in horizontal and transversal section.

The defect length and width distributions so obtained correspond to the observed defects in a weld or a structure after fabrication and before repair, and are corrected automatically in order to take into account the sample size, the accuracy of the measurement equipment, the size of acceptable defects according to the construction rules, and the reliability of thc NDT methods (probability of having undetected and correspondingly unprepared defects).

To consider all combinations among α, b and C class intervals at every stress transient would in fact mean continuously mixing the material properties, whose scattering has, on the contrary, to be applied only once to the fatigue phenomenon considered as a whole.

The modeling defined above was introduced in the COVASTOL computer code thus allowing calculation of the propagation along the two axes of an elliptical defect subjected to a periodical loading. Temperatures and stresses as a function of location and time are given as deterministic analytical functions for each situation.

The probability of onset of unstable crack propagation is calculated as the convolution of fracture toughness and stress intensity factor histograms. Its evolution is followed during the stress transients as well as the evolution of any defect.

The COVASTOL program outputs give on the one hand the crack growths and if needed the evolution of defect size distribution and on the other hand the rupture probability associated with each defect size. The crack growth and the rupture probability computation procedure for internal and surface defects as well as test cases to calculate the rupture risk of welded steel Pressure Vessels are well presented and analyzed by Dufresne *et al.* (1986, 1988) where the reader is referred for details.

The failure probability, when a sophisticated program like the COVASTOL code is not available, can be calculated by using the limit state function concept. By integrating the Paris-Erdogan law (6.12) one obtains:

$$\int_{\alpha_0}^{\alpha_N} \frac{\mathrm{d}a}{[F(\alpha)\sqrt{\pi\alpha}\,]^m} = C\sum_{i=1}^{N} S_i^m \tag{6.21}$$

where α_0 is the initial crack size and α_N is the crack size after N stress cycles. For a given critical crack size α_c , the failure occurs when α_c-$\alpha_N \leq 0$. Hence the limit state function can be expressed as:

$$g(z) = \int_{\alpha_0}^{\alpha_c} \frac{\mathrm{d}\alpha}{[F(\alpha)\sqrt{\pi\alpha}\,]^m} - C\sum_{i=1}^{N} S_i^m \tag{6.22}$$

A similar limit state function using the Forman law (6.13) can be easily evaluated. The failure probability of the structure can be evaluated using the equations of Section 6.3.2 and the statistical distributions of the parameter C (m can be considered constant) and of the load sequence S_i. All the above analysis concerns mainly FCG under static loading conditions.

Zhu and Lin, (1992), propose a new analytical procedure to predict the fatigue life and reliability of a mechanical or structural component, with random material resistance to crack growth and under random loading. The procedure is applicable when the fatigue crack growth is a slow process compared with the stress process, which is the case for the high cycle fatigue. In the special case, in which the stress is a narrow-band stationary Gaussian process and a randomized Paris-Erdogan crack growth law is applicable, analytical expressions have been obtained for the probability densities of the fatigue crack size and the fatigue life, and the reliability function. A numerical example is given for the case of a degrading system. The accuracy of the proposed analytical procedure is confirmed by comparing the theoretical and simulation results.

Quality of the fatigue life prediction and failure prognosis. From the above discussion it is clear that the quality of the prediction depends directly on the quality of the method used to process the raw FCG experimental data and to estimate the parameters of the probabilistic fracture mechanics model. A poor estimation of the parameters will lead to an inaccurate prediction of the life-time, even if sophisticated FCG prediction models are used (Stavrakakis et al., 1990).

The currently used standard method to estimate the parameters, the ASTM E647-83, has several weak points. The determination of the derivative, $d\alpha/dN$, required by this method, introduces a scatter in the FCG rate data which varies considerably with the data processing technique used. Thus, a significant variation in $d\alpha/dN$ at a given ΔK level is introduced due to the raw FCG data processing technique. This variability introduced on the estimated FCG model parameters distributions due to the FCG raw data processing technique leads to a pessimistic structural reliability assessment. Moreover, the ASTM E647-83 standard method is strictly related and thus limited to the application of the Paris law or Paris-like models to describe the FCG phenomenon.

Stavrakakis, (1992), proposes a general method to process the raw FCG data based on non-linear regression techniques. In this method, the parameters of any probabilistic FCG rate model are estimated directly form its integral form, namely $\alpha=f(N)$. It is not restricted to the application of the Paris law for FCG predictions and handles, because of its generality, any probabilistic FCG rate relationship with any number of parameters.

This method permits a significantly reduced contribution to the variation in $d\alpha/dN$ at a given ΔK level, due to the raw data processing method. The performance of the method is evaluated using the integrated computer program COVASTOL for structural reliability assessment, when the FCG rate model coefficients are determined by the currently used ASTM E647-83 standard method and the new technique proposed there.

The two methods were used to process the Virkler *et al.* experimental data. The Virkler *et al.* data comprises 68 replications with constant load amplitude cycling loading. The data consists of the number of cycles required to reach 164 crack lengths, starting at 9mm and terminating at 49.8 mm, for each replication. Center crack (CCT) aluminum 2024-T3 test specimens were employed. The 68 sample functions of time (cycles) to reach a half crack length α are plotted, statistically analyzed and discussed in Virkler *et al.*, (1979). The non-linear FCG rate model considered was the Paris law for convenience.

In order to evaluate the influence of the FCG data processing method on the results of a FCG prediction program, the COVASTOL program was run for the same initial crack and stress transients conditions as in the Virkler *et al.* experiments, namely α_0=9mm and $\Delta\sigma$ =48MPa except for the Paris law parameters, i.e. the mean value of n and the C-histograms (Stavrakakis, 1992). First, the prediction of the defect propagation after a service-life of 2×10^5 cycles as resulted from the COVASTOL program is performed, when the Paris law parameters used are those derived by the standard ASTM E647-83. Then, the defect propagation prediction after 2×10^5 cycles is calculated by the COVASTOL program for the same initial and loading conditions, but with the C-histograms, n-mean values derived by the non-linear regression method. Finally, the real defect distribution (histogram) after a service life of 2×10^5 cycles, derived directly from the Virkler *et al.* experimental data, is given.

A comparison of the predicted defect histogram for the propagated crack length after 2×10^5 cycles with the real defect histogram has shown that even if the real crack length classes are predicted, the predicted probability of the upper classes (crack-length-α between ~31mm and ~40mm) is very high (~50 per cent) compared to the reality (less than ~10 per cent). Moreover, the prediction gives a small probability (~8 per cent) for fast crack propagation and fracture (crack length $\alpha >$ ~55mm up to ~71mm) that does not exist in reality. This is a quite conservative (i.e. pessimistic) prediction.

A comparison of the predicted defect histogram after 2×10^5 cycles with the real defect histogram has shown that the predicted probabilities of the different crack-length classes differ by less than 10 per cent from those of reality, and a successful coincidence between the two histograms occurs.

Thus, it is obvious that even if the variability introduced by the raw FCG data processing techniques of the ASTM E647-83 standard does not induce a significant amount of bias in the processed results it can induce an unacceptable bias in the final FCG prediction and residual-life-time results which make them conservative and thus less realistic.

In the above experimental evaluation the Paris law was used because this is the case in the COVASTOL program. This is not restrictive in any case. An analysis of examining the applicability of the unified fatigue crack propagation (FCP) approach proposed earlier for the FCP in engineering plastics such as PMMA and PVC is described by Chow and Wond (1987).

A Paris-like formulation is proposed to characterize FCP in polymeric materials and it is found, using measurements, that it is able to assess satisfactorily the FCP in both PMMA and PVC materials.

In this way, all the considerations of this section can be easily extended using this formulation to assess in-time the FCP phenomenon in polymeric materials and plastic pipes.

6.3.4 *Stochastic process approach for FCG life prediction*

In general both the loading actions and the resistance degradation mechanisms have the characteristics of stochastic processes. They can thus be defined as random variables which are functions of time. The particular load history which affects a component is one of the possible realizations of the stochastic load process and the same applies for the environmental condition or for the evolution of the dimensions of a defect inside the component.

The prediction of the component lifetime is to a large extent based on the representation of the stochastic processes which act on the component. The damage accumulation mechanisms can, in general, be represented by a positive "damage rate" function such that the measure of damage is a monotonic increasing function of time.

The physical situation to be contemplated is as follows: a structural component is in operation in a certain environment. During cyclic operation, irreversible changes occur. These irreversible changes accumulate until the component can no longer perform satisfactorily. The component is then said to have failed. The time at which the component ceases to perform satisfactorily is called the time-to-failure or the lifetime of the component.

The process by which the irreversible changes accumulate is called a cumulative damage (CD) process. Fatigue, wear, crack growth, creep are examples of physical processes in which CD takes place.

The particular damage process of interest here is the FCG as experienced for instance in failures of pressurized mechanical systems having a structure which contains defects as a result of technological operations like weldings. The defect dimensions, although continuous variables, are in fact associated with a discrete level or state which allows (without excessive restrictions) the use of well-known mathematical tools for discrete Markov processes.

The damage levels are represented by the states y=1, 2, ..., b; b being the conventional rupture state. The loading process is represented at cycle x by the transition matrix $\boldsymbol{P}_x$:

$$\boldsymbol{P}_x = \begin{bmatrix} p_1 & q_1 & 0 & . & . & 0 & 0 \\ 0 & p_2 & q_2 & 0 & . & . & 0 \\ . & . & . & . & . & . & . \\ . & . & . & . & . & . & . \\ 0 & 0 & 0 & 0 & 0 & p_{b\text{-}1} & q_{b\text{-}1} \\ 0 & 0 & 0 & 0 & 0 & 0 & 1 \end{bmatrix} \tag{6.23}$$

where p_j, q_j>0, p_j+q_j=1; j =1, 2, ..., b-1.

As the transition between the states is governed by eq. (6.23) the damage state at cycle x is linked to that at cycle x-1 by:

$$p_x = p_{x-1} \boldsymbol{P}_x \tag{6.24}$$

and thus,

$$p_x = p_0 \prod_{k}^{x} \boldsymbol{P}_k \tag{6.25}$$

which describes a *unitary jump* (*UJ*) stationary stochastic process.

The relationships (6.23)÷(6.25) represent the mathematical basis of the discrete Markov process; from them one can easily find the probability distribution of the number of cycles to failure and of the damage level at a given number of cycles x (Bogdanoff and Kozin, 1985).

The sample functions (SFs), that is the functions $\alpha(N)$ of each sample of the set form FCG experiments, are the complete, even if elementary, representation of the damage process. Starting from a set of SFs and from the first two statistical moments (mean value and variance of the cycles) related to a given value of the crack size α, the Markovian model (the above three equations) of the FCG process can be defined. This is called a *unitary-jump* (UJ) *stationary B-model* of CD. The mathematical details of this operation may be found in Bogdanoff and Kozin, 1985. The important point to be remarked here is the fundamental hypothesis of a Markovian process (that is the statistical independence of damage states).

The Markovian hypothesis characterizes a process "without memory" of the past events except those which occurred in the time immediately before. This assumption is purely theoretical because any damage state depends on the past history (Lucia *et al.*, 1987, Kozin and Bogdanoff, 1992, Bogdanoff and Kozin, 1985).

The disadvantages of the Markovian assumption are related mainly to an overestimation of the variance of the predicted time-to-failure distributions, when the initial crack population is different from the trivial case of a single crack, located at the origin of the SFs set.

The way to overcome this limitation of the B-model of CD has been suggested by Bogdanoff and Kozin (1984) which consider the propagation of a population of cracks as the superposition of many elementary propagation processes, each one starting from a particular crack size α_k, belonging to a given initial distribution. This can be done if one thinks in terms of many UJ stationary B-models, each starting from crack size α_k and considering as random variables the differences in cycles (N_j-N_k), where N_k is the cycle number corresponding to α_k. These new variables constitute a subset of the main random variable set N_j defined for every $j \neq k$.

The statistical moments of the first and second order corresponding to these variables are expressed taking into account the statistical dependence between N_j and N_k. The application of the method of statistical moments to the random variables (N_j-N_k) or/and (T_j-T_k) (T_j being the holding time in the state S_j) for the estimation of the parameters of each UJ Markovian model implicitly introduces the statistical dependence of the theoretical random variables N_j (or T_j).

A CD model having these characteristics is called a *semi-Markovian B-model*, because the fundamental assumption of an elementary Markovian model is disregarded and the dependence between N_j levels is considered. With this difference in mind the computer code *RELIEF 2·0* was developed at the Joint Research Center of EEC at Ispra-Italy which optimizes the efficiency of the Markovian scheme according to the above considerations. The calculation of the first and second order statistical moments to estimate the CD B-model parameters is now included in the code itself due to the dependence of this calculation step on the current crack size. In particular the evaluation of the covariances has now to be carried out in order to account for the statistical dependence between the number of cycles at the different crack sizes describing the process.

In their recent work, Kozin and Bogdanoff, (1992), propose and study a probabilistic macro model of FCG based upon a micro result from the reaction rate theory. A center crack panel under periodic tensile load is the basic physical situation considered. The model's explicit dependence on the temperature and the wave form of the periodic load indicates the importance of these two quantities in the evolution of the crack length. The straightforword relation of the semi-Markovian B-model parameters with the parameters of this probabilistic model illuminated many of the complexities that are experimentally observed in the FCG process.

The simplicity and flexibility of models based on Markov schemes is the reason for their frequent appearance in the literature. In the case in which the emphasis is rather upon the stochastic process of the loads and environmental conditions than upon the mechanism of damage accumulation, the traditional techniques for the treatment of processes of this type become more important. It is in this context that the Caldarola and Bolotin methods are described representatively by Lucia (1985). Many others can by found in the literature.

The structural reliability, in its most stringent formulation, can be defined as the probability that the largest of the loads envisaged is smaller than the smallest of the resistances hypothesized. This means that what one needs to know is the distributions of the extreme values of the loads and of the resistances, rather than their effective distributions. This observation, together with the fact that the possible distributions of the extreme values of a random variable are asymptotically independent of the distribution of the variable itself, leads to the consideration of the extreme values theory as a fundamental ingredient of structural reliability.

Some methods, all based on the hypothesis that the lowest resistance has a Weibull distribution, have been proposed by Freudenthal, Ang and Talreja and presented by Lucia, (1985).

A comparative study of the probabilistic fracture mechanics and the stochastic Markovian process approaches was performed by Stavrakakis *et al.*, (1990). The two computer codes COVASTOL and RELIEF based respectively on the randomization of a differential crack growth law and on the theory of discrete Markov processes were applied for fatigue crack growth predictions using the Virkler *et al.* (1979) and Ghonem and Dore (1987) sets of data of crack propagation curves from specimens. The results are critically analyzed and an extensive discussion is cited there on the merits and limitations of each approach.

The analysis and the presented results of the last two sections permits one to deduce some propositions on the applicability of the two codes (and consequently of the corresponding theoretical approaches) for real-time structural reliability assessment.

The COVASTOL code can be used for any real structure FCG prediction and reliability safety assessment given that any stress transient corresponding to some defined real conditions can be applied. Code outputs giving the predicted defect distributions could be more realistic - less pessimistic - if the scatter introduced by the treatment of the raw (α, N) data could be reduced, i.e. by using the method proposed in Stavrakakis (1992).

For the COVASTOL code numerical experimentation, raw data on defects (number; positions; dimensions) and material properties (FCG law parameters; static fracture toughness; crack arrest toughness; etc.) have to be converted into histograms by preliminary processing. Temperatures and stresses as a function of location and time are given as deterministic analytical functions for each situation. An important advantage of the COVASTOL code is that it is data-base independent, that is it can be applied for FCG predictions for stress transients other than those catalogued in the component material fatigue properties databank. Code outputs give the evolution of defect size distribution and the rupture probability associated with each defect size. Thus the COVASTOL code has a very large applicability to problems involving real structures.

The RELIEF code can be used in real situations where a representative databank exists for well-defined stress transients and environment conditions. The possibility to apply catalogued crack-growth processes, obtained in certain elementary loading conditions, to other more complex situations occurring in real structures, is a delicate problem which should be carefully examined in the particular context, with the approximations decided by the analyst. This necessity restrains the applicability of the code on structures which are loaded (even accidentally) with a limited small number of stress transients during their life.

The RELIEF code is based on the representation of the damage accumulation as a stochastic process by omitting any analytical fracture mechanics relation. It is important to recall that the markovian approach (B-model), as implemented in the RELIEF code, may be applied to any type of damage process (creep, corrosion, material embrittlement, etc.) when a representative description of the process is given in terms of a SFs set. Even though the structural reliability assessment performed by the RELIEF code is more precise than that performed by the COVASTOL code (smaller scatter), its applicability is limited with respect to complex real situations.

FCG predictions allowed by the RELIEF code are those concerning the SF sets (same material, environment conditions, type of load) corresponding to the different loading intensities (stress transients) which have been catalogued in the databank.

6.3.5 *Time series analysis approach for FCG prediction*

The principle underlying this methodology is that the fatigue crack growth data (N, α) occur in a form of a time series where observations are dependent. This dependency is not necessarily limited to one step (Markov assumption) but it can extend to many steps in the past of the series. Thus, in general, the current value N_α (=number of cycles at crack size α) of the process N can be expressed as a finite linear aggregate of previous values of the process and the present and previous values of a random shock u (Solomos and Moussas, 1991), i.e.

$$N_\alpha=\varphi_1 N_{\alpha-1}+\varphi_2 N_{\alpha-2}+...+\varphi_p N_{\alpha-p}+u_\alpha-\theta_1 u_{\alpha-1}-\theta_2 u_{\alpha-2}-...-\theta_q u_{\alpha-q} \tag{6.26}$$

In eq. (6.26) N_α, $N_{\alpha-1}$, $N_{\alpha-2}$, ... and u_α, $u_{\alpha-1}$, $u_{\alpha-2}$, ... represent respectively the number of cycles and the value of the random shock at the indexing equally spaced crack sizes α, α-1, α-2, ... The random shock u is modeled as a white noise stochastic process, whose distribution is assumed to be Gaussian with zero mean and standard deviation σ_u (specified by the structure random loading conditions).

Defining the *autoregressive operator of order p* by,

$$\varphi(B) = 1 - \varphi_1 B - \varphi_2 B^2 - ... - \varphi_p B^p$$

and the *moving-average operator of order q* by,

$$\theta(B) = 1 - \theta_1 B - \theta_2 B^2 - ... - \theta_q B^q$$

Eqn. (6.26) can be rewritten compactly, as,

$$\varphi(B)N_\alpha = \theta(B)u_\alpha$$

It is recalled that B stands for the backward shift operator defined as $B^s N_\alpha=N_{\alpha-s}$. Another closely related operator, to be used below, is the backward difference operator ∇ defined as $\nabla N_\alpha = N_\alpha - N_{\alpha-1}$ and thus equal to 1-B.

In an attempt to physically interpret the above equations and connect them to the observed inhomogeneous crack propagation properties, one could associate the autoregressive terms to the mean behavior of each individual test curve and the moving-average terms to the non smoothness within it, which is due to the inhomogeneity of the material ahead of the crack tip. In this manner, this spatial irregularity is approximated by the homogeneous random field u.

The *autoregressive moving-average model* (*ARMA*) as formulated above, is limited to modeling phenomena exhibiting stationarity, i.e., broadly speaking, fluctuating about a fixed mean. Clearly, this is not the case for the fatigue crack growth curves for which nonstationary processes will have to be employed. It is possible though that, even under these circumstances, the processes still possess a homogeneity of some kind. It is usually the case that the *d*th difference of the original time series (or, a nonlinear transformation of it) exhibits stationary characteristics. The previous ARMA model could than be applied to the new stationary process $\nabla^d N$ and the eq. (6.26) will correspondingly read,

$$\varphi(B)\nabla^d N_\alpha = \theta(B)u_\alpha \qquad (6.27)$$

This equation represents the general model used here. Clearly, it can describe stationary (d=0) or nonstationary (d≠0), purely autoregressive (q=0) or purely moving-average (p=0) processes. It is called *autoregressive integrated moving-average* (*ARIMA*) process of order (p, d, q). It employs $p+q+1$ unknown parameters $\varphi_1, \ldots, \varphi_p$; $\theta_1, \ldots, \theta_q$; σ_u , which will have to be estimated from the data.

Expecting that the fatigue crack growth curves would eventually reveal some stationary characteristics, the task of estimating the aforementioned unknown parameters is undertaken below. A phenomenological theoretical model will thus be built identifying the mechanism of crack propagation under certain loading and geometrical conditions. An outcome of direct practical importance will evidently be the possibility of forecasting the future behavior of the series N_α from its current and past values. This, of course, will be expressed in a probabilistic manner, in the form of a distribution.

Elaborating briefly on the terminology, if the values of N are known up to a current crack size α and a prediction of N is desired for ℓ steps ahead (i.e., at crack size $\alpha+\ell$), then one refers to "origin α", "lead time ℓ" and "forecasted value $N_\alpha(\ell)$". The methodology employed is capable of providing, beyond a "best" value of the forecast, probability limits on either side of it for a set of convenient values, for example 50%, 95%. If $\alpha+\ell$ is chosen to represent a critical value of the crack size, these forecasted results will obviously yield the distribution of the time-to-failure.

In the FCG type of series one expects relationships to exist (i) between observations of successive number of cycles in a particular record (the previously tackled problem); and (ii) between observations for the same crack size in successive records. Starting from the ARIMA model, it can be deduced that a seasonal series can be mathematically represented by the general multiplicative model (Solomos and Moussas, 1991)

$$\varphi_p(B)\Phi_p(B^s)\nabla^d\nabla_s^D N_A = \theta_q(B)\Theta_Q(B^s)u_A \qquad (6.28)$$

In this equation the parameters p, d, q and the operators $\varphi_p(B)$ and $\theta_q(B)$ are exactly as those defined for the ARIMA model and refer to the aforementioned point (i), while ∇ =1-B^s $\Phi_p(B^s)$ and $\Theta_Q(B^s)$ are proper polynomials in B^s of degrees P and Q, respectively, representing relationships of point (ii) above. This multiplicative process is said to be of order (p, d, q)×(P, D, Q)$_s$.

The building of the model for a specific physical problem is composed again the same steps: identification, estimation, diagnostic checking. The general scheme for determining a model includes three phases, which are:

- Model identification, where the values of the parameters p, d, q are defined.
- Parameter estimation, where the $\{\varphi\}$ and $\{\theta\}$ parameters are determined in some optimal way, and
- Diagnostic checking for controlling the model's performance.

As is stated however by Box and Jenkins (1976), there is no uniqueness in the ARIMA models for a particular physical problem. In the selection procedure, among potentially good candidates, one is aided by certain additional criteria. Among them are ***Akaike's information criterion*** (*AIC*) and ***Schwartz's Bayesian criterion*** (*SBC*). If $L=L(\varphi_1, \ldots, \varphi_p, \theta_1, \ldots, \theta_q, \sigma_u)$ represents the likelihood function, formed during the parameter estimation, the AIC and SBC are expressed, respectively, as,

$$AIC = -2\ln L + 2k$$
$$SBC = -2\ln L + \ln(n)k \tag{6.29}$$

where k, the number of free parameters ($=p+q$) and n, the number of residuals, that can be computed for the time series. Proper choice of p and q calls for a minimization of the AIC and SBC. Last, in the overall efficiency of the model, the principle of parsimony should be observed. Inclusion of an excessive number of parameters might give rise to numerical difficulties (ill-conditioning of matrices etc.), and might render the model too stiff and impractical.

Model building. It is well known that the ARMA model (6.26) can be written for identification purposes in the form of an observation equation as follows (see also Chapter 3):

$$y \overset{\Delta}{=} N_\alpha = \left[N_{\alpha-1}\; N_{\alpha-2} \cdots N_{\alpha-p}\; u_\alpha\; -u_{\alpha-1} \cdots\; -u_{\alpha-q}\right] \begin{bmatrix} \varphi_1 \\ \cdot \\ \cdot \\ \cdot \\ \varphi_p \\ 1 \\ \theta_1 \\ \cdot \\ \cdot \\ \cdot \\ \theta_q \end{bmatrix} \overset{\Delta}{=} \boldsymbol{u}^{\mathrm{T}} \boldsymbol{\theta} \tag{6.30}$$

An ARIMA model can also be written easily in a similar form, considering the new stationary process $\nabla^d N$ in the place of the process N.

On the other hand, as mentioned earlier in Section 6.3.3, the Paris-Erdogan and Forman logarithmic FCG equations (6.12) and (6.13) are the most suitable for accurate FCG prediction purposes because they can model satisfactorily the curves of fig. 6.17. The experimental points of fig. 6.17 do not form exactly a straight line. However, straight lines modeled by the Paris-Erdogan and Forman logarithmic FCG equations (6.14) and (6.15) of Section 6.3.3. can adequately represent large portions of them. The logarithmic equations (6.14) and (6.15) can also be rewritten in an observation form as follows:

$$Y = [1 \quad X]\begin{bmatrix} \log(C) \\ m \end{bmatrix} \stackrel{\Delta}{=} \boldsymbol{u}^{\mathrm{T}}\boldsymbol{\theta} \qquad (6.31)$$

The same considerations are obviously valid for the FCG laws of crack length as an exponential function of the number of accumulated cycles, presented before.

It can therefore be claimed that quite efficient linear regression models for the fatigue crack growth phenomenon have been constructed. In addition, they have the advantage of being compact, easily presentable and implementable. They can thus serve in practical situations, as they can readily furnish updated predictions of a component's residual lifetime after periodic inspections.

Every such model is built based on the primary form of information of the crack growth, i.e. the (N, α) sample functions, and consequently is suitable for a specific set of geometric and loading conditions. The possibility of utilizing the same model under different conditions, or of attaching physical significance to its parameters, can also be envisaged.

In particular, if one considers moving windows of data of appropriate length, iterative regression techniques can be used to track the varying conditions. In this way an adaptive prediction method is introduced by Stavrakakis and Pouliezos, (1991), which is especially desirable in such cases, since the parameters of the logarithmic Paris-Erdogan (6.14), logarithmic Forman (6.15), ARMA, ARIMA and logarithmic exponential FCG models (6.17) change with time (number of cycles), due to the continuous variation of the conditions related with the FCG condition (stress transients, random overloads, temperature, material properties, inspection technique variability, etc.).

To denote explicitly the dependence of the various regression models (6.14)÷(6.17) estimated parameters on the number of cycles, the observation equation derived before for the various model cases may be written more accurately as,

$$y = \boldsymbol{u}^{\mathrm{T}}\, \boldsymbol{\theta}\,(N) \qquad (6.32)$$

For n pairs of ($\alpha(N)$, N) experimental points, the well known linear least-squares regression formula gives,

$$\hat{\boldsymbol{\theta}} = [\boldsymbol{U}^{\mathrm{T}}\boldsymbol{U}]^{-1}\boldsymbol{U}^{\mathrm{T}}\boldsymbol{y} \qquad (6.33)$$

where $\boldsymbol{U}$, $\boldsymbol{y}$ hold the information for the whole set of data.

Iterative methods that update the estimate whenever new information is available can also be used. For accurate detection purposes, a moving window regression formula is more appropriate, since it is more sensitive to parameter changes during the variation of the thermomechanical conditions of the structures. As shown in Stavrakakis and Pouliezos, (1991), a moving window estimate is given by the following recursive equations (see also Appendix 3.A):

$$\hat{\theta}(k+1) = \hat{\theta}(k) - \boldsymbol{P}(k+1)[\boldsymbol{\Gamma}(k+1)\hat{\theta}(k) - \boldsymbol{\delta}(k+1)]$$

$$\boldsymbol{P}^{-1}(k+1) = \boldsymbol{P}^{-1}(k) + \boldsymbol{\Gamma}(k+1) \qquad (6.34)$$

where,

$$\Gamma(k+1) = \boldsymbol{u}(k+1)\boldsymbol{u}^{\mathrm{T}}(k+1) - \boldsymbol{u}(k-n_w+1)\boldsymbol{u}^{\mathrm{T}}(k-n_w+1)$$

$$\delta(k+1) = \boldsymbol{u}(k+1)y(k+1) - \boldsymbol{u}(k-n_w+1)y(k-n_w+1)$$

and n_w is the window length.

The proposed FCG prediction algorithm consists of the following steps:

Step 1: Compute $\hat{\theta}(n_w)$ for the first n_w pairs of ($\ell n\, a$, N) data, using the one-shot linear least-squares regression formula and one of the FCG linear regression models proposed before. If an ARMA or ARIMA model is adopted the AIC and/or SBC criteria (6.29) must be applied off-line using a big amount of raw FCG data in order to determine the appropriate structure of the model for the specific case.

Step 2*:* Process the pair of data coming from the next inspection using the moving window regression formulae.

Step 3*:* Estimate the one step ahead predicted value for $a(N)$ using the adopted model equation(s) and $\hat{\theta}(N)$. The value of N (number of cycles) used in this one step ahead predictor must be the number of cycles for the next inspection according to the inspection-maintenance schedule of the structure.

Step 4*:* The predicted value of $\hat{\alpha}(N)$ is checked against the predetermined critical crack length threshold a_c. If $\hat{\alpha}(N) \geq \alpha_c$ an emergency condition is declared and appropriate action should be taken, otherwise go to step 2.

The estimator variance $\boldsymbol{P}(k+1)$ is automatically updated by the above moving window procedure, thus the predicted defect variance can be easily calculated. It is then possible to determine in real time, i.e. at any number of cycles N, the probability of structural failure, i.e. the probability $U(N)$ that an inadmissible failure state will occur

$$U(N) = P[\hat{\alpha}(N) \geq \alpha_c] = 1 - P[\alpha(N) < \alpha_c] = 1 - R(N) \qquad (6.35)$$

where $R(N)$ is the structural reliability function.

The FCG law $\ell n\alpha(N)=C^*+m_3N$ (see Section 6.3.3, eqs. (6.17)) is fitted by Stavrakakis and Pouliezos, (1991), into the Virkler *et al.*, (1979) data, using the linear moving window regression technique described before. The "deterministic" value of the parameter m_3 is estimated to be $6{\cdot}89\times10^{-6}$ and the mean value and variance of the parameter C^* are estimated as 1·94 and $7{\cdot}67\times10^{-3}$ respectively.

In this case, the failure probability of the structure or component can be calculated in closed form as follows (see for details Stavrakakis and Pouliezos, 1991):

$$U(N)=\varphi\left(\frac{N-A}{B}\right) \tag{6.36}$$

where the parameters A, B are:

$$A=\frac{\ln\alpha_c-\mathrm{E}\{C^*\}}{\overline{m}_3}, \qquad B^2=\frac{[\mathrm{var}\{C^*\}]}{\overline{m}_3^2}$$

Parameter A represents an estimation of the mean number of cycles in order to attain the critical crack length α_c.

Parameter B determines the role of the quality of the product, i.e. variability of the properties of the material and of the loading and thermal conditions, or the measurement error introduced by the crack detection method.

The failure probability of a cracked aluminum-2024-T3 structure constructed using the aluminum corresponding to the loading conditions of Virkler *et al.*, (1979), experiment can be calculated using the above equations. The failure probability for the Virkler's experiment at $N=2\times10^5$ cycles and for a critical crack length α_c=32.68mm, is found, by applying the above equations, to be $\hat{U}$(200000)=0.0274. Parameters A, B were found to be 224958.15 and 12715.5 cycles respectively. From the propagated crack length histogram at $N=2\times10^5$ cycles derived directly from the Virkler *et al.*, (1979), experiment the same probability is evaluated as U(200000)=0·0294. This represents a discrepancy of 6.8%.

The usefulness of the moving window method is illustrated using one set of Virkler's data. Simulation runs for the one-step ahead predictor indicated that the optimum window length was n_w=4. This produced a maximum absolute prediction error of 0.23 over the whole range of data. If predictions of longer horizon are required, simulation runs could establish the corresponding optimum window length.

In cases where crack length measurements are available on-line using appropriate hardware equipment, the recursive nature of the method makes it suitable for an integrated automatic safety alarm system.

Autoregressive integrated moving-average processes have been employed by Solomos and Moussas (1991) for the modeling of the number of cycles over the crack size for the fatigue crack propagation phenomenon. Even though no perfect stationarity conditions have been obtained in the treatment of the Virkler *et al.* (1979) records, an overall good performance of the derived models has been observed. It has been found that a single record can be reproduced satisfactorily by an ARIMA process of order $(p, d, q)=(2, 3, 1)$. The quality of the forecasts depends upon the origin; an early origin allows for short forecasts while a later origin yields unconditionally good forecasts. A multiplicative ARIMA process of order $(p, d, q)\times(P, D, Q)_s=(1, 2, 1)\times(0, 1, 1)_{89}$ has been found to represent very efficiently the whole set of the fatigue crack records. Its forecasting capabilities are excellent both at reproducing existing data, and at the monitoring and prediction of new experiments.

6.3.6 Intelligent systems for in-time structural damage assessment

As it has already been discussed, the mathematical FCG models available for representing the relevant physical processes are only approximate representations of the physical reality, having peculiar, but often ill-defined characteristics of precision, sensitivity and range of validity. Furthermore, they do not constitute an exhaustive representation of the reality. The knowledge to be used, related to various fields, is not fully representable by algorithms or mathematical tools but contains also qualitative and heuristic parts. Any a priori estimate of the life span distribution of a structure shows therefore, quite a large scatter which can be progressively reduced by using proper updating techniques.

Traditional algorithmic approaches are unable to cope with such a complex context. Expert systems are potentially, the breakthrough. Expert systems, roughly consisting of a procedure for inferring intermediate or definitive conclusions on structural damage and remnant lifetime, using the domain knowledge and the accumulating service data, can deal with real world problems by properly incorporating all the knowledge which may become available. An expert system for structural reliability assessment must have the ability to analyse and interpret large quantities of information in order to achieve the following goals:

- Identification of the actual state of the structure and of the damage process actually taking place.
- Prediction of the future behavior of the structure.
- Decision and planning of appropriate actions.

The bone of the expert system can be thought of as a coordinator and manager of operators which mutually collaborate and supply the information the system needs. Each step of the assessment procedure (e.g. defect population identification, material properties selection, microdamage analysis, macrodamage analysis, etc.) can constitute one operator or be subdivided into more specialized operators. The user can exploit interactively the functions performed by the operators. Rules and decision criteria can be modified under a set of metarules. The modular array allows an easier representation of the base of knowledge and an incremental construction of the system (see also Chapter 4 and Jovanovic *et al.* (1989)).

An expert system for assessing damage states of structures will consist of an interpreter, data-base and rule-base. All the rules involved are described through production rules with certainty factors. The inspection results are used as the input data. The inspection results regarding cracks are firstly input into the system, rules concerning their damage degree, cause and expanding speed are implemented to provide a solution for the damage assessment. This inference procedure is performed as shown in fig. 6.18.

The uncertainties involved in the input data and rules can be taken into account by introducing certainty factors. Damage pattern, damage cause and deterioration speed are employed to interpret the inspection data from the multi-aspect point of view.

Certainty factor. Most of the data available in the damage assessment generally include certain kinds of uncertainty and experience-based knowledge may be vague and ambiguous. Thus, an expert system should, have the ability to treat these uncertainties in a logical manner. The certainty factor is calculated hereafter. Input data and production rules are written as follows, with certainty factors:

Data 1: C_1; Data 2: C_2; . . ., Data p: C_p
IF Ant. 1, Ant. 2, . . ., Ant. m
THEN Con. 1: C'_1, Con. 2: C'_2, ..., Con. n: C'_n

where Ant. and Con. denote *antecedent* and *conclusion*, respectively, and C_p and C'_j are certainty factors. p, m and n are the numbers of input data, antecedents and conclusions, respectively. At execution of the inference procedure using the rules, including the certainty factors, the following must be done:

1. Calculate the certainty factor for the resultant antecedent.
2. Calculate the certainty factor for the resultant conclusion.
3. Determine the final conclusion and calculate its certainty factor for more than two rules which provide the same conclusion.

One can employ the following calculation methods corresponding to the items above:

1. $C_{in}= \min(C_1, C_2, \ldots, C_m)$, where C_{in} is the certainty factor for the resultant antecedent.
2. $C_{out,k}=C_{in}\times C'_k$ where C'_k is the original certainty factor for the *k-th* conclusion and $C_{out,k}$ is the certainty factor of the *k-th* output.
3. The certainty factor C for the final conclusion is calculated as follows, using $C_{out,k}$

$$C = \max\,(C_{out,1}, C_{out,2}, \ldots, C_{out,k})$$

Suppose that inspection data are as given in Table 6.2.

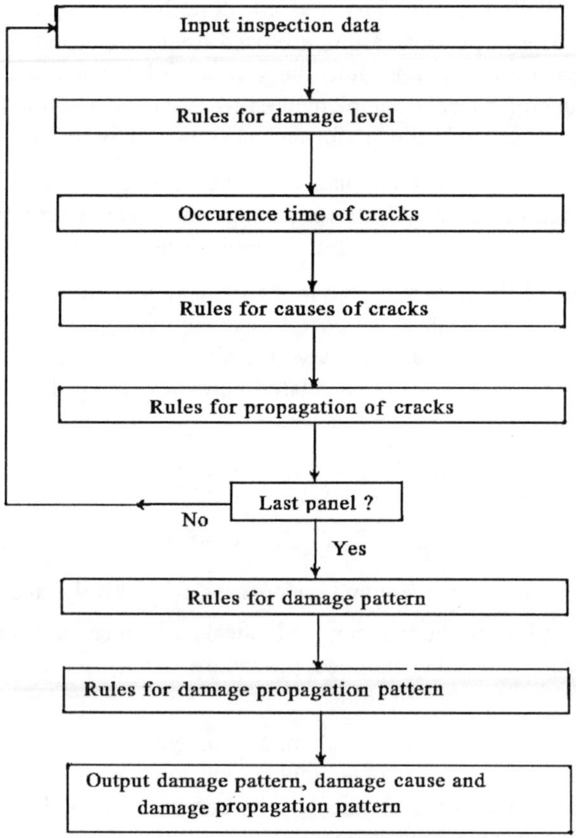

Figure 6.18 The expert structural damage assessment inference process.

Table 6.2 Example of inspection data

Inspection item	Result	CF
Direction of cracks	2 directions	0.9
Width of cracks	Middle	0.5
Interval of cracks	Small	0.7
Fracture	Large	0.5

By dividing the data-base and rule-base into several groups, it becomes possible to reduce the execution time which is proportional to the number of available rules.

Fig. 6.19 shows examples of rules for the inference process.

```
(damage-degree-2-1
 if
        (direction-of-cracks 2-directions=CF1)
        (width-of-cracks middle=CF2)
        (interval-of-cracks small=CF3)
 then
        (*deposit (damage-degree A (*times 1.0 (*min=CF1=CF2=CF3)))))
(damage-degree-4-1
 if
        (fracture large=CF1)
 then
        (*deposit (damage-degree A(*times 1.0 (*min=CF1)))))
```

Figure 6.19 Examples of rules for the damage degree of reinforced concrete bridge decks.

In practice, the values of the certainty factors involved in the input data and production rules are given by an expert who has been engaged in maintenance work for more than 20 years. First, matching succeeds in the rule (damage degree 2-1), where 0.9, 0.5 and 0.7 are prescribed for $=CF1$, $=CF2$, and $=CF3$, respectively. The symbol = denotes CF_i is a variable. Second, C_{in} is calculated as 0.7, using step (1), i.e., C_{in}=min($=CF1$, $=CF2$, $=CF3$). According to step (2), $C_{out,k}$ is obtained as 0.7 from 0.7×1.0. This leads to the conclusion that damage state is A with CF=0.7. Similarly, the rule (damage degree 4-1) leads to another conclusion that the damage state is A with CF=0.5. From these two conclusions, the final conclusion is that damage state is A with CF=0.7, using step (3).

In the MYCIN approach (see Chapter 4), the certainty factors are formally defined, extensively tested and correct results/diagnoses have been obtained in many circumstances.

Evaluation method. Usual damage state evaluation is based only on the information obtained from visual inspection. If one desires a high accuracy in its evaluation, the damage degree ought to be classified into several categories, where too many categories may often induce a contradiction among them derived by each individual and make classification meaningless. To increase the evaluation accuracy, one can introduce three damage measures; damage pattern, damage propagation pattern and damage cause. An appropriate damage pattern is chosen among prescribed basic damage patterns. Similarly, the most probable damage propagation pattern is determined by using the inference results of the crack occurrence time, crack pattern, cause of crack, and serviceability of the concrete deck. Basic damage patterns are determined by considering the following:

- Pattern 1: Severe damage is seen all over the structure.
- Pattern 2: Severe damage is concentrated at the structure edges.
- Pattern 3: Severe damage is concentrated at both ends of a structure component.

- Pattern 4: Severe damage is concentrated at the overhang portions of the structure (if these portions exist).
- Pattern 5: Severe damage is concentrated in the structure's center region.
- Pattern 6: Severe damage is not seen all over the structure.

To demonstrate the usefulness of the expert system in FCG real-time assessment, a plate-girder bridge with four main girders and seven cross beams is employed by Shiraishi *et al.* (1991).

A large number of rules useful for the damage assessment could be acquired through an intensive interview with well-experienced engineers on repair and maintenance work. The use of certainty factors can lead to a reliable conclusion using vague and ambiguous data and rules. Introducing the three damage measures such as damage pattern, damage propagation pattern and damage cause, it is possible to give useful information to predict the change of structural durability in the future.

The damage causes are estimated on the basis of damage degree, damage pattern, and loss of serviceability and the estimation is important to clarify the occurrence mechanism of damage as well as useful for establishing an efficient repair and maintenance program (see Lucia and Volta, 1991).

Recently Vancoille *et al.*, (1993), have developed a new module that explicitly deals with corrosion troubleshooting. During the development of this module it was observed that expert systems are not always suited to carry out part of the tasks involved in corrosion troubleshooting. Therefore, the possibilities of neural networks were investigated. It was realized that they have some potential that might open completely new perspectives in dealing with problems where expert systems tend to fail. The combination of expert systems and neural network techniques gives rise to powerful architectures that can be used to solve a wide range of problems.

In cases where conventional analytical techniques cannot provide a useful means for the evaluation of system reliability, techniques based on expert opinions may be used until such time that either performance data can be obtained and/or mathematical modeling of system reliability along with adequate field or laboratory data can be used. The expert opinion technique can also be used in conjunction with an analytical approach in cases where the performance data are sparse but system failure modes are well known (Mohammadi *et al.*, 1991).

Specific examples of engineering systems for which the expert opinion approach can be used in lieu of acquiring data from conventional sources are given next.

Bridge inspection. In this problem, the evaluation of bridge components, i.e., determination of their levels of deterioration and extent of damage is conducted by experts (bridge inspection personnel). The results of an inspection are then verified, analyzed and used along with structural analyses to arrive at a specific rating for a given bridge. The rating is indicative of the level of structural integrity of the bridge.

Interior gas piping systems. Interior gas piping systems operate under low pressure, 1.75 to 14.00 kPa (0.25 to 2.0 psi). Under normal operating conditions, the internal stresses are low and do not impose any safety problems. However, there are many factors (such as poor installation practice, component malfunction, loose joints due to external factors, etc.) that can contribute to system failure resulting in a leak. An expert opinion approach can effectively be used (i) to identify components' modes of failure; and (ii) to compile system performance data for reliability evaluation purposes (Mohammadi et al., 1991, Sandberg et al., 1989).

Human error. The impact of human error on the reliability of an engineering system is another problem that may be investigated using the expert opinion approach. One typical example is fabrication errors occurring during construction of a facility. Identification of factors that may ultimately promote structural failure and evaluation of the likelihood of occurrence of such factors can be done using the expert opinion approach.

In the above three examples the objective is well defined, i.e., the objective is to acquire information on the performance of a system and to determine its reliability. In certain non-engineering areas, however, the objectives may be unknown or not clear. Thus a separate expert opinion survey may be used only to arrive at a set of objectives and attributes to the problem being investigated.

In engineering problems, because the objectives are often well known, the expert opinion approach becomes simply a data collection process that can be used for one or more of the following tasks:

- Identification of failure modes in terms of component or system performance.
- Establishment of statistics or occurrence rates for individual modes of failure.
- Fault-tree and event-tree analyses and identification of the sequence of events (scenarios) whose occurrence would lead to the formation of a top event (in fault tree analysis) or a series of consequences (in event tree analysis).

The general process of the expert opinion method is very much dependent on the type of problem. As described earlier, in cases where the problem's objectives are well defined and the parameters influencing these objectives are also known, the procedure degenerates to a data collection scheme for ranking or scaling the objectives and their associated parameters. Many engineering problems fall under this category and represent cases each with a limited number of well defined objectives. Each objective may then be expressed with a performance level and a series of attributes. In other extreme cases where uncertainties exist in specific objectives and their attributes, the expert opinion approach may become very complicated. Generally problems associated with societal or economics issues fall under this category. In such cases the method may have to be repeated for several rounds before a final decision on the objectives can be made.

The following list presents the basic elements of the method and can well be expanded for certain cases.

1. Discuss why the expert opinion approach is employed instead of other methods.

2. Identify a series of objectives in the study. If the objectives are not well defined, a separate expert opinion approach may be used to arrive at definite objectives.
3. Solicit expert opinions for ranking or scaling these objectives. At this stage the final refinement of the rankings may be done in more than one round if time and money permit and especially if a somewhat large discrepancy in the opinions is observed.
4. Summarize the findings in a form that can be used as a mathematical tool for the system risk analysis or merely as a support document. The findings may also be evaluated using statistical methods. Of course, prior to these steps, experts must be identified.

A case study is presented by Mohammadi *et al.*, (1991), to demonstrate the applicability of the expert opinion approach in system reliability evaluation. In this case study, the risk associated with leak development in several interior gas piping system is evaluated and the results are presented. The structure considered in the case study is a simple system made of components with binary modes of failure. For more complicated structures with multiple independent and/or dependent modes of failure the reliability formulation and evaluation of results require additional analyses including the translation of the expert opinion data into numerical values that can be used in the formulation of the individual modes of failure. One objective of the case study presented there was to compare an existing system (black steel piping system) with a new product (corrugated stainless steel tubing). In the absence of reliable performance data on these systems the expert opinion approach was employed. As demonstrated in this example, the approach offers an effective method in the analysis of system reliability of each system and the evaluation and comparison of the performance of the two systems.

To treat the uncertainty and ambiguity involved in the expression in terms of natural languages, it is useful to introduce the concept of fuzzy sets.

Garribba *et al.*, (1988), present a specific application of fuzzy measures relevant to structural reliability assessment for the treatment of imperfections in ultrasonic inspection data.

Looking from a general point of view at the problem of combining multiple non-homogeneous sources of knowledge, whilst the structure of the composition problem can differ from one case to another, the preservation of a general pattern may be suppossed. Thus, the investigation and characterization of this pattern can help to highlight the nature of the dependencies between the different sources.

Assessment of damaged structures is usually performed by experts through subjective judgments in which linguistic values are frequently used.

The fuzzy set concept is then used to quantify the linguistic values of the variables of damage criteria and to construct the rules. Assessments from the same group of experts may result in rules with the following cases:

1. Similar antecedents and consequents.
2. Similar antecedents but different consequents.

3. Similar consequents but different antecedents.
4. Different antecedents and consequents.

In the case of similar antecedents and consequents (1), fuzzy set operations need not be used. The total number of similar rules determines the weight of damage levels in the rules. In the case where several rules have similar antecedents but different consequents, these rules can be combined. For example, consider a case in which there are five rules with similar antecedents but three of the consequents indicate that the damage level is very severe or "DL is VSE" an two others indicate that the damage level is severe or "DL is SEV". The combined consequents of rules 1 and 2 can be represented by the following:

CONS 1: DL is VSE (0.6) AND
CONS 2: DL is SEV (0.4)

where CONS denotes "consequent" and where (0.6) and (0.4) are obtained from 3/5 and 2/5, indicating the weights of CONS 1 and CONS 2, respectively.

Two rules can have similar consequents but different antecedents, as is shown below:

Rule (expert 1):	Rule (expert 2):
ANT 1: DEQ is VSE AND	ANT 1: DEQ is SEV AND
ANT 2: IOC is VSE	ANT 2: IOC is VSE

where ANT denotes the antecedent, DEQ, is equipment damage level, and IOC is the injury level of the occupants; then these rules can be combined through the use of an OR-gate as follows:

ANT 1: DEQ is (VSE OR SEV) AND
ANT 2: IOC is VSE

The antecedents and consequents of the rules may be different:

Rule (expert 1):	Rule (expert 2):
ANT 1: AR is MAJ AND	ANT 1′: AR is VSB AND
ANT 2: RT is VLO AND	ANT 2′: RT is VLO AND
ANT 3: RC is EXP AND	ANT 3′: RC is VEX AND
ANT 4: RA is ABD	ANT 4′: RA is ABD
CONS 1: DL is VSE	CONS 1′: DL is SEV

where AR, RT, RC and RA are the amount of repair, repair time, repair cost, and resource availability, respectively. The linguistic values MAJ, VSB, VLO, EXP, VEX, and ABD denote major, very substantial, very long, expensive, very expensive, and abundant, respectively.

Then if the rules do not conflict with each other, they will stay as they are. But if conflicting rules occur as in the ANT 1 and ANT 1′, ANT 3 and ANT 3′, and CONS 1 and CONS 1′ in the above example, a combined rule should be sought through the use of a fuzzy relation such that R11=MAJ×VSE, R31=EXP×VSE, R1′1′=VSB×SEV, and R3′1′ =VEX×SEV, where R_{ij} is the fuzzy relation between ANT i and CONS j; R11 and R1′1′ are contained in the classes of all fuzzy sets of (AR×DL); and R31 and R3′1′ are contained in the classes of all fuzzy sets of (RC×DL).

The combined relations of R11 and R1′1′ can be obtained through the use of the modified combined fuzzy relation method introduced by Boissonnade which is an extension of Mamdani's approach which combined all relations though fuzzy disjuctions. The method uses modified Newton iterations to reach an optimal solution for the combined fuzzy relations. Details of these techniques can be found in Chapter 4 and Hadipriono and Ross (1987). Through the use of this method, the combined relations of R11 and R1′1′ yield R111′1′. A similar procedure is performed for R31 and R3′1′ to yield R313′1′. The fuzzy composition between R111′1′ and R313′1′ results in R131′3′, contained in the classes of all fuzzy sets of (AR×RC). The fuzzy set value for AR and RC, is the projection of R131′3′ on planes AR and RC, respectively. The result now yields two rules with similar antecedents but different consequents. Hence, similar procedures can be applied as in cases (2), (3) and (4).

A complete rule may require the participation of the three damage criteria. Therefore, the rules should also be combined to incorporate the functionality, repairability, and structural integrity of the damage structure. Zadeh developed the extension principle to extend the ordinary algebraic operations to fuzzy algebraic operations. One method based on this principle is the DSW technique introduced by Dong, Shah, and Wong (see Hadipriono and Ross, 1987). The technique uses the lambda-cut representations of fuzzy sets and performs the extended operations by manipulating the lambda-intervals. For brevity, further details of these techniques can be obtained in the above references.

In order to accommodate the effect of each damage criteria on the total damage, in this study, one can include the weighting factor of each criterion. For example, if the weights of the damage level assessed, based on the above three damage criteria, are assumed to be "high" (HIH), "fairly high" (FHI), and "moderate" (MOD), respectively, and the values of the damage level are DL1, DL2, and DL3, respectively, then the overall combined damage level becomes,

$$DL_{tot} = \frac{(HIH \times DL1) + (FHI \times DL2) + (MOD \times DL3)}{HIH + FHI + MOD} \tag{6.37}$$

Based on the complete rules, new or intermediate rules can be constructed through partial matching.

Consider the following production rule: "IF deformation (DF) is very severe (VSE), THEN damage level (DL) is severe (SEV)". When a fact shows that DF is VSE, the consequent is then realized. However, when the value of DF does not match exactly, e.g., the fact shows that "DF is SEV", then partial matching is in order.

This can be performed by the following fuzzy logic operations: ***truth functional modification*** (*TFM*), ***inverse truth functional modification*** (*ITFM*), and ***modus ponens deduction*** (*MPD*). Brief descriptions of these operations follow:

TFM, first introduced by Zadeh, is a logic operation that can be used to modify the membership function of a linguistic value in a certain proposition with a known truth value. Suppose that damage level (DL) is "negligible" or NNE and is believed to be "false", or FA. This proposition can be expressed as

$$P\text{: (DL is NNE) is FA; NNE} \subset \text{DL, FA} \subset T$$

where DL is a variable (universe of discourse), T is the truth space, and NNE and FA are the values of DL and T, respectively. The symbol $\subset$ denotes "a subset of ". Modification of this proposition yields,

$$P'\text{: (DL is DL1): DL1} \subset \text{DL}$$

where DL1 is a value of DL. A graphical solution is shown in fig. 6.21 where the fuzzy set NNE and FA are represented by Baldwin's model, (1980), and plotted in figs 6.21.b and 6.21.a, respectively.

Note that the axes of fig. 6.21.a are rotated 90° counterclockwise from fig. 6.21.b. Since the elements of FA are equal to the membership values of NNE, they are represented by the same vertical axis in fig. 6.21. This means that for any given element of NNE, one can obtain the corresponding element of FA. Also, since the membership values of FA and DL1 are the same, the membership values of DL1 can be found as shown by the arrowheads and plotted in fig. 6.21.b.

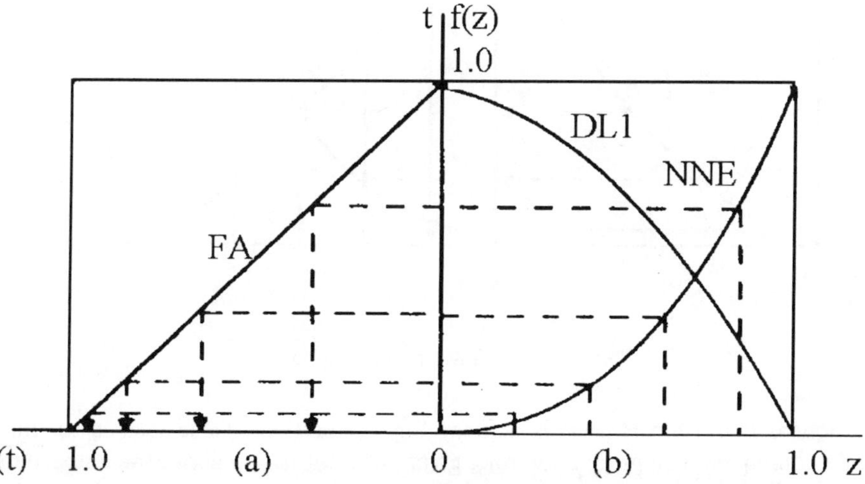

Figure 6.21 TFM graphical solution.

ITFM is a logic operation that can be used to obtain the truth values of a conditional proposition. Suppose a proposition, *P*, is expressed as "damage level is negligible given damage level is severe"; then the proposition can be rewritten as,

$$P: \text{(DL is NNE)} \mid \text{(DL is SEV); NNE, SEV} \subset \text{DL}$$

The ITFM reassesses the truth of (DL is NNE) by modifying this proposition to yield,

$$P': \text{(DL is NNE) is T1; T1} \subset T$$

where T1 is the new truth value for (DL is NNE). The truth value, T1 can also be obtained through the graphical solution shown in fig. 6.22. Suppose NNE and SEV are again represented by Baldwin's model. The values NNE and SEV are first plotted as shown in fig. 6.22.b. Since the truth level is equal to the membership value of NNE they lie on the same vertical axis. Hence, for each membership value of NNE, the corresponding element of T1 is also known. Then too, since the membership value of T1 equals that of SEV, for any given element of both NNE and SEV, one can find the corresponding element and membership value of T1. The truth value, T1, in fig. 6.22.a is constructed by successively plotting the membership values of SEV (*d*1, *d*2, etc.) from fig. 6.22.b at each truth level. Note that the axes in fig. 6.22.a are rotated 90° counterclockwise from fig. 6.22.b.

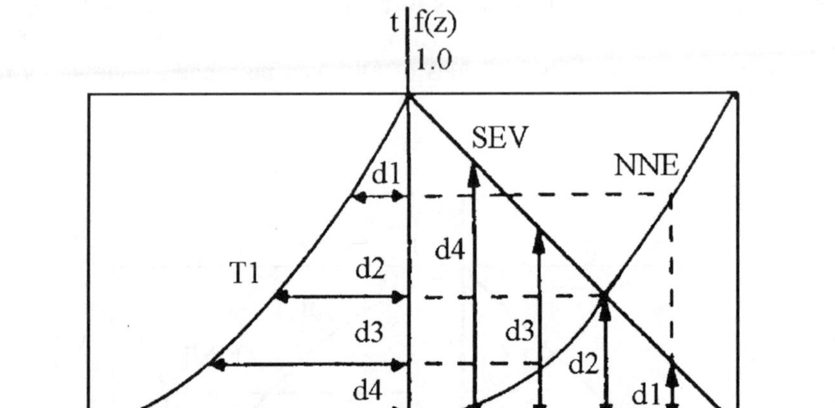

Figure 6.22 ITFM graphical solution.

Modus ponens deduction (MPD) is a fuzzy logic operation whose task is to find the value of a consequent in a production rule, given the information about the antecedent. A simple MPD is: *A* implies *B* and given *A*, then the conclusion is *B*. Consider again the proposition: "if deformation is very severe, then damage level is severe", (IF DF is

VSE, THEN DL is SEV). However, suppose further information is available, i.e., "deformation is severe" (DF is SEV). These propositions can be represented by the following:

P : (DF is VSE) ⊃ (DL is SEV)

P ′: (DF is SEV)

where the symbol ⊃ represents the implication relation between (DF is VSE) and (DL is SEV). This example can be conveniently solved through the following graphic representation. Through the ITFM, P and P ′ can be combined:

P ″: (DF is VSE) is T1 ⊃ (DL is SEV)

One can obtain the truth value of (DL is SEV), i.e., T2, through the use of the implication relation operation introduced by Lukasiewicz (Hadripriono and Ross, 1987). He incorporated the truth relation, denoted as I, of "if P1 then P2" or "P1⊃P2". The parameters of the truth relations, I, are the elements of T2 and T1. These relations, for different values of the elements of T2, are shown in fig. 6.23.a as parallel lines.

The intersections of I and T1 yield the membership values of T2. Subsequently, the truth value, T2, can be found and plotted as in fig. 6.23.a. Now (DL is SEV) is T2 can be modified through the TFM process to give (DL is DL1) in fig. 6.23b which concludes that DL is "close to fairly severe".

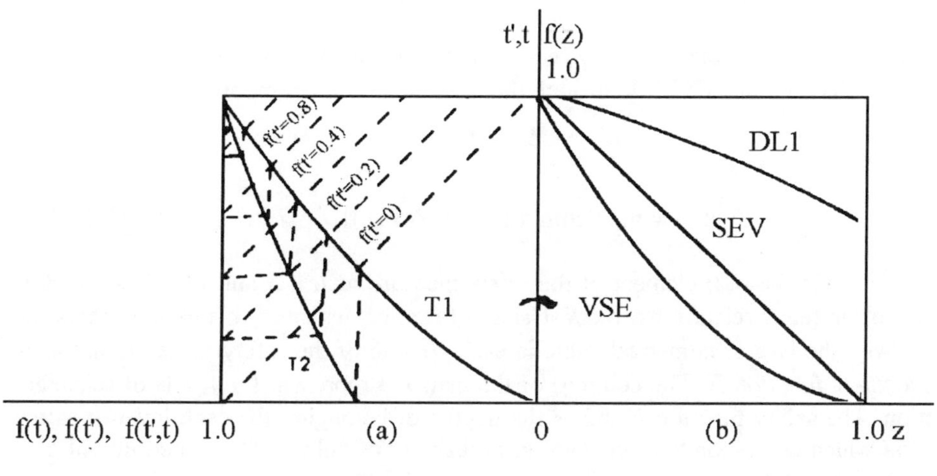

Figure 6.23 MPD graphical solution.

Brown and Yao, (1983), developed an algorithm to illustrate the effect of qualitative parameters in existing structures. In their analysis, *quality* Q_i is used to describe the condition, such as good, fair, poor, etc., of the i-th parameter or structural component.

This description is based on the inspector's observation. Associated with each parameter's quality is its *consequence* $\boldsymbol{C}_i$ which describes the consequence that this parameter's quality has on the structure. For example, connections in "poor" condition may lead to a "catastrophic" consequence. The total *effect* $\boldsymbol{T}$ is a union of all the parameters inspected along with their consequences. It can be calculated as

$$\boldsymbol{T} = \bigcup_i (\boldsymbol{Q}_i \cap \boldsymbol{C}_i) \tag{6.38}$$

and,

$$T(j,k) = \max_i [\min[Q_i(j), C_i(k)]] \tag{6.39}$$

in which $Q_i(*)$ and $C_i(*)$ are the membership or degree of belonging at the numerical rating * of quality $\boldsymbol{Q}$ and consequence $\boldsymbol{C}$, respectively, for the i-th parameter; $T(j,k)$ is the (j,k) element of the total effect matrix $\boldsymbol{T}$; and the symbols $\cup$ and $\cap$ represent, respectively, the relations union and intersection between two fuzzy events. A *fuzzy relation* $\boldsymbol{R}$ is then developed relating the consequence to the *safety reduction* $\boldsymbol{N}$. The safety reduction $\boldsymbol{N}$ describes the level of resistance reduction, verbally, according to the type of the resulting consequence. For example, a "catastrophic" consequence may lead to a "very large" safety reduction. The fuzzy relation $\boldsymbol{R}$ can be calculated in the same manner as the total effect $\boldsymbol{T}$,

$$\boldsymbol{R} = \bigcup_i (\boldsymbol{C}_i \cap \boldsymbol{N}_i) \tag{6.40}$$

Once the total effect $\boldsymbol{T}$ and the fuzzy relation $\boldsymbol{R}$ are obtained, a safety measure $\boldsymbol{S}$ can be computed by combining $\boldsymbol{T}$ with $\boldsymbol{R}$ through the operation called composition.

$$\boldsymbol{S} = \boldsymbol{T} \cdot \boldsymbol{R}$$

and,

$$S(j,\ell) = \max_k [\min[T(j,k), R(k,\ell)]] \tag{6.41}$$

in which $S(j,\ell)$ is the (j,ℓ) element of the safety measure matrix $\boldsymbol{S}$ and $R(k,\ell)$ is the (k,ℓ) element of the fuzzy relation matrix $\boldsymbol{R}$. Using a fuzzifier, which in this case extracts the element with the largest numerical value in each column of the safety measure matrix $\boldsymbol{S}$, yields a *safety function* F. The columns of the matrix $\boldsymbol{S}$ represent the levels of reliability reduction. The safety function F shows the degree of belonging for each level of safety reduction which corresponds to increase in probability of failure. This function will give engineers some idea of the possible reductions for the design reliability after an inspection was done. The engineers may use the results to assist them in deciding on the priority of their actions or in allocating resources in order to maintain the current usage of the structure.

An illustration of the Brown and Yao algorithm for structural damage assessment using fuzzy relation approach can be found in Chou and Yuan, (1992), where a typical rigidly connected plane frame was analyzed.

The fuzzy relation approach presented by Brown and Yao, (1983), incorporating qualitative parameters in assessing existing structures was modified by applying a filter to the total effect T.

Since the fuzzy relation approach presented by Brown and Yao, failed to differentiate the importance of various levels of consequences from the total effect T, a filtering process was presented by Chou and Yuan (1992) which is used to emphasize the more critical effects over the minor effects. The total effect T can be modified to

$$T^f(j,k) = \max_i \{(k/m)T_i(j,k)\} \tag{6.42}$$

in which $T^f(j,k) = (j,k)$ element of the filtered total effect matrix T^f; ($T_i(j,k)$=min[$Q_i(j)$, $C_i(k)$] and m is the total number of numerical ratings used to define consequence C. Note that this filtering equation assumes that the numerical rating is in an ascending order of seriousness. That is, 0, 1, 2 and 3 are the numerical rating for an "insignificant" consequence while 15, 16, 17 and 18 are the rating for a "catastrophic" consequence. Due to the filtration, the membership values for the less serious consequence will be reduced substantially. This reduction may lead to a low membership value in the safety function. However, if one is only interested in the relative degree of belonging in the safety reduction that the existing structure may have, it would be more appropriate to normalize the safety function with the highest membership being 1.

A filtering process is applied to the total effect T because the focus here is at the overall consequence of the structure. Six different filtering processes were considered in order to determine if the results would alter significantly. The filtering processes are shown graphically in fig. 6.24. In each process, m is the same as defined for the filtering equation and k is the numerical rating used for consequence C. The discontinuous filtering functions (figs. 6.24.b and 6.24.c) yielded unsatisfactory results. The reason is that the zero slope region of the filtering function has the same effect as a no filter. The results from the remaining filtering processes (figs. 6.24.d, 6.24.f and 6.24.g) were similar to that obtained from the linear filtering function of fig. 6.24.a.

It was suggested that perhaps a modified membership function for consequence C_i would be fundamentally more sound. A membership function reflects the degree of belonging a numerical rating for a verbal description. An individual consequence C_i is not intended to represent the integrated effect of a structure. It only contributes to the overall effect (that is the function of a total effect T). Thus, modifying C_i in general is not desirable. In addition, if the unanimous professional opinion of a bad consequence has a numerical rating of 12 and a catastrophic consequence has a rating of 18, then these opinions should not be altered just because a catastrophic consequence is more serious than a bad consequence.

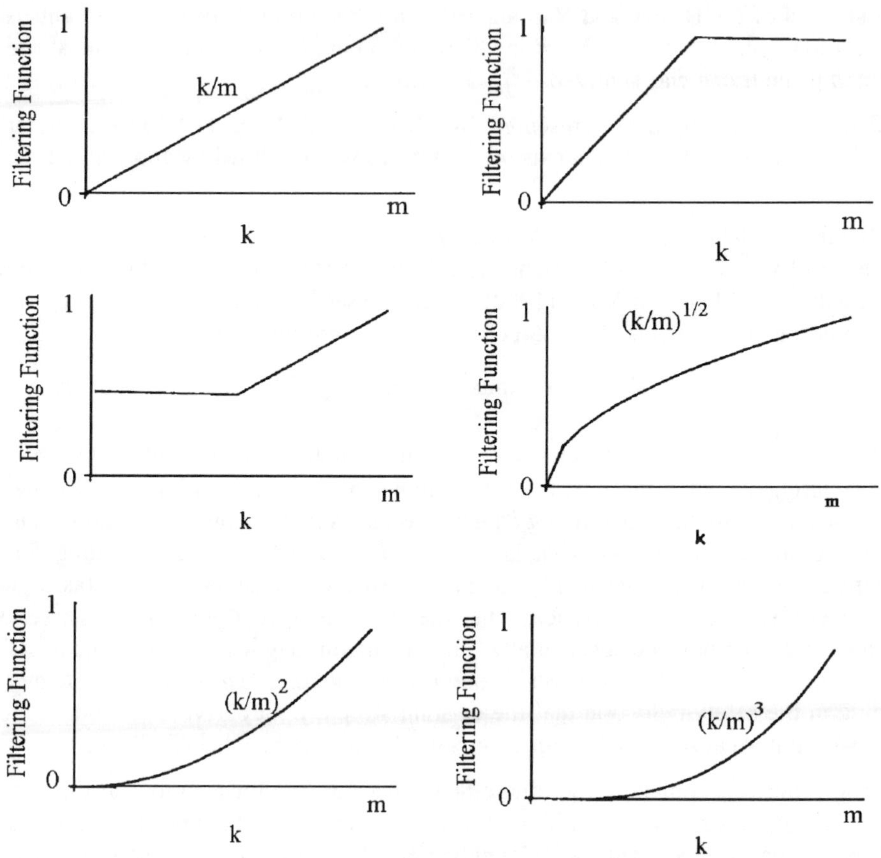

Figure 6.24 Types of filtering processes.

Based on the representative cases studied, the filtered fuzzy relation algorithm yields safety functions *F* which are in tune with intuition. The concept will enhance the current practice of relying heavily on the inspector's experience to analyze the qualitative information. Although a rigid frame was used to illustrate the algorithm presented, the application of the concept is not constrained to buildings only. It can apply to any structural system. The information required for the analysis is the condition (through inspection) of the parameters within the system, and the consequence and effect on the overall performance of the system associated with the condition of each parameter.

A sensitivity analysis on the shapes, degrees of curvature, shifting, gradient and maximum membership values was performed. It was found that all, but two of the factors examined have no effect, or only very insignificant effects, on the safety function *F*. The most pronounced effects are those due to changes in the gradient and due to shifting. In the gradient study, the gradient will influence the range of the safety reductions. The lower gradient value will yield a wider range of the safety reductions with a high membership value. In the shifting study, the levels of safety reduction having a high membership value will shift in the same direction as the membership function for consequence $\boldsymbol{C}$ or safety reduction $\boldsymbol{N}$.

Based on the results of membership function sensitivity analysis, the safety function $\boldsymbol{F}$, in general, is not significantly affected by the preciseness of membership functions developed for every parameter considered. Thus, the algorithm presented by Chou and Yuan, (1992), has practical applications in assessing aging infrastructure system with minimal expert input to establish the necessary membership functions.

The methods presented heretofore are among the most important for the representation of reasoning under uncertainty. These methods, which the artificial intelligence community refers to as *uncertain inference schemes*, contain also Bayesian decision and causal network approaches. Bayesian decision theory may be described as statistical inference using the Bayesian position, i.e. a personalistic view of probability. If probabilities are thought to describe orderly opinions, Bayes' theorem describes how the opinions should be updated in the light of new information. Two major problems have been identified with the implementation of Bayesian inference schemes: (1) combinatorial explosion for realistic networks and (2) the estimation of prior probabilities. Because it has been shown that humans display characteristics such as representativeness, conservatism and the gambler's fallacy when dealing with probability, it has been assumed that a subjective estimation of probability is not meaningful. These results may be thought of as an impetus for the development of other uncertain inference schemes, in the sense that alternative methods for modeling the "true" meaning of what humans term "belief" have been sought. Others have attempted to develop methods for easier assessment in order to use the Bayesian approach.

The causal network approach is a computationally tractable form of Bayesian decision theory. A causal network is defined as an acyclical directed graph in which probabilistic nodes are connected in a causal manner. "Directed" means that the nodes are connected by arrows and "acyclical" means that the arrows cannot form a circle or cycle in the graph or network. The connection in a causal manner allows for an easier assessment of probabilities. In real-world usage, assessment is frequently made in the direction of,

$$\text{observable} \rightarrow \text{unobservable}$$

or,

$$\text{effect} \rightarrow \text{cause}$$

which is generally more difficult to assess than going from,

$$\text{effect} \leftarrow \text{cause}$$

The latter assessment is made in the causal direction.

A causal network is shown in fig. 6.25. In this figure, the conditional independence assumption is illustrated.

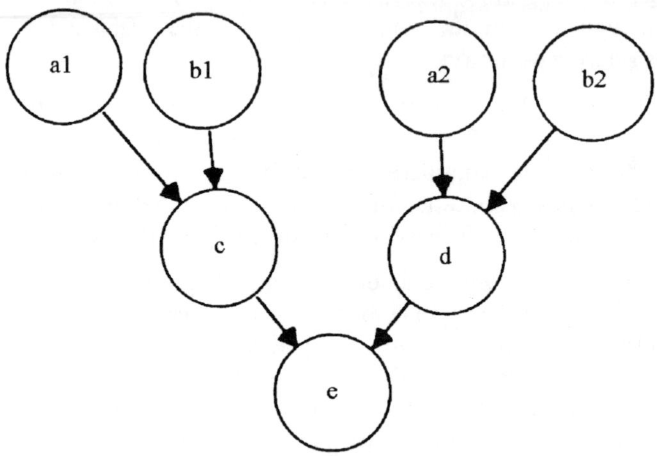

Figure 6.25 A typical causal network.

The following equations are used to evaluate nodes *e*, *c* and *d:*

$$p(e_k) = \sum_i^n \sum_j^n p(e_k|c_i,d_j)p(c_i,d_j) \tag{6.43}$$

where $p(e_k|c_i,d_j)$ is the probability of e_k conditioned upon c_i and d_j. If the events represented by nodes c_i and d_j are independent, then $p(c_i, d_j) = p(c_i)p(d_j)$.

$$p(c_i) = \sum_m \sum_n p(c_i|a_{1_m}, b_{1_n})p(a_{1_m})p(b_{1_n}) \tag{6.44}$$

$$p(d_j) = \sum_m \sum_n p(d_j|a_{2_m}, b_{2_n})p(a_{2_m})p(b_{2_n})$$

It can be seen from the above equations that node *e* is conditionally independent of nodes a_1, a_2, b_1 and b_2 (because there are no arrows connecting these nodes) given that nodes *c* and *d* are updated using nodes a_1, a_2, b_1 and b_2. The condition for the graph to be acyclical, i.e. it cannot contain any cycles, means that the node can never become conditional upon itself. It is noted that a consequence of the Bayes theorem is that it holds in both (arrow) directions.

In some instances, a node *y* may be conditional upon several, say, *n* nodes, x_r, where r $=1, 2, 3, \dots, n$. In order to reduce the assessment of 2^n probability values to *n* values, a technique called the *noisy OR gate* was developed (Reed, 1993). In the noisy OR gate, the probability of *y* conditional on *n* nodes, x_r, $r=1, 2, 3, \dots, n$, is estimated as

$$p(y|x_1,x_2,x_3,\ldots,x_n)=1-\prod_{r=1}^{n}(1-p(y|x_r)) \qquad (6.45)$$

In this equation, $p(y|x_1)$, $p(y|x_2)$, $p(y|x_3)$, ..., $p(y|x_n)$, are assessed and then used to estimate $p(y|x_1, x_2, x_3, \ldots, x_n)$.

Dubois and Prade, (1986), discuss specific features of probability and possibility theory with emphasis on semantical aspects. They fig. out how statistical data and possibility theory could be matched. As a result, procedures for constructing weak possibilistic substitutes of probability measures and for processing imprecise statistical data are outlined. They provide new insights on relationship between fuzzy sets and probability theory.

In this way, fuzzy causal networks can be constructed to improve reasoning with uncertainty in structural damage assessment. Probability, fuzzy set theory and the Dempster-Shafer theory have been combined in *SPERIL* developed by Yao, (1985), and Brown and Yao, (1983), for evaluating the post-earthquake structural system safety.

The Dempster-Shafer Theory has been developed primarily to model measures of belief when the probability distribution is incompletely known. This theory enables one to include the consideration of lack of evidence. Dempster's rule for combining evidence from different sources is provided in the method. The interval notation used in this method suggests bounded limits on probability values. Although this method has been combined successfully with others, the main criticism of it is that many consider it a generalization of probability.

It should be obvious from the previous presentation that it is not simple to undertake a definitive comparative study of the various uncertain inference schemes. First of all, it is clear that the modeling approaches are different in the definitions and assessments of uncertainty. Secondly, the manner in which the uncertainty (ies) is (are) combined is different. Converting from one scheme to the other numerically for all cases is not trivial, if at all possible. However, measures for comparison were defined (i.e. clarity, completeness, hypothetical conditioning, complementarity, consistency, etc.) and comparison results of fundamental properties of uncertain inference schemes are summarized in Reed, (1993).

6.4 Application examples

6.4.1 Nuclear reactor safety assessment using the probabilistic fracture mechanics method

Research on incidents occurring with conventional pressure vessels has shown that in 90% of cases, the initial defects were located in a weld. For this reason, the present analysis is primarily concerned with defects in welds; under cladding defects have also been considered in order to evaluate their harmfulness.

Data were collected from 3 European manufacturers: BREDA (Italy), FRAMATOME (France) and Rotterdam Nuclear (Netherlands). Each manufacturer filled in, for each weld, a standard form, giving complete information on NDT results (US or X-ray) before and after repair: instrument calibration, weld size and description and position of the defect in azimuth, in depth and in relation to the axis of symmetry of the weld. All of this information was sent in confidence to the Ispra JRC of EEC-Italy which processed and harmonized the data.

A total of 338 meters of PWR and BWR shell were analyzed. The main conclusions are as follows (Dufresne *et al.*, 1986, 1988):

- Density of defects: the number of defects per weld varies from 0 to 50 (mean value is 13).
- Position of the defects in the weld: there is no clear distribution of the defects according to their depth and to their position in relation to the axis of symmetry of the weld, but, for a given weld, defects are frequently gathered in some limited areas of the weld, this probably being due to maladjustment of a parameter during the welding process.
- Length of the defects: the cumulative distribution function before repair shows that, for defects larger than 20mm, the log-normal distribution is a good approximation.

With regard to the width distribution of the defects, unfortunately no data has been obtained from manufacturers. After discussion with experienced welding operators, it seems that a defect larger than a single pass is very unlikely. Therefore the number and distribution of defects wider than one pass have been calculated by estimating the probability for two or more defects to overlap, both in azimuthal and transversal section. This probability is calculated using the Monte Carlo method.

The defect length and width distributions so obtained correspond to the observed defects in a weld after fabrication and before repair, but the distribution to be incorporated in the code must be processed in order to take into account the following factors: the sample size, the accuracy of the measurement equipment, the reliability of the NDT methods and equipment, and the size of acceptable defects according to the construction rules.

An overall statistical interpretation of all the available (da/dN) vs. ΔK measurement points has been made for SA 508 and SA 533 steels, using four laws: Paris, Forman, Priddle and Walker (see Hoeppner and Krupp, 1974). Numerical coefficients have been calculated for each of these formulas by linear regression from experimental results and by several partitions of the measurement range. Paris' law: $da/dN=C(\Delta K)^m$ has been found to be the most suitable for this first application to reactor pressure vessels. The values of coefficients C and m have been defined, in 4 different domains, as a function of the value of ΔK and for two types of environment: air and water. The values of the m coefficient giving the best determination coefficient have been chosen and the corresponding "C" values have been determined in histogram forms. The rate of propagation of defects under water is higher than in air above a given threshold ΔK_t. It has been assumed that for the values of $\Delta K<\Delta K_t$, the propagation rates in air and water are identical. The values of ΔK_t are correlated with R ($R=\sigma_{min}/\sigma_{max}$).

When an internal defect emerges at the surface, one can assume that it becomes semi-elliptic, the major axis does not change, and the small axis doubles with respect to its initial value.

Loading of the vessel has been computed for 22 observable and incidental conditions (ANS Conditions I and II and second category conditions according to the French Standard Order dated February 26, 1974). The frequency of occurrence of these conditions is determined on the basis of statistical data, rather than from "envelope data" which are used in safety assessments. Fatigue crack growth was calculated according to the above mentioned assumptions for all defect sizes and for all positions in the thickness.

Failures can occur in different situations: either under normal, upset, emergency of faulted conditions or during pressure tests. In the first case, it can be due to a through-crack propagated by fatigue or in an instability when the stress intensity factor becomes larger than toughness. Evaluation of those phenomena has shown that for crack distribution computed at the end of the plant's life the probability of rupture was negligible.

The second category encompasses emergency or faulted conditions (ANS Conditions III and IV, and Third and Fourth Category Conditions according to the French Order dated February 2, 1974). The probability of occurrence of these later conditions during the life of the power plant is very low, and they are therefore not taken into consideration for fatigue crack propagation. Identification and quantification of these conditions have been performed using event tree analysis, followed by classification into a limited number of "envelope" faulted conditions to which a probability of occurrence is assigned. Faulted conditions so defined concern:

- LOCA with different break sizes, different temperatures of the safety injection water (storage pool water), and different times to operate the core cooling recirculation system.

- Steam break, with different break sizes, with and without electrical power, and different times, to depressurize the primary circuit.
- Hot and cold overpressure during operation or start-up of the plant.

Concerning the pressure tests, the rupture probability has been computed for a test pressure of 206 bars and at different temperatures between 40°C and 100°C.

For all the situations thus defined, a thermoelastic analysis is made at different crack depths. Temperature and the three mains stresses are computed at different steps of the transient, and the stress intensity factor is computed with plastic zone correction.

Different approaches for rupture criteria have been considered: LEFM, EPFM and plastic instability. After comparison of these three criteria with 141 representative experimental data it has been concluded that the criterion proposed by Dowling and Townley gives the best evaluation of fracture conditions.

Toughness distribution as a function of the RT_{NDT} temperature of a 508 Cl 3 steel has been calculated from experimental results. The shift in transition temperature (ΔRT_{NDT}) due to irradiation has been calculated as a function of the neutron flux and of the phosphorus and copper content, according to the RG 1.99 formula. From these data, it is then possible to compute the toughness distribution at any point of the pressure vessel and at each step of the transient.

The fracture probability is found by searching for the intersection of toughness and stress intensity coefficient histograms at each transient step. The probability of initiating unstable crack growth for the considered defect is equal to the maximum value of the various probabilities thus obtained during the various steps of the transient.

The modeling of the initiation of unstable crack growth during a transient has been determined according to the following assumptions:

- When a internal defect becomes a through-wall defect, it is transformed into a semi-elliptical surface crack having the same excentricity and keeping its inner front at the same position as the initial crack.
- When a defect becomes unstable at the crack tip of the major axis, the defect becomes an infinite length crack and the crack arrest criterion is applied at its inner front.
- Every defect becoming unstable at its inner front is considered as leading to vessel failure; the code does not consider the possibility of subsequent arrest in warmer and less irradiated zones through the wall.

Calculations made with the COVASTOL code from the data compiled on French PWRs, operated according to EDF, rules allow the following conclusions to be drawn:

- Rupture probabilities at 40 years vary from 10^{-8} to 10^{-12} according to the weld location and the safety injection water temperature; final results are presented in fig. 6.26.

- The evolution of the rupture probability of various welds as a function of time is calculated. It is shown that the belt weld is more sensitive to the time than the others. This is due to the fact that, for this weld, time has two effects: crack growth and steel embrittlement.
- The main rupture risk is due to a LOCA; the contribution of the different break size is calculated, the associated conditional probability being of the order of 10^{-5}. The rupture probability of a PWR pressure vessel is therefore less than that of a LOCA by several orders of magnitude.
- The protective effect of the cladding decreases markedly the fatigue growth of defects and lowers the overall rupture probability by several orders of magnitude.
- The fatigue growth of defects located in the shell during normal and incidental transients is very limited for the belt line weld; it is greater for the nozzle shell.
- The fatigue growth of defects located in the nozzles and their welds with the nozzle shell is markedly larger than that of the defects located in the core shell.
- The rupture probability of the most irradiated area of the belt line weld is 10^5 times greater than that of the least irradiated area (neutron shielding is not homogeneous) over the whole periphery of the vessel).
- The most severe conditions for fatigue propagations are: cooling, load increase between 15% and 100%, and steady state fluctuations (for the chosen modelization).
- The harmfulness of under cladding defects located in the nozzles depends greatly on their location. The most harmful defects are located in the vertical crossection in the vicinity of the internal angle.
- Residual stresses, when large, can increase markedly the fatigue growth of defects, and the rupture probability (up to a factor of 5).
- Taking the pressure vessel life duration into account, the rupture probability during a pressure test is greater than the overall rupture probability, when the pressure vessel temperature is less than 60°C during the test.

After a new NDT inspection, the data can be exploited to update the FCG model parameters and consequently to update the above risk estimations. The adaptive models of Section 6.3.5 could be used as a complementary prediction tool, especially for individual defects evolution and the related risk assessment.

6.4.2 Marine structures safety assessment using the probabilistic fracture mechanics method

In dealing with the quality control and/or the reliability of marine structures, attention will be focused mainly on ships and offshore structures, leaving aside other types of marine structures such as submersible vehicles, subsea installations and pipelines.

The most frequently techniques for the detection of cracks in weld seams of underwater constructions are the magnetic particle test methods and the ultrasonic technique (see Section 6.2.2). The advantage of the magnetic particle testing technique is the high detection capability in the indication of cracks in the weld seam and the simple evaluation of the results. Disadvantageous is the fact that the area to be inspected has to be cleaned accurately. The realization of the inspection with a heavy magnetizing yoke is a very hard work for the divers, for in most cases it is not possible to have a secure support at the structure. In low water depths the sun light hinders an evaluation of the indication, for the fluorescence of the particles cannot be recognized. The inspection must be postponed to the evening or night. The most important disadvantage of the magnetic particle testing technique however is that it is very difficult to integrate into manipulator systems.

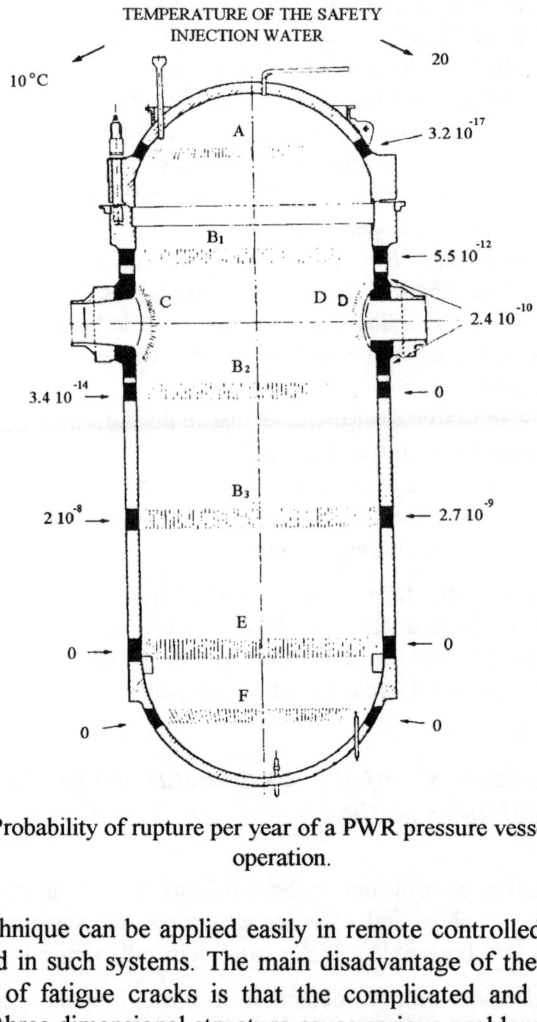

Figure 6.26 Probability of rupture per year of a PWR pressure vessel after 40 years of operation.

The ultrasonic technique can be applied easily in remote controlled manipulators and is actually often used in such systems. The main disadvantage of the ultrasonic technique for the detection of fatigue cracks is that the complicated and permanent changing geometries of the three dimensional structure cause serious problems in the evaluation of the signals. A further problem is the fact that ultrasonic inspection requires a very exact sensor movement which can be difficult for the diver, so that a mechanical guide or a manipulator is necessary. But it can be stated that for the detection of the depth of cracks with known position the ultrasonic technique can be applied with good success.

The eddy current testing method, which is a traditional technique for the detection of surface cracks, is applied offshore for a very short time. The reason is that the signals of weld seam roughness, and the changes of magnetic permeability and electric conductivity caused by the welding process, superpose the crack signal which reduces the detectability of cracks. The fact that the eddy current technique is now more interesting for the inspection of welds, can be explained by the development of sensors and evaluation techniques. The eddy current method has very low demands of surface cleaning, so that inspections without removing dirty layers are possible. The signal evaluation is independent from ambient conditions and the eddy current method can be applied easily in remote controlled manipulators (Camerini *et al.*, 1992).

Ship and offshore steel structures are designed to withstand the cyclic stresses created by sea waves. Ship welding is a critical technology. It is difficult to manufacture welds in which the defects are few and far between. Some smaller defects will always be present in the welds and there is a risk of having large defects. The welds frequently determine the strength and usability of a welded product, more than any other single factor. Therefore, the presence of severe stress in as-fabricated structures may cause fatigue cracking during service. If a corrosion or a fatigue crack is detected in the steel or the welds of the ship structure during in-service inspection, it is necessary to assess its significance with respect to the safety of the facility. The key to the assessment of fatigue cracks in marine structures is a validated analytical model that predicts the growth rate of fatigue cracks and the corresponding probability of failure for the anticipated service conditions (Stavrakakis, (1990), (1993)).

Fatigue crack growth behavior in metals is best described using the fracture mechanics approach, in which fatigue crack growth rates, $\mathrm{d}\alpha/\mathrm{d}N$, are correlated to the applied stress intensity factor range ΔK (see Section 6.3.3).

Cyclic stresses caused by sea-wave loading are random in nature, and fatigue analysis of structures subjected to random loading is complicated. An equivalent-stress-range scheme will be presented here, used to correlate fatigue crack growth results for structures subjected to random loadings with those of structures subjected to constant-amplitude loading.

If the crack growth per cycle is much less than the crack length ($\mathrm{d}\alpha/\mathrm{d}N << \alpha$), and there are no load-sequence interaction effects, then the total increment of crack growth in N successive cycles, for a specific ΔK range is,

$$\sum_{i=1}^{N} \Delta\alpha_i = C[(\pi\alpha)^{1/2} F]^m \sum_{i=1}^{N} \sigma_i^m$$

This result is derived by summing the crack-length increment per stress cycle calculated using the Paris-Erdogan equation (see Section 6.3.3), for the N successive cycles, when the applied stress-range is σ_i, i=1, 2, ..., N for the 1st, 2nd, ...Nth cycle.

The average FCGR per stress cycle is then,

$$\frac{\mathrm{d}\alpha}{\mathrm{d}N} = C[(\pi\alpha)^{1/2}F]^m \frac{1}{N}\sum_{i=1}^{N}\sigma_i^m \quad \text{or} \quad \frac{\mathrm{d}\alpha}{\mathrm{d}N} = C[\sigma_{eq}(\pi\alpha)^{1/2}F]^m$$

where,

$$\sigma_{eq} = \left[\frac{1}{N}\sum_{i=1}^{N}\sigma_i^m\right]^{1/m}, \qquad i = 1,2,\ldots,N$$

is the equivalent-stress range.

Here N should be large enough for the equivalent-stress range, σ_{eq}, to be representative of a stress spectrum. The task then becomes the computation of σ_{eq}, from a random load history. In the above derivation, load-sequence interaction effects are ignored. This means that the well known Miner's rule is also applicable. This is adequate for sea-wave loading, Cheng, (1985, 1988).

The equivalent-stress range for a random sea-wave load history, σ_{eq}, can be computed directly from the exceedances spectrum or power spectral density function (PSDF) of the anticipated (or assumed) sea-wave stress spectrum.

The exceedances spectrum expresses the accumulated number of stress cycles at each normalized (or actual) stress range (normalized to the maximum stress range) over the design lifetime, see fig. 6.27.a. For a given stress spectrum, the exact exceedances spectrum varies, depending upon the cycle-counting method used, except in the case of the constant-amplitude loading condition (Cheng, 1985, 1988).

For a complete description of the random load history, frequency and irregularity have to be specified along with the exceedances spectrum. For a given stress spectrum, the exact exceedances spectrum varies, depending upon the cycle-counting methods used, except in the case of constant-amplitude loading.

A histogram, or probability distribution, is used to calculate the σ_{eq} of a random load history expressed as an exceedances spectrum. The histogram corresponding to an exceedances spectrum is usually readily available. If not, a histogram can be constructed directly from the corresponding exceedances spectrum by dividing the axis of normalized stress range into intervals and calculating the frequency of occurrence for each interval. For example, the ordinate in fig. 6.27.a is divided into 20 equal intervals. Then the statistical data needed to construct a histogram can be extracted from this exceedances spectrum. The resultant histogram obtained from fig. 6.27.a is presented in fig. 6.27.b.

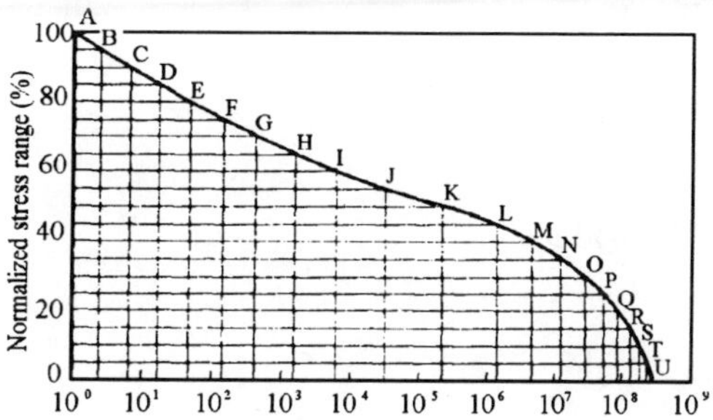

Figure 6.27.a
Exceedances spectrum divided for construction of histogram

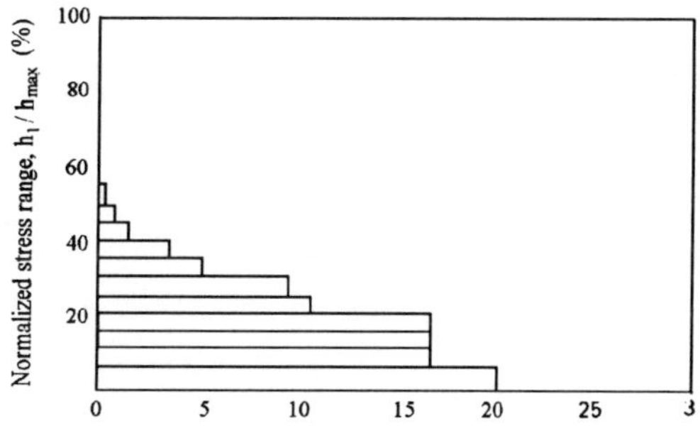

Figure 6.27.b
Stress-range histogram corresponding to the exceedances spectrum shown in fig. 6.27.a.

From the histogram of fig. 6.27.a, σ_{eq} is evaluated from the equation,

$$\sigma_{eq} = \left[\sum \gamma_i \sigma_i^m\right]^{1/m}$$

where σ is the stress range, γ is the frequency of its occurrence, and m is the exponent in the Paris-Erdogan FCG rate equation.

The other way to describe the random load history, is to use the PSDF, as shown in fig. 6.28. The PSDF is a result obtained from a spectral analysis of the original random load history. If the random load history is a stationary Gaussian process, as is commonly assumed, then a PSDF exists, $G(\omega)$, which possesses all the statistical properties of the original load history (Cheng, 1988).

The two most important parameters in the random-loading fatigue analysis that can be retrieved from the PSDF, are the root-mean-square (rms) value of the load amplitude and the irregularity factor, a, of the random load history. The rms value is the square root of the area under the PSDF.

The irregularity factor is defined as the ratio of the number of positive-slope zero crossings, N_0, to the number of peaks, F_0 per unit time in a load history:

$$a = N_0/F_0$$

The exact value of N_0 and F_0 can be evaluated from $G(\omega)$ as follows:

$$N_0 = (M_2 / M_0)^{1/2}$$

and,

$$F_0 = (M_4 / M_2)^{1/2}$$

where M_0, M_2 and M_4 are the zeroth, second and fourth moments of $G(\omega)$ about the origin (zero frequency) and are defined as,

$$M_0 = \int_0^{\infty} G(\omega)\mathrm{d}\omega$$

$$M_2 = \int_0^{\infty} \omega^2 G(\omega)\mathrm{d}\omega$$

$$M_4 = \int_0^{\infty} \omega^4 G(\omega)\mathrm{d}\omega$$

Thus,

$$a = M_2 / (M_0 M_4)^{1/2}$$

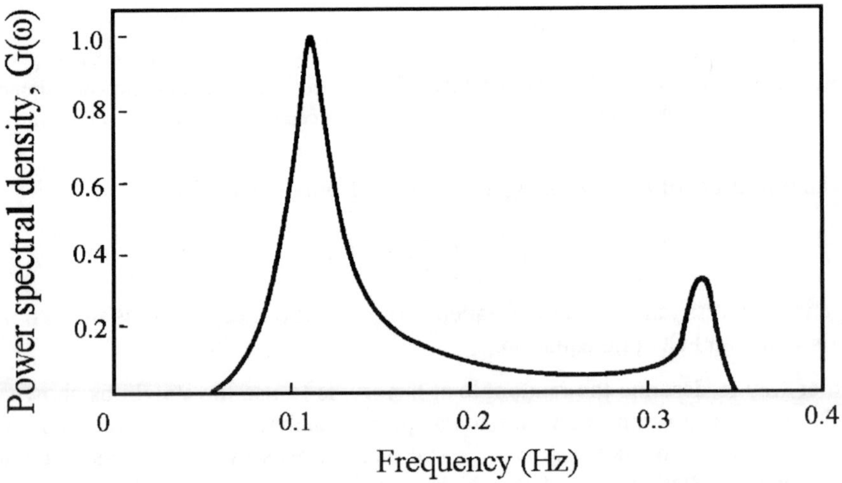

Figure 6.28 Example of power spectral density function (double peaked spectra).

The free surface elevation of the sea can be modeled by an ergodic Gaussian process for adequately short periods of time. This short-term description implies that the process is homogeneous in time and in space, that is, its probabilistic properties do not change with time nor with location. Thus, it is equivalent to estimate those properties from several sea surface elevation records made at different times in the same point or made at the same time in different points.

Each sea state is completely described by a wave spectrum. These spectra result from physical processes and are therefore amenable to theoretical modeling. Various proposed mathematical expressions to represent average sea spectra have appeared in the past. The one that has become generally accepted and that has been commonly used in response analysis is due to Pierson and Moskowitz, although it is most commonly seen in the parametric form proposed in the International Ship Structures Congress (ISSC), see Guedes-Soares (1984) and Hogben *et al.* (1976). A sea state is often characterized by an average wave period T_z and a significant wave height H_s , which are related to the spectral moments by:

$$H_s = 4\sqrt{m_0} \quad \text{and} \quad T_z = \frac{1}{2\pi}\frac{m_0}{m_1}$$

where $m_i = \int_0^{\infty} \omega^i S(\omega)\mathrm{d}\omega$, is the moment of order i, of the sea spectrum $S(\omega)$. Developing seas have a more peaked spectrum, as has been demonstrated during the *JONSWAP* project by Hasselmann *et al*., (1970). They proposed a spectral form that accounted for the dependence on wind speed and fetch.

The JONSWAP spectrum has also been recommended by the ISSC (Hogben *et al*., 1976), where a parameterization in terms of significant wave height and average period was proposed.

Recent evidence on the adequacy of the ISSC and JONSWAP models to represent the average shape of measured spectra has shown that both of these formulations represent single peaked spectra while many of the measured spectra exhibits two peaks.

A spectrum with two peaks is expected to occur whenever the sea state contains two wave systems with different average periods. Often this will be a combination of a swell component and a wind sea component.

It must be noticed that the main feature of a double peaked spectrum is the partition of the energy about two distinct peak frequencies. For a sea state of given H_s and T_z , a double peaked spectrum has the energy more spread along the frequency axis than in the case of a single peak spectrum.

The main idea behind the model of double-peaked spectra described hereafter is that they can be represented by two spectral components of the JONSWAP type. The wind sea part of the spectrum will often be in a developing situation, to which the JONSWAP formulation is appropriate. As to the swell part, it is mostly its narrowness that justifies the choice (Guedes - Soares, 1984).

Having defined the shape of the two spectral components, a double peaked spectrum becomes completely described by the ratios of the peak frequencies and of the spectral peaks. If the sea spectrum S is represented by the sum of a swell S_s and a wind sea S_w component,

$$S(\omega) = S_s(\omega) + S_w(\omega)$$

its moments must be equal to the sum of the moments of the components:

$$m_0 = m_{0s} + m_{0w}$$

$$m_1 = m_{1s} + m_{1w}$$

where m_{ik}, i=0,1 and k=s,w, is the moment of order i of spectrum k.

Having a sea state defined by the four parameters H_s, T_z, H_R and T_R, where H_R and T_R are the ratios of significant wave height and average period of the two spectral components, each spectral component S_k, k=s,w can be given by the equation:

$$S_k(f) = S_{PM}(f)\gamma^q, \ (\mathrm{m}^2 \cdot \mathrm{sec})$$

where S_{PM} is the ISSC spectrum:

$$S(f) = 0.11H^2T(Tf)^{-5}\exp\{-0.44(Tf)^{-4}\}, \ (\mathrm{m}^2 \cdot \mathrm{sec})$$

and γ is the peak enhancement factor of the JONSWAP spectrum,

$$q = \exp\{-(1.296Tf - 1)^2/2\sigma^2)\}$$

$$f = \omega/2\pi$$

$$T = T_z \cdot F_2$$

$$H = H_s / \sqrt{F_1}$$

The JONSWAP parameter σ is used at its mean values:

$$\sigma_a = 0.07 \text{ for } f \leq 1/1.296T$$

$$\sigma_b = 0.09 \text{ for } f > 1/1.296T$$

The quantities F_1 and F_2 are two constants that correct for the difference in peak period and area between a Pierson-Moskowitz (P-M) and a JONSWAP spectrum. The values of these parameters depend on γ as shown in Table 6.3.

The two additional parameters that define the double peaked spectrum are H_R and T_R. The last one is easily determined from a measured spectrum, as the ratio of the spectral peak frequencies. Other easily obtainable quantity is the ratio between the two spectral ordinates S_R. To relate this ratio with H_R, it is necessary to obtain the expression of the spectral ordinate at the peak frequency.

The peak frequency is determined by equating to zero the derivative of S_k with respect to f. If this is done, it follows that the ratio of the two spectral peaks S_{sp} and S_{wp} is given by:

$$S_R = \frac{S_{sp}}{S_{wp}} = \frac{H_s^2 T_s}{H_w^2 T_w} = H_R^2 T_R$$

A correction must be introduced to account for the unsymmetry of the component spectra about the peak frequency. The fact that the spectra have higher energy above the peak frequency implies that adding two spectral components of equal energy produces a double peaked spectrum with the high frequency peak larger than the low frequency one.

This effect decreases with increasing separation between the two spectral peaks, being thus of less importance in the case of two well defined peaks. However, it is easy to correct for that effect, so as to make the procedure applicable regardless of the distance between spectral peaks.

Table 6.3 Values of the parameters F_1 and F_2 on the JONSWAP spectrum

γ	$F_1 = \dfrac{m_0[JONSWAP]}{m_0[P-M]}$	$F_2 = \dfrac{(m_1/m_0)[JONSWAP]}{(m_1/m_0)[P-M]}$
1	1.0	1.0
2	1.24	0.95
3	1.46	0.93
3.3	1.52	0.92
4	1.66	0.91
5	1.86	0.90
6	2.04	0.89

Having a double peaked spectrum defined by its 4 parameters, one determines first the ordinates of the theoretical spectrum at the two peak frequencies, and their estimated ratio S_R. This value is larger or equal to the value of the spectral parameter S_R. The value of H_R to be used in the above equations is thus corrected by the factor $k_R = \hat{S}_R / S_R$, which upon substitution in the last equation results in:

$$H_R = \left(\frac{k_R S_R}{T_R} \right)^{0.5}$$

If a better accuracy is desired the procedure can be repeated until convergence is obtained in the value of k_R. This generally occurs with two iterations.

It should be noticed that when S_R is equal to zero this double peaked spectral representation reduces to the single peaked JONSWAP spectrum. If in addition γ is set equal to one, the ISSC spectrum results.

Ship responses, regardless of whether the quantities of interest are motions or loads, are basically inertia dominated, implying that except in extreme situations they are proportional to wave height. It is the linearity of this relationship that justifies the adoption of the spectral approach.

The response spectrum $G(\omega)$ (i.e. short-term wave induced load effects) from which the σ_{eq} is calculated, can be determined from the knowledge of the structure transfer function $H(\omega)$ and the input sea spectrum $S(\omega)$ as:

$$G(\omega) = S(\omega) \cdot H^2(\omega)$$

where ω is the frequency of the wave components.

If the sea spectrum describes a stationary Gaussian process, the assumption that the transfer function is linear implies that the response is also a stationary Gaussian process, thus completely described by the response spectrum.

These theoretical formulations describe average spectral shapes expected to occur in the presence of a given wind or in a sea state of known characteristic wave height and period. There is however considerable uncertainty in the shape of individual spectrum due to the large variability of the generation process and of the estimation methods.

The sources of uncertainty in the spectral shape definition are discussed by Guedes-Soares, (1984), and a method is proposed there to model them and to determine their effect on the uncertainty of the response parameters. This treatment accounts for both fundamental and statistical uncertainties in the spectral shape. The results are given in terms of the response quantities predicted by the standard method of calculation of ship responses. They indicate the bias and the coefficient of variation of the standard predictions, being thus a representation of the model uncertainty of that response calculation method.

The main feature of the standard response method is, in this context, the use of the ISSC spectrum to represent all sea states. Thus the results can also be interpreted as the model uncertainty of the ISSC spectrum in representing all types of sea spectra.

In addition to the uncertainties related with the shape of the spectrum, ship responses are also subjected to other uncertainty sources such as relative course and speed of the ship. Thus it is often more meaningful to operate with expected values of responses than with responses to specific sea states. Different possible formulations of mean response and mean response and mean bias are examined there.

If the load history under consideration is in the form of a narrow-band random process (irregularity factor α=1), the value of the equivalent stress range σ_{eq} can be calculated from the following closed-form expression (Cheng, 1988):

$$\sigma_{eq} = 2(rms)\left[\frac{\Gamma(1+m/2)}{2(1/2)^{1+m/2}}\right]^{1/m}$$

where $\Gamma(...)$ is the gamma function and m is the exponent in the Paris-Erdogan FCG rate equation. There are no closed-form solutions for wide-band (α<0.99) random loads.

For wide-band random loads, solutions derived from numerical analysis to convert PSDF to σ_{eq} are presented in graphical form in Cheng (1988).

FCG rates under sea-wave loading can thus be calculated using the probabilistic fracture mechanics approach described in Section 6.3.3 and the equivalent-stress-range concept described above.

The procedures start with an anticipated (or assumed) stress spectrum acting on a component of interest. The equivalent-stress-range for the sea-wave loading is then calculated.

The COVASTOL computer program, presented in sections 6.3.3, 6.3.4 and 6.4.1, or any similar FCG prediction program, can be applied in a straightforward manner to predict at any time the residual life-time and to estimate using inspection data the failure probability of marine structures under sea-wave action. The main objective of applying this program is to provide information about the availability of marine structures, which is important for their efficient design, for defining optimal after-service inspection and maintenance policies and for estimating the loss of marine structures financial risk.

6.4.3 Structural damage assessment using a causal network

In this section, a simple causal network is built for post-earthquake structural damage assessment. "Uncertainty" in this context includes the description and prediction of loading conditions, material properties of structural components, the difference between the simplistic mathematical modeling of the structure and the actual behavior and loading path, and the imprecision involved in construction. Evaluation of structural damage by an expert involves all of these "uncertainties" in some implicit fashion which is assumed to be appropriately characterized as a degree of belief in the severity of damage from observations. For example, assigning a degree of belief to an event such as "global loss of strength" as "moderate" is assumed to be an accurate characterization of damage assessment.

The causal form of the network corresponds to the physical "pathway" of damage, i.e. the two causes of structural failure are inadequate stiffness and insufficient strength. Creating a causal network in which the component damage "caused" global damage which "caused" a level of structure failure extent corresponds to the physical model of damage which is familiar to structural engineers. It would also allow engineers to identify in a concise manner how the damage would be evaluated at each stage.

It is cognitively easier to make (subjective) probability assessments in the causal direction. If necessary, the noisy OR gate (eq. (6.45)) provides a method for generating conditional probabilities.

The structural damage assessment network will be evaluated using fig. 6.25. Each node or event can take on the value of *none*, *slight*, *moderate* and *severe*, i.e. the indices in the equations (6.44), will all be defined on the range from unity to four. In fig. 6.25, the events that the nodes represent, are as follows:

Node	*Meaning*
e	Structure failure extent
c	Global loss of strength
d	Global loss of stiffness
a_1	Component loss of strength
a_2	Component loss of stiffness
b_1	Global damage to strength
b_2	Global damage to stiffness

The network is interpreted as follows. The probability of "structure failure extent" taking on the values of "none", "slight", "moderate" or "severe" is denoted as $p(e_k)$, where k=1, 2, 3, 4, represents the four values. The event "structure failure extent" is "caused" by global loss of strength and stiffness, nodes c and d, respectively. The events these nodes represent are "caused" by "component loss of strength" and "component loss of stiffness", respectively, and "global structural changes" such as a change in the natural frequency. It could be argued that component loss of stiffness is influenced by component loss of strength; however, this influence is assumed to be small enough that the two should be treated as different, distinct causes of failure. Preliminary conditional probability values can be assessed on the basis of experience and limited consultation with colleagues; additional values are estimated on the basis of the noisy OR gate. It is necessary to normalize the probabilities generated in this manner.

This aspect of the evaluation process is the most complex, and the most uncertain. Given the input or marginal probabilities for these nodes, the probability p (structure failure extent) may be calculated using eqns. (6.43)÷(6.45) of Section 6.3.6. First, marginal probability values are assessed for nodes a_1, a_2, b_1 and b_2. These values are input into eqns. (6.44). Multiplication and summation of the marginal probabilities yield $p(c_i)$ and $p(d_j)$, where i, j=1, ..., 4. These values are input into eqn. (6.43) for $p(e_k)$, where multiplication and summation yield $p(e_k)$, k=1, ..., 4.

For purposes of illustration, numerical examples are given in Table 6.4. It can be seen that the degrees of belief in the structure failure extent, seem reasonable for the given inputs. For example, extreme inputs yield extreme results. For mixed inputs on the component level, the structure failure extent appears to have the greatest degree of belief associated with moderate damage, as would be expected. Given this information, a structural engineer would be able to decide whether structural rehabilitation was required. Structures for which the degree of belief in the structure failure extent being moderate or severe was high, would be required to undergo rehabilitation.

A more realistic causal network would include damage observations for each component and individual types of global structural changes. The total component loss of strength and stiffness would therefore be conditional upon the type and severity of damage observed. Another extension of the present efforts should be to evaluate diagrams which have horizontal arrows connecting nodes at each level. Any set of influences is permitted in a causal network as long as the network is acyclic.

Table 6.4 Numerical example for network of fig. 6.25

$p(a_1)$	$p(a_2)$	$p(b_1)=p(b_2)$	$p(c)$	$p(d)$	$p(e)$
[0	[0	[0	[0	[0	[0
0.33	0	0.33	0.29	0.14	0.19
0.34	0.5	0.34	0.41	0.48	0.5
0.33]	0.5]	0.33]	0.3]	0.38]	0.32]

e=structure failure extent, c=global loss of strength, d=global loss of stiffness, a_1=component loss of strength, a_2=component loss of stiffness; b_1=global damage to strength, b_2=global damage to stiffness.

Reed, (1993), presents a complete multiply-connected network for post-earthquake damage assessment of a reinforced concrete building, see fig. 6.29. In this figure, the final node represents "building failure extent". It can be seen that the two basic causes of structural failure, global loss of strength and global loss of stiffness, are related to the framing type and component loss of strength and loss of stiffness. Observations of permanent drift, and stiffness and damping changes are "caused" by the global loss of strength and global loss of stiffness, so the arrows point into these variables. The component damage, which is broken down into the strength and stiffness includes observed damage to structural as well as non-structural components. This damage includes observations of spalling, cracking and buckling. Although the diagram shows only one component, in reality there are many components, which would be shown in a 3-D diagram.

At the present time, the only other structural damage assessment programs are *SPERIL* (Yao, 1985) and *RAMINO* (Lucia and Volta, 1991). They both employ the uncertain inference approach described in Section 6.3.6.

This formulation considers both global and local damage, but not in an explicitly causal manner. An experimental comparison can only be meaningful when the knowledge is extracted from the same source(s) and the degree of detail is equivalent. Although the input will be different in the sense that, say, fuzzy set theory will require membership function, as opposed to subjective probability assessment, the type of input, e.g. "what is the damage to this beam", must be similar.

Of course, the true test will be to evaluate actual damage cases. However, it does not seem premature to recommend the causal network approach as one in which the Bayesian updating is accomplished in an effective and reasonable manner.

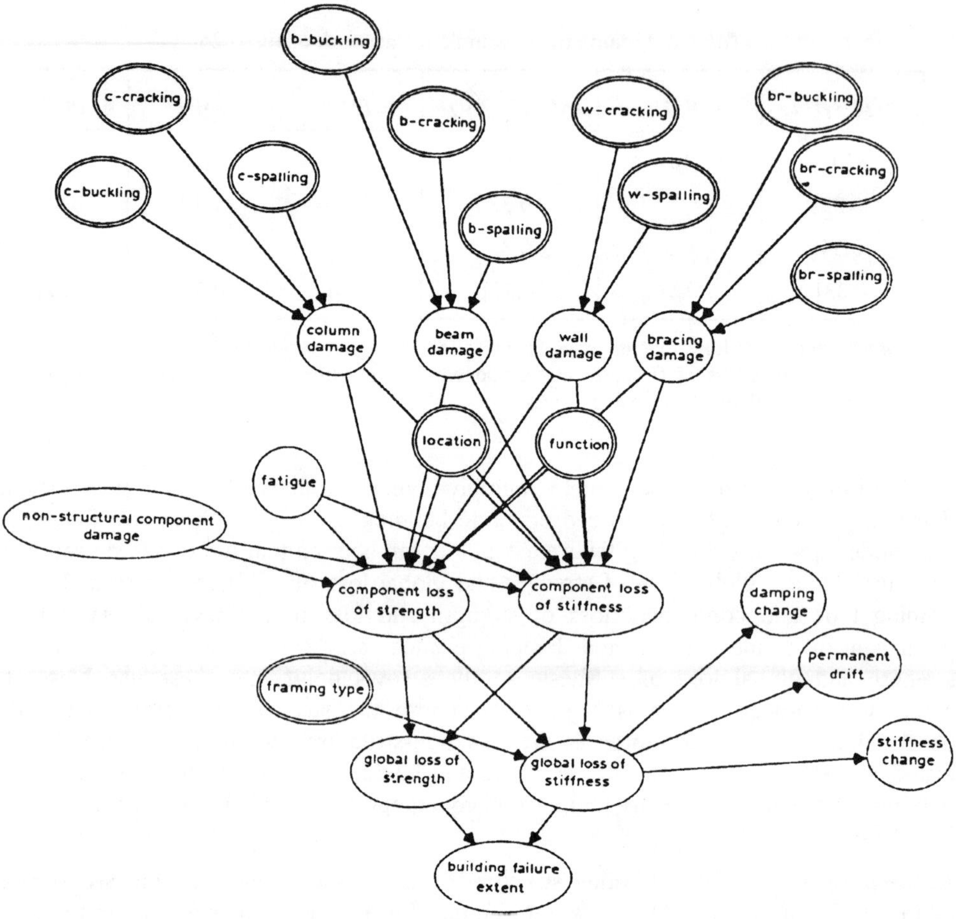

Figure 6.29 Network for post-earthquake damage assessment of a reinforced concrete building.

References

Akyurek T. and O.G. Bilir (1992). A survey of fatigue crack growth life estimation methodologies. *Engineering Fracture Mechanics*, **42**, 5, p. 797.

Al-Obaid Y.F. (1992). Fracture toughness parameter in a pipeling. *Engineering Fracture Mechanics*, **43**, 3, p. 461.

Ben-Amoz M. (1992). Prediction of fatigue crack initiation life from cumulative damage tests. *Engineering Fracture Mechanics*, **41**, 2, p. 247.

Bhargava V. *et al.* (1986). Analysis of cyclic crack growth in high strength roller bearings. *Theoretical and Applied Fracture Mechanics*, **5**, p. 31.

Bladwin J.F., and B.W. Pilsworth (1980). Axiomatic approach to implication for approximate reasoning with fuzzy logic. *Fuzzy Sets and Systems*, **3**, p. 193.

Bogdanoff J.L. and F. Kozin (1985). Probabilistic models of cumulative Damage. John Wiley and Sons, N.Y.

Bogdanoff J.L. and F. Kozin (1984). Probabilistic models of fatigue crack growth. *Engineering Fracture Mechanics*, **20**, 2, p. 225.

Box G.E.P. and G.M. Jenkins (1976). Time Series Analysis, Forecasting and Control. Holden-Day, San Francisco, CA.

Brown C.B. and J.T.P. Yao (1983). Fuzzy Sets and Structural Engineering. *Journal of Structural Engineering, ASCE*, **109**, 5, p. 211.

Camerini *et al.* (1992). Application of automated eddy current techniques for off-shore inspection. In C. Hallai-P. Kulcsar (Eds.), "Non-Destructive testing '92", Elsevier Science Publishers.

Cheng Y.W. (1985). The fatigue crack growth of a ship steel in saltwater under spectrum loading. *International Journal of Fatigue*, **7**, 2, p. 95.

Cheng Y.W. (1988). Fatigue crack growth analysis under sea-wave loading. International Journal of Fatigue, **10**, 2, p. 101.

Chou K. C. and J. Yuan (1992). Safety assessment of existing structures using a filtered fuzzy relation. *Structural Safety*, **11**, p. 173.

Chow C. L. and K.H. Wong (1987). A comparative study of crack propagation models for PMMA and PVC. *Theoretical and Applied Fracture Mechanics*, **8**, p. 101.

Cortie M. B. and G.G. Garrett (1988). On the correlation between the C and m in the Paris equation for fatigue crack propagation. *Engineering Fracture Mechanocs*, **30**, 1, p. 49.

D'Attelis C. *et al.* (1992). A bank of Kalman filters for failure detection using acoustic emission signals. In C. Hallai - P. Kulcsar (Eds.), "Non-destructive testing '92", Elsevier Science Publishers.

Dubois D. and H. Prade (1986). Fuzzy sets and statistical data. *European Journal of Operational Research*, **25**, p. 345.

Dufresne J., Lucia A., Grandemange J. and A. Pellissier-Tanon (1986). The COVASTOL program. *Nuclear Engineering and Design*, **86**, p. 139.

Dufresne J., Lucia A., Grandemange J. and A. Pellissier-Tanon (1988). Probabilistic vessels-study of the failure of pressurized water reactor (PWR) vessels. *Report EUR No 8682*, JRC-Ispra (Italy), Commission of the European Communities.

Fukuda T. and T. Mitsuoka (1986). Pipeline inspection and maintenance by applications of computer data processing and Robotic technology. *Computers in Industry*, 7, p. 5.

Garribba S. *et al.* (1988). Fuzzy measures of uncertainty for evaluating non-destructive crack inspection. *Structural Safety*, **5**, p. 187.

Georgel B. and R. Zorgati (1992). EXTRACSION: a system for automatic Eddy Current diagnosis of steam generator tubes in nuclear power plant. In C. Hallai - P. Kulcsar (Eds.), "Non-Destructive testing 92", Elsevier Science Publishing.

Ghonem H. and S. Dore (1987). Experimental study of the constant - probability crack growth curves under constant amplitude loading. *Engineering Fracture Mechanics*, **27**, 1, p. 1.

Fukuda T. and T. Mitsuoka (1986). Pipeline inspection and maintenance by applications of computer data processing and Robotic technology. *Computers in Industry*, 7, p. 5.

Godfrey M.W., Mahcwood L.A. and D.C. Emmony (1986). An improved design for point contact transducer. *NDT International*, **19**, 2.

Grangeat P. *et al.* (1992). X-Ray 3D cone beam tomography application to the control of ceramic parts. In C. Hallai - P. Kulcsar (Eds.), "Non-Destructive testing '92", Elsevier Science Publishers.

Guedes-Soares C. (1984). Probabilistic models for load effects in ship structures. *Report UR-84-38*, Marine Technology Dept., The Norwegian Institute of Technology, Trondheim, Norway.

Hadipriono F. and T. Ross (1987). Towards a rule-based expert system for damage assessment of protective structures. *Proceedings of International Fuzzy Systems Association (IFSA) Congress*, Tokyo, Japan, July 20-25.

Hagemaier D.J., Wendelbo A.H. and Y. Bar-Cohen (1985). Aircraft Corrosion and detection methods. *Materials Evaluation*, **43**, p. 426.

Halford *et al.* (1989). Fatigue life prediction modeling for turbine hot section materials. *ASME Journal of Engineering for Gas Turbines and Power*, **11**, 1, p. 279.

Hasselmann K. *et al.* (1976). A parametric wave prediction model. *Journal of Physical Oceanography*, **6**, p. 200.

Hoeppner D.W. and W.E Krupp (1974). Prediction of component life by application of fatigue crack growth knowledge. *Engineering Fracture Mechanics*, **6**, p. 47.

Hogben H. *et al.* (1976). Environmental conditions. *Report of Committee I.1, Procedings of 6th International Ship Structures Congress*, Boston.

Hull B. and J. Vernon (1988). Non-Destructive Testing. MacMillan Education, London.

Journet B.G. and R.M. Pelloux (1987). A methodology for studying fatigue crack propagation under spectrum loading: application to rail steels. *Theory and Applications of Fracture Mechanics*, **8**, p. 117.

Jovanovic A.S. *et al.* (1989). Expert Systems in Structural Safety Assessment. Springer-Verlag, Berlin, 1989.

Kalyanasundaram P. *et al.* (1991). *Brit. Journal of NDT*, **33**, 5, p. 221.

Komatsu H. *et al.* (1992). Basic study on ECT data evaluation method with neural network. In C. Hallai and P. Kulcsar (Eds.), "Non-Destructive Testing '92", Elsevier Science Publishers.

Kozin F. and J.L. Bogdanoff (1992).Cumulative damage model for fatigue crack growth based on reaction rate theory. *Engineering Fracture Mechanics*, **41**, 6, p. 873.

Landez J.P. *et al.* (1992). Ultrasonic inspection of vessel closure head penetrations. In C. Hallai - P. Kulsar (Eds.), "Non-Destructive testing 92", Elsevier Science Publishers.

Lankford J. and S.J. Hudak Jr. (1987). Relevance of the small crack problem to lifetime prediction in gas turbines. *International Journal of Fatigue*, **9**, 2, p. 87.

Lucia A.C. (1985). Probabilistic structural reliability of PWR pressure vessels. *Nuclear Engineering and Design*, **87**, p. 35.

Lucia A.C., Arman G. and A. Jovanovic (1987). Fatigue crack propagation: probabilistic models and experimental evidence. In *Trans. 9th SMiRT Conf.*, Vol. **M**, Lausanne, p. 313.

Lucia A.C. and G. Volta (1991). A knowledge-based system for structural reliability assessment. *Trans. SMiRT 11*, Vol. **SD1**, Tokyo, pan.

Ludwing and Roberti (1989). A nondestructive ultrasonic imaging system for detection of flaws in metal blocks. *IEEE Transactions on Instruments and Measurements*, **38**, 1.

Madsen H.O., Krenk S. and N.C. Lind (1986). Methods of structural safety. Prentice-Hall, N.J., USA.

Marci G. (1992). A fatigue crack growth threshold. *Enginneering Fracture Mechanics*, **41**, 3, p. 367.

Mohammadi J. *et al.* (1991). Evaluation of system reliability using expert opinions. *Structural Safety*, **9**, p. 227.

Namioka T. *et al.* (1992). Development and experience pipeline inspection robots by TV camera. In C. Hallai - P. Kulcsar (Eds.), "Non-Destructive testing '92", Elsevier Science Publishers.

Nielsen N. (1981). P-scan system for ultrasonic weld inspection. *British Journal of NDT*, March 1981, p. 63.

Nisitani H., Goto M. and N. Kawagoishi (1992). A small-crack growth law and its related phenomena. *Enginering Fracture Mechamics*, **41**, 4, p. 499.

Parpaglione M.C. (1992). Neural networks applied to fault detection using acoustic emission. In C. Hallai - P. Kulcsar (Eds.), "Non-Destructive testing '92", Elsevier Science Publishers.

Raj, B., (1992) "Reliable solutions to engineering problems in testing through acoustic signal analysis", in "Non-Destructive testing 92", C. Hallai and P. Kulcsar (Eds.), Elsevier Science Publishers.

Reed D.A. (1993). Treatment of uncertainty in structural damage assessment. *Reliability Engineering and Systems Safety*, **39**, p. 55.

Sandberg G. *et al.* (1989). The application of a continuous leak detection system to pipelines and associated equipment. *IEEE Transactions in Industry Applications*, **25**, 5, p. 906.

Schicht A. and A. Zhirabok (1992). The integrated Expert Systems for NDT in quality control systems. In C. Hallai - P. Kulcsar (Eds.), "Non-Destructive testing 92", Elsevier Science Publishing.

Shiraishi N. *et al.* (1991). An expert system for damage assessment of a reinforced concrete bridge deck. *Fuzzy Sets and Systems*, **44**, p. 449.

Singh G.P. and S. Udpa (1986). The role of digital signal processing in NDT. *NDT International*, **19**, 3, p. 125.

Solomos G.P. and V.C. Moussas (1991). A time series approach to fatigue crack propagation. *Structural Safety*, **9**, p. 211.

Stavrakakis G.S. (1990). Quality assurance of welds in ship structures. *Quality and Reliability Engineering International*, **6**, p. 323.

Stavrakakis G.S. (1992). Improved structural reliability assessment using non-linear regression techniques to process raw fatigue crack growth test data. *Quality and Reliability Engineering International*, **8**, p. 341.

Stavrakakis G.S. (1993). An efficient computer program for marine structures reliability and risk assessment. *The Naval Architect*, July/August '93, p. E342.

Stavrakakis G.S., Lucia A.C. and G. Solomos (1990). A comparative study of the probabilistic fracture mechanics and the stochastic markovian process approaches for structural reliability assessment. *International Journal Pres. Ves. Piping*, **41**, p. 25.

Stavrakakis G.S. and A. Pouliezos (1991). Fatigue life prediction using a new moving window regression method. *Mechanical Systems and Signal Processing*, **5**, 4, p. 327.

Stavrakakis G.S. and S.M. Psomas (1993). NDT data interpretation using Neural Networks. In "*Knowledge based system applications in power plant and structural engineering*", SMiRT 12 post-conference Seminar no. 13, August 23-25, Konstanz, Germany.

Thoft-Christensen P. and J.D. Sorensen (1987). Optimal strategy for inspection and repair of structural systems. *Civil Engineering Systems,* **4**, p. 17.

Van Dijk G.M. and J. Boogaard (1992). NDT reliability - a way to go. In C. Hallai and P. Kulcsar (Eds.), "Non-Destructive testing '92", Elsevier Science Publishers.

Vancoille M.J.S., Smets H.M.G. and F.L. Bogaerts (1993). Intelligent corrosion management systems. In "*Knowledge based system applications in power plant and structural engineering*" SMiRT 12 post-conference Seminar no. 13, August 23-25, Konstantz, Germany.

Verreman Y. *et al.* (1987). Fatigue life prediction of welded joints - a reassessment. *Fatigue Fracture Engineering and Materials Structure*, **10**, 1, p. 17.

Virkler D.A., Hillberry B.M. and P.K. Goel (1979). The statistical nature of fatigue crack propagation. *ASME Journal Enginnering Materials Technology*, **101**, p. 148.

Yanagi C. (1983). Robotics in material inspection. *The NDT Journal of Japan*, **1**, No 3, p. 162.

Yao J.T.R. (1985). Safety and Reliability of Existing Structures. Pitman Publishing, Marshfield.

Zhu W.Q. and Y.K Lin (1992). On fatigue crack growth under random loading. *Engineering Fracture Mechanics*, **43**, 1, p.1.

Author index

D

E

F

G

H

Z

Subject index

G

S